# Plant Signaling Molecules in Regulation of ROS-Scavenging System

Noushina Iqbal • Nafees A. Khan
Antonio Ferrante
Editors

# Plant Signaling Molecules in Regulation of ROS-Scavenging System

## Adapting to Climate Change

Springer

*Editors*
Noushina Iqbal
Department of Botany
School of Chemical and Life Sciences
Jamia Hamdard
New Delhi, India

Nafees A. Khan
Department of Botany
Aligarh Muslim University
Aligarh, Uttar Pradesh, India

Antonio Ferrante
Department of Agricultural
and Environmental Sciences
University of Milan
Milan, Italy

ISBN 978-981-95-3376-3 ISBN 978-981-95-3377-0 (eBook)
https://doi.org/10.1007/978-981-95-3377-0



This Springer imprint is published by the registered company Springer Nature Singapore Pte Ltd.
The registered company address is: 152 Beach Road, #21-01/04 Gateway East, Singapore 189721, Singapore

# Preface

The twenty-first century is witnessing a convergence of climate change, environmental degradation, and population growth that places unprecedented pressure on global agriculture. Crops across the world-whether fruit-bearing trees, staple cereals, or vegetables—are increasingly exposed to an array of abiotic stresses, including drought, salinity, nutrient deficiencies, temperature extremes, heavy metal toxicity, and intense light. These environmental challenges disrupt plant metabolism and, in many cases, trigger the excessive generation of reactive oxygen species (ROS).

ROS such as superoxide radicals, hydrogen peroxide, hydroxyl radicals, and singlet oxygen play a dual role in plants. At controlled levels, they act as critical signaling molecules, orchestrating defense responses, and adaptive growth patterns. In excess, however, they inflict oxidative damage on proteins, lipids, nucleic acids, and membranes, impairing growth and productivity. To survive, plants have evolved intricate antioxidant defense systems—comprising both enzymatic and nonenzymatic components—that work alongside osmolyte accumulation, hormonal regulation, nutrient balancing, and beneficial microbial partnerships to restore redox equilibrium.

This book brings together recent advances in understanding the molecular, biochemical, and physiological interplay between ROS, antioxidants, and stress adaptation in plants. By spanning multiple crops, stress types, and intervention strategies—from nutrient and hormone modulation to genetic engineering and microbial inoculation—it provides a comprehensive reference for researchers, students, and practitioners aiming to strengthen plant resilience and promote sustainable agriculture in a rapidly changing climate.

## Chapter-by-Chapter Overview

Chapter 1 examines how woody and fruit trees respond to environmental stresses such as drought, salinity, nutrient deficiency, and extreme temperatures. It explores the dynamics of ROS accumulation, the activation of antioxidant defenses, and specific physiological adjustments like osmolyte buildup and ionic regulation, while highlighting strategies to improve tolerance.

Chapter 2 focuses on redox signaling as a master regulator of plant adaptation, explaining how ROS and antioxidants interact to maintain cellular balance and how targeted manipulation of these pathways could enhance resilience to climate-induced stresses.

Chapter 3 delves into the regulation of amino acid metabolism and transport under stress, revealing how transcriptional control, protein modifications, and metabolic feedback loops influence nitrogen use and plant adaptation.

Chapter 4 explores the central role of osmolytes—such as proline, glycine betaine, and trehalose—in scavenging ROS, protecting macromolecules, and stabilizing redox balance under abiotic stress, alongside insights into the regulatory networks controlling their synthesis.

Chapter 5 investigates the synergistic effects between isoprenoid-derived signaling molecules (e.g., gibberellins, cytokinins, brassinosteroids) and mineral nutrients in enhancing thermotolerance, suggesting new opportunities for nutrient–hormone cross-talk in stress mitigation.

Chapter 6 highlights ROS-mediated signaling cascades, including MAPK pathways and transcription factor activation, which integrate environmental cues and trigger gene expression patterns essential for oxidative stress tolerance.

Chapter 7 uses proteomics to unravel the ROS signaling landscape in fruit vegetables during development and ripening, identifying antioxidant enzymes, metabolic adjustments, and the protective roles of compounds like melatonin and benzothiadiazole.

Chapter 8 presents experimental evidence on how salicylic acid and potassium, applied individually or together, alleviate salt stress in mustard. Enhanced antioxidant activity, osmotic adjustment, nitrogen assimilation, and secondary metabolite accumulation emerge as key mechanisms.

Chapter 9 reviews the diverse ways in which plant growth-promoting rhizobacteria (PGPR) help mitigate oxidative stress—through nutrient mobilization, water regulation, and activation of stress-response genes—while supporting sustainable agriculture.

Chapter 10 analyzes the intricate link between nutrient availability and ROS homeostasis, showing how balanced nutrition not only prevents oxidative damage but also strengthens stress signaling and tolerance.

Chapter 11 identifies the main sources of ROS under normal and stressed conditions, outlining their role in growth regulation and stress responses, and the necessity of efficient antioxidant systems to prevent cellular injury.

Chapter 12 revisits osmolytes as dual protectors against osmotic and oxidative stress, detailing their hormonal regulation, signal transduction pathways, and potential for genetic engineering to enhance stress resilience.

Chapter 13 investigates how PGPR enhance both enzymatic and nonenzymatic antioxidant defenses, protecting plants against drought, salinity, temperature fluctuations, and heavy metal toxicity.

Chapter 14 explores the synergistic role of PGPR and plant growth regulators in mitigating ROS damage, improving nutrient uptake, and sustaining crop productivity under varied abiotic stresses.

Chapter 15 examines the interplay between abscisic acid (ABA) and ROS in stress perception and adaptation, from stomatal regulation to antioxidant activation, providing insights into cultivar-specific differences in stress responses.

Chapter 16 explores the significance of Ascorbate–Glutathione (AsA–GSH) cycle as a central antioxidant system against heat-induced oxidative stress. The concerted action of ascorbate, glutathione, and enzymes such as APX, DHAR, MDHAR, and GR is focused together with the influence of various phytohormones and other metabolites in regulating this cycle for stress tolerance.

New Delhi, India — Noushina Iqbal
Aligarh, Uttar Pradesh, India — Nafees A. Khan
Milan, Italy — Antonio Ferrante

# Contents

# Editors and Contributors

## About the Editors

**Noushina Iqbal** is an accomplished researcher with extensive experience in plant biology, currently serving as an Assistant Professor at Jamia Hamdard's Department of Botany, New Delhi. Her research focuses on plant stress physiology, with notable contributions to understanding the role of phytohormones in abiotic stress tolerance. She has held prestigious positions such as Post-Doctoral Fellow under SERB-NPDF and DS Kothari Fellowship, receiving several accolades, including the ESDA Young Scientist Award and being listed among the top 2% of scientists globally. She has published 70 research papers, 12 book chapters, and 5 books, amassing over 9836 citations, an h-index of 43, and an i10-index of 71. Her work has earned her recognition for publishing in high-impact journals. A life member of ISPP, she continues to advance sustainable agriculture through integrative stress physiology. She is review editor for *Frontiers in Plant Science* and has acted as special issue editor for various journals. Her work centrally explores the synergy of phytohormones (e.g., salicylic acid, ethylene) and key nutrients (notably potassium, nitrogen, sulfur) in enhancing plant resilience under abiotic stress. She has contributed to understanding mechanisms that improve photosynthesis, nutrient use efficiency, and stress tolerance in crop systems.

**Nafees Ahmad Khan** is a distinguished Professor of Plant Physiology at the Department of Botany, Aligarh Muslim University. With a PhD and DSc from AMU, he is globally recognized for his extensive research on abiotic stress tolerance in plants, phytohormone signaling, and sustainable agriculture. A Highly Cited Researcher (Clarivate Analytics, 2019–2023), he has over 200 research publications, an H-index of 83 and numerous prestigious awards, including the UGC-Research and Mid-Career Awards, the Teaching Excellence Award, and the Research Excellence Citation Award 2023 (Clarivate Analytics). He is a fellow of multiple esteemed academies, including the National Academy of Sciences, the National Academy of Agricultural Sciences, the Linnean Society of London, the Indian Botanical Society, and the Indian Society for Plant Physiology. He is a member of the editorial board of various international journals.

His research focuses on responses of growth and photosynthetic processes, the antioxidant system to abiotic stress factors, and the identification of mechanisms for developing abiotic stress-tolerant crop plants through phytohormone signaling and its crosstalk with primary nutrient elements, nitrogen, and sulfur. In recognition of his impactful work, the Ministry of Education, Government of India, has highlighted AMU's dedication to research and excellence, specifically acknowledging his contributions to sustainable solutions for food security. He has supervised 22 PhDs and received research grants from the University Grants Commission, Council of Scientific and Industrial Research, UP-CST, Department of Science & Technology and Department of Biotechnology.

**Antonio Ferrante** is a full Professor of Vegetables and Ornamental Crops at Scuola Superiore Sant'Anna, Pisa, Italy. He is a leading horticultural scientist with a PhD in Advanced Technologies in Horticultural Science. He previously held key research and teaching roles at the University of Milan. His authoritative work on biostimulants, nutrient use efficiency, postharvest senescence, and fresh-cut vegetable physiology has been widely recognized, with over 200 publications and more than 8,000 citations. He has contributed influential reviews—such as on ethylene in plant senescence and biostimulants in stress physiology—that have been cited hundreds of times globally. He also plays prominent roles in professional societies, including as President of the Horticulture Section of the Italian Society for Horticultural Science.

## Contributors

**Gholamreza Abdi** Department of Biotechnology, Persian Gulf Research Institute, Persian Gulf University, Bushehr, Iran

**Ubaida Akbar** Department of Food Technology and Nutrition, Lovely Professional University, Phagwara, Punjab, India

**Mohd Ali** Department of Zoology, Guru Nanak Dev University, Amritsar, Punjab, India

**Ameena Fatima Alvi** Plant Physiology and Biochemistry Laboratory, Department of Botany, Aligarh Muslim University, Aligarh, India

**Hamzeh Amiri** Department of Biology, Lorestan University, Khorramabad, Iran

**Ananya Baidya** Department of Plant Physiology, Bidhan Chandra Krishi Viswavidyalaya, Mohanpur, Nadia, West Bengal, India

**Nanya Baidya** Department of Plant Physiology, Bidhan Chandra Krishi Viswavidyalaya, Mohanpur, Nadia, West Bengal, India

**Tamanna Bhardwaj** Department of Botanical and Environmental Sciences, Guru Nanak Dev University, Amritsar, Punjab, India

**Renu Bhardwaj** University Centre for Research and Development, Chandigarh University, Mohali, Punjab, India

**Anilava Chatterjee** Department of Agricultural Biochemistry, Bidhan Chandra Krishi Viswavidyalaya, Mohanpur, Nadia, West Bengal, India

**Kamini Devi** Department of Botanical and Environmental Sciences, Guru Nanak Dev University, Amritsar, Punjab, India

**Padmanabh Dwivedi** Department of Plant Physiology, Institute of Agricultural Sciences, Banaras Hindu University, Varanasi, India

**Antonio Ferrante** Institute of Crop Science, Sant'Anna School of Advanced Studies of Pisa, Pisa, Italy

**Alessandra Francini** Institute of Crop Science, Sant'Anna School of Advanced Studies of Pisa, Pisa, Italy

**Giulia Franzoni** Department of Agricultural and Environmental Science, University of Milan, Milan, Italy

**Ritu Gill** Centre for Biotechnology, Maharshi Dayanand University, Rohtak, Haryana, India

**Sarvajeet Singh Gill** Centre for Biotechnology, Maharshi Dayanand University, Rohtak, Haryana, India

**Akankhya Guru** Department of Agriculture Botany, Janta Vedic College, Baraut (Baghpat); affiliated to CCS University, Meerut, Uttar Pradesh, India

**Akash Hidangmayum** Department of Plant Physiology, Institute of Agricultural Sciences, Banaras Hindu University, Varanasi, India

**Noushina Iqbal** Department of Botany, School of Chemical and Life Sciences, Jamia Hamdard, New Delhi, India

**Parul Jain** Department of Botany, School of Chemical and Life Sciences, Jamia Hamdard, New Delhi, India

**Divya Jaiswal** Department of Plant Physiology, Institute of Agricultural Sciences, Banaras Hindu University, Varanasi, India

**Radha Jangra** Centre for Biotechnology, Maharshi Dayanand University, Rohtak, Haryana, India

**Anshu Jha** Department of Botany, School of Chemical and Life Sciences, Jamia Hamdard, New Delhi, India

**Gopal Kalwan** ICAR-National Institute for Plant Biotechnology, New Delhi, India

PG School, ICAR-Indian Agricultural Research Institute, New Delhi, India

**Nitika Kapoor** Department of Botanical and Environmental Sciences, Guru Nanak Dev University, Amritsar, Punjab, India

**Shruti Kaushik** ICAR – Central Soil Salinity Research Institute, Karnal, India

**Nafees A. Khan** Plant Physiology and Biochemistry Laboratory, Department of Botany, Aligarh Muslim University, Aligarh, India

**Varisha Khan** Department of Botany, School of Chemical and Life Sciences, Jamia Hamdard, New Delhi

**Jaspreet Kour** Department of Botanical and Environmental Sciences, Guru Nanak Dev University, Amritsar, Punjab, India

**Aman Kumar** College of Biotechnology, CCS HAU, Hisar, Haryana, India

**Nahida Rehman Mir** Department of Botanical and Environmental Sciences, Guru Nanak Dev University, Amritsar, Punjab, India

**Md Sabir Ahmed Mondol** Department of Agricultural Biochemistry, Bidhan Chandra Krishi Viswavidyalaya, Mohanpur, Nadia, West Bengal, India

**Aritra Mukherjee** Department of Seed Science and Technology, Bidhan Chandra Krishi Viswavidyalaya, Mohanpur, Nadia, West Bengal, India

**Ashima Nehra** Centre for Biotechnology, Maharshi Dayanand University, Rohtak, Haryana, India

**Nidhi** Plant Physiology and Biochemistry Laboratory, Department of Botany, Aligarh Muslim University, Aligarh, India

**Puja Ohri** Department of Zoology, Guru Nanak Dev University, Amritsar, Punjab, India

**Parkirti** Department of Zoology, Guru Nanak Dev University, Amritsar, Punjab, India

**Amisha Rani** Department of Forest Products, Dr. Yashwant Singh Parmar University of Horticulture and Forestry, Nauni, Solan, India

**Alok Ranjan** Department of Biotechnology, Patna Women's College, Patna University, Patna, Bihar, India

**Deepika Ranjan** Department of Plant Physiology, Institute of Agricultural Sciences, Banaras Hindu University, Varanasi, India

**Fabio Scotto Di Covella** Institute of Crop Science, Sant'Anna School of Advanced Studies of Pisa, Pisa, Italy

**Luca Sebastiani** Institute of Crop Science, Sant'Anna School of Advanced Studies of Pisa, Pisa, Italy

**Manu Sharma** Department of Botanical and Environmental Sciences, Guru Nanak Dev University, Amritsar, Punjab, India

**Pooja Sharma** D.A.V College Amritsar, Amritsar, Punjab, India

**Amandeep Singh** Department of Botanical and Environmental Sciences, Guru Nanak Dev University, Amritsar, Punjab, India

**Ragini Singh** Department of Botany, School of Chemical and Life Sciences, Jamia Hamdard, New Delhi, India

**Tadela Susmitha** Department of Agricultural Biochemistry, School of Agricultural Sciences, Mallareddy University, Hyderabad, Telangana, India

**Diksha Taneja** Centre for Biotechnology, Maharshi Dayanand University, Rohtak, Haryana, India

**Narendra Tuteja** International Centre for genetic Engineering and Biotechnology (ICGEB), Aruna Asaf Ali Marg, New Delhi, India

**Shahid Umar** Department of Botany, School of Chemical and Life Sciences, Jamia Hamdard, New Delhi, India

**Vikram** Department of Zoology, Guru Nanak Dev University, Amritsar, Punjab, India

**Priya Yadav** Department of Botany, Zakir Husain Delhi College, University of Delhi, New Delhi, India

**Mero Yangfo** North Eastern Regional Institute of Science and Technology, Nirjuli, Arunachal Pradesh, India

# Reactive Oxygen Species in Woody Crops Under Abiotic Stress

1

Alessandra Francini and Luca Sebastiani

**Abstract**

Woody and fruit trees undergo various abiotic stresses throughout their life cycle, which can lead to the accumulation of reactive oxygen species (ROS) and subsequent oxidative damage. ROS, including superoxide anion radical, hydrogen peroxide, hydroxyl radical, and singlet oxygen, are byproducts of oxygen metabolism and can cause severe damage to cell membranes and organelles. Plants have evolved enzymatic and non-enzymatic antioxidant systems to scavenge ROS and maintain cellular homeostasis. However, severe and long-term nutrient deficiency, drought, high-temperatures, and salinity can overwhelm plant defense mechanisms, leading to oxidative stress, which impairs growth and development. Nutrient deficiencies, like nitrogen, phosphorus, iron, and magnesium, induce ROS formation and activate antioxidant enzymes such as superoxide dismutase, catalase, and peroxidases. Drought stress leads to photosystem imbalance, accumulation of osmolytes, such as proline, and the activation of antioxidant defenses. High-temperatures increase transpiration rates and decrease water use efficiency, while salinity increases toxic ions ($Na^+$ and $Cl^-$) and ROS. Understanding the mechanisms of ROS production and scavenging in woody and fruit trees is crucial for developing strategies to enhance their tolerance to these abiotic stresses. This review discusses the current knowledge of ROS accumulation and antioxidant defense systems in woody and fruit trees under various abiotic stresses, as well as potential strategies for improving their resilience to environmental challenges.

A. Francini (✉) · L. Sebastiani
Institute of Crop Science, Sant'Anna School of Advanced Studies of Pisa, Pisa, Italy
e-mail: a.francini@santannapisa.it



N. Iqbal et al. (eds.), *Plant Signaling Molecules in Regulation of ROS-Scavenging System*, https://doi.org/10.1007/978-981-95-3377-0_1

**Keywords**

Detoxification · Drought · Salinity · High temperatures · Mineral nutrient deficiency

## 1.1 Introduction

Plants during their life are continuously exposed to environmental stress, and they cannot escape from adverse conditions. For this reason, their survival relies on appropriate defense strategies that could be slightly different in herbaceous, shrub, and woody species. Plant stressors can originate from anthropogenic activities (i.e., pollution) or natural environmental conditions (Francini and Sebastiani 2019). Stressed plants undergo physiological adjustments and very often reduced growth and yield. The most common and rapid marker of plants stress is the increase of reactive oxygen species (ROS). ROS includes superoxide anion radical ($O_2^-$), hydrogen peroxide ($H_2O_2$), hydroxyl radical ($OH^•$), and singlet oxygen ($^1O_2$) and are continuously produced in plant cells as an inevitable byproduct of processes associated with oxygen metabolism (Trivellini et al. 2017; Ferrante and Francini 2023; Francini et al. 2006). ROS are produced in mitochondria during respiration and also in chloroplasts and peroxisomes during photosynthesis and photorespiration. Plants have highly efficient enzymatic and non-enzymatic antioxidant systems that remove ROS immediately after their formation. Moreover, ROS play a crucial role in plant responses to environmental stressors (Ferrante and Francini 2023) regulating the cell redox status, but too high and uncontrolled variation in ROS concentrations, causes damage to cells components, especially to the plasmalemma and other membranes. Prolonged exposure to abiotic stress and then ROS determines cells death and organ senescence (Trivellini et al. 2017).

Woody crops have a long cultivation cycle and stay in the same environment for decades. Differently from cultivated herbaceous annual plants that could escape critical environmental conditions for the vegetative periods by appropriate selection of the sowing date, woody perennials cannot. Moreover, woody plants could integrate the previous year's stress and activate epigenetic responses that persist for a longer time, an aspect that still deserves more scientific studies. As in other species, ROS accumulation in woody plants leads to oxidative damage and impairs cellular activities. Understanding the mechanisms of ROS production and scavenging is essential for developing strategies to enhance plant tolerance to abiotic stress (Sebastiani and Gucci 2023).

In plants, primary abiotic stressors include a series of factors: ultraviolet radiation (UV), intense light, extreme temperatures (both high and low), salinity, water scarcity (drought) or flooding, xenobiotics, and heavy metals (Francini and Sebastiani 2019) and woody crops can be exposed to all these abiotic stresses. All these stresses can lead to the accumulation of ROS and subsequent oxidative damage to cell membranes and organelles. Plants have evolved enzymatic and non-enzymatic antioxidant systems to scavenge ROS and maintain cellular homeostasis.

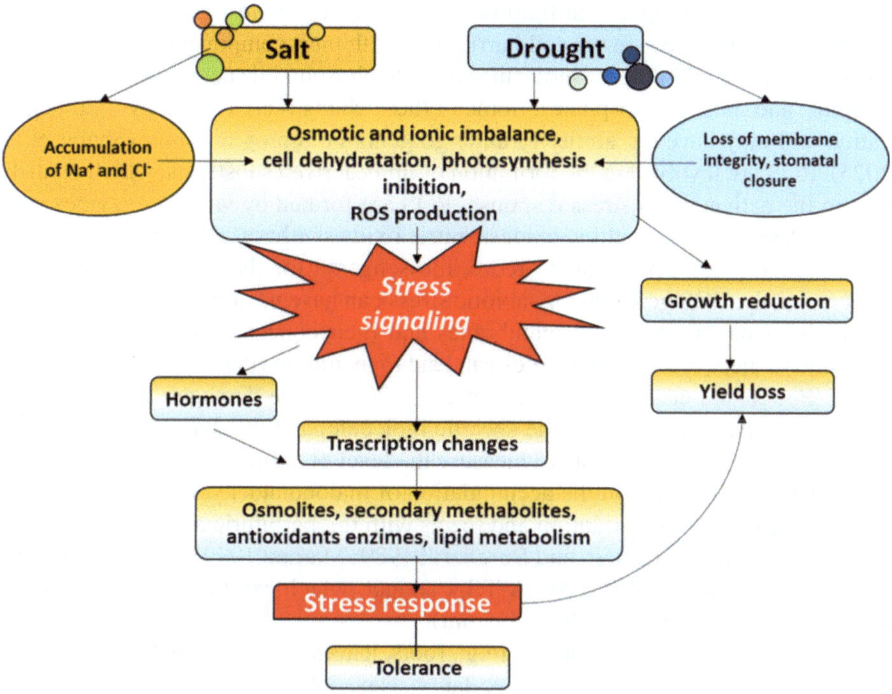


**Fig. 1.1** Schematic overview of the common mechanism between salinity and drought stress

However, prolonged exposure to abiotic stresses can overwhelm these defense mechanisms, leading to impaired plant growth and development. Nutrient deficiencies, particularly nitrogen (N), phosphorus (P), iron (Fe), and magnesium (Mg), can induce ROS formation and activate antioxidant enzymes. Salinity and drought share common osmotic stress pathways and lead to the accumulation of osmolytes, such as proline, and the activation of antioxidant defenses (Fig. 1.1).

## 1.2 ROS Production, Membrane Destabilization and Damage Assay

The cell is the smallest fundamental unit of structure and function of any plant and is organized into various compartments separated by membranes. Plant cells are more complex than animal cells because of the cell walls, the external rigid layer that protects and ensures tissue stability, the vacuole, and the plastids. However, both types of cells are delimited by plasmalemma, and the inside organelles have specific cell membranes, such as the tonoplast of the vacuole (Clément 2007). The cell has several compartments, and the cell membranes are responsible for the separation of specific substances and enzymes. Membranes are formed by double phospholipid layers containing several channels, proteins, and other functional elements

(Shajar et al. 2025). Important lipid-derived substances have signaling and sensing functions. Membranes are the cell barrier and with their components regulate the in and out traffic. Lipid-derived molecules include phosphatidic acid, oxylipins, sphingolipids, and inositol phosphates. Some of these changes are caused by lipid degradation and may have the ability to undergo genomic reprogramming (Shajar et al. 2025). ROS are involved in the formation of lipid-derived substrates, which contributes to the activation of stress response. ROS are formed by various enzymes such as NADPH oxidases, xanthine oxidase, nitric oxide synthase, and peroxisomal constituents. They are also named "redox messengers" for their signaling function. Unfortunately, ROS generated by abiotic stress can have negative effects by damaging the plasmalemma or cell membranes and their concentration, and prolonged exposure to these reactive molecules can lead to membrane destabilization or breakdown (Podgórska et al. 2017; Francini et al. 2006). The damage caused by ROS depends on the concentration, the specific molecule, and the duration of the exposure. Classical laboratory tests to measure the level of membrane damage and lipid peroxidation level relay on the accumulation of malondialdehyde (MDA). MDA is the main product of peroxidation and reacts with thiobarbituric acid (TBA) to form a colored chromogen (Janero and Burghardt 1989; Marnett 1999). MDA reacts with TBA to form a pink-colored MDA-TBA adduct, which can be measured spectrophotometrically (typically at 532–535 nm).

Measuring MDA in olive organs (e.g., roots, leaves) is a widely used approach to assess oxidative stress and lipid peroxidation, especially in studies related to abiotic stress (e.g., drought, salinity, temperature). Salt-sensitive cultivars of olive typically exhibit significant increases in MDA, reflecting higher oxidative membrane damage under salinity. Other ways salt-tolerant varieties (e.g., "Picual," "Frantoio") often maintain stable MDA levels, showing better oxidative control. The comparative analysis of three olive cultivars ("Ayvalık," "Gemlik," and "Kilis") under salinity stress showed that MDA levels were affected by increasing salt stress during the experimental period. The most significant effect of salt stress on membrane leakage values was seen in "Gemlik" and "Ayvalık" cultivars under 300 mM NaCl-treated groups after 30 days of exposure (Ayaz et al. 2021). In their study, Ben Abdallah et al. (2018) examined how salt stress impacts oxidative damage in the leaves of the "Chétoui" cultivar, finding a 50% increase in lipid peroxidation, as indicated by MDA levels, compared to the control group. They suggested that olive trees under salt stress might accumulate reactive oxygen species and electrolytes, resulting in increased oxidative damage. The comparison of biochemical responses of *Populus davidiana* and its hybrid with *Populus alba* × *davidiana* under drought stress showed that *P. alba* × *P. davidiana* had a gradual but definite increase in MDA levels at the end of the treatments compared to *P. davidiana*. Results suggest that *P. davidiana*, with better antioxidant maintenance, displayed better drought tolerance in the initial response (Kim et al. 2022). In apples, the physiological responses of Malus species under salt stress at the seedling stages were evaluated and the results showed that *M. zumi* and *M. sieversii* had a lower MDA content than *M. baccata* in the NaCl treatment groups (Wang et al. 2022). All the studies reported clearly demonstrated that MDA accumulation is used as a quantitative measure of oxidative stress severity in plants.

### 1.2.1 Drought and High Temperatures

Water scarcity is a significant challenge in many geographic regions worldwide. The decrease in water availability constrains the variety of crop species that can be cultivated. This issue is particularly critical for fruit trees, which are characterized by extended turnover periods and substantial establishment costs. Conducting drought stress research on fruit trees in open field conditions presents significant challenges, particularly when the plants are well-established with deep root systems capable of accessing extensive soil volumes. Consequently, potted experiments have become a prevalent method for investigating drought or salinity stress in fruit trees.

Drought exerts multifaceted impacts on trees, primarily through the concentration of solutes that induces osmotic stress, thereby damaging cellular components and resulting in a loss of turgor. Additionally, the closure of stomatal pores reduces $CO_2$ uptake and fixation, leading to a carbon imbalance and a depletion of carbohydrate reserves. This stomatal closure also diminishes transpiration, thereby limiting the cooling effect of leaf evaporation and subsequently causing heat stress, which adversely affects metabolic processes, particularly photosynthesis. A further common consequence of drought stress is the accumulation of reactive oxygen species, which damages proteins and membranes. However, trees have mechanisms to mitigate this, including the synthesis of antioxidants and other protective compounds, in fact, drought induces the biosynthesis and accumulation of osmotic molecules. In *Catalpa bungei* C. A. Mey. trees, drought stress induces ABA and proline accumulation (Shi et al. 2017). Proline is a proteogenic amino acid that can accumulate under stressful conditions, especially drought and salinity conditions (Verslues and Sharma 2010). Hosseinifard et al. (2022) recently reviewed its function as an osmolyte and a signaling molecule, which triggers the production of abscisic acid, thereby serving as a strong antioxidant. Proline is primarily synthesized in the roots and then transported through the phloem to the leaves and stems, resulting in a higher concentration that aids the plant in coping with water stress (Renzetti et al. 2024). The process of synthesizing proline starts with the formation of glutamate, which is governed by the genes pyrroline-5-carboxylate synthetase (P5CS), pyrroline-5-carboxylate reductase (P5CR), and ornithine-δ-aminotransferase (OAT). Enzymes involved in the breakdown of proline, such as proline dehydrogenase (PDH) and pyrroline-5-carboxylate dehydrogenase (P5CDH), have been discovered and isolated (Kavi Kishor et al. 2015). Drought determined the loss of turgor do to stomatal closure and subsequent reduction in net photosynthesis. Research by Gallé et al. (2007) on the leaf gas exchange parameters of oak (*Quercus pubescens* Willd.) under conditions of water stress and subsequent re-watering revealed that both photosynthetic rates and stomatal conductance decreased during periods of stress and reverted to their initial levels following re-watering. Associated with a decline in net photosynthesis, the maximum efficiency of PSII is also reduced. The reduction in net photosynthesis is the result of stomatal closure that occurs under water shortages. Stomatal closure reduces leaf gas exchange and represents the last step of a crop's response to drought (Wang et al. 2018; Liu et al. 2023). The progressive reduction of water availability is linked to the activation of osmolyte biosynthesis and the accumulation of antioxidants. Osmolytes function to decrease water potential and enhance water uptake. Unfortunately, prolonged stress conditions cannot be compensated for by osmolytes, and the plants close their stomata to reduce water loss. The investigation into poplar plants examined the function of carbohydrates as both osmolytes and energy providers. The findings revealed that under conditions of water scarcity, soluble sugars, originating from newly synthesized carbon and the breakdown of starch, were mainly directed to the bark. This rise in carbohydrate levels facilitated the preservation of a water reserve, which was utilized throughout the day to avert a reduction in water within the xylem (Traversari et al. 2020). Two poplar genotypes, *Populus alba* "Marte" and *Populus deltoides* "Dvina," exhibited differing growth performances and strategies for managing water and carbon balance. These poplar species offered a mechanistic insight into their responses to water scarcity. "Marte" was identified as more anisohydric compared to "Dvina," a trait linked to its ability to redistribute carbohydrates during periods of water shortage (Traversari et al. 2018). Tolerance to drought stress has also been studied in relation to age. These results indicate that older plants have better tolerance, and this response can be attributed to the architecture of the roots. In older apple plants, ROS, as well as enzymatic and non-enzymatic scavengers, increased with age (Jia et al. 2025). Woody and fruit trees may develop thicker leaves, deeper root systems, and an altered leaf orientation to reduce water loss and enhance stress tolerance. Indicators of oxidative damage, such as lipid peroxidation, have been utilized to evaluate the rise in ROS production during drought stress. Nonetheless, the absence of symptoms does not mean that there is no increase in ROS production. Rather, the absence of symptoms is likely to lead to a simultaneous enhancement of cellular antioxidant defenses.

Higher temperatures exacerbate drought conditions by increasing evaporation and reducing water availability. The global mean temperature has been estimated to increase by approximately 2.8 ± 0.7 °C by the end of this century. The frequency and amplitude of heat events are expected to increase as a consequence of global warming (Dai et al. 2025). An increase in temperature has a negative effect on fruit tree cultivation and fruit production, with a reduction in fruit quality. This can have an impact on the economies of specific regions and social consequences. In short periods, some agronomic tools, such as biostimulants, can have mitigation action, and in the long term, genetic improvement has to find tolerant cultivars. High temperatures in well-watered trees increase transpiration rates, whereas net photosynthesis remains unaffected. Plant transpiration can modulate leaf temperature and maintain the optimal range for the activities of assimilation enzymes. Water use efficiency (WUE) in well-watered trees decreased to 60% at 33 °C compared to that at 23 °C. High temperatures above the crop threshold for several days are called heatwaves and dramatically affect crop performance. However, the extent of heat stress damage depends on the availability of water. The transpiration rate in well-watered trees promptly returned to pre-stress levels, highlighting the significant atmospheric influence on transpiration. The schematic overview highlighting the

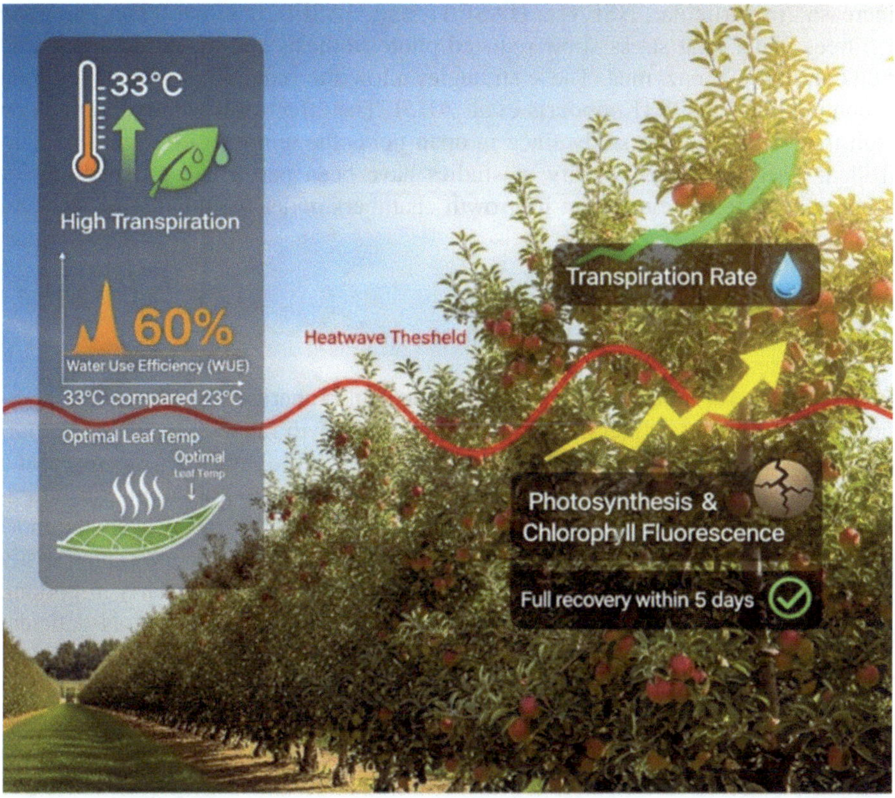


**Fig. 1.2** Schematic overview highlighting the delicate balance between temperature, water, and plant health, and emphasizing the critical role of irrigation in a changing climate (Image generated with Gemini AI tools)

delicate balance between temperature, water, and plant health, and emphasizing the critical role of irrigation in a changing climate is reported in Fig. 1.2.

Heat stress can disrupt cellular homeostasis, leading to the overproduction of reactive oxygen species (ROS) and damage to cell membranes. Melatonin's protective effects may also include reducing oxidative stress, they increasing antioxidant enzyme activity, and promoting photosynthesis, which are all important for plant survival under stress. When MdASMT9-OE apple plants that overexpress melatonin biosynthesis were exposed to heat stress showed an overaccumulation of melatonin, with a positive effect on the preservation of membrane integrity, as demonstrated by the lower concentration of MDA (Gao et al. 2024). The transgenic apple had lower ROS accumulation ($H_2O_2$ and $O_2^-$) and higher SOD, POD, and CAT enzyme activities. At the plant hormone level, apples overexpressing MdASMT9 showed increased ABA content and reduced stomatal aperture (Gao et al. 2024). ABA accumulation is confirmed by the increase in MdNCED1 and MdNCED3 gene expression. Among the transcriptional changes, heat shock factors increased in particular, HSFA1a, HASFB1, and HASFB2b. Olive (*Olea europaea* L.) trees under heat stress show reduced photosynthesis and increased activity of SOD and GPOX enzymes. These strategies allow the reduction of water loss and accumulation of ROS (Koubouris et al. 2015). The effect of high temperatures on fruit trees is difficult to study since in open fields the temperature cannot be controlled. Therefore, the majority of studies have been performed on seedlings or young plants for placing them in growth chambers under modulation of the temperature stress.

### 1.2.2 Mineral Nutrient Deficiency

Essential mineral nutrient deficiency can occur in orchards with inadequate agronomic management. Mineral deficiency induces growth reduction and physiological disorders, which can be visually observed by chlorophyll loss, necrosis, and stunted growth. This stress can be induced by a single or multiple nutrient limitations, as stated in Liebig's law. Among mineral elements, nitrogen (N) is the most labile nutrient that in nitrate form is not absorbed by soil colloids and can be easily lost by leakage. In orchards, agronomic management of N is particularly important, and its availability should be synchronized with N crop requirements. N-deficient trees produce fewer fruits and are of lower quality (Locascio et al. 1984). At the physiological level, chlorophyll concentrations can decline, especially under severe N deficiency, and induce leaf senescence (Carillo and Ferrante 2025). In olive trees, "Meski" and "Koroneiki" cultivars, N limitation reduced the maximum quantum efficiency of photosystem II by 40% or 20% N (Boussadia et al. 2015). Nitrogen limitation induces ROS formation and activates enzymatic and non-enzymatic scavenger systems. Rootstock can mitigate the nutritional deficiency provided to grafted plants with better root architecture and functionality. In citrus plants, tetraploid rootstocks can improve plant tolerance to nutrient deficiency. A study conducted on nine citrus genotypes with diploid (2x) or tetraploid (4x) seedlings with different nutrient availabilities demonstrated a 4x increased tolerance to nutrient deficiency (Oustric et al. 2019). In citrus, magnesium (Mg) deficiency in the 20-year-old citrus, *Citrus reticulata* cv. Ponkan showed an increase in membrane damage, as measured by MDA accumulation due to lipid peroxidation (Tang et al. 2012). The defense of plant was obtained by the activation of ascorbate peroxidase (APX, EC 1.11.1.11), glutathione reductase (GR, EC 1.6.4.2), guaiacol peroxidase (GPX, EC 1.11.1.7), monodehydroascorbate reductase (MDAR, EC 1.6.5.4), and superoxide dismutase (SOD, EC 1.15.1.1). The detoxification enzymatic system plays an important regulatory role in increasing fruit tree tolerance to abiotic stress. Mg is a key element in chlorophyll molecules, and Mg deficiency impairs light harvesting and subsequent photosynthesis.

In "sour pummelo" (*Citrus grandis*) seedlings under phosphorus (P) deficiency enhanced ascorbate concentration, and the enzyme activities of SOD, CAT, APX, DHAR, and GR in leaves (Meng et al. 2021). Growth reduction was observed in shoots, and no change was found in roots. An increase in hydrogen peroxide and MDA was found in the shoots, but not in the roots (Meng et al. 2021). No typical increase in anthocyanin content was observed. Blue color is often associated with P deficiency and is commonly observed in herbaceous plants. Flavonoids are non-enzymatic antioxidant compounds that play crucial roles in scavenging ROS and protecting cells from oxidative stress (Gechev and Petrov 2020; Rudenko et al. 2023). Nutrient deficiency can occur because of soil characteristics, such as high or low pH. Iron (Fe) deficiency is typical in calcareous or alkaline soils. Plants have difficulty in Fe uptake and have negative effects on growth and development. In *Malus xiaojinensis*, the Rho GTPase protein (ROP) plays an important signaling role. The overexpression of this gene in model plants demonstrated that ROS accumulation increased, suggesting a signaling role in response to Fe deficiency (Zhai et al. 2021).

### 1.2.3 Salinity and ROS

In the Mediterranean region, especially along the coast, orchards are often exposed to salinity. Fig trees are commonly cultivated in regions where salinity is an issue. An increase in salinity led to a 5.6% rise in total phenols and moderate salt stress seems to enhance the quality of fig fruits due to its beneficial impact on nutrients and antioxidant compounds like epicatechin (Francini et al. 2021). The olive (*Olea europaea* L.) is widespread in saline environments. However, many cultivars have varying degrees of tolerance. Several studies have demonstrated that the cv. "Frantoio" is salt-tolerant, and cv. Leccino is also sensitive to salinity (Rossi et al. 2015, 2016). The ability to isolate and manage $Na^+$ in roots distinguishes tolerant cultivars from those that are sensitive (Sodini et al. 2022). Additionally, "Frantoio" appears to have developed a more effective approach to mitigate the adverse effects of salt by enhancing the movement of $K^+$, $Mg^{2+}$, and $Ca^{2+}$ (Sodini et al. 2022, 2023).

As observed in other studies, biostimulants, such as the Plant Growth Promoting Bacteria (PGPB), can improve olive tolerance to salinity. Ten potential PGPB were evaluated to mitigate the salinity stress in "Arbosana" and "Arbequina" cultivars, grown in soil with 6.7 ds $m^{-1}$. The results showed differences between cultivars and PGPB showed mitigation effects. However, these studies must be conducted several times, especially in open fields, to confirm their effectiveness (Galicia-Campos et al. 2022).

Silicon (Si) is a non-essential element for plants and nowadays its application in crops as biostimulant is proposed (Franzoni et al. 2022). Fidalgo-Illesca et al. (2025) demonstrated that the Si priming effect in olive trees varies depending on the cultivar. In the "Frantoio" variety, Si triggers an early and rapid reduction in stomatal conductance, which enhances intrinsic water use efficiency, a response not seen in "Leccino" plants (Fig. 1.3). For "Leccino," the primary effect of Si was a significant decrease in $Na^+$ accumulation in new leaves (−58%) and the stabilization of $K^+$ levels under saline conditions. The interactions between Si and NaCl, as well as the number of affected polyphenols, were more pronounced in "Frantoio" compared to "Leccino." It can be concluded that photosynthesis and $Na^+$ uptake are the main mechanisms involved in the Si-mediated salt stress tolerance in "Frantoio" and "Leccino" cultivars.

**Fig. 1.3** Pre-treated with Si, the Frantoio and Leccino plant shown different responses under salinity but both cultivars activated the phenylpropanoid pathway (Image generated with Gemini AI tools)

Si contributes to enhances the antioxidant defence in plants under salinity through modulation of polyphenol compounds and in Leccino and Frantoio cultivars salinity stress activated the phenylpropanoid pathway (Fidalgo-Illesca et al. 2025).

In mandarin trees (*Citrus Clementina* cv. "Orogrande") grafted on Carrizo citrange (*Citrus sinensis* [L.] Osb. × *Poncirus trifoliata* [L.]), salinity stress was alleviated by increasing abscisic acid and salicylic acid. Date palm (*Phoenix dactylifera* L.) exposed to salinity with 80 μg $mL^{-1}$ biogenic selenium was distributed with nanoparticles that increased yield and improved oxidative stability, avoiding ROS accumulation (El-Tarabily and Abu Qamar 2025). Strawberries are extremely sensitive to salinity, and ROS reduction is avoided by Cu/Zn-SOD, Fe-SOD, and Mn-SOD (Zhang et al. 2024).

Overexpression of *VvLBD39*—lateral organ boundaries domain (LBD)—in grapes exposed to water and NaCl stress showed a higher sensitivity to drought and salt stress. Moreover, transgenic plants overexpressing *VvLBD39* were insensitive to ABA. Membrane integrity, measured as malondialdehyde, $H_2O_2$ content, and superoxide anion productivity, was higher. Moreover, *VvLBD39* expression negatively affected stomatal closure, ABA biosynthesis, and signalling under drought and salt treatments. The expression of genes related to ROS scavenging and stress response was significantly downregulated in *VvLBD39*. These results indicated that

*VvLBD39* functions as a negative regulator of salt and drought tolerance (Chen et al. 2024). The tolerance of glycophytic plants can be determined by the translocation of sodium and chloride in older leaves. Specific transporters are involved in the transport of toxic ions. The accumulation of ROS induces damage to the membrane, and transmembrane channels are impaired, with a negative effect on plant tolerance against salinity. The response of woody plants of *Dalbergia odorifera* to salinity was studied at the seedling stage. Exposure to 200 mM NaCl induced the accumulation of ROS, such as $H_2O_2$ and $O_2^-$. Seedlings showed a reduction in growth as a consequence of reduced photosynthesis. Stressed seedlings showed activation of phenylalanine ammonia-lyase (PAL E.C. 4.3.1.5) and polyphenol oxidase (PPO E.C. 1.14.18.1 and EC 1.10.3.1) enzymes. These enzymes induce the accumulation of antioxidant compounds such as flavonoids. Among detoxification enzymes, SOD showed the highest activity (Cisse et al. 2024) (Table 1.1).

**Table 1.1** Studies have been performed on woody and fruit trees. Plant responses to mitigate stress and ROS accumulation under various stress conditions

| Stress | Species | Function | References |
|---|---|---|---|
| Mg deficiency | *Citrus reticulate* cv. Ponkan | Activation of detoxification enzymes (APX, GPX, GR, MDAR, and GR) | Tang et al. (2012), Oustric et al. (2019) |
| Fe deficiency | *Malus xiaojinensis* | Increase of ROP genes and stress signaling | Zhai et al. (2021) |
| Fe deficiency | *Peach rootstock Mr. S 2/5* | *Physiological, biochemical, and molecular effects of* in vitro *induced iron deficiency* | Lombardi et al. (2003) |
| P deficiency | *Citrus grandis* | Enhance APX, GPX, GR, MDAR, and GR. | Meng et al. (2021) |
| Drought | *Catalpa bungei* | Induce proline and ABA accumulation | Shi et al. (2017) |
| Drought | *Malus domestica* | ROS accumulation and tolerance mediated by cytokinins | Jia et al. (2025) |
| Drought | *Malus hupenensis* | Reduce the accumulation of MDA | Zhang et al. (2020) |
| Drought | *Vitis vinifera* | *VvLBD39* | Chen et al. (2024) |
| Heat stress | *Malus domestica* | MdASMT9-mediated melatonin biosynthesis enhances basal thermotolerance | Gao et al. (2024) |
| Heat stress | *Olea europapea* L., cv. Koroneiki | Reduction of photosynthesis and increase of antioxidant enzyme activities | Koubouris et al. (2015) |
| Salinity | *Phoenix dactylifera* L. | Reduced the oxidative stress | El-Tarabily and Abu Qamar (2025) |
| Salinity | *Olea europaea* L. | Increase the antioxidant compounds and detoxification enzyme | Galicia-Campos et al. (2022) |
| Salinity | *Olea europapea* L. | | |
| Salinity | *Fragaria x ananassa* | Activation of different SOD families | Zhang et al. (2024) |
| Salinity | *Dalbergia odorifera* | Under salinity PAL, PPO, and SOD increased, as well as flavonoids. | Cisse et al. (2024) |
| Salinity | *Ficus carica* | Increase the antioxidant compounds | Francini et al. (2021) |

## 1.3 Conclusion

Understanding the mechanisms used for preventing ROS damage will help in developing strategies to improve the resilience of woody and fruit trees to environmental stress. Fruit trees can adjust their metabolic processes, such as increasing the production of osmoprotectants and stress-responsive proteins, to better cope with abiotic stresses. These adaptations allow fruit trees to maintain cellular homeostasis and protect vital cellular components during periods of stress. Additionally, fruit trees can modify their root architecture and enhance water uptake efficiency to mitigate the effects of drought stress. By studying these adaptive mechanisms, researchers can develop targeted approaches to enhance stress tolerance in fruit trees, potentially leading to improved crop yields and food security in the face of changing climatic conditions.

## References

Ayaz M, Varol N, Yolcu S, Pelvan A, Kaya Ü, Aydogdu E, Bor M, Özdemir F, Türkan İ (2021) Three (Turkish) olive cultivars display contrasting salt stress-coping mechanisms under high salinity. Trees 35:1283–1298. https://doi.org/10.1007/s00468-021-02115-w

Ben Abdallah M, Trupiano D, Polzella A, De Zio E, Sassi M, Scaloni A, Zarrouk M, Youssef NB, Scippa GS (2018) Unraveling physiological, biochemical and molecular mechanisms involved in olive (Olea europaea L. cv. Chétoui) tolerance to drought and salt stresses. J Plant Physiol 220:83–95. https://doi.org/10.1016/j.jplph.2017.10.009

Boussadia O, Steppe K, Van Labeke MC, Lemeur R, Braham M (2015) Effects of nitrogen deficiency on leaf chlorophyll fluorescence parameters in two olive tree cultivars 'Meski'and 'Koroneiki'. J Plant Nutr 38(14):2230–2246. https://doi.org/10.1080/01904167.2015.1069339

Carillo P, Ferrante A (2025) Decoding the intricate metabolic and biochemical changes in plant senescence: a focus on chloroplasts and mitochondria. Ann Bot:1–13. https://doi.org/10.1093/aob/mcaf003

Chen L, Ji X, Luo C, Song X, Leng X, Ma Y et al (2024) VvLBD39, a grape LBD transcription factor, regulates plant response to salt and drought stress. Environ Exp Bot 226:105918

Cisse EHM, Jiang BH, Yin LY, Miao LF, Li DD, Zhou JJ, Yang F (2024) Physio-biochemical and metabolomic responses of the woody plant *Dalbergia odorifera* to salinity and waterlogging. BMC Plant Biol 24(1):49

Clément P (2007) Introducing the cell concept with both animal and plant cells: a historical and didactic approach. Sci Educ 16:423–440

Dai Q, Wu L, Jia B, Wang L, Zhang X, Zhang L et al (2025) Accelerated warming of air and soil temperatures: evidence from observations in Fuyang from 1985 to 2024. Theor Appl Climatol 156(6):1–17

El-Tarabily KA, Abu Qamar SF (2025) Enhancement of yield, physiological characteristics, and oxidative stability of date palm (*Phoenix dactylifera* L.) trees under salt stress using biogenic selenium nanoparticles. Sci Hortic 342:114016

Ferrante A, Francini A (2023) Cross-talk of NO between RNS, ROS, and RSS under abiotic stress. In: Nitric oxide in developing plant stress resilience. Academic Press, pp 125–134. Elsevier.

Fidalgo-Illesca C, Francini A, Raffaelli A, Sebastiani L (2025) Silicon priming triggers differential physiological, ionomic and metabolic responses in olive (*Olea europaea* L.) cultivars with different tolerance to salinity. Plant Physiol Biochem 223:109809. https://doi.org/10.1016/j.plaphy.2025.109809

Francini A, Sebastiani L (2019) Abiotic stress effects on performance of horticultural crops. Horticulture 5(4):67

Francini A, Galleschi L, Saviozzi F, Pinzino C, Izzo R, Sgherri C, Navari-Izzo F (2006) Enzymatic and non-enzymatic protective mechanisms in recalcitrant seeds of Araucaria bidwillii subjected to desiccation. Plant Physiol Biochem 44(10):556–563. https://doi.org/10.1016/j.plaphy.2006.09.002

Francini A, Sodini M, Vicario G, Raffaelli A, Gucci R, Caruso G, Sebastiani L (2021) Cations and phenolic compounds concentrations in fruits of fig plants exposed to moderate levels of salinity. Antioxidants 10(12):1865. https://doi.org/10.3390/antiox10121865

Franzoni G, Cocetta G, Prinsi B, Ferrante A, Espen L (2022) Biostimulants on crops: their impact under abiotic stress conditions. Horticulture 8:189. https://doi.org/10.3390/horticulturae8030189

Galicia-Campos E, García-Villaraco Velasco A, Montero-Palmero MB, Gutiérrez-Mañero FJ, Ramos-Solano B (2022) Modulation of photosynthesis and ROS scavenging response by beneficial bacteria in *Olea europaea* plantlets under salt stress conditions. Plants 11(20):2748

Gallé A, Haldimann P, Feller U (2007) Photosynthetic performance and water relations in young pubescent oak (*Quercus pubescens*) trees during drought stress and recovery. New Phytol 174(4):799–810. https://doi.org/10.1111/j.1469-8137.2007.02047.x

Gao T, Zhang D, Shen W, Xu S, Jia X, Liu X et al (2024) MdASMT9-mediated melatonin biosynthesis enhances basal thermotolerance in apple plants. Plant Cell Environ 47(3):751–764

Gechev T, Petrov V (2020) Reactive oxygen species and abiotic stress in plants. Int J Mol Sci 21(20):7433

Hosseinifard M, Stefaniak S, Ghorbani Javid M, Soltani E, Wojtyla Ł, Garnczarska M (2022) Contribution of exogenous proline to abiotic stresses tolerance in plants: a review. Int J Mol Sci 23(9):5186. https://doi.org/10.3390/ijms23095186

Janero DR, Burghardt B (1989) Thiobarbituric acid-reactive malondialdehyde formation during superoxide-dependent, iron-catalyzed lipid peroxidation: influence of peroxidation conditions. Lipids 24(2):125–131. https://doi.org/10.1007/BF02535249

Jia X, Xu S, Wei J, Wang F, Qing Y, Zhang Z et al (2025) Cytokinin mediates age-dependent drought response by regulating the removal reactive oxygen species in apple. Plant J 121(4):e70023

Kavi Kishor PB, Hima Kumari P, Sunita MSL, Sreenivasulu N (2015) Role of proline in cell wall synthesis and plant development and its implications in plant ontogeny. Front Plant Sci 6:544. https://doi.org/10.3389/fpls.2015.00544

Kim TL, Lee K, Hwang HS, Oh C, Lee IH, Lim H (2022) Comparison of physiological and biochemical responses of two poplar species under drought stress. Plant Breed Biotechnol 10(3):145–162

Koubouris GC, Kavroulakis N, Metzidakis IT, Vasilakakis MD, Sofo A (2015) Ultraviolet-B radiation or heat cause changes in photosynthesis, antioxidant enzyme activities and pollen performance in olive tree. Photosynthetica 53:279–287

Liu X, Gao T, Liu C, Mao K, Gong X, Li C, Ma F (2023) Fruit crops combating drought: physiological responses and regulatory pathways. Plant Physiol 192(3):1768–1784. https://doi.org/10.1093/plphys/kiad202

Locascio SJ, Wiltbank WJ, Gull DD, Maynard DN (1984) Fruit and vegetable quality as affected by nitrogen nutrition. In: Nitrogen in crop production, pp 617–626. https://doi.org/10.2134/1990.nitrogenincropproduction.c42

Lombardi L, Sebastiani L, Vitagliano C (2003) Physiological, biochemical, and molecular effects of in vitro induced iron deficiency in peach rootstock Mr.S 2/5. J Plant Nutr 26(10–11):2149–2163. https://doi.org/10.1081/PLN-120024271

Marnett LJ (1999) Lipid peroxidation—DNA damage by malondialdehyde. Mutat Res Fundam Mol Mech Mutagen 424(1–2):83–95

Meng X, Chen WW, Wang YY, Huang ZR, Ye X, Chen LS, Yang LT (2021) Effects of phosphorus deficiency on the absorption of mineral nutrients, photosynthetic system performance and antioxidant metabolism in *Citrus grandis*. PLoS One 16(2):e0246944

Oustric J, Morillon R, Luro F, Herbette S, Martin P, Giannettini J et al (2019) Nutrient deficiency tolerance in citrus is dependent on genotype or ploidy level. Front Plant Sci 10:127. https://doi.org/10.3389/fpls.2019.00127

Podgórska A, Burian M, Szal B (2017) Extra-cellular but extra-ordinarily important for cells: apoplastic reactive oxygen species metabolism. Front Plant Sci 8:1353

Renzetti M, Funck D, Trovato M (2024) Proline and ROS: a unified mechanism in plant development and stress response? Plants (Basel) 14(1):2. https://doi.org/10.3390/plants14010002

Rossi L, Francini A, Minnocci A, Sebastiani L (2015) Salt stress modifies apoplastic barriers in olive (Olea europaea L.): a comparison between a salt-tolerant and a salt-sensitive cultivar. Sci Hortic 192:38–46

Rossi L, Borghi M, Francini A, Lin X, Xie DY, Sebastiani L (2016) Salt stress induces differential regulation of the phenylpropanoid pathway in Olea europaea cultivars Frantoio (salt-tolerant) and Leccino (salt-sensitive). J Plant Physiol 204:8–15. https://doi.org/10.1016/j.jplph.2016.07.014

Rudenko NN, Vetoshkina DV, Marenkova TV, Borisova-Mubarakshina MM (2023) Antioxidants of non-enzymatic nature: their function in higher plant cells and the ways of boosting their biosynthesis. Antioxidants 12(11):2014

Sebastiani L, Gucci R (2023) Responses to abiotic stresses. In: The olive: botany and production, pp 510–528

Shajar F, Nabi AU, Manzoor A, Saleem S, Mushtaq NU, Manzoor S, Rehman RU (2025) Understanding lipid-mediated defense mechanisms in plant responses to abiotic stress. Nucleus:1–14. https://doi.org/10.3390/ijms26073450

Shi H, Ma W, Song J, Lu M, Rahman SU, Bui TTX et al (2017) Physiological and transcriptional responses of *Catalpa bungei* to drought stress under sufficient-and deficient-nitrogen conditions. Tree Physiol 37(11):1457–1468

Sodini M, Astolfi S, Francini A, Sebastiani L (2022) Multiple linear regression and linear mixed models identify novel traits of salinity tolerance in *Olea europaea* L. Tree Physiol 42(5):1029–1042. https://doi.org/10.1093/treephys/tpab158

Sodini M, Francini A, Sebastiani L (2023) Early salinity response in root of salt sensitive *Olea europaea* L. cv Leccino. Plant Stress 10:100264. https://doi.org/10.1016/j.stress.2023.100264

Tang N, Li Y, Chen LS (2012) Magnesium deficiency–induced impairment of photosynthesis in leaves of fruiting Citrus reticulata trees accompanied by up-regulation of antioxidant metabolism to avoid photo-oxidative damage. J Plant Nutr Soil Sci 175(5):784–793

Traversari S, Francini A, Traversi ML, Emiliani G, Sorce C, Sebastiani L, Giovannelli A (2018) Can sugar metabolism in the cambial region explain the water deficit tolerance in poplar? J Exp Bot 69(16):4083–4097. https://doi.org/10.1093/jxb/ery195

Traversari S, Neri A, Traversi ML, Giovannelli A, Francini A, Sebastiani L (2020) Daily osmotic adjustments in stem may be good predictors of water stress intensity in poplar. Plant Physiol Biochem 146:13–22. https://doi.org/10.1016/j.plaphy.2019.10.039. Epub 2019 Nov 2

Trivellini A, Cocetta G, Francini A, Ferrante A (2017) Reactive oxygen species production and detoxification during leaf senescence. In: Reactive oxygen species and antioxidant systems in plants: role and regulation under abiotic stress. Springer Singapore, pp 115–128

Verslues PE, Sharma S (2010) Proline metabolism and its implications for plant-environment interaction. Arabidopsis Book 8:e0140. https://doi.org/10.1199/tab.0140

Wang Z, Li G, Sun H, Ma L, Guo Y, Zhao Z et al (2018) Effects of drought stress on photosynthesis and photosynthetic electron transport chain in young apple tree leaves. Biol Open 7(11):bio035279

Wang D, Gao Y, Sun S, Lu X, Li Q, Li L, Wang K, Liu J (2022) Effects of salt stress on the antioxidant activity and malondialdehyde, solution protein, proline, and chlorophyll contents of three *malus* species. Life (Basel) 12(11):1929. https://doi.org/10.3390/life12111929

Zhai L, Sun C, Li K, Sun Q, Gao M, Wu T et al (2021) MxRop1-MxrbohD1 interaction mediates ROS signaling in response to iron deficiency in the woody plant Malus xiaojinensis. Plant Sci 313:111071

Zhang D, Luo XY, Zhao T, Cheng L, Zhang R, Zhang XY et al (2020) A study on the physiological responses of *malus hupenensis* to exogenous silicon treatment under drought stress. J Fruit Sci 37(10):1506–1517

Zhang Y, Chen Y, Gao W, Tian S, Lin B, Gu X et al (2024) Characterization of superoxide dismutase (SOD) gene family and their responses to salinity stress and fruit development in octoploid strawberry. Agronomy 14(11):2514

# 2 Redox Signaling: A Master Regulator of Plant Responses to Climate Stress

Gholamreza Abdi, Tadela Susmitha, Md Sabir Ahmed Mondol, Mero Yangfo, Anilava Chatterjee, Ubaida Akbar, Aritra Mukherjee, and Amisha Rani

**Abstract**

Significant challenges to plant survival and production are being brought about by the phenomena of climate change. Heavy metals, high heat or cold, drought, or salt can have a significant impact on plant growth indices. Redox signaling, mediated by antioxidants and reactive oxygen species (ROS), plays a crucial role in controling how plants adapt to environmental challenges. Many metabolic reactions produce reactive oxygen species (ROS), which include superoxide,

G. Abdi (✉)
Department of Biotechnology, Persian Gulf Research Institute, Persian Gulf University, Bushehr, Iran
e-mail: abdi@pgu.ac.ir

T. Susmitha
Department of Agricultural Biochemistry, School of Agricultural Sciences, Mallareddy University, Hyderabad, Telangana, India

M. S. A. Mondol · A. Chatterjee
Department of Agricultural Biochemistry, Bidhan Chandra Krishi Viswavidyalaya, Mohanpur, Nadia, West Bengal, India

M. Yangfo
North Eastern Regional Institute of Science and Technology, Nirjuli, Arunachal Pradesh, India

U. Akbar
Department of Food Technology and Nutrition, Lovely Professional University, Phagwara, Punjab, India

A. Mukherjee
Department of Seed Science and Technology, Bidhan Chandra Krishi Viswavidyalaya, Mohanpur, Nadia, West Bengal, India

A. Rani
Department of Forest Products, Dr. Yashwant Singh Parmar University of Horticulture and Forestry, Solan, India



N. Iqbal et al. (eds.), *Plant Signaling Molecules in Regulation of ROS-Scavenging System*, https://doi.org/10.1007/978-981-95-3377-0_2

hydrogen peroxide, and hydroxyl radicals. In addition to causing oxidative damage when present in excess, reactive oxygen species (ROS) act as signaling molecules that set in motion a complex cascade of reactions. Redox homeostasis and cellular oxidative stress mitigation are mostly accomplished by antioxidants such ascorbate, glutathione, and a variety of enzymes generated inside plants. In this chapter, we will look at how antioxidants and reactive oxygen species (ROS) work together to keep redox homeostasis in check. Furthermore, the chapter explores the possibility of modifying redox signaling pathways to increase climatic stress tolerance in plants.

**Keywords**

ROS · Redox Signaling Pathways · Abiotic stress

## 2.1 Introduction

Natural ecosystems are threatened by global climate change. For example, changes in temperature, the intensity of rainfall, and $CO_2$ concentration are some altered environmental conditions that lead to new pests and pathogens. It reduces plant growth, causes damage to the photosynthesis process, and leads to a reduction in crop productivity, thus affecting food security (Chaudhry and Sidhu 2022). Global warming tendencies have risen to disturbing proportions in view of the impacts of human actions and the accompanying natural influences. The change in the climate has generally been associated with a rise in heat waves, increasing their frequency and intensity, harming agricultural production (IPCC 2021). One kind of heat stress is heat shock, which occurs when the body's temperature rises by 10–15 °C in a short amount of time (minutes to hours), while another form is gradual warming, which occurs over days or weeks and is characterized by a 2–5 °C increase. Heat stress has detrimental effects on plant development, growth, and reproduction on a molecular and morpho-physiological level, despite the fact that plants react to such pressures in varied ways (Zhao et al. 2020; Wang et al. 2020a). The greatest limitation this increased temperature imposes is the great restraint it places on crop yields. During gametogenesis, heat stress causes structural deformations in the ultrastructure of the ovary and pollen, thus negatively affecting its viability and reducing fertilization rates in the end. The length of time it takes for grains to fill up, the availability of nutrients, the activity of enzymes involved in starch metabolism, and the ratio of abscisic acid to ethylene are all affected by heat stress. Thermal stress, like cold, floods, or nutrient deprivation, affects plant development, breaks down relationships between plants and microbes, and has a major impact on harvest yields. (Ullah et al. 2022; Peck and Mittler 2020; Ahammed and Yu 2016; Jagadish et al. 2015). Ortiz-Bobea et al. (2021) reported that climate change because of human activities has caused around a 21% decrease in global agriculture productivity since 1961. Changes over warmer and dryer regions were the maximum; Africa, for example, faced around a 34% decrease in agriculture productivity, while cooler

Europe and Central Asia have dropped by 7.1%. In the next decades, climate change would continue to impact the global food supply system. In addition, the frequency and severity of extreme weather events will be highly impacted, which also includes increased atmospheric $CO_2$, fluctuating temperatures, supercells, cyclones, tides, hail, and altered patterns of rainfall. Most of the studies have explained that the rise in atmospheric $CO_2$ alters both the biomass and biochemistry of most plant species, but individual species respond differently to elevated $CO_2$ conditions. Overall, it is very evident that carbon dioxide emissions and other greenhouse gases have been major factors in shaping the current world conditions (Wang et al. 2018; Nogia et al. 2016). Heat stress interferes with several molecular and metabolic processes in plants, including water potential, electron transport chain, cell differentiation, photosynthetic activity, nutrient intake and transport, respiration, and nutrient transfer. Extreme heat triggers photosynthesis, which in turn alters membrane fluidity and thylakoid permeability and amplifies microscopic structural deformation in chloroplasts. The oxygen-evolving complex, the cytochrome b6f complex, the mechanisms of the ATP-generating process, and photosystem II (PSII) are the main targets of heat damage in chloroplasts. Heat stress also affects carbon absorption processes, chloroplast biosynthesis, and RUBISCO activase activity, which manifests as a yellowing of the leaves. Injuries such as protein denaturation, cytoskeleton and membrane damage, and enzyme inactivation are brought about by the cellular level increase in reactive oxygen species generation that is stimulated by heat stress. The reference is from Fortunato et al. (2023). Since a result, stress responses are impacted, since low or abrupt temperatures harm plant metabolism and development. The accumulation of harmful metabolites causes the membrane to become stiff and proteins to become unstable. The efficiency of scavenging ROS is lowered due to this, and this lowering of photosynthesis is through cold stress. When severe temperature conditions prevail, then the temperature may be lesser than 0 °C inside the plant tissues; hence, it damages the water potential inside a cell besides cellular processes owing to the development of ice crystals inside the tissues. Alteration in membrane lipid composition, accumulation of protective proteins and sugars, and induction of cold-responsive signaling pathways, including the CBF-COR pathway, for improved freezing tolerance by gene regulation of ROS detoxification and hormone metabolism. In rice, chilling tolerance is dependent on the induction of gene expression and the upregulation of antioxidants by ROS. Levels of MDA and $H_2O_2$ are found to increase at low temperatures. Redox signaling also governs cold stress response genes. Recent proteomics analysis in Brassica napus indicates that improved cold tolerance was associated with ROS scavenging pathways (Devireddy et al. 2021). The photosynthetic machinery is one of the most important targeted locations of heat stress produced damage stress, and plants have developed sophisticated signaling mechanisms to detect and react to heat (Sharma et al. 2020). How heat shock proteins (HSPs) and heat shock factors (HSFs) interact with one another is a key topic of research into the HSR. Heat-stress proteins (HSPs) are a special class of chaperone proteins that help keep cells in a stable state by refolding proteins that have become denatured, misfolded, or aggregated. This allows metabolic activities to continue even when plants are exposed to bad circumstances, such as heat stress. An intricate interplay between HSFs, HSPs, and ROS has been demonstrated, and recent research has placed a greater emphasis on reactive oxygen species (ROS) as important HSR signaling molecules (Suzuki 2023; Mittler et al. 2022; Haslbeck and Vierling 2015; Driedonks et al. 2015). This chapter examines the effects of elevated $CO_2$ levels, increasing temperatures, drought, and salt on the structure and function of plants, as well as their biochemical reactions, with an emphasis on reactive oxygen species (ROS) and the mechanisms by which plants are able to withstand stress.

## 2.2 Types of ROS

When molecular oxygen ($O_2$) undergoes partial reduction or activation, it forms very reactive oxygen species, or ROS. Hydrogen peroxide ($H_2O_2$), hydroxyl radicals (OH), and singlet oxygen ($^1O_2$) are the most common reactive oxygen species (ROS) in plants. $O_2\cdot\dot{:}\cdot$ and hydroxyl radicals are examples of radicals, while ozone and hypochlorous acid are examples of non-radical species. Each form of radical has its own unique oxidative properties and affects various physiological processes. (Huang et al. 2019). Superoxide anions ($O_2^-$) are constantly synthesized within chloroplasts as an by-product of energy transfer and partial $O_2$ reduction, predominantly in photosystem I in the course of noncyclic electron transport in the thylakoid membrane. Molecular oxygen is also critical during seed release of dormancy via cytochrome c oxidase reactions (Ma et al. 2016). Superoxide anions and reactive ROS change thiol groups and lead to the increase of the total pool of glutathione, which plays a critical role in mobilization of nitrogen and carbohydrates, leading to germination and early plant development (Bykova et al. 2011). As a cellular element present in cells naturally, hydrogen peroxide ($H_2O_2$) behaves as a constitutive and stress-inducible molecule. $H_2O_2$ is formed mainly by univalent protonation and partial $O_2$ reduction in different organelles such as the endoplasmic reticulum, mitochondria, chloroplasts, and in fatty acid oxidation, photorespiration, and in membrane-associated processes. Optimal concentrations of $H_2O_2$ are significant as a signal molecule controlling vital physiological processes like stomatal dynamics, photosynthesis, senescence, and growth of the plant. Overexpression, however, is cytotoxic and results in cell damage (Noctor et al. 2002). While electron transfer to $O_2$ contributes to the formation of one kind of ROS, chlorophyll reactions within the antenna system results in production of different ROS. The overall intracellular $CO_2$ concentration is lowered through heavy metals, salt, and dryness and leads to stomatal closure: damaging both photosystems I and II and increasing the overproduction of $^1O_2^-$ with the final damage to a number of cellular targets. In addition, ROS damage plant pigments, proteins, lipids, and nucleic acids and triggers the last effect-cell death. The most reactive and toxic species is the hydroxyl radical, or $OH^-$. They are generated under neutral pH conditions by the Fenton reaction that is initiated by $H_2O_2$ coupled with transition metals, such as Fe. Due to the absence of removal processes, lipid peroxidation and protein degradation follow, which leads to further membrane damage and cell death. Previously, $OH^-$ only caused destruction, but now it has

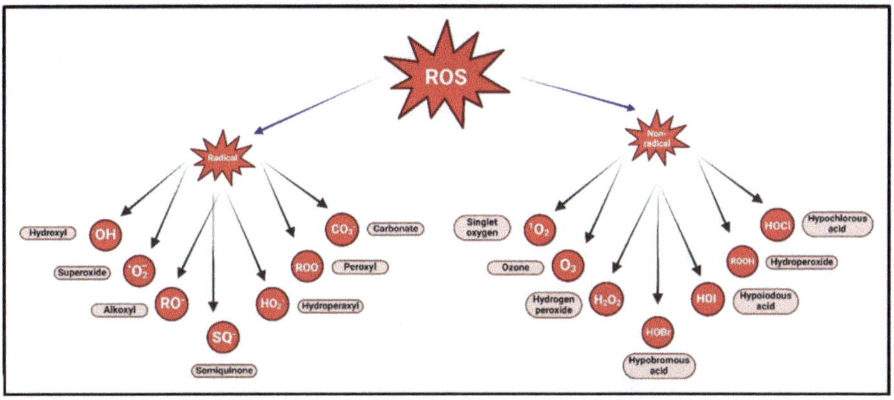


**Fig. 2.1** Types of ROS in plants. (Tarfeen et al. 2022)

been observed to exhibit signaling roles in seed germination, growth, stomatal closure, and in acclimating to stress. While they catalyze lipid peroxidation through generation of $ROO^-$ and $RO^-$ radicals affecting lipid oxidation, DNA damages, and protein modifications, alkoxyl radicals produced during lipid degradation also cause DNA damage that subsequently leads to apoptosis (Bano et al. 2021; Khan et al. 2023). The different types of ROS are shown in Fig. 2.1 and generation of ROS by energy transfer in Fig. 2.2.

## 2.3 Mechanism of ROS Production and Mitigation in Plants

Climate change has imposed augmented abiotic stress on plants across a vast area of geographical space, and 96.5% of global cultivated land is affected, according to FAO, 2019. With the increasing levels of greenhouse gases, the global surface temperature has increased by 0.85 °C, and rainfall patterns have altered. A plant water requirement tends to increase by 4–4.5% with the increase in temperature by 1 °C, which further makes cultivation difficult, particularly in drought-prone areas. Higher temperatures also convert chemical pollutants to more toxic forms, thereby increasing the environmental problem and consequently affecting the health of the plants. Fluctuation in temperature and rainfall can temporarily check growth in plants, but extreme climate events increase such stresses, severely affecting the performance of the plants and challenges to fullfill the growth and nutritional needs of the present and future populations (Jarsjö et al. 2020; Zheng et al. 2020; Sachdev et al. 2021). Metabolic processes in living cells unavoidably produce reactive oxygen species (ROS) as by-products, commonly termed as oxidative bursts. ROS in plants have a dual function—excessive accumulation is toxic, yet they are crucial signaling molecules that control growth, programmed cell death (apoptosis), and systemic responses to biotic and abiotic stresses. Changes in ROS levels and redox homeostasis disruption trigger multiple pathways of oxidative stress, depending on ROS type, uptake, and heavy metal content. Plants utilize complex survival and

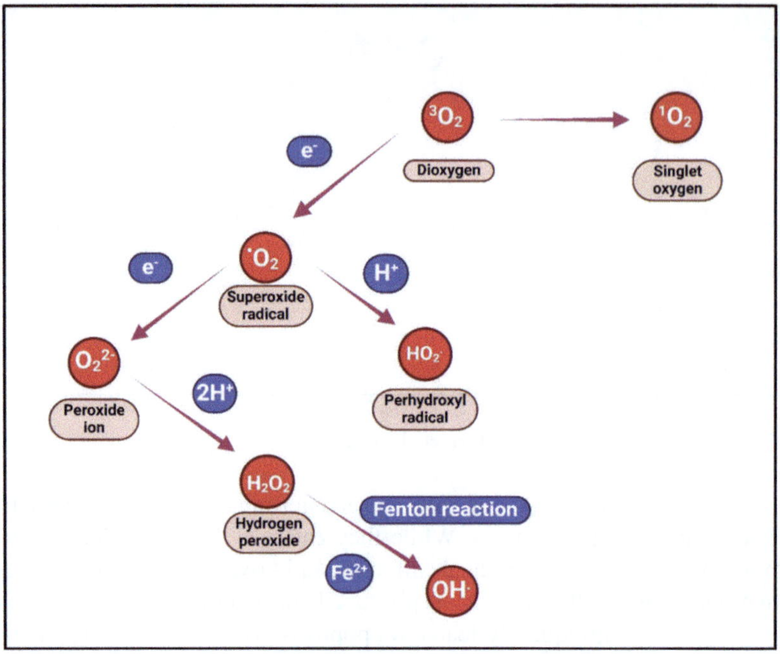


**Fig. 2.2** Generation of ROS by energy transfer. (Abouzari and Fakheri 2015)

adaptation mechanisms against environmental stress factors to limit damage and preserve cellular function (Mansoor et al. 2023). Endoplasmic reticulum, chloroplasts, mitochondria, peroxisomes, and cell walls are among the many cellular compartments that can create reactive oxygen species (ROS) in response to stress or as a normal metabolic byproduct. Chloroplasts and peroxisomes are the primary locations for reactive oxygen species (ROS) synthesis when illumination is abundant, whereas mitochondria are the primary locations for ROS generation when darkness is present. Abiotic stress causes the production of various reactive oxygen species (ROS) in various organelles, leading to the establishment of intricate signaling pathways. The presence of two unpaired electrons with the same spin quantum number renders molecular oxygen ($O_2$) a free radical in and of itself. The successive uptake of single electrons by $O_2$ causes ROS to be formed in steps due to spin restriction. (Display Unit 2). Chloroplasts are the site of primary production of ROS, being the chief location where light and chlorophyll interact. Triplet chlorophyll and the electron transport chain, comprising PSI and PSII, produce ROS. In the Mehler reaction, SOD reduces $O_2$ to $H_2O_2$, which with $Fe^{2+}$, reacts to form reactive OH. Stress conditions such as stomatal closure reduce the levels of $CO_2$, raising chloroplastic ROS levels further. They also produce harmful ROS, such as $O_2^-$ and $H_2O_2$. This is produced by less frequent energetic electrons in the mitochondrial electron transport chain, mainly produced by complexes I and III. The ROS defense mechanism includes antioxidant machinery that helps reduce oxidative stress-induced damage. It has two components: enzymatic and non-enzymatic antioxidants (Table 2.1).

**Table 2.1** Antioxidant defense mechanisms in plants—enzymatic and non-enzymatic components (Das and Roychoudhury 2014; Buchanan et al. 2015)

| Category | Antioxidant type | Function | Subcellular location |
|---|---|---|---|
| Enzymatic antioxidants | Superoxide Dismutase (SOD) | Converts superoxide anion ($O_2^-$) into hydrogen peroxide ($H_2O_2$) and molecular oxygen ($O_2$) | Peroxisomes, mitochondria, cytosol, chloroplast |
| | Catalase (CAT) | Breaks down $H_2O_2$ into water ($H_2O$) and oxygen ($O_2$) | Peroxisomes, mitochondria |
| | Ascorbate Peroxidase (APX) | Uses ascorbate to neutralize $H_2O_2$ | Peroxisomes, mitochondria, cytosol, chloroplast |
| | Monodehydroascorbate Reductase (MDHAR) | Regenerates ascorbate from monodehydroascorbate (MDHA) | Mitochondria, cytoplasm, chloroplast |
| | Dehydroascorbate Reductase (DHAR) | Restores ascorbate from dehydroascorbate (DHA) using glutathione | Mitochondria, cytoplasm, chloroplast |
| | Glutathione Reductase (GR) | Recycles glutathione disulfide (GSSG) back to reduced glutathione (GSH) | Mitochondria, cytoplasm, chloroplast |
| | Guaiacol Peroxidase (GPX) | Reduces $H_2O_2$ using phenolic compounds | Mitochondria, cytoplasm, chloroplast, ER |
| Non-enzymatic antioxidants | Ascorbic Acid (Vitamin C) | Scavenges reactive oxygen species (ROS) and regenerates other antioxidants | Cytosol, chloroplast, mitochondria, peroxisomes, vacuole, apoplast |
| | Glutathione (GSH) | Detoxifies ROS and heavy metals, supports antioxidant enzyme function | Cytosol, chloroplast, mitochondria, peroxisomes, vacuole, apoplast |
| | α-Tocopherol (Vitamin E) | Protects membranes by neutralizing lipid peroxidation products | Membranes (chloroplast, mitochondria) |
| | Carotenoids | Absorbs excess light energy and quenches singlet oxygen ($^1O_2$) | Chloroplasts, non-green plastids |
| | Flavonoids | Directly scavenges ROS, including hydroxyl radicals (OH•) and singlet oxygen ($^1O_2$) | Vacuole |

Enzymes like Mn-SOD and APX convert $O_2$ into $H_2O_2$ in the mitochondrial matrix. The production of ROS highly increases under abiotic stress (Khan et al. 2023). $O_2^-$ is the most abundantly produced under abiotic stress due to molecular oxygen gaining an electron. Through redox reactions in the mitochondrial and photosynthetic electron transport chains, a proton gradient is established for generating ATP. A state close to being fully reduced results when input of energy is high but intermediate carriers such as cytochrome c and ubiquinone are restricted. NADPH is an electron source. $O_2$ production occurs at different parts, for instance, photosystem I and mitochondrial complex due to NADPH oxidase. $O_2$ is also released by NADPH oxidase into the apoplast that enables local and systemic signaling under stress conditions (Navrot et al. 2007; Dvořák et al. 2021). The hydrogen peroxide ($H_2O_2$), that is produced through dismutation catalyzed by the enzyme superoxide dismutase (SOD) of the intermediate peroxide ($O_2^{2-}$) from $O_2^-$. $O_2^{2-}$ captures two protons to form $H_2O_2$. More abundant $H_2O_2$ is produced because it is the least reactive and more stable. Whereas low concentration of $H_2O_2$ acts as an intracellular messenger that elevates tolerance of the cell to abiotic stress, its high concentration may cause programmed cell death. It is mainly produced in the process of photorespiration in peroxisomes and can also be formed from the endoplasmic reticulum by oxidoreductase enzymes. The hydroxyl radical (OH-), the second ROS on the list, is the most reactive and poisonous of the bunch. It is produced from hydrogen peroxide by means of the Fenton reaction when redox-active metal ions like $Fe^{2+}$ and $Cu^+$ are present. Although its high reactivity limits detoxification possibilities, proline, and other antioxidants can help reduce the levels of this radical during stress. Antennae system components and the reaction centers of photosystem II are potentially potential locations for the creation of $^1O_2$, the excited state of $O_2$, which is extremely reactive and often formed within chloroplasts during photosynthesis (Singh et al. 2022). To facilitate photosynthesis, the region immediately surrounding the plant cell membrane changes the incoming $CO_2$ into a form that can enter the cytoplasm. Hydrogen peroxide ($H_kO_k$) is produced in the apoplast by signaling chemicals and abscisic acid when stress is present. These players aren't the only ones that can produce reactive oxygen species (ROS); oxidases connected to cell walls, oxidases of polyamines, and peroxidases that are pH-dependent are also involved. The survival of every plant cell depends on its ability to communicate with its surroundings through its plasma membrane. During its involvement in the transit of electrons from cytosolic NADPH to $O_2$, it creates $H_2O_2$. Additionally, it produces $O_2$ through SOD. In order for plants to fight against diseases and abiotic stress, the NADPH oxidase enzyme must be active. $O_k^-$, $OH^-$, ¬$O_2$, and $H_2O_2$ are ROS that were produced in the cell wall by the catalyzed hydroperoxidation of PUFA by lipoxygenase (LOX). In addition, cell-wall diamine oxidases that catalyze the oxidation of diamines or polyamines to generate ROS. Furthermore, $H_2O_2$ pathways entail extreme cross-linking and the production of recombinant lignin when a pathogen attacks lignin precursors (Khan et al. 2023).

Since glycolytic enzymes and the TCA cycle are very sensitive to redox alterations, it stands to reason that redox perturbations can cause rapid metabolic adaptations even in the presence of hypoxia. The reduction in fumarate and malate, two

TCA intermediates, in this condition may indicate that the fumarase and succinate dehydrogenases, which are controlled by the mitochondrial TRX system, are inactivated. In addition, fermentative enzymes like ADH are designed with resistance to oxidative stress, and therefore, for instance, $H_2O_2$ represses ADH activity through oxidation of essential cysteine residues, and high NADH maintains it in a reduced state under hypoxic conditions. However, post-re-oxygenation $H_2O_2$ could degrade ADH rapidly. Understanding the redox regulation of the metabolic enzymes under hypoxic conditions may be an invaluable source of information, including but not limited to examining lines of transgenic redox-sensitive and insensitive fumarase or succinate dehydrogenase (Sasidharan et al. 2021). Plant receptor perception of temperature changes leads to complex networks of signaling, including lower membrane thermostability, increased malondialdehyde (MDA) levels, and the production of reactive oxygen species (ROS) such as $H_2O_2$, $1O_2$, $O_2^-$, and hydroxyl radicals. However, uncontrolled production of ROS causes oxidative stress, damaging proteins, biomolecules, and membranes, thereby impairing photosynthesis and overall leaf physiology. For instance, 32/20 °C Day/Night heat stressed chickpea or rice plants showed a 6.5-fold increase in $H_2O_2$ contents, which caused oxidative damage compared with controls (Awasthi et al. 2017; Liu et al. 2019). The most recent studies report that miRNAs like miR398 are required for the regulation of transcription factors involved in heat-induced responses by repressing several ROS-scavenging enzymes, which allow the accumulation of ROS as well as the activation of HSFA1 to turn on heat stress responses. Moreover, the study carried out on mutants of ROS biosynthesis, such as NADPH oxidases (atrbohB and atrbohD), has proved that the produced thermotolerance is associated with reduced levels of ROS, suggesting that an optimal level of ROS is necessary in order to obtain thermotolerance. In addition, heat priming has increased the antioxidant capacity, decreased the ROS level, and inhibited cell death through delaying the effects of ROS. Heat treatment also increases photosynthetic activity with an improvement in the stomatal conductance. Also, the development of secondary compounds with potential antioxidant properties helps in conserving the integrity of membranes of the leaves exposed to the stress of high-temperature. In general, all of these reports indicate that the ROS play critical roles in maintaining the balancing act between growth, acclimation, and the defense response to fluctuating temperatures (Devireddy et al. 2021).

## 2.4 ROS as a Signal Molecule

At low quantities, reactive oxygen species (ROS) serve a dual purpose: they stimulate the stress and acclamatory responses in response to abiotic stress. They are produced by stress sensors, such as cyclic nucleotide-gated calcium channels (CNGCs) and activate NADPH oxidases at the plasma membrane through phosphorylation or $Ca^{2+}$ fluctuations. This ROS signaling affects gene expression and a number of subsequent processes including $Ca^{2+}$ mobilization and protein phosphorylation, which in turn decrease metabolic ROS and thereby reduce stress impacts on plants. Moreover, certain forms of ROS and redox state alterations contribute significantly to an effective defense against abiotic stress (Singh et al. 2022). Typically, reactive oxygen species (ROS) are considered harmful waste products of aerobic metabolism that need to be removed. Simultaneously, they are involved in every facet of plant cellular functions, including control of stomatal openings, growth, development, defense against pathogens, programmed cell death, and abiotic and biotic stressors. The production of ROS in chloroplasts and mitochondria affects nuclear gene expression, implying that ROS signals are indeed intracellularly transduced. ROS can be sensed by two-component histidine kinases and redox-sensitive transcription factors. It is able to activate calcium signaling through its binding to the calcium-binding proteins, such as calmodulin and phospholipases that produce phosphatidic acid. The latter affects the activity of many transcription factors and activates mitogen-activated protein kinase cascade and protein kinase OXI1. Furthermore, ROS detection-based phosphatase inhibition may further activate OXI1 and MAPKs. At last, reactive oxygen species (ROS) signals trigger defense systems by tipping the production/scavenging ratio of ROS in favor of scavenging. One to two percent of the genes in Arabidopsis and tobacco are responsive to hydrogen peroxide exposure, suggesting that exposure to reactive oxygen species (ROS) like $H_2O_2$ may significantly change gene expression. These regulate the expression of genes associated with kinases, antioxidant enzymes, and defense proteins through interactions with a number of transcription factors, including HSFs, NPR1, WRKYs, and Myb-type factors (Fig. 2.3). Plants' antioxidant defense proteins and ascorbate-glutathione cycle enzymes are strengthened when exposed to reactive oxygen species (ROS) in organelles. It is paradoxical that plants have compensating mechanisms in response to oxidative stress, which is itself caused by a decline in antioxidant activity. Scientists have recently been devoting a lot of time

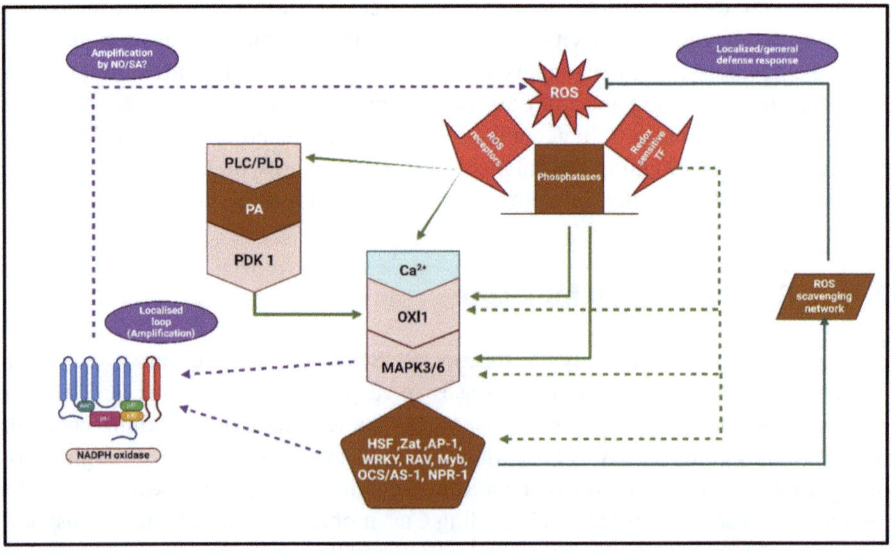


**Fig. 2.3** Proposed model of the ROS signal transduction pathway. (Buchanan et al. 2015)

and energy to deciphering the "ROS gene network" that plants have, which consists of around 150 genes encoding enzymes like peroxidases and catalases that are involved in this process. As the network evolves, more mutations or gene deletions will clarify how biotic and abiotic stimuli interact to shape the oxidative stress response (Buchanan et al. 2015).

There are several locations in plant cells where reactive oxygen species (ROS) detection and signal transduction can take place. These locations can be classified as either intrinsic (in the cytosol and nucleus) or organellar (in chloroplasts, mitochondria, peroxisomes, etc.). These pathways' ability to work together or independently under stress is a relatively new discovery. Enzymes that generate or degrade reactive oxygen species (ROS), as well as antioxidants that do not involve enzymes, are found in the apoplast and cell wall in extrinsic ROS signaling. RBOHs, AQPs, and peroxidases are important in apoplastic ROS signaling. RBOHs, which are regulated by $Ca^{2+}$, and other posttranslational modifications of regulation through phosphorylation, generate $O_2^-$ through cytosolic NADPH and have been demonstrated to influence ROS signature following stress. Peroxidases contribute to the regulation of ROS in the apoplast. Although ROS has the potential to react with receptors such as HPCA1, oxidize antioxidants, and modulate ion channels, accumulated ROS cannot enter cells on their own and therefore must be transported into cells through AQPs. A range of modifications modulates the function of AQPs. RBOHs and ROS complexes can also be taken into the cell through endocytosis to impact cytosolic ROS levels. This modulates and can exert active control over ROS levels and signaling functions of ROS in response to stress. The apoplast–cytosol interface is emerging as a critical interface for signaling of ROS-related pathways during stress (Smirnoff and Arnaud 2019; Kimura et al. 2020; Lee et al. 2022; Shen et al. 2020; Lee et al. 2020; Wang et al. 2020b; Mittler et al. 2022). While in the case of intrinsic ROS signaling, cytosol has a different array of mechanisms for scavenging ROS and some enzymes that are ROS producing and, thus regulates both cytosolic as well as transported from apoplast or organelles to the nucleus ROS. It is also equipped with hundreds of signaling hubs such as MAPK cascades and calcium-dependent protein kinases, integrating ROS signals with other molecules such as $Ca^{2+}$ and hormones. AQP enables bidirectional transport of $H_2O_2$, thereby allowing cytosolic concentration to further regulate other regions of the cell. This role that the cytosol might assume in the taking of control of ROS signaling is very important under stress, as its scavenging capacity can modulate acclimatization and defense mechanisms. The fact that there exist ROS gradients within cells suggests that scavenging processes also modulate ROS signals. Many transcriptional regulators of plant stress responses are activated in the cytosol prior to entering the nucleus. Though it is relatively well-known what regulates ROS in the cytosol, nuclear ROS regulation remains unknown though proteins like GRXs and TRXs exist to regulate the redox levels and modify transcription factors (Ding et al. 2019; Exposito-Rodriguez et al. 2017; Jabłońska and Tawfik 2021; Nietzel et al. 2020; Mittler et al. 2022). Unlike their adverse effects and contribution to thermotolerance, optimal ROS function as signaling molecules produced after the perception of stress by sensors. Signaling ROS are found generally in the metabolically active tissues of the plant and their functions are controlled by a set of genes that are

ROS-related. The production of ROS occurs, specifically in the chloroplast, upon heat stress of the temperature stresses. But signaling ROS are mainly mediated at the plasma membrane by Ca or activated NADPH oxidases themselves, and act as the transducers of heat shock signals. $Ca^+$ plus ROS and all other such signals play long-distance systemic signals for systemically inducing acclimatization from abiotic and heat stress conditions. For instance, under heat stress, sensors have been identified to stimulate the synthesis of ROS through plant hormones such as abscisic acid (ABA) and jasmonic acid (JA). The RBOH proteins in neighboring cells are activated by the ROS that are subsequently produced. This sets off a systemic signal, commonly called the "ROS wave," and raises ROS levels even higher. This kind of hormone-induced reactive oxygen species (ROS) spreads from cell to cell, setting up a feed-back loop that activates acquired acclimation systemically. Research has shown that, depending on the ROS wave, both local and systemic transcriptional responses are comparable to those induced by localized applications of heat or cold. ROS provides further activation of the calcium channels, enabling the opening of the vacuolar calcium channel Two Pore Channel1, which is involved in the transfer of calcium from the vacuoles. The respiratory burst oxidase homolog D proteins are activated by this cascade, which sets in motion a feedback process that increases ROS and calcium, resulting in a comprehensive acclimation response in the plant. (Devireddy et al. 2021).

## 2.5 Changes in ROS Levels in Various Cell Compartments and Integration of Such Signals During Stress

A plant's ability to sense changes in light, temperature, and osmotic pressure is enhanced by its many photoreceptors, osmoreceptors, and thermoreceptors. Photoreceptors such as PHYB and cryptochrome, as well as cyclic nucleotide-gated channels that sense changes in heat stress and osmotic pressure, as well as receptor-like kinases, have been identified as $Ca^{2+}$-permeable channels. Afterwards, chloroplasts emit ROS and $Ca^{2+}$ as part of retrograde signaling for light stress detection. Here, PHYB also detects the increase in temperature. This proximity of these receptors to RBOHs may enhance the rate of ROS produced during initial stress sensing. Changes in $Ca^{2+}$ signaling and phosphorylation directly related to abiotic stress sensing, with their speed, may be directly connected with the produced ROS. In particular, as deletion of specific RBOHs can abolish fast ROS accumulation under light stress, this implies that the chloroplast manages its internal ROS levels, and that RBOHs mainly drive the ROS signal. Perhaps these different functions will be fulfilled by two populations of chloroplasts—near the nucleus to support chloroplast-to-nucleus signaling and at the plasma membrane to initiate the activation of RBOH-driven ROS signals (Mittler et al. 2022). Accumulation of $O_2^-$, $H_2O_2$, or $^1O_2$ in chloroplasts following exposure to light stress may trigger the retrograde signal perceived by RBOHs, while the action of PHYB here is that of a light sensor initiating this process (Devireddy et al. 2020) (Fig. 2.4). The environment often places simultaneous stresses on plants—such as drought combined with heat and

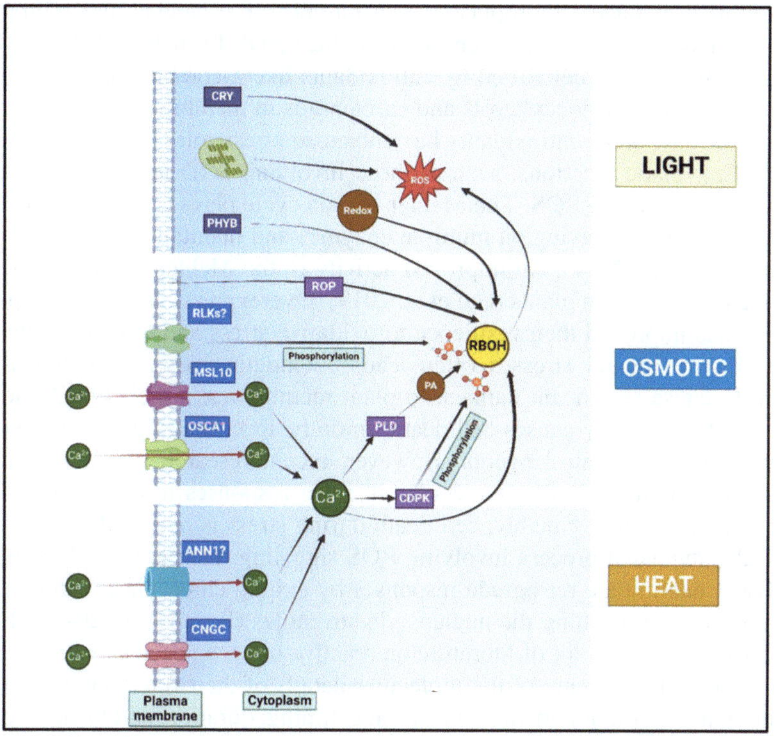


**Fig. 2.4** Different plant sensors for changes in environmental conditions. (Mittler et al. 2022)

light—that will activate several pathways involved in the "stress combination." Acclimation to such conditions is centrally focused on ROS. As an example, mutants lacking the APX1 enzyme are sensitive to drought stress and to heat stress, while mutants lacking the RBOHD gene are sensitive to multifactorial stress. On the other hand, the cascades of MPK3/MPK6 and MPK4 regulated by ROS under cold stress have contrasting functions in defense and acclimatization under pathogen responses. Different types of stresses can have unique ROS signatures, thus the integration of which may influence the overall response to dynamic environmental conditions. Interaction of stress-heat with developmental stages, such as reproduction or even the plant microbiome with ROS, determines whether a plant is tolerant or susceptible toward stress (Koussevitzky et al. 2008; Zandalinas et al. 2021; Jagodzik et al. 2018; Muhlemann et al. 2018; Pfeilmeier et al. 2021).

## 2.6 Improving Tolerance to Abiotic Stresses in Plants

In response to biotic and abiotic challenges, chloroplasts, mitochondria, and peroxisomes produce more ROS, which are primarily produced during photosynthesis, respiration, and photorespiration. To reduce reactive oxygen species (ROS) buildup and oxidative damage, chloroplasts use a number of strategies, the primary one being the dissipation of surplus energy to reduce ROS formation. When ROS are produced, they can be detoxified by antioxidants like ascorbate and glutathione in the cytosol, as well as tocopherols and carotenoids in membranes. Metabolic engineering to boost these antioxidants has enhanced stress tolerance in several plant species. Enzymatic reactions, such as those involving SOD and catalases, further neutralize less reactive ROS. The Mehler–Asada cycle plays a key role in reducing superoxide radicals, relying on multiple enzymes and maintaining antioxidant levels (Lodeyro et al. 2021). An improved activity of the Mehler-Asada pathway has been noted in mountain plants (Cui et al. 2019). Overexpression of key components in plants has improved their resilience to oxidative stress and environmental challenges. Environmental stresses often lead to oxidative stress, which was once thought to cause significant damage to plant membranes, proteins, and photosystems, which eventually causes cell death. Initially, ROS scavengers were seen primarily as damage control agents. However, recent research indicates that ROS mainly function as signaling molecules in plant responses to developmental and environmental stimuli. Notably, cell death during stress is now understood to be a genetically regulated process involving ROS signaling. Increases in ROS, particularly $H_2O_2$, can initiate retrograde responses by exiting chloroplasts through aquaporins or directly reaching the nucleus via stromules (Foyer et al. 2017; Mielecki et al. 2020). There is a lot of literature on reactive oxygen species (ROS) signaling and metabolism, but many of the molecular details of their functions in stress and development are still unknown. Type, dosage, timing, duration, and location of ROS production all influence the cellular response (Crawford et al. 2018). The effects of $H_2O_2$ are dependent on its source, which can be either the chloroplasts or the peroxisomes; however, signals from many sources can combine to produce coordinated effects. Recent reviews have shown the manner in which ROS from chloroplasts interact with signaling pathways in other parts of the cell. More research is needed to understand the molecular mechanisms, but it seems counterintuitive that enhancing tolerance would be the result of blocking the propagation of reactive oxygen species (ROS) during stress (Mielecki et al. 2020; Lodeyro et al. 2021). Instead, it was found that strengthened ROS-scavenging systems increased stress tolerance. Through stromal and thylakoid-bound APX isoforms and SOD variations containing Fe or CuZn, the Mehler-Asada cycle removes $O_2^-$ and $H_2O_2$. Several reductants and enzymes work together to recycle ascorbate from monodehydroascorbate (MDHA) and dehydroascorbate (DHA), with NADPH acting as the electron donor for MDHAR. Peroxiredoxin (Prx) is involved in a second pathway that converts hydrogen peroxide ($H_2O_2$) to water and then produces NADPH, Trxf1, and Fd-dependent Trx reductase (FTR) through ferredoxin. When $H_2O_2$ is scavenged, it forms highly reactive HO· radicals by reducing with $Fe^{2+}$. Detoxifying $^1O_2$ are tocopherols and carotenoids. This does not contain any other chemical scavengers. For further details, refer to the text (Fig. 2.5). (Lodeyro et al. 2021). Advances in the microarray technologies and next-generation sequencing have provided a detailed study of model organisms at the level of the genome and the transcriptome to reveal

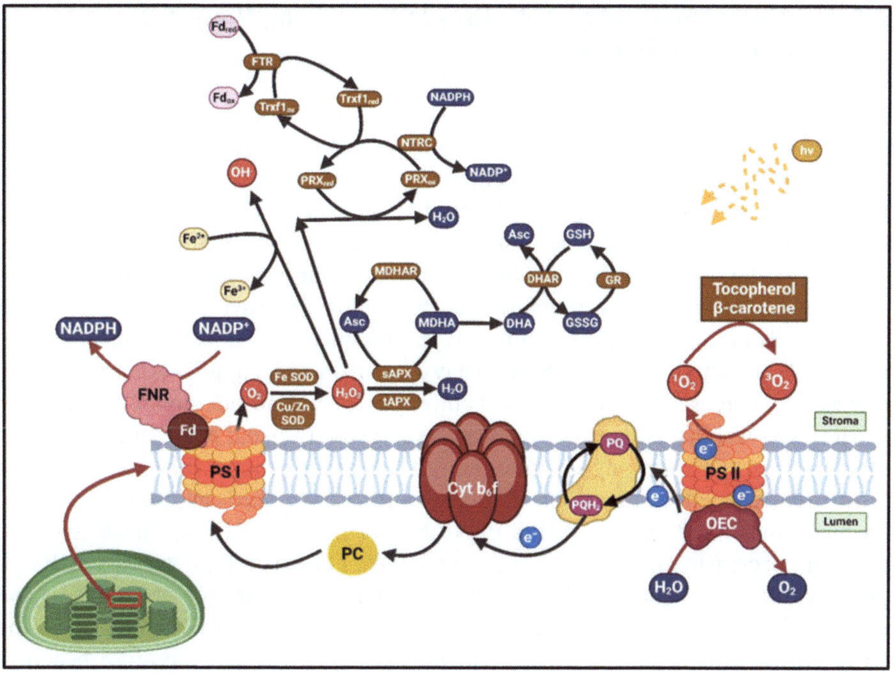


**Fig. 2.5** Summary of chloroplast ROS detoxification systems. (Lodeyro et al. 2021)

changes in the transcript behavior due to raised ROS levels. Before translation, mRNAs are processed, and they bind to RBPs of different RNA-binding domains. Amongst the factors involved in oxidative tolerant mechanisms in the Arabidopsis, several polyadenylation factors and CPSF30, among others, are recognized. Oxidative stress also impacts alternative splicing; for instance, it was possible to observe two isoforms of GORK mRNA during ozone stress.

In addition, organellar RNA maturation-related RNA editing factors encode PPR proteins. Oxidative stress regulates RNA stability; for example, SOS1 mRNA of the $Na^+/H^+$ antiporter is stabilized under oxidative stress, but AtFER1 transcripts are destabilized. Oxidative stress also affects translation and translation initiation factors like eIF4A/eIFiso4A function as redox detectors that can repress translation through disulfide bond formation.

Oxidative stress regulates several PTMs in proteins, including phosphorylation and ubiquitinylation. One such protein in Arabidopsis, CAT3, has been reported to be an mRNA-binding partner of AtGRP7, which is involved in redox homeostasis. WRKY proteins in Capsicum annuum have shown differential regulation of thermotolerance through the modulation of ROS detoxification. More research on the mRNA-bound proteome may help decipher PTMs in response to ROS (Singh et al. 2022).

### 2.6.1 Role of Several Metabolites in Improving Plant Tolerance to Abiotic Stresses

Proline is among the most studied osmoprotectants in plants. It was reported to accumulate in wilting rye grass, maintaining protein integrity and redox states. The potency differs across species, concentration, and timing. For example, proline has been known to enhance growth in stressed crops such as rice; however, very high concentrations, such as those in beans, may be toxic at or above 50 mM. Proline also caused structural changes in roots and shoots that improved the adaptation capability of plants toward stressful conditions (Szabados and Savouré 2010; Teh et al. 2016; Mansour and Ali 2017). Similarly, glutathione (GSH) is also an important detoxification activity with ROS and heavy metals. For its application, it has been reported to increase water content and chlorophyll levels in salt-stressed mungbean seedlings as well as reducing stress markers such as $H_2O_2$ and MDA and enhancing the activities of scavenging enzymes of ROS. Similar benefits for GSH treatment have been realized in a number of crops by increasing seed and pod numbers in soybeans under salt stress. GSH is important for phytochelatin synthesis and also assists in decreasing the toxicity of heavy metals. For example, it appears to revive photosynthetic activities in Pb-treated cotton plants and enhance germination and growth of different species under Hg stress. Moreover, using GSH can make the stress response of plants at higher temperatures better, which therefore becomes useful in agricultural production as well. Optimization of the applications of proline and GSH will enhance crop resistance to abiotic stresses (Nahar et al. 2015; Zhou et al. 2017). The most studied of all the plant hormones has been auxin, and L-tryptophan (TRP) applied as a treatment is usually employed because it is a precursor for the biosynthesis of auxin and reduces abiotic stresses (Mustafa et al. 2018). TRP has been found to reduce damage caused in crops like red pepper and sugar beet (Jamil et al. 2018a; Hozayn et al. 2020) by salts; thus, growth and fruit development could occur even at higher salinities. Growth and yield enhancement in wheat increased the application of TRP. It also diminishes heavy metal stress, therefore, improves the growth parameter in rice plants that had been treated with cadmium. Some findings indicated that TRP stimulates germination of seeds exposed to temperature stress and although a suitable dose depends upon specific species, it generally was reported to promote growth in plants and productivity in numerous stresses; cost-effective measures at the field level remain to be looked into (Jamil et al. 2018b; Farooq et al. 2015). Citric acid (CA) being a weaker antioxidant and chelate metal, is known for salinity as well as stress due to heavy metals. For instance, it improved nutrient uptake and growth in tomato and maize plants under salinity stress; it also induced antioxidant responses in mustard plants stressed by Cd (Pérez-Labrada et al. 2016; El-Hawary and Nashed 2019; Faraz et al. 2020). In Pb-stressed castor beans (Mallhi et al. 2019), application of CA restored growth and photosynthetic parameters. In Cu-stressed canola and peas (BenMassoud et al. 2019; Parveen et al. 2020), CA helped recover growth and biomass. CA overall presents hope in developing resistant plants for salinity as well as heavy metal stresses. Polyols are the sugar alcohols that are obtained by the reduction of carbonyl groups (CO) from their molecular structure, transforming into alcohol groups (OH). There are two general forms: linear and cyclic. The linear types include compounds like mannitol, which is a reduction of mannose and sorbitol as a derivative of glucose. Cyclic forms, called inositols, include inositol itself, which is produced from glucose-6-phosphate by dephosphorylation. Several mechanisms use D-glucose, L-galactose, uronic acid, L-glucose, and myo-inositol, or recycle its oxidized state, to create ascorbic acid (AA), another essential molecule. In addition to its role as an enzyme cofactor, ascorbic acid aids in the plant's ability to regulate cell division, growth, development, and stress tolerance; it also assists in the neutralization of reactive oxygen species (ROS) and the repair of oxidized organic molecules. (Godoy et al. 2021). Eggplants are moderately sensitive to salinity stress. Many reports show that foliar applications of sorbitol, in concentrations of 5 and 10 mg L at a concentration level for plants grown in a glasshouse, treated with saline solution of NaCl to concentrations of 1.5, 3, and 6 dS m/, may help prevent or alleviate salt damage. Among the given concentrations, 5 mg/L sorbitol proved to be most efficient as it improved the salt-stressed plant height, fruit number, fruit weight, fruit diameter, chlorophyll a and b, and carotenoid levels of treated plants as compared with their untreated controls (Issa et al. 2020). Concentrations of 15 and 30 mM mannitol offered protective effects in maize when treated with salt stress; the osmotic conditions were supplied with 100 mM NaCl. This combined treatment showed increases in total plant DW, RWC, and levels of chlorophyll a and b, suggesting that mannitol not only protects against the adverse effects of saline stress but also increases biomass production, most probably because of its contribution to cellular osmotic adjustment. Field studies have further shown the positive effects of mannitol application, where foliar treatment with 2–4 g/L mannitol improved productivity in cabbage (Brassica oleracea) under saline conditions. In addition, its foliar application at a dose of 50 and 100 mg/L concentration was applied also to the pot-grown maize wherein this study has found that treatment relieved the stress of Cr through increase of all of the parameters studied here: plant height, leaf area, root length, and even the count of leaves per plant and the DW of the respective parts of the plants which possessed phenotypes almost alike those of the control plants, which are not treated by Cr. Additional evidence supporting mannitol's protective role under heavy metal stress comes from studies conducted by Kaya et al. (2013), Ramadan and Shalaby (2018), and Habiba et al. (2019) that found that it reduced oxidative stress in roots and leaves, specifically in more tolerant cultivars. These studies also found that it enhanced the function of antioxidant enzymes such as SOD, POD, CAT, and APX, as well as chlorophyll and carotenoid contents. Similar to how mannitol inhibits wheat stomatal closure at a dose of 100 mM, this action becomes dominant. Alternatively, according to Adrees et al. (2015), the antioxidant systems in both leaves and roots were stimulated by combining the low concentrations of Cr (0.25 and 0.5 mM) with a high concentration of mannitol (100 mM), which improved shoot and root DWs as well as husk and kernel weights. Abiotic stress tolerance is an area where cyclic polyols like myo-inositol have shown promise. As an example, in a hydroponic system for crab apples (Malus hupehensis), pretreatment with 50 μM myo-inositol reduced symptoms of salt stress (200 mM), while simultaneously improving plant height, DW, root length, volume, and surface area. On the other hand, it lowered electron leakage, $H_kO_2$ generation, and $Na^+$ buildup. By keeping the leaf water, status constant, exogenous myo-inositol applications also reduce drought stress. In pepper plants, foliar treatment of 5, 15, and 25 μM myo-inositol enhanced root water content (RWC) and lowered levels of $H_2O_2$, proline, and malondialdehyde (MDA) when subjected to drought conditions (i.e., 7 days without watering) (Hu et al. 2018; Yildizli et al. 2018). Lipoic acid has antioxidant properties and may increase harvest yields. As an example, in canola plants subjected to salinity stress (<150 mM NaCl), Yıldız et al. (2015) found that applying 100 μM LA topically decreased lipid peroxidation, raised cysteine content, and enhanced POD and CAT activities, all of which contributed to better yields and a degree of resistance to salt stress. Although natural metabolites are useful for increasing crop resistance to abiotic stress, there is a lack of variety in their chemical makeup. Discovering novel compounds that improve plant development in stressful conditions is one goal of chemical genomics, which aims to increase the available compound repertoire.

## 2.7 Conclusion

During the last 50 years, human activities have put pollutants into the environment that altered the balance in nature and heavily affected crop production on the earth. There is a pressing need to determine the problems that will face plants and how they are going to react to the climate. Generally, stress-generated ROS is not only a mark of cellular disorder but also reflects changes in cellular mechanisms. Too little detoxification of excess and too much detoxification of insufficient ROS lead to toxicity and death of cells. Even so much has been gained into the dynamics of ROS; many questions still linger over how ROS plays the mediation role in plant responses under stress, particularly in combined stress conditions. The accurate measurement of ROS dynamics was targeted in future studies using advanced biosensors and fluorescent dyes that are more sensitive and specific than traditional ones. It is possible that in vivo ROS level monitoring may be enhanced by the development of novel protein-based ROS sensors that aim at subcellular compartments. The research relationships between ROS and other reactive species will be continued, as well as that between ROS, melatonin, and phytohormones. Stress tolerance crop varieties such as rice and wheat will be under study in order to develop deeper insights into the tolerance mechanisms. Improving abiotic stress tolerance requires the identification of key regulators and interactors utilizing multi-omics technology. To ensure long-term crop production, it is essential to breed premier cultivars with genes that reduce reactive oxygen species (ROS). Transgenic approaches, coupled with advanced genome editing methods like CRISPR/Cas, will greatly enhance tolerance by targeting ROS-regulating genes. Finally, improved irrigation and accurate weather forecasting will monitor crop performance under increased environmental stresses toward sustainable farm lines. In understanding ROS, even while there is better current knowledge about their roles, there remain formidable issues in elucidating their signaling and stress-related functions that require more interrogations through functional genomics and proteomics.

## References

Abouzari A, Fakheri BA (2015) Reactive oxygen species: generation, oxidative damage, and signal transduction. Int J Life Sci 9:3–17. https://doi.org/10.3126/ijls.v9i5.12699

Adrees M, Ali S, Iqbal M, Aslam Bharwana S, Siddiqi Z, Farid M, Ali Q, Saeed R, Rizwan M (2015) Mannitol alleviates chromium toxicity in wheat plants in relation to growth, yield, stimulation of anti-oxidative enzymes, oxidative stress and Cr uptake in sand and soil media. Ecotoxicol Environ Saf 122:1–8. https://doi.org/10.1016/j.ecoenv.2015.07.003

Ahammed GJ, Yu JQ (2016) Plant hormones under challenging environmental factors (pp. 1–21). Dordrecht: Springer. doi:https://doi.org/10.1007/978-94-017-7758-2

Awasthi R, Gaur P, Turner NC, Vadez V, Siddique KH, Nayyar H (2017) Effects of individual and combined heat and drought stress during seed filling on the oxidative metabolism and yield of chickpea (Cicer arietinum) genotypes differing in heat and drought tolerance. Crop Pasture Sci 68(9):823–841. https://doi.org/10.1071/CP17028

Bano A, Gupta A, Rai S, Fatima T, Sharma S, Pathak N (2021) Mechanistic role of reactive oxygen species and its regulation via the antioxidant system under environmental stress. In: Plant stress physiology—perspectives in agriculture, pp 1–18. https://doi.org/10.5772/intechopen.101045

BenMassoud M, Sakouhi L, Chaoui A (2019) Effect of plant growth regulators, calcium and citric acid on copper toxicity in pea seedlings. J Plant Nutr 42(10):1230–1242. https://doi.org/10.1080/01904167.2019.1609506

Buchanan BB, Gruissem W, Jones RL (2015) Biochemistry and molecular biology of plants. Wiley

Bykova NV, Hoehn B, Rampitsch C, Hu J, Stebbing JA, Knox R (2011) Thiol redox-sensitive seed proteome in dormant and non-dormant hybrid genotypes of wheat. Phytochemistry 72(10):1162–1172. https://doi.org/10.1016/j.phytochem.2010.12.021

Chaudhry S, Sidhu GPS (2022) Climate change regulated abiotic stress mechanisms in plants: a comprehensive review. Plant Cell Rep 41(1):1–31. https://doi.org/10.1007/s00299-021-02759-5

Crawford T, Lehotai N, Strand Å (2018) The role of retrograde signals during plant stress responses. J Exp Bot 69(11):2783–2795. https://doi.org/10.1093/jxb/erx481

Cui G, Ji G, Liu S, Li B, Lian L, He W, Zhang P (2019) Physiological adaptations of Elymus dahuricus to high altitude on the Qinghai–Tibetan Plateau. Acta Physiol Plant 41:1–9. https://doi.org/10.1007/s11738-019-2904-z

Das K, Roychoudhury A (2014) Reactive oxygen species (ROS) and response of antioxidants as ROS-scavengers during environmental stress in plants. Front Environ Sci 2:53. https://doi.org/10.3389/fenvs.2014.00053

Devireddy AR, Liscum E, Mittler R (2020) Phytochrome B is required for systemic stomatal responses and reactive oxygen species signaling during light stress. Plant Physiol 184(3):1563–1572. https://doi.org/10.1104/pp.20.01084

Devireddy AR, Tschaplinski TJ, Tuskan GA, Muchero W, Chen JG (2021) Role of reactive oxygen species and hormones in plant responses to temperature changes. Int J Mol Sci 22(16):8843. https://doi.org/10.3390/ijms22168843

Ding X, Jimenez-Gongora T, Krenz B, Lozano-Duran R (2019) Chloroplast clustering around the nucleus is a general response to pathogen perception in Nicotiana benthamiana. Mol Plant Pathol 20(9):1298–1306. https://doi.org/10.1111/mpp.12840

Driedonks N, Xu J, Peters JL, Park S, Rieu I (2015) Multi-level interactions between heat shock factors, heat shock proteins, and the redox system regulate acclimation to heat. Front Plant Sci 6:999. https://doi.org/10.3389/fpls.2015.00999

Dvořák P, Krasylenko Y, Zeiner A, Šamaj J, Takáč T (2021) Signaling toward reactive oxygen species-scavenging enzymes in plants. Front Plant Sci 11:618835. https://doi.org/10.3389/fpls.2020.618835

El-Hawary MM, Nashed ME (2019) Effect of foliar application by some antioxidants on growth and productivity of maize under saline soil conditions. J Plant Prod 10(2):93–99. https://doi.org/10.21608/jpp.2019.36238

Exposito-Rodriguez M, Laissue PP, Yvon-Durocher G, Smirnoff N, Mullineaux PM (2017) Photosynthesis-dependent $H_2O_2$ transfer from chloroplasts to nuclei provides a high-light signalling mechanism. Nat Commun 8(1):49. https://doi.org/10.1038/s41467-017-00074-w

Faraz A, Faizan M, Sami F, Siddiqui H, Hayat S (2020) Supplementation of salicylic acid and citric acid for alleviation of cadmium toxicity to Brassica juncea. J Plant Growth Regul 39:641–655. https://doi.org/10.1007/s00344-019-10007-0

Farooq H, Asghar HN, Khan MY, Saleem M, Zahir ZA (2015) Auxin-mediated growth of rice in cadmium-contaminated soil. Turk J Agric For 39(2):272–276. https://doi.org/10.3906/tar-1405-54

Fortunato S, Lasorella C, Dipierro N, Vita F, de Pinto MC (2023) Redox signaling in plant heat stress response. Antioxidants 12(3):605. https://doi.org/10.3390/antiox12030605

Foyer CH, Ruban AV, Noctor G (2017) Viewing oxidative stress through the lens of oxidative signalling rather than damage. Biochem J 474(6):877–883. https://doi.org/10.1042/BCJ20160814

Godoy F, Olivos-Hernández K, Stange C, Handford M (2021) Abiotic stress in crop species: improving tolerance by applying plant metabolites. Plants 10(2):186. https://doi.org/10.3390/plants10020186

Habiba U, Ali S, Rizwan M, Ibrahim M, Hussain A, Shahid MR, Alamri SA, Alyemeni MN, Ahmad P (2019) Alleviative role of exogenously applied mannitol in maize cultivars differing in chromium stress tolerance. Environ Sci Pollut Res 26:5111–5121

Haslbeck M, Vierling E (2015) A first line of stress defense: small heat shock proteins and their function in protein homeostasis. J Mol Biol 427(7):1537–1548. https://doi.org/10.1016/j.jmb.2015.02.002

Hozayn M, Abd-Elmonem AA, Samaha GM (2020) The physiological effect of pre-soaking with tryptophan on sugar beet (Beta vulgaris L.) productivity under different levels of salinity stresses. Bull Natl Res Cent 44:1–9. https://doi.org/10.1186/s42269-020-00324-w

Hu L, Zhou K, Li Y, Chen X, Liu B, Li C et al (2018) Exogenous myo-inositol alleviates salinity-induced stress in Malus hupehensis Rehd. Plant Physiol Biochem 133:116–126. https://doi.org/10.1016/j.plaphy.2018.10.037

Huang H, Ullah F, Zhou DX, Yi M, Zhao Y (2019) Mechanisms of ROS regulation of plant development and stress responses. Front Plant Sci 10:800. https://doi.org/10.3389/fpls.2019.00800

IPCC (2021) Climate change 2021: the physical science basis. In: Masson-Delmotte VP, Zhai A, Pirani SL, Connors C, Péan S, Berger N, Caud Y, Chen L, Goldfarb MI (eds) Contribution of Working Group I to the sixth assessment report of the Intergovernmental Panel on Climate Change. Cambridge University Press, Cambridge, UK; New York, NY, USA, In Press.

Issa DB, Alturki SM, Sajyan TK, Sassine YN (2020) Sorbitol and lithovit-guano 25 mitigates the adverse effects of salinity on eggplant grown in pot experiment. https://doi.org/10.15159/AR.20.075

Jabłońska J, Tawfik DS (2021) The evolution of oxygen-utilizing enzymes suggests early biosphere oxygenation. Nat Ecol Evol 5(4):442–448. https://doi.org/10.1038/s41559-020-01386-9

Jagadish SVK, Murty MVR, Quick WP (2015) Rice responses to rising temperatures–challenges, perspectives and future directions. Plant Cell Environ 38(9):1686–1698. https://doi.org/10.1111/pce.12430

Jagodzik P, Tajdel-Zielinska M, Ciesla A, Marczak M, Ludwikow A (2018) Mitogen-activated protein kinase cascades in plant hormone signaling. Front Plant Sci 9:1387. https://doi.org/10.3389/fpls.2018.01387

Jamil M, Ahamd M, Anwar F, Zahir ZA, Kharal MA, Nazli F (2018a) Inducing drought tolerance in wheat through combined use of L-tryptophan and pseudomonas fluorescens. Pak J Agric Sci 55(2):331–337. https://doi.org/10.21162/PAKJAS/18.4980

Jamil M, Kharal MA, Ahmad M, Abbasi GH, Nazli F, Hussain A, Akhtar MFUZ (2018b) Inducing salinity tolerance in red pepper (Capsicum annuum L.) through exogenous application of proline and L-tryptophan. Soil Environ 37(2):160–168. https://doi.org/10.25252/SE/18/31052
Jarsjö J, Andersson-Sköld Y, Fröberg M, Pietroń J, Borgström R, Löv Å, Kleja DB (2020) Projecting impacts of climate change on metal mobilization at contaminated sites: controls by the groundwater level. Sci Total Environ 712:135560. https://doi.org/10.1016/j.scitotenv.2019.135560
Kaya C, Sonmez O, Aydemir S, Ashraf M, Dikilitas M (2013) Exogenous application of mannitol and thiourea regulates plant growth and oxidative stress responses in salt-stressed maize (Zea mays L.). J Plant Interact 8:234–241
Khan M, Ali S, Al Azzawi TNI, Saqib S, Ullah F, Ayaz A, Zaman W (2023) The key roles of ROS and RNS as a signaling molecule in plant–microbe interactions. Antioxidants 12(2):268
Kimura S et al (2020) CRK2 and C-terminal phosphorylation of NADPH oxidase RBOHD regulate reactive oxygen species production in Arabidopsis. Plant Cell 32:1063–1080
Koussevitzky S et al (2008) Ascorbate peroxidase 1 plays a key role in the response of Arabidopsis thaliana to stress combination. J Biol Chem 283:34197–34203
Lee DH et al (2020) Regulation of reactive oxygen species during plant immunity through phosphorylation and ubiquitination of RBOHD. Nat Commun 11:1838
Lee J, Nguyen HH, Park Y, Lin J, Hwang I (2022) Spatial regulation of RBOHD via AtECA4-mediated recycling and clathrin-mediated endocytosis contributes to ROS accumulation upon salt stress response but not flg22-induced immune response. Plant J 109:816–830
Liu J, Hasanuzzaman M, Wen H, Zhang J, Peng T, Sun H, Zhao Q (2019) High temperature and drought stress cause abscisic acid and reactive oxygen species accumulation and suppress seed germination growth in rice. Protoplasma 256:121
Lodeyro AF, Krapp AR, Carrillo N (2021) Photosynthesis and chloroplast redox signaling in the age of global warming: stress tolerance, acclimation, and developmental plasticity. J Exp Bot 72(16):5919–5937
Ma Z, Marsolais F, Bykova NV, Igamberdiev AU (2016) Nitric oxide and reactive oxygen species mediate metabolic changes in barley seed embryo during germination. Front Plant Sci 7:138
Mallhi ZI, Rizwan M, Mansha A, Ali Q, Asim S, Ali S, Hussain A, Alrokayan SH, Khan HA, Alam P et al (2019) Citric acid enhances plant growth, photosynthesis, and phytoextraction of lead by alleviating the oxidative stress in Castor beans. Plants 8:525
Mansoor S, Ali A, Kour N, Bornhorst J, AlHarbi K, Rinklebe J et al (2023) Heavy metal induced oxidative stress mitigation and ROS scavenging in plants. Plants 12(16):3003. https://doi.org/10.3390/plants12163003
Mansour MMF, Ali EF (2017) Evaluation of proline functions in saline conditions. Phytochemistry 140:52–68. https://doi.org/10.1016/j.phytochem.2017.04.016
Mielecki J, Gawroński P, Karpiński S (2020) Retrograde signaling: understanding the communication between organelles. Int J Mol Sci 21(17):6173. https://doi.org/10.3390/ijms21176173
Mittler R, Zandalinas SI, Fichman Y, Van Breusegem F (2022) Reactive oxygen species signalling in plant stress responses. Nat Rev Mol Cell Biol 23(10):663–679. https://doi.org/10.1038/s41580-022-00499-2
Muhlemann JK, Younts TL, Muday GK (2018) Flavonols control pollen tube growth and integrity by regulating ROS homeostasis during high-temperature stress. Proc Natl Acad Sci U S A 115(47):E11188–E11197. https://doi.org/10.1073/pnas.1811492115
Mustafa A, Imran M, Ashraf M, Mahmood K (2018) Perspectives of using l-tryptophan for improving productivity of agricultural crops: a review. Pedosphere 28(1):16–34. https://doi.org/10.1016/S1002-0160(18)60002-5
Nahar K, Hasanuzzaman M, Alam MM, Fujita MJBP (2015) Roles of exogenous glutathione in antioxidant defense system and methylglyoxal detoxification during salt stress in mung bean. Biol Plant 59:745–756. https://doi.org/10.1007/s10535-015-0542-x
Navrot N, Rouhier N, Gelhaye E, Jacquot JP (2007) Reactive oxygen species generation and antioxidant systems in plant mitochondria. Physiol Plant 129(1):185–195. https://doi.org/10.1111/j.1399-3054.2006.00777.x

Nietzel T, Mostertz J, Ruberti C, Née G, Fuchs P, Wagner S et al (2020) Redox-mediated kick-start of mitochondrial energy metabolism drives resource-efficient seed germination. Proc Natl Acad Sci 117(1):741–751. https://doi.org/10.1073/pnas.1910501117

Noctor G, Veljovic-Jovanovic SONJA, Driscoll S, Novitskaya L, Foyer CH (2002) Drought and oxidative load in the leaves of C3 plants: a predominant role for photorespiration? Ann Bot 89(7):841–850. https://doi.org/10.1093/aob/mcf096

Nogia P, Sidhu GK, Mehrotra R, Mehrotra S (2016) Capturing atmospheric carbon: biological and nonbiological methods. Int J Low Carbon Technol 11(2):266–274. https://doi.org/10.1093/ijlct/ctt077

Ortiz-Bobea A, Ault TR, Carrillo CM, Chambers RG, Lobell DB (2021) Anthropogenic climate change has slowed global agricultural productivity growth. Nat Clim Chang 11(4):306–312. https://doi.org/10.1038/s41558-021-01000-1

Parveen A, Saleem MH, Kamran M, Haider MZ, Chen JT, Malik Z et al (2020) Effect of citric acid on growth, ecophysiology, chloroplast ultrastructure, and phytoremediation potential of jute (Corchorus capsularis L.) seedlings exposed to copper stress. Biomolecules 10(4):592. https://doi.org/10.3390/biom10040592

Peck S, Mittler R (2020) Plant signaling in biotic and abiotic stress. J Exp Bot 71(5):1649–1651. https://doi.org/10.1093/jxb/eraa051

Pérez-Labrada F, Benavides-Mendoza A, Valdez-Aguilar LA, Robledo-Torres V (2016) Citric acid in the nutrient solution increases the mineral absorption in potted tomato grown in calcareous soil. Pak J Bot 48(1):67–74

Pfeilmeier S, Petti GC, Bortfeld-Miller M, Daniel B, Field CM, Sunagawa S, Vorholt JA (2021) The plant NADPH oxidase RBOHD is required for microbiota homeostasis in leaves. Nat Microbiol 6(7):852–864. https://doi.org/10.1038/s41564-021-00929-5

Ramadan ME, Shalaby OA (2018) Effect of salicylic acid and mannitol on white cabbage plants under saline conditions. J Plant Prod 9(4):397–402. https://doi.org/10.21608/jpp.2018.35766

Sachdev S, Ansari SA, Ansari MI, Fujita M, Hasanuzzaman M (2021) Abiotic stress and reactive oxygen species: generation, signaling, and defense mechanisms. Antioxidants 10(2):277. https://doi.org/10.3390/antiox10020277

Sasidharan R, Schippers JH, Schmidt RR (2021) Redox and low-oxygen stress: signal integration and interplay. Plant Physiol 186(1):66–78. https://doi.org/10.1093/plphys/kiaa081

Sharma A, Kumar V, Shahzad B, Ramakrishnan M, Singh Sidhu GP, Bali AS et al (2020) Photosynthetic response of plants under different abiotic stresses: a review. J Plant Growth Regul 39:509–531. https://doi.org/10.1007/s00344-019-10018-x

Shen J, Zhang J, Zhou M, Zhou H, Cui B, Gotor C et al (2020) Persulfidation-based modification of cysteine desulfhydrase and the NADPH oxidase RBOHD controls guard cell abscisic acid signaling. Plant Cell 32(4):1000–1017. https://doi.org/10.1105/tpc.19.00826

Singh A, Mehta S, Yadav S, Nagar G, Ghosh R, Roy A et al (2022) How to cope with the challenges of environmental stresses in the era of global climate change: an update on ROS stave off in plants. Int J Mol Sci 23(4):1995. https://doi.org/10.3390/ijms23041995

Smirnoff N, Arnaud D (2019) Hydrogen peroxide metabolism and functions in plants. New Phytol 221(3):1197–1214. https://doi.org/10.1111/nph.15488

Suzuki N (2023) Fine tuning of ROS, redox and energy regulatory systems associated with the functions of chloroplasts and mitochondria in plants under heat stress. Int J Mol Sci 24(2):1356. https://doi.org/10.3390/ijms24021356

Szabados L, Savouré A (2010) Proline: a multifunctional amino acid. Trends Plant Sci 15(2):89–97. https://doi.org/10.1016/j.tplants.2009.11.009

Tarfeen N, Nisa Q, Kahlief K (2022) Antioxidant defense system in plants against biotic stress. In: Antioxidant defense in plants: molecular basis of regulation. Springer, Singapore, pp 383–395. https://doi.org/10.1007/978-981-16-7981-0_17

Teh CY, Shaharuddin NA, Ho CL, Mahmood M (2016) Exogenous proline significantly affects the plant growth and nitrogen assimilation enzymes activities in rice (Oryza sativa) under salt stress. Acta Physiol Plant 38:1–10. https://doi.org/10.1007/s11738-016-2163-1

Ullah A, Nadeem F, Nawaz A, Siddique KH, Farooq M (2022) Heat stress effects on the reproductive physiology and yield of wheat. J Agron Crop Sci 208(1):1–17. https://doi.org/10.1111/jac.12572

Wang J, Vanga SK, Saxena R, Orsat V, Raghavan V (2018) Effect of climate change on the yield of cereal crops: a review. Climate 6(2):41. https://doi.org/10.3390/cli6020041

Wang P, Hsu CC, Du Y, Zhu P, Zhao C, Fu X et al (2020a) Mapping proteome-wide targets of protein kinases in plant stress responses. Proc Natl Acad Sci U S A 117(6):3270–3280. https://doi.org/10.1073/pnas.1919901117

Wang L, Ma KB, Lu ZG, Ren SX, Jiang HR, Cui JW et al (2020b) Differential physiological, transcriptomic and metabolomic responses of Arabidopsis leaves under prolonged warming and heat shock. BMC Plant Biol 20:1–15. https://doi.org/10.1186/s12870-020-2292-y

Yıldız M, Akçalı N, Terzi H (2015) Proteomic and biochemical responses of canola (Brassica napus L.) exposed to salinity stress and exogenous lipoic acid. J Plant Physiol 179:90–99. https://doi.org/10.1016/j.jplph.2015.03.006

Yildizli A, Çevik S, Ünyayar S (2018) Effects of exogenous myo-inositol on leaf water status and oxidative stress of Capsicum annuum under drought stress. Acta Physiol Plant 40:1–10. https://doi.org/10.1007/s11738-018-2690-z

Zandalinas SI, Sengupta S, Fritschi FB, Azad RK, Nechushtai R, Mittler R (2021) The impact of multifactorial stress combination on plant growth and survival. New Phytol 230(3):1034–1048. https://doi.org/10.1111/nph.17232

Zhao J, Lu Z, Wang L, Jin B (2020) Plant responses to heat stress: physiology, transcription, noncoding RNAs, and epigenetics. Int J Mol Sci 22(1):117. https://doi.org/10.3390/ijms22010117

Zheng W, Wang S, Tan K, Lei Y (2020) Nitrate accumulation and leaching potential is controlled by land-use and extreme precipitation in a headwater catchment in the North China plain. Sci Total Environ 707:136168. https://doi.org/10.1016/j.scitotenv.2019.136168

Zhou Y, Wen Z, Zhang J, Chen X, Cui J, Xu W, Liu HY (2017) Exogenous glutathione alleviates salt-induced oxidative stress in tomato seedlings by regulating glutathione metabolism, redox status, and the antioxidant system. Sci Hortic 220:90–101. https://doi.org/10.1016/j.scienta.2017.02.021

# Regulation of Amino Acid Metabolic Enzymes and Transporters in Plants Under Abiotic Stress

3

Kamini Devi, Mohd Ali, Parkirti, Vikram, Amandeep Singh, Manu Sharma, Nahida Rehman Mir, Nitika Kapoor, Puja Ohri, and Renu Bhardwaj

**Abstract**

Amino acids play vital roles in various aspects of plant function, serving as crucial components for protein synthesis and essential metabolites that engage with diverse metabolic pathways. Additionally, these molecules function as key carriers of organic nitrogen within the plant. Given their pivotal role in nitrogen metabolism, the processes of amino acid biosynthesis, degradation, and transport are intricately regulated to meet the demands dictated by nitrogen and carbon availability. While our understanding of the feedback regulation within the branched biosynthesis pathways involving amino acids is substantial, the regulatory mechanisms at the transcriptional, post-transcriptional, and protein levels are still awaiting identification. This chapter primarily delves into the current comprehension of how enzymes and transporters are regulated at the transcript level under abiotic stress. Existing findings emphasize the impact of transcription factors and protein modifications, offering a fragmented yet insightful perspective that suggests a multitude of intricate regulatory layers governing the coordination of transport and enzyme activities. Notably, the regulation of amino acid metabolism, transport, and integration of stress signals appears to exhibit a complex interplay that, as of now, remains challenging to predict with certainty.

K. Devi · A. Singh · M. Sharma · N. R. Mir · N. Kapoor
Department of Botanical and Environmental Sciences, Guru Nanak Dev University, Amritsar, Punjab, India

M. Ali · Parkirti · Vikram · P. Ohri
Department of Zoology, Guru Nanak Dev University, Amritsar, Punjab, India

R. Bhardwaj (✉)
University Centre for Research and Development, Chandigarh University, Mohali, Punjab, India



N. Iqbal et al. (eds.), *Plant Signaling Molecules in Regulation of ROS-Scavenging System*, https://doi.org/10.1007/978-981-95-3377-0_3

**Keywords**

Abiotic stress · Amino acids · Metabolism · Biosynthesis

## 3.1 Introduction

Amino acids can serve as building blocks for the production of proteins linked to stress. Plants frequently need to create certain proteins in response to stress in order to help them deal with the stressor (Khan et al. 2020). The building blocks required for the creation of these stress-responsive proteins are amino acids. Amino acids such as glutamine and proline, for instance, are needed for the production of stress-related proteins that shield plants from excessive salinity and dehydration (Trovato et al. 2021; Zulfiqar and Ashraf 2023). Furthermore, amino acids have the ability to control the synthesis and signaling of plant hormones. Plant stress reactions depend heavily on plant hormones like jasmonic acid (JA) and abscisic acid (ABA). Plant adaptation to stressful circumstances can be regulated by amino acids through their influence on the synthesis and signaling of these hormones (Pratelli and Pilot 2014). Methionine, for instance, is a necessary amino acid for the production of ethylene, a hormone that controls how plants react to both biotic and abiotic stressors (Zemanová et al. 2014). Many studies reveal that a significant level of methionine ameliorates salt stress in *Zea mays* L. and also promotes crop productivity in stress conditions (Shahid et al. 2023). Another essential element in amino acid metabolism is ammonium, which is derived from nitrate absorption or the breakdown of amino acids (Cánovas et al. 2007). Glutamate and glutamine are produced by ammonium's absorption, and these two molecules are the building blocks for the production of additional amino acids. Ammonium's availability is necessary for the synthesis of certain amino acids (Schjoerring et al. 2002). Sometimes proteins can get damaged by high temperatures, unfolding and clumping together, which can result in a loss of function or even cell death. By functioning as molecular chaperones, which support appropriate protein folding and inhibit aggregation, amino acids can aid in the prevention of protein aggregation (Goyal et al. 2005; Housmans et al. 2023). Additionally, they can maintain the normal functioning of high-temperature-sensitive enzymes like lactate dehydrogenase by stabilizing their structure and activity. Plants need amino acids to survive and respond to heat stress because they can shield proteins and enzymes from thermally induced damage (Mogk et al. 2018).

Moreover, protein synthesis and secondary metabolism are the main processes that drive the need for amino acids. A steady supply of amino acids is needed for protein synthesis in order to construct polypeptide chains (Ali et al. 2019). Furthermore, amino acids are precursors for the production of several secondary metabolites, including hormones, neurotransmitters, and pigments, in addition to being needed for the synthesis of proteins (Batista-Silva et al. 2019). On the other hand, compounds from the metabolism of carbohydrates are used in amino acid production. For instance, the pyruvate and phosphoenolpyruvate intermediates produced by the glycolytic process serve as precursors for the production of certain amino acids (Khan et al. 2020). By producing ribose-5-phosphate, a precursor for the synthesis of nucleotide bases and some amino acids, the pentose phosphate pathway also aids in the biosynthesis of amino acids (Pratelli and Pilot 2014).

Conversely, the breakdown of amino acids produces a number of metabolites that the citric acid cycle can employ as energy sources. For example, ammonium ions are produced during the breakdown of most amino acids and are then transformed into glutamate and finally α-ketoglutarate. Energy in the form of ATP can be produced by directly entering the citric acid cycle and oxidizing α-ketoglutarate (Ingrisano et al. 2023). Overall, the demand for protein synthesis, secondary metabolism, ammonium metabolism, and glucose metabolism are closely related to amino acid metabolism, which ensures an appropriate supply of amino acids for different cellular functions.

## 3.2 Biosynthesis and Classification of Amino Acids

The significance of amino acids in plant development and stress defense has emerged during the past few decades, sparking a surge of interest in both basic and practical plant research (Parthasarathy et al. 2021). Many amino acids, even those not involved in protein synthesis, have been shown to play active roles in plant development and to aid in the plant's response to environmental stressors in addition to serving as building blocks for the synthesis of proteins (Kishor et al. 2020). Furthermore, amino acids have critical functions in human nutrition as either a source of nutraceutical substances or as necessary dietary components. They also act as precursors for a variety of primary and secondary metabolites (Lopes et al. 2023). However, the biosynthesis pathways of aromatic amino acids (AAA) in bacteria have been studied intensively due to their utilization in the food and pharmaceutical industries (Dickey et al. 2021). In plants, bacteria and fungi the three AAA are produced from the same precursor metabolite chorismate, which comes from the shikimate pathway (Yokoyama et al. 2021). Although the AAA pathway is almost entirely used in bacteria for the synthesis of proteins, however, in plants, it also serves as a precursor for a number of distinct plant hormones, including auxin and salicylate, as well as for a very large variety of aromatic secondary metabolites with a variety of biological roles and biotechnological applications (Wu et al. 2024). Moreover, the shikimate pathway enzyme 5-enolpyruvylshikimate-3-phospate synthase (EPSPS), is the target of the glyphosate herbicide and non-plant EPSPS confers the herbicide-resistance trait in several commercialized transgenic crops (Ruszkowski and Forlani 2022). However, defining the regulation of the shikimate and AAA biosynthesis pathways in plants is mainly inspired by these significant features. The synthesis of aromatic amino acids (AAAs), or L-tyrosine, L-phenylalanine, and L-tryptophan (Tyr, Phe, and Trp, respectively), as well as a variety of aromatic natural products, is facilitated by the shikimate pathway, which also controls carbon flow from central carbon metabolism (Brey et al. 2020). Since the shikimate route is absent in animals and AAAs are necessary for protein

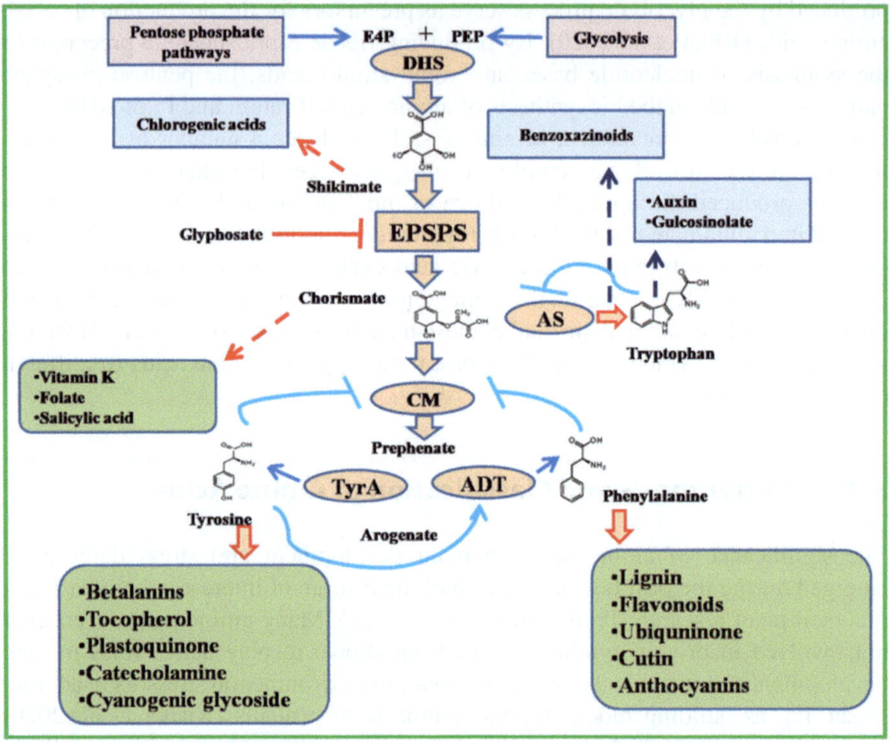


**Fig. 3.1** Biosynthesis and shikimate pathway of amino acids in plants. The shikimate pathway results in the biosynthesis of AAAs, which are needed for protein synthesis as well as being used as precursors of many AAA-derived natural products in plants. DHS 3-deoxy-D-arabino-heptulosonate 7-phosphate synthase, EPSPS 5-enolpyruvylshikimate-3-phosphate synthase, AS anthranilate synthase, CM chorismate mutase, TyrA arogenate dehydrogenase, ADT arogenate dehydratase, E4P D-erythrose 4-phosphate, PEP phosphoenolpyruvate

synthesis in all species, glyphosate, the most commonly used herbicide, targets the shikimate pathway, making AAAs vital for human nutrition (Shende et al. 2024). Plant natural products derived from the AAA and shikimate pathways are widely employed in biomaterials, medicines, and nutraceuticals. These compounds are important to plant physiology and adaptability (Fig. 3.1).

### 3.2.1 Classification of Amino Acids in Plants

As building blocks of proteins, amino acids are categorized as either essential or non-essential depending on production in humans. While both plants and humans can synthesize non-essential amino acids (AAs), only plants can synthesize essential amino acids (EAAs). The EAAs must be obtained from external sources in the

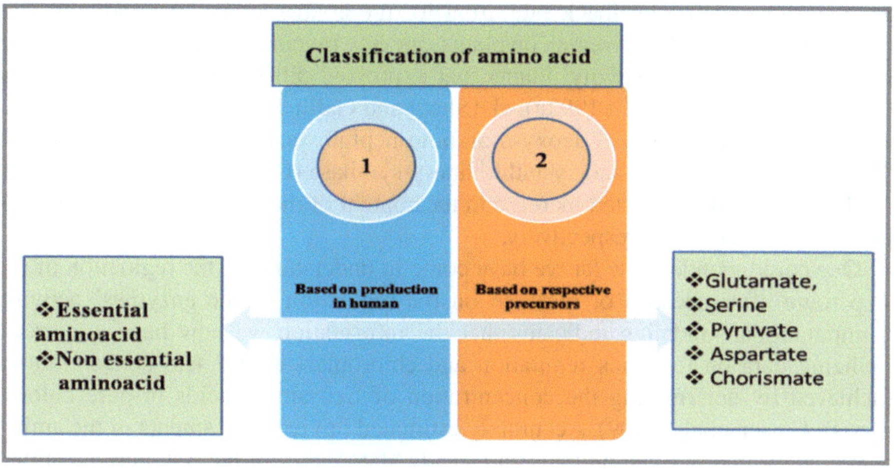


**Fig. 3.2** Classification of amino acids in plants

form of a plant or animal diet (Kumar et al. 2019; Farooq et al. 2023; Chen and Zhang 2024).

However, based on their respective precursors amino acids are classified into five groups based glutamate, serine, pyruvate, aspartate, or chorismate (Trovato et al. 2021; Zhang et al. 2024) Animal food is more nutrient-dense than plant food, but the production of plant food has a far smaller environmental impact than that of animal food (Smith et al. 2021). The environmental footprint measures how much of the planet's resources are required for human use (Lenzen et al. 2020) (Fig. 3.2).

## 3.3 Regulation of Enzymes Involved in Amino Acids Metabolism

First, the biochemical separation of the enzymes from plant tissues and their subsequent in vitro characterization provided insight into the control of the functions of the metabolic enzymes (Dotson et al. 1989). Later, the relevant cDNAs could be isolated and expressed in Escherichia coli thanks to molecular cloning, which allowed for accurate characterization of the purified enzymes (Curien et al. 2005). These methods demonstrated that S-adenosylmethionine and amino acids, which are the end products of different pathways, regulate feedback on many enzymes.

Numerous amino acids are hazardous to cells (Bonner and Jensen 1997) and to plant development (Pratelli et al. 2010) as a result of feedback inhibition. For example, adding Lys and Thr to plants suppresses the Asp pathway's aspartate kinase (AK) activity. This deprives the plant of Met and therefore stunts growth. Through genetic screens, this inhibition has been used to find plants resistant to harmful amino acid combinations or analogues. It has also been used to find mutations that render enzymes resistant to feedback inhibition (Sarrobert et al. 2000). The necessity of enzyme feedback inhibition for the control of pathway activity was clearly demonstrated by research on plants carrying mutant genes or *E. coli* enzymes with low feedback sensitivity. Plants that expressed dihydrodipicolinate synthase (DHDPS; Shaul and Galili 1992a), ak (Shaul and Galili 1992b), chorismate mutantase (Tzin et al. 2009), 3-deoxy-d-arabino-heptulosonate 7-phosphate synthase (Tzin et al. 2012, 2013), and cystathionine-γ-synthase (CGS; Hacham et al. 2006a, b) all exhibited the expected increase in the content of the downstream amino acids Lys, Thr, Trp, and Met, respectively.

One could wonder how far we have come in understanding the regulation of the Asp route after decades of research on the regulation of the enzymes' activity. Computational modeling indicates that the information we now have is correct: Utilizing data on feedback regulation and chloroplast amino acid concentrations (achieved by determining the concentration of free amino acids in pure chloroplasts), Curien et al. (2009). Accurately estimated in vivo assessments of the amino acid content in described mutants and steady-state fluxes of the Asp-derived amino acid pathway. It has been demonstrated that the mRNA accumulation of the transcripts encoding the amino acid pathway's enzymes varies depending on a number of factors. Numerous studies have addressed the subject of how perturbations affect the accumulation of transcripts and, less frequently, associated proteins, since the first cloning of the genes encoding metabolic enzymes. A large number of the genes involved in the production and breakdown of aromatic amino acids are induced by biotic and abiotic stressors as well as stress hormones as well as the genes connected to the S-adenosylmethionine and Met pathways for instance, injury, drought, free radicals, jasmonic acid, elicitors, pseudomonas syringae, and $AgNO_3$ can all stimulate the anthranilate synthase (AS) gene, which is implicated in the synthesis of Trp. A thorough analysis looked into variations in the mRNA content of the enzymes involved in the metabolic pathways that produce amino acids. Utilizing microarray data that is accessible to the public in reaction to stress (Less and Galili 2008, 2009). The genes of the aromatic amino acid system were the most responsive to the stress circumstances, according to the first study's authors. Additionally, they came to the conclusion that, in comparison to the enzymes engaged in biosynthesis, the genes of the enzymes involved in degradation were often the most sensitive (Less and Galili 2008). In the second study, Less and Galili (2009) used a novel approach to find genes that exhibit co-regulated expression, or gene coordination. Discovered gene modules that react to certain situations in a comparable way. Three major gene modules were found: a catalytic module that is suppressed by active growth and stimulated by most stresses; an aromatic amino acid module that responds to most stresses; and a Met metabolism module that responds positively to situations of active growth. The presence of these modules indicates the existence of a shared signaling and regulatory mechanism that regulates the expression of genes engaged in related or unrelated pathways. A co-expression study of metabolic enzymes and transcription factor genes tests the idea that common transcription factors are involved (Jander and Joshi 2010). Additionally, it has been demonstrated that a number of transcription factors regulate the expression of DAHPS, CM, 5-enolpyruvylshikimate 3-phosphate synthase (EPSPS), and phenylalanine ammonia lyase (PAL), which are all involved in the synthesis and breakdown of Phe. A rise in the phenylpropanoid concentration metabolite detection has always been associated with an increase in the activity of the upstream Phe pathway. Metabolite production is a component of secondary metabolism that occurs downstream from Phe. The majority of transcription factors that have been demonstrated to affect the expression of the phenylpropanoid pathway's genes also control, albeit indirectly, the expression of the aromatic amino acid pathway's genes, which is in good agreement with this result (Maeda and Dudareva 2012). The regulation of δ-pyrroline-5-carboxylate reductase (P5CR) and CGS has been characterized at the post-transcriptional/translational level. P5CR plays a role in stress-induced pro synthesis, while CGS plays a role in met synthesis. The investigation of the P5CR mRNA's translation efficiency under salt stress was spurred by an intriguing disparity between P5CR mRNA and protein quantities (Hare et al. 1999). The authors discovered that the 5′-untranslated region of P5CR is involved in simultaneous mRNA stability and translation inhibition, as well as surprisingly the regulation of transcription efficiency (Hua et al. 2001). It is unclear what these somewhat at odds regulatory paths actually signify. The regulation of CGS mRNA translation and degradation is attributed to S-adenosylmethionine. S-adenosylmethionine causes ribosome stalling, which lowers mRNA accumulation and protein synthesis by binding the developing CGS protein at the just translated so-called MTO domain (Onouchi et al. 2005) In plants without the MTO region that are not inhibited by S-adenosylmethionine, a secondary CGS transcript has been found (Hacham et al. 2006a, b). The mRNA region encoding the MTO domain is thought to be cut, most likely forming a hairpin, to produce the shorter form of the CGS transcript from the full-length transcript (Hacham et al. 2006a, b). Even in the presence of significant amounts of S-adenosylmethionine and Met, Met synthesis would be permitted by the transcript's natural production.

## 3.4 Role of Amino Acid in Response to Abiotic Stress in Plants

Plants encounter various environmental stressors because they are sessile in nature. Among various stresses, abiotic stresses such as high and low temperatures, salinity, drought, and heavy metals are one of the chief reasons for directly changing agronomical production and yield. These abiotic stressors have detrimental effects on crop development, growth, and productivity (Haghighi et al. 2022). Plants interpret a temperature rise above optimal levels as heat stress because temperature is one of the key elements in the maintenance of cellular metabolism. Heat can cause disruptions to the homeostasis of cells, which can result in stunted growth or even death. Because of variations in the ideal temperature for crop maturity, rises in temperate area climates may cause a shift in the geographical distribution of major agriculturally significant crops as well as modifications to their growing seasons in a similar way, the degree of stress and length of exposure determine how plants respond toward cold stress (Imran et al. 2021). The effects of drought stress on crop species related to their physiology and growth have been thoroughly studied. Drought has several main effects on plants, such as poor germination, decreased growth of seedlings, decreased availability of nutrients, decreased photosynthetic rate, declined number and size of each leaf, reduced fresh and dry weight of the plants, etc. (Begna 2020). Another abiotic factor that may restrict crop productivity is salinity in the water or soil that causes some general effects such as reduced root length, smaller and fewer leaves, and a slower rate of growth in plants particularly in arid or semi-arid areas, (Kul et al. 2020). Similarly, in the case of heavy metal stress, the heavy metals are non-biodegradable that is, they cannot be eliminated from the environment naturally. Heavy metal also has several effects on plants. The enzymes are negatively impacted by heavy metal stress in plants through denaturation or inactivation. It also disrupts the integrity of membranes by interfering with the substitution reaction of necessary metallic ions with biomolecules, which alters respiration, photosynthetic capacity, and other processes (Abd Elnabi et al. 2023). Plants typically respond to abiotic stresses in a number of ways and various changes in amino acid metabolism are among the ways in which plants respond to abiotic stress. Under ideal circumstances, amino acids are well-known bio stimulant and they aid in the growth and development of plants. Extensive research has been conducted on the impact of amino acids on various plants. According to studies, amino acids can enhance a variety of plants' ability to absorb nutrients and water, assimilate fertiliser, and engage in photosynthesis, which increases flower production, fruit set, and fruit yield. For example- In a study the effect of L-Arginine was observed on maize plant and the result showed an increase in chlorophyll content of plants, it also reduced the ill-effects of hydrogen peroxide and alleviated MDA content in drought affected plants. Moreover, it effectively increased catalase, glutathione reductase, superoxide dismutase, ascorbate peroxidase, glutathione S-transferase and peroxidase levels in plants by 108.5%, 181.1%, 46.1%, 11.6%, 104.4%, and 18.3% (Sun et al. 2023). In another study, the foliar application of poly-γ-glutamic acid (γ-PGA) was introduced to wheat seedlings under salinity stress, the results favored increased growth of the plant and enhanced activities of antioxidant enzymes such as catalase, superoxide dismutase and peroxidase (Guo et al. 2017). In one more study, the iron chelated lysine combination was used to treat chromium toxicity in *Brassica napus* plants. The treatments effectively alleviated Cr toxicity, improved growth, and enhanced antioxidant enzymatic activity in plants (Zaheer et al. 2020). Some other examples of different amino acids effective against various abiotic stressors are given in the Table 3.1 below:

## 3.5 Transporters of Amino Acids

The distribution of nitrogen throughout the plant body is vital for growth and development, and amino acid transporters play a crucial role in this process. Within the plant kingdom, there exist two families of amino acid transporters: the Amino Acid-Polyamine-Organocation (APC) family and the amino acid/auxin permease (AAAP) family, which is also known as the amino acid transporter (ATF) family

**Table 3.1** Role of amino acid in response to abiotic stress in plants

| Abiotic stress type | Plant species | Amino acid | Concentration of amino acid | Evaluation type (pot/field/aqueous) | Metabolic effects | Reference |
|---|---|---|---|---|---|---|
| Drought stress | *Vicia faba* L. | β-aminobutyric acid (BABA) | 0.5, 1 and 1.5 mM of BABA | Hydroponic system | Reduced MDA level, reduced levels of electrolyte leakage, increased activity of antioxidant enzymes such as CAT, APX, SOD, GPX, enhanced leaf photosynthesis rate | Abid et al. (2020) |
| Heavy metal stress | *Zea mays* | Proline | 10 mM and 5 mM | Pot study | Enhanced chlorophyll content and carotenoid content | Adejumo et al. (2020) |
| Drought stress | *Triticum aestivum* L. | – | 5 ml/L | Pot trials | Improved biomass yield, grain yield, number of spikes also increased | Ahmad et al. (2019) |
| Salt stress | *Solanum lycopersicum* | Arg, Met, Glu, Pro, and Trp | 5% (v/v) | Hydroponic system | Improved vegetative growth, enhanced relative water and mineral nutrient content | Alfosea-Simón et al. (2020) |
| Arsenic stress | *Oryza sativa* L. | L-glutamic acid | 0, 2.5, 5 and 10 μM | Lab study | Enhanced activities of nitrite reductase, glutamine synthetase, glutamate synthase and nitrate reductase, enhanced growth attributes | Asgher et al. (2022) |

(continued)

**Table 3.1** (continued)

| Abiotic stress type | Plant species | Amino acid | Concentration of amino acid | Evaluation type (pot/field/aqueous) | Metabolic effects | Reference |
|---|---|---|---|---|---|---|
| Mercury stress | *Zea mays* | Cysteine | 200 μM | Hydroponic experiment | Improved heavy metal stress, enhanced chlorophyll content, enhanced root and shoot length | Aysin et al. (2020) |
| Salt-stress | *Glycine max* | Glycine (2.5%), Arginine (2.4%), Aspartic acid (1.6%), and Tryptophane (0.5%) | 0.4 mL $L^{-1}$, 0.8 mL $L^{-1}$, and 1.2 mL $L^{-1}$ | Pot experiment | Improved growth, chlorophyll content enhanced, improved transpiration (*E*), stomatal conductance | Peña Calzada et al. (2022) |
| Water-deficit stress | *Phaseolus vulgaris* L. | γ-aminobutyric acid (GABA) | 0.5, 1 and 2 mM | Field study | Improved growth, antioxidant defence system enhanced, better nutrient acquisition | Abd El-Gawad et al. (2021) |
| Cadmium toxicity | *Lens culinaris* | L-glutamic acid | 10 mM | Petri dish | Improved plant growth, enhanced activity of CAT, GPOX, APOX and monodehydroascorbate reductase, reduced cd uptake | Fardus et al. (2021) |

| | | | | | | |
|---|---|---|---|---|---|---|
| Drought stress | *Brassica oleracea* | L-amino acids, specifically: L-aspartic acid, L-cysteine, L-glutamic acid, L-glycine, L-proline, L-tyrosine, L-tryptophan, L-histidine, L-arginine, L-aspartic acid, L-valine, L-threonine, L-lysine, L-phenylalanine, L-methionine, and L-serine | 0, 150 and 300 mg/L | Field study | Increased protein, phenolic, and proline concentration | Haghighi et al. (2020) |
| Drought | *Triticum aestivum* | Arginine | 0.25, 0.5, and 1 mM | Greenhouse study | Lowered plant growth and chlorophyll content, enhanced sugar and flavonoid content in plants | Hussein et al. (2022) |
| Heat stress | *Lolium* | Aspartate | 0.1, 1, 2, 5, 10, and 15 mM | Controlled conditions | Increased activity of antioxidant enzymes (CAT, SOD, APX and POD), increased chlorophyll content, reduced MDA content | Lei et al. (2022) |

(continued)

**Table 3.1** (continued)

| Abiotic stress type | Plant species | Amino acid | Concentration of amino acid | Evaluation type (pot/field/aqueous) | Metabolic effects | Reference |
|---|---|---|---|---|---|---|
| Temperature stress | *Zea mays* | Glycine (Gly) and L-Arginine (L-Arg) | 0.006 mM, 0.003 mM, 0.06 mM, 0.2 mM, 0.3 mM, 0.6 mM, 2 mM, 3 mM, 6 mM, and 9 mM | Test plates and pot experiments | Enhanced length of radicals and number of lateral roots, increased root and shoot weight | Matysiak et al. (2020) |
| Salt stress | *Vigna radiata* L. | γ-aminobutyric acid (GABA) | 1.5 mM, 200 ml; 100 ml GABA/pot | Pot study | Improved fresh weight and dry weight of plant, increased levels of Chl a, Chl b and carotenoid content, significant increase in enzymatic antioxidants | Ullah et al. (2023) |
| Salt treatment | *Solanum lycopersicum* L. | Gamma-aminobutyric acid (GABA) | 5 mmol·$L^{-1}$ GABA | Pot study | Reduced salt accumulation in leaves, increased fresh weight and dry weight of plants | Wu et al. (2020) |
| Chromium stress | *Brassica napus* L. | cysteine (Cys) | 0.5 mM | – | Improved activities of antioxidant enzymes, reduced lipid peroxidation and Cr accumulation | Yıldız and Terzi (2021) |
| High temperature stress | *Gracilariopsis lemaneiformis* | L-Arginine | 0, 0.1, 0.5, and 1 mM | Lab conditions | Alleviated heat stress, improved MDA content, elevated activities of CAT, POD, SOD | Zhang et al. (2021) |

(Ortiz-Lopez et al. 2000). The APC family is made up of five sub-families, four of which are amino acid transporters: the hydroxy/aromatic amino acid permease (HAAAP) family, the alanine or glycine cation symporter (AGCS) family, the AAAP (auxin/amino acid permease) family, and the cation-chloride cotransporter (CCC) family (Jack et al. 2000). The cationic amino acid transporters (CATs), amino acid/choline transporters (ACTs), and polyamine $H^+$-symporters (PHSs) of the APC family, along with the amino acid permeases (AAPs), lysine and histidine transporters (LHTs), proline transporters (ProTs), γ-aminobutyric acid transporters (GATs), auxin transporters (AUXs), and aromatic and neutral amino acid transporters (ANTs) of the AAAP family are of particular interest (Pratelli and Pilot 2014) as shown in Fig. 3.3. According to Tegeder and Rentsch (2010), most of the well-researched amino acid transporters in the APC superfamily have importer traits, which clearly indicates their involvement in long-distance transport and the transfer of amino acids into the cytosol.

There is also a family of transporters called USUALLY MULTIPLE ACIDS MOVE IN AND OUT TRANSPORTERS (UMAMITs) that have been discovered to play a role in transporting amino acids (Ladwig et al. 2012). UMAMITs belong to the DMT superfamily, which is mainly composed of transporters for triose phosphate and nucleotide-sugar complex (Jack et al. 2001). There are currently only two known members of this family, AtSiAR1/AtUMAMIT18 (Ladwig et al. 2012) which mediates the transport of amino acids, and WAT1/AtUMAMIT05 which mediates the transport of auxin (Ranocha et al. 2013). Furthermore, the APC superfamily's phylogenetic tree also includes AtBAT1/AtGABP, a transporter that is distantly related and was described as a bidirectional transporter, mediating γ-aminobutyric acid (GABA) transport to the mitochondrion (Dündar and Bush 2009; Michaeli et al. 2011).

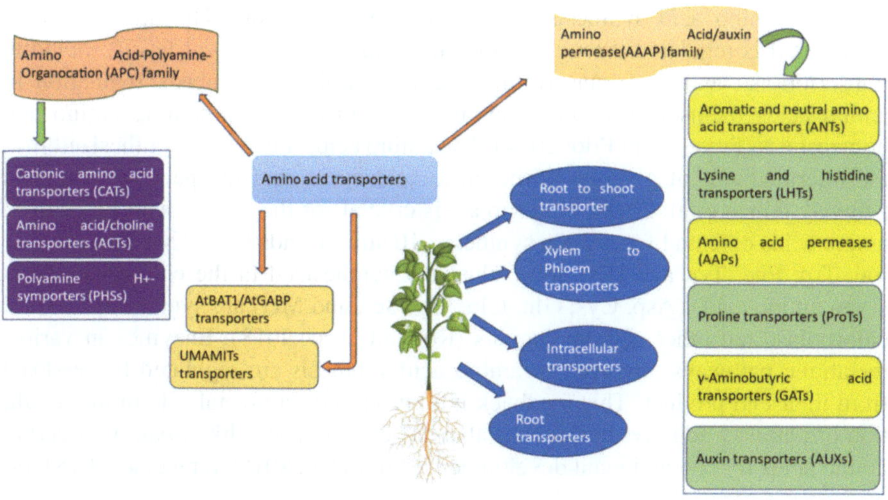


**Fig. 3.3** Amino acid transporter in plants

The terms "low affinity" and "high affinity" are used to describe transporters. The high-affinity ammonium transporter (AMT) family, particularly AMT1;1 and AMT1;2, found in plants, absorb a significant amount of ammonium from the soil (Yao et al. 2020). In the roots of Arabidopsis, there are transporters from three families, AAPs, LHTs, and ProTs, that help in the transport of amino acids. AAPs and LHTs are responsible for neutral and acidic amino acids, while ProTs facilitate proline, glycine, and γ-aminobutyric acid (GABA) transport (Fischer et al. 2002; Okumoto et al. 2002; Grallath et al. 2005). LHT1 and AAP5 play a crucial role in amino acid absorption when soil concentrations are below 50 μM. The phloem of aap6 mutant plants has lower amounts of amino acids, indicating that Arabidopsis AAP6 is located in the xylem parenchyma, where it facilitates nitrogen exchange between the xylem and phloem (Hunt et al. 2010). Amino acids produced inside cells need to travel across different organelles and long distances, which requires intracellular transporters. Researchers have found that several amino acid transporters localize to organellar membranes instead of the plasma membrane. They used fusion proteins between transporters and the green fluorescent protein (GFP) to demonstrate this (Yao et al. 2020). In Arabidopsis, CAT2 and CAT4 are cationic amino acid transporters that localize to the vacuolar membrane (tonoplast). CAT2 regulates amino acid levels in leaves, while AAP1 acquires glutamate and neutral amino acids from the soil at ecologically relevant concentrations. LHT6, on the other hand, aids in the import of the acidic amino acids alanine and aspartate by roots at both low and high concentrations (Perchlik et al. 2014).

## 3.6 Potential Mechanisms Governing the Regulation of Amino Acid Homeostasis

In plants, the amino acid homeostasis is a complex process which involves the regulation of amino acid synthesis, degradation, and transport. The homeostasis of amino acid is regulated through various mechanisms at different levels, including transcriptional and post-transcriptional levels. Similarly, feedback regulation by amino acids themselves is an important mechanism in maintaining amino acid homeostasis (Pratelli and Pilot 2014). The amino acids are mainly synthesized from the intermediates of glycolysis, the citric acid cycle, and the pentose phosphate pathway (Chen et al. 2018). Chloroplast is crucial for the metabolism of 17 of 20 amino acids essential for protein synthesis. 10 amino acids (Arg, Lys, Thr, Leu, Ile, Val, Trp, Phe, Tyr, and His) are exclusively synthesized in the chloroplast while seven amino acids (Asp, Cys, Gln, Glu, Gly, Ser, and Met) are synthesized in both chloroplast and other cellular regions (Rolland et al. 2018). Enzymes in various metabolic pathways, involved in amino acid synthesis are regulated by feedback from their end product. This feedback inhibition plays a pivotal role in the regulation of pathway activity; however, it also has consequences like toxicity of certain amino acids for cell and plant development (Amir et al. 2018; Yang et al. 2018). For instance, plants with certain amino acids like lysine and threonine can inhibit the activity of the aspartate kinase enzyme in the aspartate pathway, which leads to a decrease in the production of methionine and inhibit plant growth. Similarly, plants with enzymes like dihydrodipicolinate synthase, aspartate linase, chorismite mutase, 3-deoxy-D-arabinoheptulosonate 7-phosphate synthase and cystathionine-γ-synthase, insensitive to feedback exhibit an increase in the content of amino acids like Lys, Thr, Trp, and Met respectively (Pratelli and Pilot 2014). The main mechanism used by the plant cells to control the content of amino acids is by regulating their degradation. The study of (Karchi et al. 1995; Stepansky and Galili 2003; Stepansky et al. 2005) focused on the degradation of lysine pathway, where the first step for lysine degradation was the transcript of lysine ketoglutarate reductase/saccaropine dehydrogenase enzyme which is induced by various conditions including the addition of lysine itself. The basic leucine zipper (bZIP) transcription factor family like bZIP12/DPBF4 and bZIP39/ABI5 downregulate the expression of AK during darkness and low-sugar conditions (Ufaz et al. 2011) similarly, bZIP1 and bZIP39 expression is regulated by sugar and abscisic acid (Kang et al. 2010; Brocard et al. 2002). Transcription factors (dihydrodipicolinate synthase, 5-enolpyruvylshikimate 3-phosphate synthase, chorismite mutase, and phenylalanine ammonia lyase) control the gene expression for Phe synthesis and degradation hence, indicate a complex network for amino acid homeostasis regulation (Pratelli and Pilot 2014). At the post-transcriptional/translational level δ-pyrroline-5-carboxylate reductase and cystathionine-γ-synthase has been found in the synthesis of Pro and Met (Hare et al. 1999). Not much known about the post-translational modifications of enzymes in amino acid biosynthesis pathways. Proteomic approach has identified four enzymes of Met cycle which can nitrosylate Cys close to the active site of AtSAMS1 thereby reducing its activity by 60% (Lindermayr et al. 2005, 2006; Pratelli and Pilot 2014).

## 3.7 Conclusion

The examination of existing data concerning free amino acids in cereals has unveiled their dynamic engagement in metabolic and physiological functions across various stages of plant development, alongside their substantial capability to promote resistance to stress. The regulatory pathways facilitated by external amino acids are designed to enhance defensive enzyme mechanisms and elevate the levels of natural protective substances, thereby decreasing reliance on synthetic chemicals for farming, like pesticides and fertilizers. Utilization of amino acids exogenously represents a potentially innovative strategy to enhance the efficiency of crops metabolic pathways, and boost yields. Future investigations should prioritize the examination of genes responsible for amino acid biosynthesis, unraveling the specific contributions of individual amino acids in conferring resistance to abiotic pollutants, signaling cascades, and interplay with other signaling molecules.

## References

Abd El-Gawad HG, Mukherjee S, Farag R, Abd Elbar OH, Hikal M, Abou El-Yazied A, Elhady SAA, Helal N, ElKelish A, Nahhas NE, Azab E, Ismail. IA, Mbarki S, Ibrahim MF (2021) Exogenous γ-aminobutyric acid (GABA)-induced signaling events and field performance associated with mitigation of drought stress in Phaseolus vulgaris L. Plant Signal Behav 16(2):1853384

Abd Elnabi MK, Elkaliny NE, Elyazied MM, Azab SH, Elkhalifa SA, Elmasry S, Mouhamed MS, Shalamesh EM, Alhorieny NA, Abd Elaty AE, Elgendy IM, Etman AE, Saad KE, Tsigkou K, Ali SA, Kornaros M, Mahmoud YAG (2023) Toxicity of heavy metals and recent advances in their removal: a review. Toxics 11(7):580

Abid G, Ouertani RN, Jebara SH, Boubakri H, Muhovski Y, Ghouili E, Abdelkarim S, Chaieb O, Hidri Y, Kadri S, Ayed ME, Elkahoui S, Barhoumi F, Jebara M (2020) Alleviation of drought stress in faba bean (Vicia faba L.) by exogenous application of β-aminobutyric acid (BABA). Physiol Mol Biol Plants 26:1173–1186

Adejumo SA, Awoyemi V, Togun AO (2020) Exogenous proline and hormone in combination with compost improves growth and tolerance of maize under heavy metal stress. Plants Environ 2:40–53

Ahmad A, Aslam Z, Ilyas MZ, Ameer H, Mahmood A, Rehan M (2019) Drought stress mitigation by foliar feeding of potassium and amino acids in wheat. J Environ Agric Sci 18(1):10–18

Alfosea-Simón M, Zavala-Gonzalez EA, Camara-Zapata JM, Martínez-Nicolás JJ, Simón I, Simón-Grao S, García-Sánchez F (2020) Effect of foliar application of amino acids on the salinity tolerance of tomato plants cultivated under hydroponic system. Sci Hortic 272:109509

Ali Q, Haider MZ, Shahid S, Aslam N, Shehzad F, Naseem J et al (2019) Role of amino acids in improving abiotic stress tolerance to plants. In: Plant tolerance to environmental stress. CRC Press, Boca Raton, pp 175–204

Amir R, Galili G, Cohen H (2018) The metabolic roles of free amino acids during seed development. Plant Sci 275:11–18

Asgher M, Sehar Z, Rehaman A, Rashid S, Ahmed S, Per TS, Alyemeni MN, Khan NA (2022) Exogenously-applied L-glutamic acid protects photosynthetic functions and enhances arsenic tolerance through increased nitrogen assimilation and antioxidant capacity in rice (*Oryza sativa* L.). Environ Pollut 301:119008

Aysin F, Karaman A, Yilmaz A, Aksakal Ö, Gezgincioğlu E, Kohnehshahri SM (2020) Exogenous cysteine alleviates mercury stress by promoting antioxidant defence in maize (*Zea mays* L.) seedlings. Turk J Agric For 44(5):506–516

Batista-Silva W, Heinemann B, Rugen N, Nunes-Nesi A, Araújo WL, Braun HP, Hildebrandt TM (2019) The role of amino acid metabolism during abiotic stress release. Plant Cell Environ 42(5):1630–1644

Begna T (2020) Effects of drought stress on crop production and productivity. Int J Res Studies iAgric Sci 6(9):34–43

Bonner CA, Jensen RA (1997) Recognition of specific patterns of amino acid inhibition of growth in higher plants, uncomplicated by glutamine-reversible 'general amino acid inhibition'. Plant Sci 130(2):133–143

Brey LF, Włodarczyk AJ, Thøfner JFB, Burow M, Crocoll C, Nielsen I et al (2020) Metabolic engineering of Synechocystis sp. PCC 6803 for the production of aromatic amino acids and derived phenylpropanoids. Metab Eng 57:129–139

Brocard IM, Lynch TJ, Finkelstein RR (2002) Regulation and role of the Arabidopsis abscisic acid-insensitive 5 gene in abscisic acid, sugar, and stress response. Plant Physiol 129(4):1533–1543

Cánovas FM, Avila C, Canton FR, Canas RA, de la Torre F (2007) Ammonium assimilation and amino acid metabolism in conifers. J Exp Bot 58(9):2307–2318

Chen C, Zhang J (2024) Enhancing leukemia treatment: the role of combined therapies based on amino acid starvation. Cancer 16(6):1171

Chen Y, Zhou B, Li J, Tang H, Tang J, Yang Z (2018) Formation and change of chloroplast-located plant metabolites in response to light conditions. Int J Mol Sci 19(3):654

Curien G, Ravanel S, Robert M, Dumas R (2005) Identification of six novel allosteric effectors of Arabidopsis thaliana aspartate kinase-homoserine dehydrogenase isoforms: physiological context sets the specificity. J Biol Chem 280(50):41178–41183

Curien G, Bastien O, Robert-Genthon M, Cornish-Bowden A, Cárdenas ML, Dumas R (2009) Understanding the regulation of aspartate metabolism using a model based on measured kinetic parameters. Mol Syst Biol 5(1):271

Dickey RM, Forti AM, Kunjapur AM (2021) Advances in engineering microbial biosynthesis of aromatic compounds and related compounds. Bioresour Bioprocess 8:1–17

Dotson SB, Somers DA, Gengenbach BG (1989) Purification and characterization of lysine-sensitive aspartate kinase from maize cell cultures. Plant Physiol 91(4):1602–1608

Dündar E, Bush DR (2009) BAT1, a bidirectional amino acid transporter in Arabidopsis. Planta 229:1047–1056

ElKelish A, Nahhas NE, Azab E, Ismail. IA, Mbarki S, Ibrahim MF (2021) Exogenous γ-aminobutyric acid (GABA)-induced signaling events and field performance associated with mitigation of drought stress in Phaseolus vulgaris L. Plant Signal Behav 16(2):1853384

Fardus J, Hossain MS, Fujita M (2021) Potential role of L-glutamic acid in mitigating cadmium toxicity in lentil (Lens culinaris Medik.) through modulating the antioxidant defence system and nutrient homeostasis. Notulae Botanicae Horti Agrobotanici Cluj-Napoca 49(4):12485

Farooq M, Jang YH, Kim EG, Park JR, Eom GH, Zhao DD, Kim KM (2023) Evaluation of amino acid profiles of rice genotypes under different salt stress conditions. Plants 12(6):1315

Fischer WN, Loo DD, Koch W, Ludewig U, Boorer KJ, Tegeder M et al (2002) Low and high affinity amino acid H+-cotransporters for cellular import of neutral and charged amino acids. Plant J 29(6):717–731

Goyal K, Walton LJ, Tunnacliffe A (2005) LEA proteins prevent protein aggregation due to water stress. Biochem J 388(1):151–157

Grallath S, Weimar T, Meyer A, Gumy C, Suter-Grotemeyer M, Neuhaus JM, Rentsch D (2005) The AtProT family. Compatible solute transporters with similar substrate specificity but differential expression patterns. Plant Physiol 137(1):117–126

Guo Z, Yang N, Zhu C, Gan L (2017) Exogenously applied poly-γ-glutamic acid alleviates salt stress in wheat seedlings by modulating ion balance and the antioxidant system. Environ Sci Pollut Res 24:6592–6598

Hacham Y, Schuster G, Amir R (2006a) An in vivo internal deletion in the N-terminus region of Arabidopsis cystathionine gamma-synthase results in CGS expression that is insensitive to methionine. Plant J 45:955–967

Hacham Y, Schuster G, Amir R (2006b) An in vivo internal deletion in the N-terminus region of Arabidopsis cystathionine γ-synthase results in CGS expression that is insensitive to methionine. Plant J 45(6):955–967

Haghighi M, Saadat S, Abbey L (2020) Effect of exogenous amino acids application on growth and nutritional value of cabbage under drought stress. Sci Hortic 272:109561

Haghighi TM, Saharkhiz MJ, Kavoosi G, Jowkar A (2022) Monitoring amino acid profile and protein quality of Licorice (*Glycyrrhiza glabra* L.) under drought stress, silicon nutrition and mycorrhiza inoculation. Sci Hortic 295:110808

Hare PD, Cress WA, Van Staden J (1999) Proline synthesis and degradation: a model system for elucidating stress-related signal transduction. J Exp Bot 50(333):413–434

Housmans JA, Wu G, Schymkowitz J, Rousseau F (2023) A guide to studying protein aggregation. FEBS J 290(3):554–583

Hua XJ, Van De Cotte B, Van Montagu M, Verbruggen N (2001) The 5′ untranslated region of the At-P5R gene is involved in both transcriptional and post-transcriptional regulation. Plant J 26(2):157–169

Hunt E, Gattolin S, Newbury HJ, Bale JS, Tseng HM, Barrett DA, Pritchard J (2010) A mutation in amino acid permease AAP6 reduces the amino acid content of the Arabidopsis sieve elements but leaves aphid herbivores unaffected. J Exp Bot 61(1):55–64

Hussein HAA, Alshammari SO, Kenawy SK, Elkady FM, Badawy AA (2022) Grain-priming with L-arginine improves the growth performance of wheat (*Triticum aestivum* L.) plants under drought stress. Plants 11(9):1219

Imran QM, Falak N, Hussain A, Mun BG, Yun BW (2021) Abiotic stress in plants; stress perception to molecular response and role of biotechnological tools in stress resistance. Agronomy 11(8):1579

Ingrisano R, Tosato E, Trost P, Gurrieri L, Sparla F (2023) Proline, cysteine and branched-chain amino acids in abiotic stress response of land plants and microalgae. Plants 12(19):3410

Jack DL, Paulsen IT, Saier MH (2000) The amino acid/polyamine/organocation (APC) superfamily of transporters specific for amino acids, polyamines and organocations. Microbiology 146(8):1797–1814

Jack DL, Yang NM, Saier H Jr, M. (2001) The drug/metabolite transporter superfamily. Eur J Biochem 268(13):3620–3639

Jander G, Joshi V (2010) Recent progress in deciphering the biosynthesis of aspartate-derived amino acids in plants. Mol Plant 3(1):54–65

Kang SG, Price J, Lin PC, Hong JC, Jang JC (2010) The Arabidopsis bZIP1 transcription factor is involved in sugar signaling, protein networking, and DNA binding. Mol Plant 3(2):361–373

Karchi H, Miron D, Ben-Yaacov S, Galili G (1995) The lysine-dependent stimulation of lysine catabolism in tobacco seed requires calcium and protein phosphorylation. Plant Cell 7(11):1963–1970

Khan N, Ali S, Zandi P, Mehmood A, Ullah S, Ikram M et al (2020) Role of sugars, amino acids and organic acids in improving plant abiotic stress tolerance. Pak J Bot 52(2):355–363

Kishor PK, Suravajhala R, Rajasheker G, Marka N, Shridhar KK, Dhulala D et al (2020) Lysine, lysine-rich, serine, and serine-rich proteins: link between metabolism, development, and abiotic stress tolerance and the role of ncRNAs in their regulation. Front Plant Sci 11:546213

Kul R, Ekinci M, Turan M, Ors S, Yildirim E (2020) How abiotic stress conditions affects plant roots. In: Plant roots. IntechOpen

Kumar V, Sharma A, Kohli SK, Yadav P, Bali S, Bakshi P et al (2019) Amino acids distribution in economical important plants: a review. Biotechnology Research and Innovation 3(2):197–207

Ladwig F, Stahl M, Ludewig U, Hirner AA, Hammes UZ, Stadler R et al (2012) Siliques are Red1 from Arabidopsis acts as a bidirectional amino acid transporter that is crucial for the amino acid homeostasis of siliques. Plant Physiol 158(4):1643–1655

Lei S, Rossi S, Huang B (2022) Metabolic and physiological regulation of aspartic acid-mediated enhancement of heat stress tolerance in perennial ryegrass. Plants 11(2):199

Lenzen M, Malik A, Li M, Fry J, Weisz H, Pichler PP et al (2020) The environmental footprint of health care: a global assessment. Lancet Planet Health 4(7):e271–e279

Less H, Galili G (2008) Principal transcriptional programs regulating plant amino acid metabolism in response to abiotic stresses. Plant Physiol 147(1):316–330

Less H, Galili G (2009) Coordinations between gene modules control the operation of plant amino acid metabolic networks. BMC Syst Biol 3:1–18

Lindermayr C, Saalbach G, Durner J (2005) Proteomic identification of S-nitrosylated proteins in Arabidopsis. Plant Physiol 137(3):921–930

Lindermayr C, Saalbach G, Bahnweg G, Durner J (2006) Differential inhibition of Arabidopsis methionine adenosyltransferases by protein S-nitrosylation. J Biol Chem 281(7):4285–4291

Lopes M, Coimbra MA, Costa MDC, Ramos F (2023) Food supplement vitamins, minerals, amino-acids, fatty acids, phenolic and alkaloid-based substances: an overview of their interaction with drugs. Crit Rev Food Sci Nutr 63(19):4106–4140

Maeda H, Dudareva N (2012) The shikimate pathway and aromatic amino acid biosynthesis in plants. Annu Rev Plant Biol 63:73–105

Matysiak K, Kierzek R, Siatkowski I, Kowalska J, Krawczyk R, Miziniak W (2020) Effect of exogenous application of amino acids l-arginine and glycine on maize under temperature stress. Agronomy 10(6):769

Michaeli S, Fait A, Lagor K, Nunes-Nesi A, Grillich N, Yellin A et al (2011) A mitochondrial GABA permease connects the GABA shunt and the TCA cycle, and is essential for normal carbon metabolism. Plant J 67(3):485–498

Mogk A, Bukau B, Kampinga HH (2018) Cellular handling of protein aggregates by disaggregation machines. Mol Cell 69(2):214–226

Okumoto S, Schmidt R, Tegeder M, Fischer WN, Rentsch D, Frommer WB, Koch W (2002) High affinity amino acid transporters specifically expressed in xylem parenchyma and developing seeds of Arabidopsis. J Biol Chem 277(47):45338–45346

Onouchi H, Nagami Y, Haraguchi Y, Nakamoto M, Nishimura Y, Sakurai R et al (2005) Nascent peptide-mediated translation elongation arrest coupled with mRNA degradation in the CGS1 gene of Arabidopsis. Genes Dev 19(15):1799–1810

Ortiz-Lopez A, Chang HC, Bush DR (2000) Amino acid transporters in plants. Biochim Biophys Acta Biomembr 1465(1–2):275–280

Parthasarathy A, Borrego EJ, Savka MA, Dobson RC, Hudson AO (2021) Amino acid–derived defense metabolites from plants: A potential source to facilitate novel antimicrobial development. J Biol Chem 296:100438

Peña Calzada K, Olivera Viciedo D, Habermann E, Calero Hurtado A, Lupino Gratão P, De Mello Prado R, Lata-Tenesaca LF, Martinez CA, Celi GEA, Rodríguez JC (2022) Exogenous application of amino acids mitigates the deleterious effects of salt stress on soybean plants. Agronomy 12(9):2014

Perchlik M, Foster J, Tegeder M (2014) Different and overlapping functions of Arabidopsis LHT6 and AAP1 transporters in root amino acid uptake. J Exp Bot 65(18):5193–5204

Pratelli R, Pilot G (2014) Regulation of amino acid metabolic enzymes and transporters in plants. J Exp Bot 65(19):5535–5556

Pratelli R, Voll LM, Horst RJ, Frommer WB, Pilot G (2010) Stimulation of nonselective amino acid export by glutamine dumper proteins. Plant Physiol 152(2):762–773

Ranocha P, Dima O, Nagy R, Felten J, Corratgé-Faillie C, Novák O et al (2013) Arabidopsis WAT1 is a vacuolar auxin transport facilitator required for auxin homoeostasis. Nat Commun 4(1):2625

Rolland N, Bouchnak I, Moyet L, Salvi D, Kuntz M (2018) The main functions of plastids. In: Plastids: methods and protocols. Springer, pp 73–85

Ruszkowski M, Forlani G (2022) Deciphering the structure of Arabidopsis thaliana 5-enol-pyruvyl-shikimate-3-phosphate synthase: an essential step toward the discovery of novel inhibitors to supersede glyphosate. Comput Struct Biotechnol J 20:1494–1505

Sarrobert C, Thibaud MC, Contard-David P, Gineste S, Bechtold N, Robaglia C, Nussaume L (2000) Identification of an Arabidopsis thaliana mutant accumulating threonine resulting from mutation in a new dihydrodipicolinate synthase gene. Plant J 24(3):357–368

Schjoerring JK, Husted S, Mäck G, Mattsson M (2002) The regulation of ammonium translocation in plants. J Exp Bot 53(370):883–890

Shahid S, Kausar A, Zahra N, Hafeez MB, Raza A, Ashraf MY (2023) Methionine-induced regulation of secondary metabolites and antioxidants in maize (Zea mays L.) subjected to salinity stress. Gesunde Pflanzen 75(4):1143–1155

Shaul O, Galili G (1992a) Increased lysine synthesis in tobacco plants that express high levels of bacterial dihydrodipicolinate synthase in their chloroplasts. Plant J 2(2):203–209

Shaul O, Galili G (1992b) Threonine overproduction in transgenic tobacco plants expressing a mutant desensitized aspartate kinase of Escherichia coli. Plant Physiol 100:1157–1163

Shende VV, Bauman KD, Moore BS (2024) The shikimate pathway: gateway to metabolic diversity. Nat Prod Rep 41:604

Smith NW, Fletcher AJ, Hill JP, McNabb WC (2021) Animal and plant-sourced nutrition: complementary not competitive. Anim Prod Sci 62(8):701–711

Stepansky A, Galili G (2003) Synthesis of the Arabidopsis bifunctional lysine-ketoglutarate reductase/saccharopine dehydrogenase enzyme of lysine catabolism is concertedly regulated by metabolic and stress-associated signals. Plant Physiol 133(3):1407–1415

Stepansky A, Yao Y, Tang G, Galili G (2005) Regulation of lysine catabolism in Arabidopsis through concertedly regulated synthesis of the two distinct gene products of the composite AtLKR/SDH locus. J Exp Bot 56(412):525–536

Sun Y, Miao F, Wang Y, Liu H, Wang X, Wang H, Guo J, Shao R, Yang Q (2023) L-Arginine alleviates the reduction in photosynthesis and antioxidant activity induced by drought stress in maize seedlings. Antioxidants 12(2):482

Tegeder M, Rentsch D (2010) Uptake and partitioning of amino acids and peptides. Mol Plant 3(6):997–1011

Trovato M, Funck D, Forlani G, Okumoto S, Amir R (2021) Amino acids in plants: regulation and functions in development and stress defense. Front Plant Sci 12:772810

Tzin V, Malitsky S, Aharoni A, Galili G (2009) Expression of a bacterial bi-functional chorismate mutase/prephenate dehydratase modulates primary and secondary metabolism associated with aromatic amino acids in Arabidopsis. Plant J 60(1):156–167

Tzin V, Malitsky S, Zvi MMB, Bedair M, Sumner L, Aharoni A, Galili G (2012) Expression of a bacterial feedback-insensitive 3-deoxy-d-arabino-heptulosonate 7-phosphate synthase of the shikimate pathway in Arabidopsis elucidates potential metabolic bottlenecks between primary and secondary metabolism. New Phytol 194(2):430–439

Tzin V, Rogachev I, Meir S, Moyal Ben Zvi M, Masci T, Vainstein A et al (2013) Tomato fruits expressing a bacterial feedback-insensitive 3-deoxy-D-arabino-heptulosonate 7-phosphate synthase of the shikimate pathway possess enhanced levels of multiple specialized metabolites and upgraded aroma. J Exp Bot 64(14):4441–4452

Ufaz S, Shukla V, Soloveichik Y, Golan Y, Breuer F, Koncz Z et al (2011) Transcriptional control of aspartate kinase expression during darkness and sugar depletion in Arabidopsis: involvement of bZIP transcription factors. Planta 233:1025–1040

Ullah A, Ali I, Noor J, Zeng F, Bawazeer S, Eldin SM, Asghar MA, Javed HH, Saleem K, Ullah S, Ali H (2023) Exogenous γ-aminobutyric acid (GABA) mitigated salinity-induced impairments in mungbean plants by regulating their nitrogen metabolism and antioxidant potential. Front Plant Sci 13:1081188

Wu X, Jia Q, Ji S, Gong B, Li J, Lü G, Gao H (2020) Gamma-aminobutyric acid (GABA) alleviates salt damage in tomato by modulating Na+ uptake, the GAD gene, amino acid synthesis and reactive oxygen species metabolism. BMC Plant Biol 20:1–21

Wu J, Chen Y, Huang Y, Hao B, Dai S, Zhao L et al (2024) The cytosolic aminotransferase VAS1 coordinates aromatic amino acid biosynthesis and metabolism. *Science*. Advances 10(2):eadk0738

Yang QQ, Zhao DS, Zhang CQ, Wu HY, Li QF, Gu MH et al (2018) A connection between lysine and serotonin metabolism in rice endosperm. Plant Physiol 176(3):1965–1980

Yao X, Nie J, Bai R, Sui X (2020) Amino acid transporters in plants: identification and function. Plants 9(8):972

Yıldız M, Terzi H (2021) Exogenous cysteine alleviates chromium stress via reducing its uptake and regulating proteome in roots of *Brassica napus* L. seedlings. S Afr J Bot 139:114–121

Yokoyama R, de Oliveira MV, Kleven B, Maeda HA (2021) The entry reaction of the plant shikimate pathway is subjected to highly complex metabolite-mediated regulation. Plant Cell 33(3):671–696

Zaheer IE, Ali S, Saleem MH, Imran M, Alnusairi GS, Alharbi BM, Riaz M, Abbas Z, Rizwaan M, Soliman MH (2020) Role of iron–lysine on morpho-physiological traits and combating chromium toxicity in rapeseed (*Brassica napus* L.) plants irrigated with different levels of tannery wastewater. Plant Physiol Biochem 155:70–84

Zemanová V, Pavlík M, Pavlíková D, Tlustoš P (2014) The significance of methionine, histidine and tryptophan in plant responses and adaptation to cadmium stress. Plant Soil Environ 60:426
Zhang J, Liu S, Hu C, Chen X, Sun X, Xu N (2021) Physiological and transcriptome analysis of exogenous l-arginine in the alleviation of high-temperature stress in *Gracilariopsis lemaneiformis*. Front Mar Sci 8:784586
Zhang Y, Wang Y, Guo Y, Ji N, Chen Y, Sun Y et al (2024) Effects and regulatory mechanism of exogenous $H_2O_2$ on amino acid metabolism in post-harvest broccoli: an integrated analysis of transcriptomic and Metabolomic. Food Bioprocess Technol, pp 1–14
Zulfiqar F, Ashraf M (2023) Proline alleviates abiotic stress induced oxidative stress in plants. J Plant Growth Regul 42(8):4629–4651

# Role of Osmolytes in ROS-Scavenging and Maintaining Homeostasis Under Abiotic Stress

4

Akankhya Guru, Deepika Ranjan, Divya Jaiswal, Akash Hidangmayum, and Padmanabh Dwivedi

**Abstract**

Abiotic stresses such as drought, salinity, heat, and cold pose significant challenges to plant growth and productivity by disrupting cellular homeostasis and triggering the excessive generation of reactive oxygen species (ROS). These ROS, including superoxide radicals, hydrogen peroxide, and hydroxyl radicals, lead to oxidative stress and cause damage to vital cellular components such as proteins, lipids, and nucleic acids. To mitigate these detrimental effects, plants have developed complex defense mechanisms, among which the accumulation of osmolytes plays a crucial role. This chapter explores the pivotal role of osmolytes in ROS scavenging and in maintaining cellular homeostasis under abiotic stress conditions. It begins with an overview of ROS generation and their effects on plant cells, followed by a detailed classification of osmolytes—such as proline, glycine betaine, trehalose, and polyols—and their protective functions. Emphasis is placed on the diverse mechanisms through which osmolytes contribute to oxidative stress tolerance, including direct ROS scavenging, enhancement of antioxidant defense systems, stabilization of macromolecules, and maintenance of redox balance. Furthermore, the chapter examines the regulatory pathways influencing osmolyte accumulation, highlighting transcriptional control, hormonal regulation, and genetic engineering strategies aimed at improving stress resilience in crops. Future perspectives focus on integrating omics technologies and biotechnological approaches to enhance osmolyte-mediated stress

A. Guru
Department of Agriculture Botany, Janta Vedic College, Baraut (Baghpat); affiliated to CCS University, Meerut, Uttar Pradesh, India

D. Ranjan · D. Jaiswal · A. Hidangmayum · P. Dwivedi (✉)
Department of Plant Physiology, Institute of Agricultural Sciences, Banaras Hindu University, Varanasi, India
e-mail: pdwivedi25@bhu.ac.in



N. Iqbal et al. (eds.), *Plant Signaling Molecules in Regulation of ROS-Scavenging System*, https://doi.org/10.1007/978-981-95-3377-0_4

tolerance. Through a comprehensive synthesis of current research, this chapter underscores the essential functions of osmolytes in mitigating oxidative stress and maintaining cellular stability, offering valuable insights for the development of stress-resilient crop varieties in the face of global climate change.

**Keywords**

Osmolytes · Reactive Oxygen Species · Abiotic Stress · Antioxidant Defense · Redox Homeostasis

## 4.1 Introduction

Plants are constantly exposed to environmental fluctuations that impose stress on their growth and productivity. Abiotic stress conditions such as drought, salinity, extreme temperatures (heat and cold), and heavy metal toxicity severely affect plant physiological and biochemical processes. These stress factors disturb water balance, ion homeostasis, nutrient uptake, and photosynthetic efficiency, ultimately leading to cellular damage and reduced plant vigour. One of the most severe consequences of abiotic stress is oxidative stress, which arises due to an imbalance between the production and scavenging of reactive oxygen species (ROS). Under normal physiological conditions, ROS, including superoxide radicals ($O_2^-$), hydrogen peroxide ($H_2O_2$), hydroxyl radicals (OH·), and singlet oxygen ($^1O_2$) are continuously generated as byproducts of metabolic activities and are efficiently detoxified by antioxidant systems (Choudhury et al. 2017). However, under stress conditions, the rate of ROS production exceeds the plant's scavenging capacity, leading to oxidative stress and cellular damage. If not effectively managed, oxidative stress can result in programmed cell death (PCD) and reduced plant vigor.

To counteract ROS-induced damage and maintain cellular homeostasis, plants have evolved a highly sophisticated defense system consisting of enzymatic and non-enzymatic antioxidants. Among the most crucial non-enzymatic mechanisms is the accumulation of osmolytes—low molecular weight organic compounds that help protect plants from stress-induced cellular damage. These osmolytes, including proline, glycine betaine, polyamines, sugar alcohols (mannitol, sorbitol, and trehalose), and other compatible solutes, act as osmoprotectants, ROS scavengers, and stabilizers of biomolecules under abiotic stress conditions (Ejaz et al. 2020).

This chapter provides an in-depth discussion of the role of osmolytes in ROS scavenging and homeostasis maintenance, highlighting their biosynthesis, regulatory mechanisms, and interaction with antioxidant defense systems in plants under abiotic stress.

## 4.2 Reactive Oxygen Species (ROS) and Oxidative Stress in Plants

Reactive oxygen species (ROS) are highly reactive molecules that play a dual role in plant physiology. Under normal conditions, they serve as important signaling molecules involved in stress perception, hormone signaling, and defense responses. However, when produced in excess under abiotic stress conditions, ROS become cytotoxic, causing oxidative stress and severe cellular damage (Bhattacharjee and Bhattacharjee 2019). The delicate balance between ROS production and scavenging determines whether they function as beneficial regulators or harmful agents in plants.

### 4.2.1 Generation of ROS Under Abiotic Stress

ROS generation occurs mainly in chloroplasts, mitochondria, peroxisomes, and the plasma membrane, where electron transport and metabolic activities take place. Under optimal conditions, ROS levels are tightly regulated by antioxidant defense systems (Singh 2022). However, abiotic stress factors such as drought, salinity, extreme temperatures, and heavy metal toxicity disrupt this balance, leading to overproduction of ROS and subsequent oxidative stress.

Chloroplasts are the primary site of ROS production, particularly under stress conditions that disrupt photosynthesis. When plants experience drought, high light intensity, or temperature extremes, the photosynthetic electron transport chain (PETC) becomes over-reduced, leading to the formation of superoxide radicals ($O_2^-$) and singlet oxygen ($^1O_2$). These ROS species damage thylakoid membranes, photosynthetic pigments, and D1 proteins of photosystem II (PSII), ultimately reducing photosynthetic efficiency and leading to photoinhibition (Foyer and Hanke 2022).

Mitochondria also contribute to ROS generation during cellular respiration. Under stress conditions, electron leakage from complex I and complex III of the mitochondrial electron transport chain (ETC) results in superoxide radical ($O_2^-$) formation. The generated superoxide radicals are converted into hydrogen peroxide ($H_2O_2$) by superoxide dismutase (SOD). Excess ROS in mitochondria can damage respiratory enzymes, disrupt ATP synthesis, and impair metabolic homeostasis (Zhao et al. 2019).

Peroxisomes are another significant source of ROS, particularly hydrogen peroxide ($H_2O_2$), which is generated as a byproduct of photorespiration, fatty acid β-oxidation, and enzymatic reactions (Sandalio et al. 2013). During photorespiration, oxygenation of ribulose-1,5-bisphosphate (RuBP) by Rubisco leads to the formation of glycolate, which undergoes oxidation to produce $H_2O_2$. Similarly, the breakdown of fatty acids in peroxisomes generates ROS, which can accumulate under abiotic stress conditions, leading to lipid peroxidation and metabolic disruption (Del Río and López-Huertas 2016).

The plasma membrane also plays a key role in ROS generation through NADPH oxidases (Respiratory Burst Oxidase Homologs, RBOHs). These enzymes transfer electrons from NADPH to molecular oxygen, producing superoxide radicals ($O_2^-$). While NADPH oxidase-generated ROS serve as signaling molecules in stress responses, excessive accumulation causes membrane damage, ion leakage, and cell dysfunction (Sies et al. 2022).

Thus, abiotic stress conditions lead to excessive ROS accumulation by disrupting photosynthesis, respiration, peroxisomal metabolism, and membrane-associated oxidative reactions. If not effectively managed, ROS-induced oxidative stress severely impacts plant growth and survival.

### 4.2.2 Effects of ROS on Plant Cells

While ROS are essential for normal cellular signaling, their excessive accumulation under abiotic stress results in oxidative stress, leading to severe cellular damage. High ROS levels negatively impact lipid membranes, proteins, nucleic acids, and cellular structures, disrupting metabolic homeostasis and triggering programmed cell death (PCD). The major damaging effects of ROS on plant cells include lipid peroxidation, protein oxidation, DNA damage, and cell death (Sahu et al. 2022).

Lipid peroxidation is one of the most harmful consequences of ROS accumulation, particularly affecting cellular membranes. ROS, especially hydroxyl radicals (OH•) and singlet oxygen ($^1O_2$), attack polyunsaturated fatty acids in membrane lipids, initiating a chain reaction of lipid peroxidation. This results in the formation of malondialdehyde (MDA) and other cytotoxic aldehydes, leading to membrane rigidity, increased permeability, and ion leakage. As a consequence, oxidative damage to membranes disrupts chloroplast structure, mitochondrial function, and overall cell integrity, impairing metabolic processes and stress responses (Mittler et al. 2022).

Protein oxidation occurs when ROS modify amino acid residues, leading to structural and functional alterations in proteins. Enzymes involved in photosynthesis, respiration, and metabolic pathways are particularly susceptible to oxidative damage. ROS-induced oxidation can lead to misfolding, aggregation, or degradation of proteins, rendering them non-functional (Sachdev et al. 2021). This affects crucial processes such as enzyme activity, transport mechanisms, and cellular signaling, further aggravating stress responses.

DNA damage is another major effect of ROS accumulation, primarily caused by hydroxyl radicals (OH•) attacking nucleic acids. ROS can induce strand breaks, base modifications (e.g., guanine oxidation to 8-oxo-guanine), and chromosomal aberrations, leading to genetic instability. Persistent oxidative damage impairs gene expression, cell division, and DNA replication, ultimately affecting plant growth and productivity. Severe DNA damage activates cellular stress responses, but if the repair mechanisms fail, the affected cells may undergo apoptosis or programmed cell death (PCD) (Wang et al. 2024).

Cell death is triggered when ROS levels exceed the plant's antioxidant defense capacity, leading to irreparable oxidative damage. Programmed cell death (PCD) is initiated as a protective mechanism to remove severely damaged cells and prevent further spread of oxidative stress. During PCD, mitochondrial dysfunction, cytochrome c release, and activation of proteases and nucleases lead to controlled cell degradation. While PCD is beneficial in eliminating damaged tissues, excessive ROS-induced cell death can cause tissue necrosis, premature senescence, and reduced plant vigor, ultimately compromising crop yield and survival under stress conditions (Šoln et al. 2022).

To mitigate the detrimental effects of ROS, plants have evolved antioxidant defense mechanisms, comprising enzymatic and non-enzymatic antioxidants. Enzymatic antioxidants such as superoxide dismutase (SOD), catalase (CAT), and ascorbate peroxidase (APX) neutralize ROS and prevent oxidative damage (Ye et al. 2021). Additionally, non-enzymatic antioxidants, including glutathione, ascorbic acid, flavonoids, and carotenoids, play crucial roles in ROS scavenging. Furthermore, osmolytes play a vital role in ROS detoxification and cellular protection. Osmolytes, such as proline, glycine betaine, trehalose, and polyamines function as osmoprotectants, stabilizing cellular structures and enhancing antioxidant defense (Choudhary et al. 2023). These molecules help maintain redox homeostasis, protect proteins, and membranes from oxidative damage, and regulate stress-responsive genes. In the following sections, we will explore the role of osmolytes in ROS scavenging and homeostasis maintenance, highlighting their importance in improving plant stress tolerance.

## 4.3 Osmolytes: Classification and Functions

Osmolytes are small organic molecules that play a crucial role in protecting cells against various abiotic stresses such as drought, salinity, high temperature, and cold. These stress conditions disrupt cellular homeostasis by causing osmotic imbalance and oxidative damage. Osmolytes accumulate in cells to counteract the detrimental effects of these stressors, thereby aiding in stress tolerance without interfering with normal metabolic processes. Their multifunctional role includes stabilization of proteins and membranes, scavenging of reactive oxygen species (ROS), and maintenance of turgor pressure, making them indispensable components of the plant stress response.

### 4.3.1 Classification of Osmolytes

Osmolytes are broadly classified into three major categories based on their chemical nature: *amino acids and their derivatives*, *sugars*, *and polyols*, and *quaternary ammonium compounds and betaines* (Kaur et al. 2024).

1. *Amino Acids and Derivatives*: This class includes *proline*, *glutamine*, *glycine*, and *ectoine*. Among them, *proline* is the most studied osmolyte, known for its high accumulation under stress and its role in osmoprotection, ROS scavenging,

and redox balance. These compounds not only stabilize proteins and membranes but also serve as nitrogen and carbon reservoirs.

2. *Sugars and Polyols*: Sugars like *trehalose*, *sucrose*, and *raffinose*, and polyols such as *mannitol*, *sorbitol*, and *inositol*, act as compatible solutes. They help maintain osmotic balance and protect macromolecules through vitrification and water replacement mechanisms. Trehalose, in particular, is known for its unique glass-forming properties that stabilize cellular structures during dehydration.
3. *Quaternary Ammonium Compounds and Betaines*: These include *glycine betaine*, *betaine*, and *choline-O-sulfate*. These compounds are zwitterionic and highly soluble in water, which makes them excellent osmoprotectants. *Glycine betaine* is one of the most effective osmolytes, known for stabilizing the quaternary structures of proteins and maintaining the integrity of membranes under salt and drought stress.

### 4.3.2 Functions of Osmolytes

Osmolytes perform a variety of critical roles that underpin cellular protection and enhance stress tolerance in plants subjected to abiotic challenges.

#### 4.3.2.1 Osmoregulation

Osmolytes are integral in maintaining cellular water homeostasis by modulating the osmotic gradient across cellular membranes. Under stress conditions such as drought or salinity, extracellular osmolarity increases, leading to potential cellular dehydration. Accumulation of osmolytes within the cytoplasm lowers intracellular osmotic potential, thereby facilitating water retention and sustaining turgor pressure (Ozturk et al. 2021). This osmotic adjustment preserves cell volume and ensures the continuity of physiological processes despite adverse environmental conditions.

#### 4.3.2.2 Stabilization of Proteins and Membranes

Osmolytes act as chemical chaperones that stabilize macromolecular structures. They protect proteins from denaturation induced by dehydration, high ionic concentrations, or temperature extremes through preferential exclusion from the protein hydration shell. This exclusion effect favors the native, folded state of proteins, enhancing their structural integrity and functional capacity (Choudhary et al. 2023). Similarly, osmolytes stabilize membrane bilayers by preserving lipid organization and preventing phase transitions that could compromise membrane fluidity and permeability.

#### 4.3.2.3 Reactive Oxygen Species (ROS) Scavenging and Redox Homeostasis

Certain osmolytes, notably proline and specific polyols, exhibit antioxidant properties by directly scavenging ROS such as hydrogen peroxide ($H_2O_2$), superoxide anions ($O_2^-$), and hydroxyl radicals (OH•). This mitigates oxidative damage to nucleic acids, proteins, and lipids, which commonly occurs under abiotic stress

(Anjum et al. 2023). Furthermore, osmolytes contribute to cellular redox balance by influencing key redox couples, including $NADP^+$/NADPH and glutathione (GSH/GSSG). By maintaining these redox buffers, osmolytes help regulate signaling pathways and protect enzymatic activities essential for stress adaptation.

#### 4.3.2.4 Energy Storage and Carbon/Nitrogen Reservoirs

Beyond their protective roles, osmolytes, such as sugars and amino acids serve as vital metabolic reserves. They store carbon and nitrogen resources that can be mobilized during post-stress recovery phases to support cellular repair, growth, and development (Choudhary et al. 2023). This dual function enables plants to efficiently transition from stress endurance to active growth once favorable conditions are restored.

#### 4.3.2.5 Signal Transduction

Emerging research indicates that osmolytes may function as signaling molecules that modulate stress-responsive gene expression. For example, proline accumulation has been linked to the activation of transcription factors and stress signaling cascades, suggesting a regulatory role in stress perception and response (Singh et al. 2022). This signaling function positions osmolytes as integral components in the orchestration of complex adaptive networks beyond their traditional roles in osmotic balance.

## 4.4 Mechanisms of Osmolyte-Mediated ROS Scavenging

Osmolytes are low-molecular-weight compounds, including sugars, polyamines, secondary metabolites, amino acids, and polyols, that play a vital role in helping organisms cope with environmental stress by maintaining cellular function and preventing damage (Roychoudhury and Banerjee 2013; Roychoudhury and Das 2014). These small molecules protect cells, primarily by scavenging reactive oxygen species (ROS), reducing oxidative stress, and supporting cellular health. Exposure to harsh environmental conditions causes plants to overproduce toxic ROS, which can damage biomolecules and induce oxidative stress, ultimately threatening plant productivity and global food security. As oxidative damage to biomolecules increases, the efficiency of enzymatic antioxidants in scavenging ROS decreases. However, plants counteract the harmful effects of ROS by synthesizing osmolytes, which help mitigate stress by protecting membrane structures and directly neutralizing ROS (Fig. 4.2). Certain osmolytes, like floridoside and glycine betaine, directly scavenge ROS, reducing their accumulation and minimizing cellular damage. Major osmolytes synthesized by plants, such as amino acids, quaternary amines, sugars, and sugar polyols, help regulate osmotic potential and keep ROS production and accumulation below harmful levels. Osmolytes also regulate antioxidant enzyme activity, chelate metal ions, and perform other functions that help manage oxidative stress. They maintain osmotic balance, crucial for cellular integrity (Wijayasinghe et al. 2017), and enhance survival under extreme environmental conditions (Evers

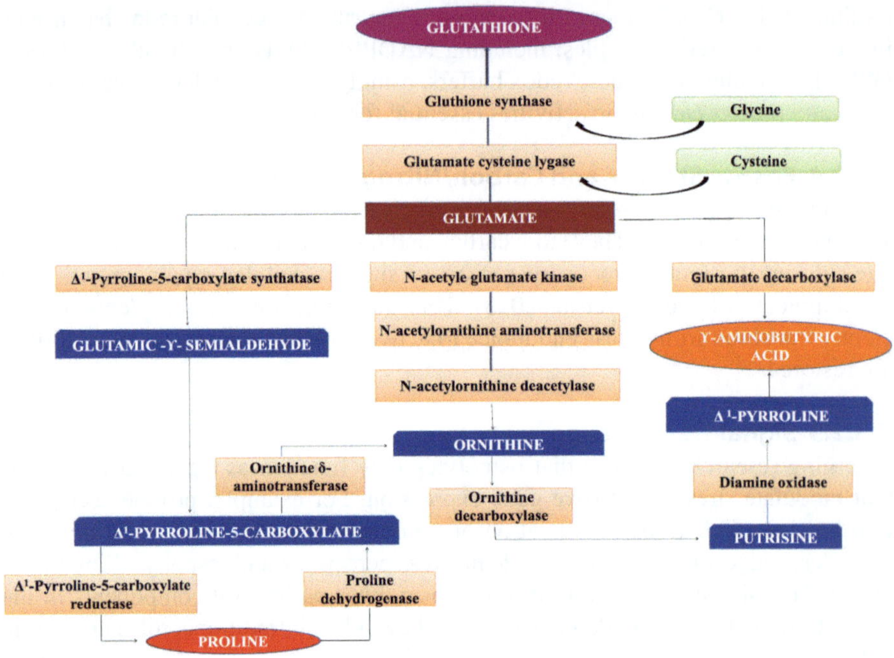


**Fig. 4.1** Biosynthesis of stress mitigating compounds from a common precursor glutamate. Glutathione, proline and γ-aminobutyric acid are biosynthetically derived from glutamate and are important in non-enzymatic defense mechanism of plants against abiotic stresses. Proline is an OH• scavenger, glutathione an antioxidant, and γ-aminobutyric acid an antioxidative enzymes inducer

et al. 2011). In addition to direct synthesis, osmolyte accumulation is often stimulated by phytohormones and micronutrient supplementation, improving the plant's tolerance to stress. Understanding the role of osmolytes in helping plants cope with environmental stresses is crucial for developing strategies to protect plant growth and productivity, particularly in the face of changing climatic conditions. The osmolyte-mediated ROS scavenging system includes:

1. Direct ROS scavenging.
2. Enhancement of antioxidant defense system.
3. Protection and stabilization of macromolecules.
4. Maintenance of cellular redox homeostasis.

Some osmolytes and their role in mitigating ROS by following all the above-mentioned processes are illustrated below:

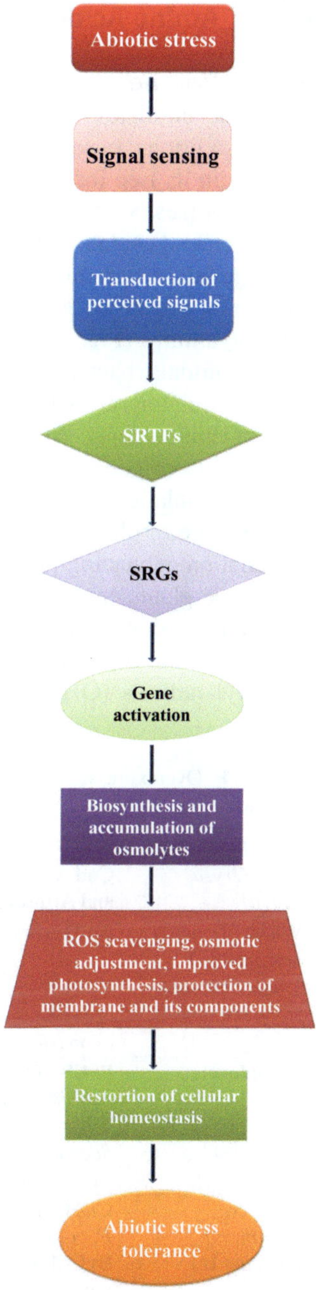


**Fig. 4.2** Schematic representation of overall relation between abiotic stress and osmolytes. Upon insight of abiotic stress signal, the related signaling pathway is initiated and results in the induction of stress-responsive transcription factors (SRTFs) which consequently upregulate stress-responsive genes (SRGs) related to biosynthesis and accumulation of osmolytes such as free amino acids and their derivatives, carbohydrates and soluble sugars, polyols, polyamines, free amines to protect and make plants tolerant of encountered abiotic stresses

### 4.4.1 Amino Acids

#### 4.4.1.1 Proline

Proline accumulates in significant amounts in higher plants as a response to various stress conditions. In addition to its well-established role as an antioxidant, proline—whether in its free form or bound to polypeptides—functions as a signaling molecule, stabilizes proteins and protein complexes, protects photosynthetic machinery, acts as a metal chelator, maintains cellular homeostasis, provides energy for growth, and regulates programmed cell death (Szabados and Savoure 2010). Proline shares a common precursor, glutamate, with other important compounds like glutathione and polyamines (Fig. 4.1). Glutathione, a key antioxidant, plays a critical role in plants, animals, bacteria, and fungi, while polyamines are ubiquitous in all living cells, modulating growth and development, and potentially enhancing cell defense mechanisms against oxidative stress by boosting antioxidant enzyme levels (Sánchez-Rodríguez et al. 2016). Due to this shared precursor, plants often shift their accumulation of antioxidants toward either proline or glutathione under stress. Furthermore, proline and polyamine biosynthesis are interconnected, with proline degradation enhancing polyamine production during stressful conditions. A list of transgenic plants engineered for osmolyte synthesis or metabolic gene expression and their associated stress tolerance is provided in Table 4.1.

Proline acts as an osmoprotectant and is particularly effective at scavenging hydroxyl radicals (OH•) in plants subjected to salinity, drought, or heavy metal

**Table 4.1** Overexpression of osmolytes biosynthetic genes in transgenic plants showing stress tolerance

| Transgenic plant | Gene and gene source | Gene product | References |
|---|---|---|---|
| Rice (*Oryza sativa*) | Cod A from spinach and Arthrobacter | Choline oxidase | Kathuria et al. (2009) |
| Wheat (*Triticum aestivum* L.) | MtlD from *E. coli* | Mannitol-1-phosphate dehydrogenase | Abebe et al. (2003) |
| Tomato (*Solanum lycopersicum*) | TPS1 from *Saccharomyces cerevisiae* | Treahalose-6- phosphate synthase | Mattioli et al. (2008) |
| Arabidopsis | P5CS1 from *Arabidopsis thaliana*<br>TPS1-TPS2 from *Saccharomyces cerevisiae* | Pyrroline-5- carboxylate synthetase<br>Treahalose-6- phosphate synthase | Miranda et al. (2007) |
| Tobacco (*Nicotiana tabacum* L.) | CDH and BADH from *E. coli*<br>BvCMO from *Beta vulgaris*<br>BADH from *Spinacia oleracea* and *Oryza sativa*<br>P5CS from *Vigna aconitifolia* | Choline dehydrogenase and betaine aldehyde dehydrogenase<br>Choline monooxygenase<br>Betaine aldehyde dehydrogenase<br>Pyrroline-5- carboxylate synthetase | Zhang et al. (2008)<br>Hasthanasombut et al. (2009)<br>Kishor et al. (1995) |

toxicity. It helps prevent ROS-induced apoptosis by neutralizing free radicals. The synthesis and accumulation of proline have been linked to improved tolerance against various abiotic stresses, especially salinity and drought. In addition to reducing cytoplasmic acidosis, increased proline production optimizes the $NADP^+$/NADPH ratio, which is crucial for maintaining cellular metabolism. Proline synthesis, which is catalyzed by enzymes in the pentose phosphate pathway, provides $NADP^+$ to support redox cycling, thereby ensuring the proper functioning of the plant's antioxidant defense system during stress (Hare and Cress 1997).

Proline in free or polypeptide bound form reacts directly with ROS ($H_2O_2$ and OH•) and forms stable free radical addition products of derivatives of proline and hydroxyproline, such as 3-hydro-xyproline and 4-hydroxyproline. Proline caused a considerable reduction in the lipid peroxidation and significantly suppressed $^1O2$ generated during light exposure time. The mechanism through which proline quenches $^1O2$ is most likely by transferring charge that leads to the return of molecular oxygen to $^3O2$ (ground triplet state) (Alia et al. 1997). Consequently, proline stabilizes DNA, proteins and membrane structures.

#### 4.4.1.2 γ-Aminobutyric Acid

γ-Aminobutyric acid (GABA) is a four-carbon non-protein amino acid that is universally synthesized across prokaryotic and eukaryotic organisms. In plants, GABA is produced through the metabolism of glutamic acid, catalyzed by the enzyme glutamate decarboxylase, which plays a central role in its biosynthesis. Under normal conditions, endogenous GABA levels in plants are relatively low. However, its concentration increases significantly in response to stress, helping plants counteract adverse environmental effects. In a study by Liu et al. (2011), systems generating $O_2•^-$, $H_2O_2$, and $1O_2$ were used to assess GABA's radical-scavenging capacity in comparison to proline. The results revealed that GABA was more effective than proline in scavenging $O_2•^-$, with inhibition rates of 41.42% and 13.95%, respectively. GABA also demonstrated a greater ability to scavenge $H_2O_2$. Further research examined GABA's protective role against oxidative stress in barley seedlings exposed to $H^+$ and $Al^{3+}$, where GABA-treated roots exhibited significantly lower levels of carbonylated proteins, as detected by anti-dinitrophenol antibodies. Additionally, these roots showed higher activities of catalase, peroxidase, and superoxide dismutase, alongside reduced levels of malondialdehyde, $H_2O_2$, and $O_2•^-$. These findings indicate that GABA mitigates oxidative damage by enhancing antioxidant enzyme activity, thereby reducing reactive oxygen species (ROS) accumulation and lipid peroxidation. This underscores GABA's role in supporting plant metabolic stability under stress conditions.

### 4.4.2 Quaternary Ammonium Compounds

Unfavorable environmental conditions force plants to accumulate quaternary ammonium compounds such as β-alanine betaine, glycine betaine, hydroxyproline betaine, and proline betaine which act as osmolytes and perform diverse functions to help plants cope with stress.

#### 4.4.2.1 Glycine Betaine

Glycine betaine is a naturally occurring compound present in plants, animals, and microorganisms. In plants, its accumulation markedly increases under various abiotic stress conditions. It enhances stress tolerance by stabilizing cellular structures such as proteins and membranes, supporting gene expression and protein synthesis, and enhancing the activity of antioxidant enzymes like catalase (Park et al. 2006). Exogenous application of glycine betaine has also been shown to increase the activities of additional antioxidant enzymes, including superoxide dismutase, peroxidase, and ascorbate peroxidase (Bharwana et al. 2014; Zhang et al. 2014). Furthermore, it boosts the performance of glutathione-dependent enzymes—such as glutathione peroxidase, glutathione S-transferase, glutathione reductase—as well as glyoxalase I and II, all of which are involved in detoxifying reactive oxygen species (ROS) and methylglyoxal. These enhancements lead to decreased hydrogen peroxide ($H_2O_2$) levels and reduced lipid peroxidation, suggesting that glycine betaine fortifies the antioxidant defense system by limiting oxidative damage.

### 4.4.3 Sugars

Carbohydrates are vital to life, serving a wide range of cellular functions including structural support, signaling, transport, and energy storage. Sugars, in particular, play a crucial role in helping cells endure various stress conditions while maintaining normal metabolic functions (Dwivedi 2000; Peshev et al. 2013). Small, soluble sugar molecules are increasingly recognized for their stress-mitigating effects, and the idea of sugars functioning as antioxidants in plants is gaining acceptance. Interestingly, sugars exhibit a dual role in relation to reactive oxygen species (ROS). On one hand, increased sugar production from photosynthesis can elevate cytosolic hydrogen peroxide ($H_2O_2$) levels. On the other hand, sugars also contribute to ROS detoxification by fuelling the oxidative branch of the pentose phosphate pathway, which produces NADPH—a key reducing agent required for the synthesis of glutathione, a major antioxidant (Bolouri-Moghaddam et al. 2010).

#### 4.4.3.1 Sucrose

Sucrose is one of the most prevalent sugars in plants and serves multiple roles, including functioning as a metabolic substrate, signaling molecule, osmotic regulator, and a key component in the transport of photosynthates. Beyond these functions, sucrose has also been shown to enhance the plant's reactive oxygen species (ROS) defense system. It promotes the accumulation of ascorbic acid—the most abundant antioxidant in plants—and anthocyanins, secondary metabolites known for scavenging various reactive oxygen and nitrogen species. Supplementing plants with exogenous sucrose leads to increased levels of ascorbic acid, anthocyanins, and polyphenols, contributing to enhanced ferric reducing antioxidant power (Guo et al. 2011). Sucrose may also provide the necessary carbon skeletons for the biosynthesis of flavonoids and polyphenols (Baque et al. 2012). Furthermore, studies in *Arabidopsis* have shown that sucrose supplementation can alleviate oxidative stress under various abiotic conditions, such as prolonged darkness (Rosenwasser et al. 2011), hypoxia and herbicide-induced stress (Sulmon et al. 2007).

#### 4.4.3.2 Fructans

Polymerized carbohydrates like fructans are considered very effective in ROS quenching, as fructans are hypothesized to be directly or indirectly stimulating antioxidative mechanism in plants (Peshev et al. 2013). Vacuolar fructans protrude deeply between tonoplastic membranes headgroups to stabilize them. Stressful conditions lead to production of excessive cytoplasmic $H_2O_2$ that enters into the vacuole via diffusion. There, peroxidases (Type III) convert $H_2O_2$ into OH• and OOH• radicals through hydroxylic cycle (Dunand et al. 2007). Vacuolar fructans are well positioned not only to stabilize tonoplast but also to scavenge these radicals in the vicinity of membranes. During this process, fructans are converted into less active fructan radicals which are proposed to be recycled back into fructans with the help of phenolic compounds (Takahama 2004). Therefore, vacuolar sugars present in tonoplast are postulated to be scavenging excessive $H_2O_2$ and, therefore, preventing peroxidation of lipids during stress.

#### 4.4.3.3 Raffinose Family Oligosaccharides

Raffinose family oligosaccharides (RFOs) are water-soluble carbohydrates derived from sucrose and include non-reducing sugars such as raffinose, stachyose, and verbascose. These oligosaccharides function not only as signaling molecules but also provide protection during abiotic stress by preventing embryo desiccation, supporting ATP generation, and aiding electron transport within chloroplast membranes—largely due to their ROS scavenging properties (Bolouri-Moghaddam et al. 2010). Elevated raffinose levels in *Arabidopsis* during heat stress have been associated with enhanced resistance to salinity and cold stress. In vitro studies show that raffinose can protect salicylate from hydroxylation mediated by hydrogen peroxide radicals, displaying a reducing capacity comparable to glutathione (Nishizawa et al. 2006, 2008). These findings indicate that raffinose, beyond serving as an osmoprotectant, also acts as an effective ROS scavenger in *Arabidopsis* leaves. In seeds, pre-existing stachyose levels are adequate to shield organelles from oxidative harm. Additionally, enzymes like galactinol synthase, involved in RFO biosynthesis, help maintain seed viability and longevity by neutralizing excessive ROS and minimizing cellular damage. Transgenic *Arabidopsis* plants overexpressing genes responsible for raffinose synthesis exhibited higher levels of raffinose, ascorbic acid, and glutathione, along with reduced ROS accumulation and lipid peroxidation (Nishizawa et al. 2008; Nishizawa-Yokoi et al. 2008). These modified plants also demonstrated improved photosystem II performance and increased resilience to high light and chilling stress, underscoring the antioxidative role of endogenous raffinose in enhancing stress tolerance.

#### 4.4.3.4 Trehalose

Trehalose is a disaccharide sugar found in small amounts in most plants, where it serves primarily as an osmoprotectant or 'compatible solute' that helps stabilize macromolecules under stress conditions. Following glucose and sucrose, trehalose is considered a key sugar signaling molecule in plant cells. Its synthesis is governed by the enzymes trehalose-6-phosphate synthase and trehalose-6-phosphate phosphatase (Karim et al. 2007). Trehalose contributes to reactive oxygen species (ROS) mitigation by enhancing the antioxidant defense system, which includes enzymes like superoxide dismutase, peroxidase, and catalase, along with non-enzymatic antioxidants such as ascorbic acid and phenolic compounds (Shafiq et al. 2015). In radish, trehalose application has been shown to elevate proline levels and increase superoxide dismutase activity (Akram et al. 2016). A positive correlation between trehalose content and ROS resistance has also been reported in *Saccharomyces cerevisiae* mutants lacking trehalose, where external application of trehalose reversed hydrogen peroxide ($H_2O_2$)-induced oxidative damage to protein amino acids, suggesting its role as an ROS scavenger. Hence, trehalose accumulation in stressed cells helps safeguard cellular components from oxidative stress (Benaroudj et al. 2001).

### 4.4.4 Sugar Alcohol (Polyols)

Sugar alcohols, also referred to as polyols, comprises of mannitol, inositol, sorbitol etc. Polyols are known to possess ROS scavenging activities. The OH• scavenging activities of mannitol, myo-inositol, and sorbitol have already been reported three decades ago (Smirnoff and Cumbes 1989).

#### 4.4.4.1 Mannitol

Shen et al. (1997) explored the development of transgenic tobacco plants that express a bacterial mannitol-1-phosphate dehydrogenase gene specifically in their chloroplasts. Their study demonstrated that targeting mannitol production to chloroplasts can enhance the plant's ability to scavenge hydroxyl radicals (OH•), thereby reducing oxidative stress within the cells. The transgenic plants accumulated high levels of mannitol in their chloroplasts, which significantly improved their tolerance to oxidative stress induced by methyl viologen. Moreover, cells containing mannitol showed an increased capacity to neutralize OH• under stress conditions. Certain plant species, such as parsley and snapdragon, naturally accumulate substantial amounts of mannitol in their chloroplasts (Moore et al. 1997), supporting the idea that mannitol plays a functional role in scavenging ROS at the chloroplast level, a primary site of ROS generation.

#### 4.4.4.2 Myo-inositol

Although myo-inositol is not a direct scavenger of reactive oxygen species (ROS), growing evidence points to its indirect role in ROS detoxification through its involvement in the biosynthesis of ascorbic acid (Lorence et al. 2004). Research has shown that overexpressing the gene encoding myo-inositol oxygenase—a crucial enzyme in the pathway converting myo-inositol to ascorbic acid—can increase ascorbic acid levels by two to three times, supporting myo-inositol's function as a precursor in this biosynthetic process. Similarly, Lisko et al. (2013) reported that transgenic Arabidopsis plants overexpressing this gene exhibited elevated ascorbic acid levels and showed reduced damage in response to pyrene, an oxidative stress-inducing compound. Additionally, myo-inositol contributes indirectly to ROS scavenging by serving as a precursor in the synthesis of raffinose, a sugar that acts as an effective ROS scavenger when it accumulates in significant amounts (Valluru and Van den Ende 2011).

### 4.4.5 Polyamines

Polyamines are nitrogen-containing compounds found in all living cells and are known to be involved in numerous cellular processes (Kusano et al. 2007). Various types of polyamines, including the diamine putrescine, triamine spermidine, and tetramine spermine, play crucial roles in helping organisms cope with extreme environmental conditions (Mustafavi et al. 2018). The accumulation of polyamines such as homospermine and cadaverine has been observed in different plant tissues both during normal growth and under abiotic stress conditions. External application of spermidine has been shown to alleviate salinity-induced damage to the plasma membrane in both salt-sensitive and salt-tolerant rice varieties (Roy et al. 2005) (Table 4.2). Additionally, spermine has been found to prevent oxidative DNA damage caused by hydroxyl radicals in *Mesembryanthemum crystallinum*, highlighting the role of polyamines as potential osmolytes and reactive oxygen species (ROS) scavengers (Kuznetsov and Shevyakova 2007).

## 4.5 Regulation of Osmolyte Accumulation

### 4.5.1 Transcriptional Control

Plants react to environmental challenges via a complex combination of sensing, signaling, and response at the molecular, biochemical, and physiological levels. These stresses, such as drought, cold, and salt, activate a variety of pathways that enable plants to adapt, tolerate, or escape harsh situations. These signals activate various transcription factors that lead to osmolyte accumulation.

#### 4.5.1.1 DREB (Dehydration-Responsive Element-Binding)

DREB transcription factors, such as ICE1 in cold stress, or MAP kinases that influence DREB activity (Joshi et al. 2016; Agarwal et al. 2017). Once activated, DREB proteins move into the nucleus where they bind to DRE sequences in the promoters of stress-responsive genes (Qin et al. 2004). This binding modulates the expression of these genes, often leading to the production of protective molecules called

**Table 4.2** Specific osmolytes and their functional roles in abiotic stress tolerance

| Osmolyte | Chemical class | Major abiotic stresses | Mechanism of action | Representative plants | References |
|---|---|---|---|---|---|
| Proline | Amino acid | Drought, salinity, heat | Osmotic adjustment, ROS scavenging, membrane and protein stabilization | *Arabidopsis thaliana*, rice | Zouari et al. (2019) |
| Glycine betaine | Quaternary ammonium | Salinity, drought, cold | Stabilizes membrane structure, protects PSII, enzyme protection | Maize, wheat, sugar beet | Pei et al. (2022) |
| Trehalose | Disaccharide sugar | Drought, heat, cold | Stabilizes proteins and membranes, regulates carbon metabolism | Rice, Arabidopsis | Ye et al. (2023) |
| Mannitol | Sugar alcohol | Salinity, drought | ROS scavenging, osmotic adjustment | Celery, olive, tobacco | Du et al. (2021) |
| Sorbitol | Sugar alcohol | Salinity, drought | Maintains osmotic balance, stabilizes enzymes | Apple, pear | Wu et al. (2015) |
| Putrescine | Polyamine | Salt, drought, cold | Stabilizes DNA and membranes, modulates ion channels, ROS scavenging | Stevia, barley, tomato | González-Hernández et al. (2022) |
| Spermidine | Polyamine | Salinity, drought | DNA/RNA stabilization, modulates stress signaling pathways | Rice, wheat | Paul and Roychoudhury (2017) |
| Betaine aldehyde | Quaternary ammonium | Salinity, drought | Precursor of glycine betaine, involved in osmoprotection | Spinach, maize | Niazian et al. (2021) |

osmolytes (Trono and Pecchioni 2022). For instance, overexpression of the OsDREB2A gene has been shown to increase the expression of GmP5CS, an enzyme involved in proline production, and to activate the BADH gene, leading to the accumulation of glycine. Studies have also demonstrated that overexpressing OsDREB2A can enhance salt tolerance in plants by increasing osmolyte accumulation and improving the expression of other stress-related genes betaine (Zhang et al. 2013). In essence, the ability of *OsDREB2A* overexpression to enhance salt tolerance by increasing osmolyte accumulation and up regulating other stress-related genes perfectly illustrates the central role that DREB transcription factors play in plant responses to abiotic stresses.

#### 4.5.1.2 AREB/ABF (ABA-Responsive Element-Binding Protein/ABA-Binding Factor) Family

It belongs to a subfamily of basic leucine zipper bZIP (Basic leucine zipper) transcription factors. They are key regulatory molecules downstream of the ABA signaling pathway regulating plant response to hormones and stresses, initiating the expression of downstream genes (Lu et al. 2025). Endogenous ABA levels increase in response to osmotic stresses such as drought and high salinity, and ABA activates the expression of many genes via ABA-responsive elements (ABREs) in their promoter regions. ABRE-binding protein/ABRE-binding factor (AREB/ABF) transcription factors (TFs) regulate the ABRE-mediated transcription of downstream target genes (Pan et al. 2023). Under salinity stress, AREB2 and ABF2 showed up-regulation in response to the increasing level of salinity in the leaf tissues, but in the root tissues, ABF2 was up-regulated in response to increasing NaCl concentration, while AREB2 was down-regulated. It seems AREB2 has tissue-specific response to salinity stress (Zandkarimi et al. 2015).

#### 4.5.1.3 Other TF Families like MYB, NAC, and WRKY Also Play Roles in Stress-Responsive Gene Expression

##### MYB

MYB transcription factors, characterized by the conserved MYB DNA-binding domain, form a large plant-specific R2R3 subfamily. Genome-wide studies show that many MYB genes are regulated by abiotic stresses like drought, suggesting their involvement in stress responses. MYBs often interact with other transcription factors like MYCs and WRKYs. Functional studies have demonstrated their roles in drought tolerance. For example, AtMYB44 in *Arabidopsis* enhances ABA sensitivity and stomatal closure, improving drought tolerance, while AtMYB96 activates cuticular wax biosynthesis to reduce water loss. MYBs are also involved in salt stress tolerance, with AtMYB20 positively regulating salt tolerance in *Arabidopsis* (Yoon et al. 2020). Expression of OsMYB2 was up-regulated by salt, cold, and dehydration stress. There was greater up-regulation of stress-related genes, including OsLEA3, OsRab16A, and OsDREB2A, in the OsMYB2-overexpressing plants (Yang et al. 2012).

##### NAC

NACs are an important class of TFs found in plants (Puranik et al. 2012). By attaching to a NAC recognition sequence (NACRS) in the gene promoters and triggering abiotic stress tolerance, NAC TFs control the transcription of downstream target genes once they are activated. Several stress-inducible genes are up-regulated in transgenic Arabidopsis plants that overexpress AtNAC019, AtNAC055, and AtNAC072, which leads to drought tolerance (Tran et al. 2004; Hickman et al. 2013).

**WRKY**

The DNA binding domain, which is 60 amino acids long and comprises a conserved N-terminal sequence of WRKYGQK together with a Zn finger-like motif, is present in the WRKY TFs (Chen et al. 2018). The production of downstream target stress-responsive genes is typically the consequence of WRKY genes binding to a particular W-box domain or W-box sequence in response to stress stimuli. Due to its partial involvement in thermotolerance, AtWRKY25 triggers the expression of genes that are sensitive to heat (Li et al. 2009). WRKY proteins may be a key regulatory node of multiple abiotic stress responses (Erpen et al. 2018).

## 4.5.2 Hormonal Regulation

Osmolytes and plant hormones have both been shown to have significant effects under stressful circumstances, and thus it is essential to correlate the two and clarify how phytohormones affect osmolyte regulation in abiotic stress (Sharma et al. 2019).

### 4.5.2.1 Brassinosteroids (BRs)

The crucial function of BRs in stress management has been investigated in several plants (Rehman et al. 2019). Plant physiological processes under stress are regulated by BR signaling, and the production of all important osmoprotectants is believed to be regulated by BRL3, a vascular-enriched member of the BR receptor family. In drought stress of Arabidopsis, overexpression of BRL3 has been found to increase the accumulation of crucial osmolytes, which provide stress resistance (Fàbregas et al. 2018).

### 4.5.2.2 Ethylene

Ethylene is a well-known stress hormone and its signaling play a key role in numerous aspects of plant growth and development (Cao et al. 2007). Iqbal et al. (2015) found that ethylene regulates abiotic stress by influencing the accumulation of osmoprotectants. The role of ethylene in osmolyte production was further validated by the study in plant responses and adaptation to water shortages, seedlings of two ethylene insensitive mutants, ein2–5, ein3–1, and wild-type Col-0, were exposed to water stress conditions mimicked by PEG-6000, a non-ionic molecule, and a non-permeating osmolyte. ET signaling promoted the accumulation of suitable solutes, such as soluble carbohydrates and proline. This resulted in lower intracellular lipid peroxidation in *Arabidopsis* during water stress (Cui et al. 2015).

### 4.5.2.3 Salicylic Acid

Salicylic acid (SA) is a plant phenolic hormone present in large quantities involved in both local and systemic response to stress (Khan et al. 2015). Under abiotic stress, SA is widely known to regulate the production of osmolytes such as GB, proline, and sugars (Yuan et al. 2014). The proline biosynthesis enzymes glutamyl kinase and pyrroline-5-carboxylate showed markedly increased activity in *Lens esculenta* plants treated with 0.5 mM SA under salt stress. In salt-stressed plants, an increased proline metabolism is linked to a stress tolerance mechanism (Misra and Saxena 2009). found that proline oxidase, which is responsible for increased proline production in response to abiotic stress, was less active and that of proline metabolism enzymes (pyrroline-5-carboxylate and glutamyl kinase) were elevated. The increase in proline levels in *Rauvolfia serpentina* plants under salt stress also suggested a substantial role for SA in the control of cell turgor.

#### 4.5.2.4 Cytokinin

Cytokinins are important for controlling the buildup of osmolytes in plants, especially when they are subjected to abiotic stressors like salt and drought (Gujjar and Supaibulwatana 2019). Proline, total soluble sugars, total phenolics, and total flavonoids are among the osmolytes that have been observed to accumulate in response to cytokinin administration. For example, cis-zeatin, a form of cytokinin, significantly increased these osmolytes in stressed maize seedlings (Sharma et al. 2019).

#### 4.5.2.5 Jasmonic Acid

Among several phytohormones, jasmonates (JAs) are regarded as one of the crucial phytohormones that play a major part in tolerance to abiotic stress. Jasmonic acid (JA) controls several plant reactions in a wide range of stressful situations, including as gene regulation, the production of specific proteins, and secondary metabolism. In both biotic and abiotic stress situations, the role of JA in plant growth and development is widely acknowledged (Sharma et al. 2018; Bali et al. 2018). According to (Anjum et al. 2011) and (Shan et al. 2015), JA has a positive impact on reducing drought stress by promoting the production of osmolytes such as proline.

#### 4.5.2.6 Abscisic Acid

Abscisic acid, a key plant hormone significantly influences plant growth and development (Hartung 2010). One of the primary physiological processes mediated by ABA under stress conditions is the regulation of osmolytes in stressed plant cells and the maintenance of osmotic adjustments (Karimi and Ershadi 2015; Ding et al. 2010; Zhang et al. 2012). ABA functions as a signaling molecule to regulate the biosynthetic pathways of osmolytes at the molecular level (Sripinyowanich et al. 2013; Pál et al. 2018) and the increased levels of this osmoprotectant through ABA stimulation further aid in the reduction of oxidative stress (Ding et al. 2010). Abscisic acid promotes the GB biosynthesis pathway and increases the amount of this osmolyte in plant cells, which eventually aids in defense against abiotic stress (Sarafraz-Ardakani et al. 2014; Burbulis et al. 2010).

### 4.5.3 Genetic Engineering Approaches in Osmolyte Accumulation

Genetic engineering provides potential techniques for enhancing osmolyte accumulation in plants, hence increasing their tolerance to abiotic challenges such as drought, salt, and severe temperatures. Introducing and overexpressing genes that encode enzymes involved in the manufacture of certain osmolytes can result in greater accumulation of these protective substances. Shirasawa et al. (2006) targeted genes include choline dehydrogenase (CDH) and betaine aldehyde dehydrogenase (BADH) in the GB production pathway. Introducing or altering genes encoding transcription factors that control the expression of many stress-responsive genes, including those involved in osmolyte production, can result in a more robust and coordinated stress response. Modifying genes encoding transporters important for osmolyte absorption, distribution, and compartmentalization within the plant allows them to be concentrated in specific cellular sites where they give the greatest protection. Overexpression of P5CS has been shown to raise proline levels while improving drought and salinity tolerance in rice, tobacco, and Arabidopsis (Yang et al. 2021).

## 4.6 Future Perspectives

Despite significant advances in understanding the roles of osmolytes in abiotic stress tolerance, several critical questions remain unanswered, offering opportunities for future research. One promising direction involves the elucidation of complex regulatory networks that govern osmolyte biosynthesis, accumulation, and cross-talk with other stress signaling pathways. High-throughput omics technologies, such as transcriptomics, proteomics, metabolomics, and integrated systems biology approaches, can be utilized to map the dynamic responses of osmolyte metabolism under multiple or combined stresses.

The development of genome editing tools like CRISPR/Cas systems presents new possibilities for precise manipulation of osmolyte biosynthetic genes to create stress-resilient crops without yield penalties. Moreover, the role of epigenetic regulation in osmolyte-mediated stress responses remains largely unexplored and represents an exciting frontier for understanding heritable stress memory in plants.

Another vital area involves the translation of laboratory findings into field applications. Most studies on osmolytes have been conducted under controlled environments, and their performance under field conditions remains variable. Therefore, breeding programs need to integrate molecular insights with traditional practices, targeting both stress tolerance and agronomic performance. Additionally, interspecies comparisons and the study of stress-resilient wild relatives may uncover novel osmolytes or regulatory mechanisms that can be harnessed for crop improvement.

Finally, exploring the synergistic effects of osmolytes with other cellular defense mechanisms—such as antioxidant systems, ion transporters, and heat shock proteins—will contribute to a more holistic understanding of plant stress physiology, facilitating the development of multi-target strategies for climate-resilient agriculture.

## 4.7 Conclusion

Osmolytes are indispensable components of the plant stress response, playing multifaceted roles in mitigating the adverse effects of abiotic stress through osmotic adjustment, protection of macromolecules, and scavenging of reactive oxygen species. Their ability to stabilize proteins and membranes, enhance antioxidant defenses, and maintain cellular redox homeostasis makes them critical for plant survival under harsh environmental conditions. This chapter has highlighted the classification, functional roles, and mechanistic insights into osmolyte-mediated ROS scavenging, along with the regulatory controls governing their accumulation. The synergistic action of osmolytes with antioxidant systems underscores their importance in the broader framework of stress tolerance mechanisms. Understanding and manipulating osmolyte pathways provide a viable route for improving plant resilience to environmental stressors. Continued research in this domain, supported by advanced molecular tools and integrative approaches, will be instrumental in addressing the challenges posed by climate change and ensuring sustainable crop production in the future.

## References

Abebe T, Guenzi AC, Martin B, Cushman JC (2003) Tolerance of mannitol-accumulating transgenic wheat to water stress and salinity. Plant Physiol 131(4):1748–1755. https://doi.org/10.1104/pp.102.003616

Agarwal PK, Gupta K, Lopato S, Agarwal P (2017) Dehydration responsive element binding transcription factors and their applications for the engineering of stress tolerance. J Exp Bot 68(9):2135–2148

Akram NA, Waseem M, Ameen R, Ashraf M (2016) Trehalose pretreatment induces drought tolerance in radish (*Raphanus sativus* L.) plants: some key physio-biochemical traits. Acta Physiol Plant 38(1):3

Alia S, Saradhi P, Mohanty P (1997) Involvement of proline in protecting thylakoid membranes against free radical-induced photodamage. J Photochem Photobiol B Biol 38:253–257

Anjum NA, Thangavel P, Rasheed F, Masood A, Pirasteh-Anosheh H, Khan NA (2023) Osmolytes: efficient oxidative stress-busters in plants. In: Global climate change and plant stress management. Wiley, New York, pp 399–409

Anjum SA, Wang L, Farooq M, Khan I, Xue L (2011) Methyl jasmonate-induced alteration in lipid peroxidation, antioxidative defence system and yield in soybean under drought. J Agron Crop Sci 197(4):296–301

Bali S, Kaur P, Sharma A, Ohri P, Bhardwaj R, Alyemeni MN et al (2018) Jasmonic acid-induced tolerance to root-knot nematodes in tomato plants through altered photosynthetic and antioxidative defense mechanisms. Protoplasma 255:471–484

Baque MA, Elgirhan A, Lee EJ, Paek KY (2012) Sucrose regulated enhanced induction of anthraquinone, phenolics, flavonoids biosynthesis and activities of antioxidant enzymes in adventitious root suspension cultures of Morinda citrifolia (L.). Acta Physiol Plant 34:405–415

Benaroudj N, Lee H, Goldberg AL (2001) Trehalose accumulation during cellular stress protects cells and cellular proteins from damage by oxygen radicals. J Biol Chem 276(26):24261–24267

Bharwana SA, Ali S, Farooq MA, Iqbal N, Hameed A, Abbas F, Ahmad MSA (2014) Glycine betaine-induced lead toxicity tolerance related to elevated photosynthesis, antioxidant enzymes suppressed lead uptake and oxidative stress in cotton. Turk J Bot 38(1):281–292

Bhattacharjee S, Bhattacharjee S (2019) ROS and oxidative stress: origin and implication. In: Reactive oxygen species in plant biology. Springer, New Delhi, pp 1–31
Bolouri-Moghaddam MR, Le Roy K, Xiang L, Rolland F, Van den Ende W (2010) Sugar signalling and antioxidant network connections in plant cells. FEBS J 2779:2022–2037
Burbulis N, Jonytiene V, Kupriene R, Blinstrubiene A, Liakas V (2010) Effect of abscisic acid on cold tolerance in Brassica napus shoots cultured in vitro. J Food Agric Environ 8(3&4):698–701
Cao WH, Liu J, He XJ, Mu RL, Zhou HL, Chen SY, Zhang JS (2007) Modulation of ethylene responses affects plant salt-stress responses. Plant Physiol 143(2):707–719
Chen YH, Cao YY, Wang LJ, Li LM, Yang J, Zou MX (2018) Identification of MYB transcription factor genes and their expression during abiotic stresses in maize. Biol Plant 62:222–230
Choudhary S, Wani KI, Naeem M, Khan MMA, Aftab T (2023) Cellular responses, osmotic adjustments, and role of osmolytes in providing salt stress resilience in higher plants: polyamines and nitric oxide crosstalk. J Plant Growth Regul 42(2):539–553
Choudhury FK, Rivero RM, Blumwald E, Mittler R (2017) Reactive oxygen species, abiotic stress and stress combination. Plant J 90(5):856–867
Cui M, Lin Y, Zu Y, Efferth T, Li D, Tang Z (2015) Ethylene increases accumulation of compatible solutes and decreases oxidative stress to improve plant tolerance to water stress in *Arabidopsis*. J Plant Biol 58:193–201
Del Río LA, López-Huertas E (2016) ROS generation in peroxisomes and its role in cell signaling. Plant Cell Physiol 57(7):1364–1376
Ding W, Song L, Wang X, Bi Y (2010) Effect of abscisic acid on heat stress tolerance in the calli from two ecotypes of Phragmites communis. Biol Plant 54:607–613
Du B, Chen N, Song L, Wang D, Cai H, Yao L et al (2021) Alfalfa (Medicago sativa L.) MsCML46 gene encoding calmodulin-like protein confers tolerance to abiotic stress in tobacco. Plant Cell Rep 40:1907–1922
Dunand C, Crévecoeur M, Penel C (2007) Distribution of superoxide and hydrogen peroxide in Arabidopsis root and their influence on root development: possible interaction with peroxi dases. New Phytol 174:332–341
Dwivedi P (2000) Regulation of root respiration and sugar-mediated gene expression in plants. Curr Sci 78(10):1196–1202
Ejaz S, Fahad S, Anjum MA, Nawaz A, Naz S, Hussain S, Ahmad S (2020) Role of osmolytes in the mechanisms of antioxidant defense of plants. Sustain Agric Rev 39:95–117
Erpen L, Devi HS, Grosser JW, Dutt M (2018) Potential use of the DREB/ERF, MYB, NAC and WRKY transcription factors to improve abiotic and biotic stress in transgenic plants. Plant Cell Tissue Organ Cult 132:1–25
Evers F et al (2011) Reduced protein adsorption by osmolytes. Langmuir 27:6995–7001
Fàbregas N, Lozano-Elena F, Blasco-Escámez D, Tohge T, Martínez-Andújar C, Albacete A et al (2018) Overexpression of the vascular brassinosteroid receptor BRL3 confers drought resistance without penalizing plant growth. Nat Commun 9(1):4680
Foyer CH, Hanke G (2022) ROS production and signalling in chloroplasts: cornerstones and evolving concepts. Plant J 111(3):642–661
González-Hernández AI, Scalschi L, Vicedo B, Marcos-Barbero EL, Morcuende R, Camañes G (2022) Putrescine: a key metabolite involved in plant development, tolerance and resistance responses to stress. Int J Mol Sci 23(6):2971
Gujjar RS, Supaibulwatana K (2019) The mode of cytokinin functions assisting plant adaptations to osmotic stresses. Plants 8(12):542
Guo R, Yuan G, Wang Q (2011) Effect of sucrose and mannitol on the accumulation of health-promoting compounds and the activity of metabolic enzymes in broccoli sprouts. Sci Hortic 128(3):159–165
Hare PD, Cress WA (1997) Metabolic implications of stress-induced proline accumulation in plants. Plant Growth Regul 21:79–102
Hartung W (2010) The evolution of abscisic acid (ABA) and ABA function in lower plants, fungi and lichen. Funct Plant Biol 37(9):806–812
Hasthanasombut S, Ntui V, Supaibulwatana K, Mii M, Nakamura I (2009) Expression of Indica rice OsBADH1 gene under salinity stress in transgenic tobacco. Plant Biotechnol Rep 4(1):75–83

Hickman R, Hill C, Penfold CA, Breeze E, Bowden L, Moore JD et al (2013) A local regulatory network around three NAC transcription factors in stress responses and senescence in *Arabidopsis* leaves. Plant J 75(1):26–39

Iqbal N, Umar S, Khan NA (2015) Nitrogen availability regulates proline and ethylene production and alleviates salinity stress in mustard (Brassica juncea). J Plant Physiol 178:84–91

Joshi R, Wani SH, Singh B, Bohra A, Dar ZA, Lone AA et al (2016) Transcription factors and plants response to drought stress: current understanding and future directions. Front Plant Sci 7:1029

Karim S, Aronsson H, Ericson H, Pirhonen M, Leyman B, Welin B, Mäntylä E, Palva ET, Van Dijck P, Holmström KO (2007) Improved drought tolerance without undesired side effects in transgenic plants producing trehalose. Plant Mol Biol 64:371–386

Karimi R, Ershadi A (2015) Role of exogenous abscisic acid in adapting of 'Sultana'grapevine to low-temperature stress. Acta Physiol Plant 37:1–11

Kathuria H, Giri J, Nataraja KN, Murata N, Udayakumar M, Tyagi (2009) Glycinebetaine-induced water-stress tolerance in cod A-expressing transgenic indica rice is associated with up-regulation of several stress responsive genes. Plant Biotechnol J 7(6):512–526

Kaur G, Sanwal SK, Kumar A, Pundir RK, Yadav M, Sehrawat N (2024) Role of osmolytes dynamics in plant metabolism to cope with salinity induced osmotic stress. Discov Agric 2(1):59

Khan MIR, Fatma M, Per TS, Anjum NA, Khan NA (2015) Salicylic acid-induced abiotic stress tolerance and underlying mechanisms in plants. Front Plant Sci 6:462

Kishor PBK, Hong Z, Miao GH, Hu CAA, Verma DPS (1995) Overexpression of [delta]-pyrroline-5-carboxylate synthetase increases proline production and confers osmotolerance in transgenic plants. Plant Physiol 108(4):1387–1394

Kusano T, Yamaguchi K, Berberich T, Takahashi Y (2007) The polyamine spermine rescues Arabidopsis from salinity and drought stresses. Plant Signal Behav 2(4):251–252

Kuznetsov VV, Shevyakova NI (2007) Polyamines and stress tolerance of plants. Plant Stress 1:50–71

Li S, Fu Q, Huang W, Yu D (2009) Functional analysis of an *Arabidopsis* transcription factor WRKY25 in heat stress. Plant Cell Rep 28:683–693

Lisko KA, Torres R, Harris RS, Belisle M, Vaughan MM, Jullian B, Chevone BI, Mendes P, Nessler CL, Lorence A (2013) Elevating vitamin C content via overexpression of myo-inositol oxygenase and L-gulono-1, 4-lactone oxidase in Arabidopsis leads to enhanced biomass and tolerance to abiotic stresses. In Vitro Cell Dev Biol Plant 49(6):643–655

Liu C, Zhao L, Yu G (2011) The dominant glutamic acid metabolic flux to produce γ-amino butyric acid over proline in Nicotiana tabacum leaves under water stress relates to its significant role in antioxidant activity. J Integr Plant Biol 53(8):608–618

Lorence A, Chevone BI, Mendes P, Nessler CL (2004) myo-Inositol oxygenase offers a possible entry point into plant ascorbate biosynthesis. Plant Physiol 134:1200–1205

Lu S, Qiao Y, Pan X, Chen X, Su W, Li A et al (2025) Genome-wide identification and expression analysis of CsABF/AREB gene family in cucumber (*Cucumis sativus* L.) and in response to phytohormonal and abiotic stresses. Sci Rep 15(1):1–16

Mattioli R, Marchese D, D'Angeli S, Altamura MM, Costantino P, Trovato M (2008) Modulation of intracellular proline levels affects flow ering time and inflorescence architecture in Arabidopsis. Plant Mol Biol 66(3):277–288

Miranda JA, Avonce N, Suárez R, Thevelein JM, Van Dijck P, Iturriaga GA (2007) Bifunctional TPS–TPP enzyme from yeast confers tolerance to multiple and extreme abiotic-stress conditions in transgenic Arabidopsis. Planta 226(6):1411–1421

Misra N, Saxena P (2009) Effect of salicylic acid on proline metabolism in lentil grown under salinity stress. Plant Sci 177(3):181–189

Mittler R, Zandalinas SI, Fichman Y, Van Breusegem F (2022) Reactive oxygen species signalling in plant stress responses. Nat Rev Mol Cell Biol 23(10):663–679

Moore BD, Palmquist DE, Seemann JR (1997) Influence of plant growth at high $CO_2$ concen trations on leaf content of ribulose-1,5-bisphosphate carboxylase/oxygenase and intracellu

lar distribution of soluble carbohydrates in tobacco, snapdragon, and parsley. Plant Physiol 115:241–248
Niazian M, Sadat-Noori SA, Tohidfar M, Mortazavian SMM, Sabbatini P (2021) Betaine aldehyde dehydrogenase (BADH) vs. flavodoxin (Fld): two important genes for enhancing plants stress tolerance and productivity. Frontiers. Plant Sci 12:650215
Nishizawa A, Yabuta Y, Shigeoka S (2008) Galactinol and raffinose constitute a novel function to protect plants from oxidative damage. Plant Physiol 147:1251–1263
Nishizawa A, Yabuta Y, Yoshida E, Maruta T, Yoshimura K, Shigeoka S (2006) Arabidopsis heat shock transcription factor A2 as a key regulator in response to several types of environmental stress. Plant J 48:535–547. https://doi.org/10.1111/j.1365-313x.2006.02889.x
Nishizawa-Yokoi A, Yabuta Y, Shigeoka S (2008) The contribution of carbohydrates including raffinose family oligosaccharides and sugar alcohols to protection of plant cells from oxidative damage. Plant Signal Behav 3(11):1016–1018
Ozturk M, Turkyilmaz Unal B, García-Caparrós P, Khursheed A, Gul A, Hasanuzzaman M (2021) Osmoregulation and its actions during the drought stress in plants. Physiol Plant 172(2):1321–1335
Pál M, Tajti J, Szalai G, Peeva V, Végh B, Janda T (2018) Interaction of polyamines, abscisic acid and proline under osmotic stress in the leaves of wheat plants. Sci Rep 8(1):12839
Pan X, Wang C, Liu Z, Gao R, Feng L, Li A et al (2023) Identification of ABF/AREB gene family in tomato (*Solanum lycopersicum* L.) and functional analysis of ABF/AREB in response to ABA and abiotic stresses. PeerJ 11:e15310
Park EJ, Jeknic Z, Chen TH (2006) Exogenous application of glycine betaine increases chilling tolerance in tomato plants. Plant Cell Physiol 47(6):706–714
Paul S, Roychoudhury A (2017) Seed priming with spermine and spermidine regulates the expression of diverse groups of abiotic stress-responsive genes during salinity stress in the seedlings of indica rice varieties. Plant Gene 11:124–132
Pei L, Li H, Zhou Y, Li W, Jiang Y, Li H (2022) Exogenous glycinebetaine application contributes to abiotic stress tolerance in maize. J Plant Biol 65(6):517–529
Peshev D, Vergauwen R, Moglia A, Hideg É, Van den Ende W (2013) Towards understanding vacuolar antioxidant mechanisms: a role for fructans? J Exp Bot 64(4):1025–1038
Puranik S, Sahu PP, Srivastava PS, Prasad M (2012) NAC proteins: regulation and role in stress tolerance. Trends Plant Sci 17(6):369–381
Qin F, Sakuma Y, Li J, Liu Q, Li YQ, Shinozaki K, Yamaguchi-Shinozaki K (2004) Cloning and functional analysis of a novel DREB1/CBF transcription factor involved in cold-responsive gene expression in *Zea mays* L. Plant Cell Physiol 45(8):1042–1052
Rehman S, Shahzad B, Bajwa AA, Hussain S, Rehman A, Cheema SA et al (2019) Utilizing the allelopathic potential of *Brassica* species for sustainable crop production: a review. J Plant Growth Regul 38:343–356
Rosenwasser S, Rot I, Sollner E, Meyer AJ, Smith Y, Leviatan N, Fluhr R, Friedman H (2011) Organelles contribute differentially to reactive oxygen species-related events during extended darkness. Plant Physiol 156:185–201
Roy P, Niyogi K, Sen Gupta DN, Ghosh B (2005) Spermidine treatment to rice seedlings recovers salinity stress-induced damage of plasma membrane and PM-bound H+-ATPase in salt-tolerant and salt-sensitive rice cultivars. Plant Sci 168(3):583–591
Roychoudhury A, Banerjee A (2013) Endogenous glycine betaine accu mulation mediates abiotic stress tolerance in plants. Trop Plant Res 3:105–111
Roychoudhury A, Das K (2014) Functional role of polyamines and poly amine-metabolizing enzymes during salinity, drought and cold stresses. In: Plant adaptation to environmental change: significance of amino acids and their derivatives. CAB International Publishers, Wallingford, pp 141–156
Sachdev S, Ansari SA, Ansari MI, Fujita M, Hasanuzzaman M (2021) Abiotic stress and reactive oxygen species: generation, signaling, and defense mechanisms. Antioxidants 10(2):277

Sahu PK, Jayalakshmi K, Tilgam J, Gupta A, Nagaraju Y, Kumar A et al (2022) ROS generated from biotic stress: effects on plants and alleviation by endophytic microbes. Front Plant Sci 13:1042936

Sánchez-Rodríguez E, Romero L, Ruiz JM (2016) Accumulation of free polyamines enhances the antioxidant response in fruits of grafted tomato plants under water stress. J Plant Physiol 190:72–78

Sandalio LM, Rodríguez-Serrano M, Romero-Puertas MC, del Río LA (2013) Role of peroxisomes as a source of reactive oxygen species (ROS) signaling molecules. Peroxisomes and their key role in cellular signaling and metabolism. Subcell Biochem 69:231–255

Sarafraz-Ardakani MR, Khavari-Nejad RA, Moradi F, Najafi F (2014) Abscisic acid and cytokinin-induced osmotic and antioxidant regulation in two drought-tolerant and drought-sensitive cultivars of wheat during grain filling under water deficit in field conditions. Notulae Scientia Biologicae 6(3):354–362

Mustafavi SH, Badi HN, Sękara A, Mehrafarin A, Janda T, Ghorbanpour M, Rafiee H (2018) Polyamines and their possible mechan isms involved in plant physiological processes and elicitation of secondary metabolites. Acta Physiol Plant 40(6):1–19

Shafiq S, Akram NA, Ashraf M (2015) Does exogenously-applied trehalose alter oxidative defense system in the edible part of radish (Raphanus sativus L.) under water-deficit conditions? Sci Hortic 185:68–67

Shan C, Zhou Y, Liu M (2015) Nitric oxide participates in the regulation of the ascorbate-glutathione cycle by exogenous jasmonic acid in the leaves of wheat seedlings under drought stress. Protoplasma 252:1397–1405

Sharma A, Kumar V, Yuan H, Kanwar MK, Bhardwaj R, Thukral AK, Zheng B (2018) Jasmonic acid seed treatment stimulates insecticide detoxification in *Brassica juncea* L. Front Plant Sci 9:1609

Sharma A, Shahzad B, Kumar V, Kohli SK, Sidhu GPS, Bali AS et al (2019) Phytohormones regulate accumulation of osmolytes under abiotic stress. Biomolecules 9(7):285

Shen BO, Jensen RG, Bohnert HJ (1997) Increased resistance to oxidative stress in transgenic plants by targeting mannitol biosynthesis to chloroplasts. Plant Physiol 113(4):1177–1183

Shirasawa K, Takabe T, Kishitani S (2006) Accumulation of glycine betaine in rice plants that overexpress choline monooxygenase from spinach and evaluation of their tolerance to abiotic stress. Ann Bot 98(3):565–571

Sies H, Belousov VV, Chandel NS, Davies MJ, Jones DP, Mann GE et al (2022) Defining roles of specific reactive oxygen species (ROS) in cell biology and physiology. Nat Rev Mol Cell Biol 23(7):499–515

Singh D (2022) Juggling with reactive oxygen species and antioxidant defense system–a coping mechanism under salt stress. Plant Stress 5:100093

Singh P, Choudhary KK, Chaudhary N, Gupta S, Sahu M, Tejaswini B, Sarkar S (2022) Salt stress resilience in plants mediated through osmolyte accumulation and its crosstalk mechanism with phytohormones. Front Plant Sci 13:1006617

Smirnoff N, Cumbes QJ (1989) Hydroxyl radical scavenging activity of compatible solutes. Pytochemistry 28:1057–1060

Šoln K, Klemenčič M, Koce JD (2022) Plant cell responses to allelopathy: from oxidative stress to programmed cell death. Protoplasma 259(5):1111–1124

Sripinyowanich S, Klomsakul P, Boonburapong B, Bangyeekhun T, Asami T, Gu H, Chadchawan S (2013) Exogenous ABA induces salt tolerance in indica rice (*Oryza sativa* L.): the role of OsP5CS1 and OsP5CR gene expression during salt stress. Environ Exp Bot 86:94–105

Sulmon C, Gouesbet G, Binet F, Martin-Laurent F, El Amrani A, Couee I (2007) Sucrose amend ment enhances phytoaccumulation of the herbicide atrazine in Arabidopsis thaliana. Environ Pollut 145:507–515

Szabados L, Savoure A (2010) Proline: a multifunctional amino acid. Trends Plant Sci 15(2):89–97

Takahama U (2004) Oxidation of vacuolar and apoplastic phenolics substrates by peroxidase: physiological significance of the oxidation reactions. Phytochem Rev 3:207–219

Tran LSP, Nakashima K, Sakuma Y, Simpson SD, Fujita Y, Maruyama K et al (2004) Isolation and functional analysis of *Arabidopsis* stress-inducible NAC transcription factors that bind to a drought-responsive cis-element in the early responsive to dehydration stress 1 promoter. Plant Cell 16(9):2481–2498

Trono D, Pecchioni N (2022) Candidate genes associated with abiotic stress response in plants as tools to engineer tolerance to drought, salinity and extreme temperatures in wheat: an overview. Plants 11(23):3358

Valluru R, Van den Ende W (2011) Myo-inositol and beyond—emerging networks under stress. Plant Sci 181:387–400

Wang P, Liu WC, Han C, Wang S, Bai MY, Song CP (2024) Reactive oxygen species: multidimensional regulators of plant adaptation to abiotic stress and development. J Integr Plant Biol 66(3):330–367

Wijayasinghe YS, Tyagi A, Poddar NK (2017) Regulation of cell volume by osmolytes. In: Cellular osmolytes: from chaperoning protein folding to clinical perspectives. Springer, Singapore, pp 195–228

Wu T, Wang Y, Zheng Y, Fei Z, Dandekar AM, Xu K et al (2015) Suppressing sorbitol synthesis substantially alters the global expression profile of stress response genes in apple (Malus domestica) leaves. Plant Cell Physiol 56(9):1748–1761

Yang A, Dai X, Zhang WH (2012) A R2R3-type MYB gene, OsMYB2, is involved in salt, cold, and dehydration tolerance in rice. J Exp Bot 63(7):2541–2556

Yang D, Ni R, Yang S, Pu Y, Qian M, Yang Y (2021) Functional characterization of the *Stipa purpurea* P5CS gene under drought stress conditions. Int J Mol Sci 22(17):9599

Ye C, Zheng S, Jiang D, Lu J, Huang Z, Liu Z et al (2021) Initiation and execution of programmed cell death and regulation of reactive oxygen species in plants. Int J Mol Sci 22(23):12942

Ye N, Wang Y, Yu H, Qin Z, Zhang J, Duan M, Liu L (2023) Abscisic acid enhances trehalose content via OsTPP3 to improve salt tolerance in rice seedlings. Plants 12(14):2665

Yoon Y, Seo DH, Shin H, Kim HJ, Kim CM, Jang G (2020) The role of stress-responsive transcription factors in modulating abiotic stress tolerance in plants. Agronomy 10(6):788

Yuan Z, Cong G, Zhang J (2014) Effects of exogenous salicylic acid on polysaccharides production of *Dendrobium officinale*. S Afr J Bot 95:78–84

Zandkarimi H, Ebadi A, Salami SA, Alizade H, Baisakh N (2015) Analyzing the expression profile of AREB/ABF and DREB/CBF genes under drought and salinity stresses in grape (*Vitis vinifera* L.). PLoS One 10(7):e0134288

Zhang J, Tan W, Yang X-H, Zhang H-X (2008) Plastid-expressed choline monooxygenase gene improves salt and drought tolerance through accumulation of glycine betaine in tobacco. Plant Cell Rep 27(6):1113–1124

Zhang LX, Lai JH, Liang ZS, Ashraf M (2014) Interactive effects of sudden and gradual drought stress and foliar-applied glycine betaine on growth, water relations, osmolyte accumulation and antioxidant defence system in two maize cultivars differing in drought tolerance. J Agron Crop Sci 200(6):425–433

Zhang L, Gao M, Hu J, Zhang X, Wang K, Ashraf M (2012) Modulation role of abscisic acid (ABA) on growth, water relations and glycinebetaine metabolism in two maize (*Zea mays* L.) cultivars under drought stress. Int J Mol Sci 13(3):3189–3202

Zhang XX, Tang YJ, Ma QB, Yang CY, Mu YH, Suo HC et al (2013) OsDREB2A, a rice transcription factor, significantly affects salt tolerance in transgenic soybean. PLoS One 8(12):e83011

Zhao RZ, Jiang S, Zhang L, Yu ZB (2019) Mitochondrial electron transport chain, ROS generation and uncoupling. Int J Mol Med 44(1):3–15

Zouari M, Hassena AB, Trabelsi L, Rouina BB, Decou R, Labrousse P (2019) Exogenous proline-mediated abiotic stress tolerance in plants: possible mechanisms. In: Osmoprotectant-mediated abiotic stress tolerance in plants: recent advances and future perspectives. Springer, Cham, pp 99–121

# Interaction of Isoprenoid Derivative Signaling Molecules with Mineral Nutrients in Thermotolerance Response

# 5

Ameena Fatima Alvi, Nidhi, and Nafees A. Khan

**Abstract**

Both internal and external variables influence plant development and growth. One of the limiting parameters that determines crop productivity is temperature. The function of intrinsic components, such as phytohormones and signaling molecules, strengthens plant defense responses in high-temperature environments, increasing plant survival rates. Recent studies suggest that isoprenoid derivative molecules regulate resilience responses, which explains why plants develop with such fantastic plasticity. Gibberellins (GA), abscisic acid (ABA), cytokinins (CK), strigolactones (SLs), and brassinosteroids (BR) are isoprenoid derivative signaling molecules crucial in thermotolerance responses. Although the exact mechanism of action is still unknown, mineral nutrients may be essential in stimulating the signaling cascade, or these signaling molecules may enhance nutrient absorption, leading to thermal tolerance. This chapter focuses on the interaction between isoprenoid derivative signaling molecules and mineral nutrients in heat stress conditions.

**Keywords**

Crop productivity · HEAT stress · Isoprenoid · Phytohormone · Signaling cascade · Signaling molecule

Authors Ameena Fatima Alvi and Nidhi share first authorship.

A. F. Alvi · Nidhi · N. A. Khan (✉)
Plant Physiology and Biochemistry Laboratory, Department of Botany, Aligarh Muslim University, Aligarh, India





N. Iqbal et al. (eds.), *Plant Signaling Molecules in Regulation of ROS-Scavenging System*, https://doi.org/10.1007/978-981-95-3377-0_5

## 5.1 Introduction

Plants are susceptible to temperatures above their thresholds, and even prolonged exposure to ideal temperatures can cause irreversible damage (Wahid et al. 2007). Heat stress can cause protein denaturation, membrane fluidity, oxidative damage, and the inactivation of chloroplast and mitochondrial enzymes. These factors can ultimately result in cell death (Moore et al. 2021; Alvi et al. 2024). A deeper understanding of the extent of heat stress that plants experience is necessary due to the drastic changes in climate and the predicted rise in global temperatures. Profitable crop production is directly proportional to the photosynthetic efficiency of plants, and photosynthesis is associated with major heat-sensitive processes like electron transport, $CO_2$ assimilation, photophosphorylation, thylakoid membrane fluidity, and light-induced chemical reactions (Sun and Guo 2016; Iqbal et al. 2021). Plant systems, however, have naturally developed a variety of strategies to fend off heat stress, such as expressing protective chaperones like Heat Shock Protein 101 (HSP101) and HSP100, whose mutants in Arabidopsis and tomatoes, respectively, displayed excruciating signs of heat stress (Hong and Vierling 2000; Yang et al. 2006). The primary functions of these chaperons are maintaining photosystem II (PSII) activity and thylakoid protection. Moreover, exogenous application of phytohormones, in addition to HSPs, aids in developing thermotolerance responses. These phytohormones function as potent signaling molecules that can modulate the expression patterns of other regulatory genes and molecular chaperones to mitigate the consequences of heat stress. (Jha et al. 2021; Iqbal et al. 2022). Likewise, isoprenoids, small, lipophilic volatile compounds, are widely generated in plants, primarily in the chloroplast. For a plant to synthesize one isoprene molecule, it needs fourteen NADPH and 21 ATP (Pollastri et al. 2021). This makes the process expensive. Notwithstanding the high expense of production, isoprenoids and their derivatives control physiological and biochemical processes in both standard and stressed environments. Holstein and Hohl (2004) support the processes of carbon absorption, electron transport, signal transduction, membrane construction, and gene expression.

Recently, methylerythritol cyclodiposphate (MEcPP), an intermediate in the methylerythritol phosphate (MEP) pathway, was identified (Xiao et al. 2012), as a signal to modify genes relevant to stress. Sessile plants must endure daily and seasonal fluctuations in their environment. High temperatures are one of the unavoidable effects of climate change. Significant issues related to heat stress in plants include increased respiratory burst, pigment loss, slowed growth, enzyme denaturation, protein degradation, and excessive free radical formation (Alvi et al. 2024). The protective role of isoprenoids on stress responses is mediated by directly reacting with reactive oxygen species (ROS), indirectly altering the ROS signal cascade, and stabilizing lipid membranes. These processes help plants become more capable of handling oxidative stress within their cells (Pollastri et al. 2021). Several isoprenoid derivatives serve as signaling molecules, viz. abscisic acid (ABA), gibberellic acid (GA), cytokinins (CK), strigolactones (SLs), and brassinosteroids (BRs). Apart from their normal physiological functions, these isoprenoid-derived phytohormones regulate plant performance under heat stress. For instance, the foliar application of GA boosted antioxidant activities, mineral absorption, and photosynthesis in tomatoes under heat stress (Guo et al. 2022), and the application of CK or BRs improved the downregulatory effect of heat on rice physiology during different phenological stages (Pantoja-Benavides et al. 2022). The stomatal closure response during high transpiration is regulated by a precise balance between SL-ABA crosstalk at the signaling level. The increased SL content in shoots possibly stimulates ABA sensitivity in plants, consequently reducing stomatal conductance and promoting stomatal closure, thereby increasing plant survival rate (Korek and Marzec 2023). Synergistic implementation of silicon (Si) and GA in *Phoenix dactylifera* helped mitigate heat stress responses through ABA signaling upregulation of *glutathione peroxidase 2 (GPX2), catalase (CAT), Cyt-Cu/Zn superoxide dismutase (Cyt-Cu/Zn SOD), and Glyceraldehyde-3-phosphate dehydrogenase (GAPHD)* genes and activation of heat shock factor related gene (*HsfA3*) (Khan et al. 2020).

Under heat stress, mineral nutrient accumulation and translocation dynamics are considerably disrupted. Heat stress could also influence ion mobility, thereby increasing the potassium (K)/calcium (Ca) ratio along with lower mineral content in wheat (*Triticum aestivum*) (Cabral et al. 2016). In soybean (*Glycine max L.*), untargeted metabolome profiling of leaves showed differently produced metabolites such as carbohydrates, lipids, amino acids, peptides, cofactors, and secondary metabolites under heat stress. Such upregulation of sugar and nitrogen (N) metabolism can help mitigate plant heat stress damage (Das et al. 2017). Availability of zinc (Zn), K, and boron (B) helped efficaciously alleviate heat-induced perturbations in terms of high chlorophyll content, antioxidant enzyme activity, net photosynthesis seed yield, and boll weight in cotton compared to untreated plant under heat stress (Sarwar et al. 2019). Moreover, exposure to heat stress during anthesis significantly changed the elemental concentration of 20 mineral nutrients in quinoa (*Chenopodium quinoa*) (Tovar et al. 2022). Thus, the application or presence of mineral nutrients is equally vital in diminishing heat stress impairment in plants. Thus, both isoprenoid derivatives and mineral nutrients are essential in maintaining plant vigor against heat stress. Up here, we will focus on the potential mechanism driven by isoprenoid and mineral nutrients as well as instances of their possible interaction in limiting heat stress damage.

## 5.2 Biosynthesis of Isoprenoid Derivatives

Isoprenoids encompass over 20,000 known chemical compounds, exhibiting a broad spectrum of structural diversity. This includes straightforward linear hydrocarbon chains and intricate ring structures, some of which are considered among the most complex known. Investigations into isoprenoid biosynthesis began in the early twentieth century. It was discovered that the carbon structure of terpenoid molecules is composed of $C_5$ units joined together in a head-to-tail fashion. Isoprenoids are organic compounds derived from the basic isoprene (2-methyl-1,3- butadiene) unit (Tarkowská and Strnad 2018). This diverse chemical class is widely distributed across all living organisms, from simple prokaryotes to complex eukaryotes (Khan et al. 2023). Plants are known to have two distinct biosynthetic pathways for producing terpenoids: the methylerythritol 4-phosphate/deoxyxylulose 5-phosphate pathway (MEP/DOXP) and the mevalonate (MEV or MVA) system (Shi et al. 2010). The isoprenoid biosynthetic pathway generates a diverse array of secondary and primary metabolites, utilizing the $C_5$ precursors isopentenyl diphosphate (IPP) and its isomer dimethyl allyl diphosphate (DMAPP). These key intermediates are produced through MVA and MEP pathways (Pu et al. 2021). While the MEP pathway is limited to plastids, the MVA pathway is distributed across the cytoplasm, endoplasmic reticulum, and peroxisomes (Noushahi et al. 2022). A detailed account of the pathways is given below.

## 5.3 Mevalonate (MVA) Pathway

According to Yeh et al. (2018), the mevalonate pathway is an essential metabolic route that plays a key role in the production of isoprenoids in a variety of cellular functions. The MVA route, discovered and recognized in the 1960s, was the first known pathway for producing isoprenoids (Liao et al. 2016). It was believed to be the sole route generating IPP/DMAPP for many years, leading to the assumption that it was ubiquitous across living organisms. While other organisms exhibit flexibility in their isoprenoid biosynthesis, elucidating the MVA route in eukaryotic organisms is considered the "canonical" mechanism. Acetoacetyl-CoA (AcAc-CoA) and 3-hydroxy-3-methylglutaryl-CoA (HMG-CoA) are the products of the earliest stages of the canonical or eukaryotic MVA pathway. These steps entail the sequential condensation of three acetyl-CoA molecules, which ultimately results in the synthesis of AcAc-CoA. The enzymes acetoacetyl-CoA thiolase (AACT) and HMG-CoA synthase, respectively, are responsible for catalyzing these conversions (Nagegowda and Gupta 2020). This mechanism is the first and most important stage in the route that is promoted by the enzyme HMG-CoA reductase (Vranova et al. 2013). First, HMG-CoA is transformed into MVA via two irreversible reduction processes. Each of these steps needs NADPH as the reducing equivalent. MVA undergoes a series of three enzymatic steps to become IPP, the biological counterpart of isoprene. This process involves two ATP-dependent phosphorylations and one ATP-driven decarboxylation (Tarkowská and Strnad 2018). The phosphorylation steps occur first by mevalonate kinase (MVK) yielding 5-phosphomevalonate, followed by a reaction with 5-phosphomevalonate kinase (PMVK) to produce 5-diphosphomevalonate, and lastly, this molecule is subsequently decarboxylated by the enzyme 5-diphosphomevalonate decarboxylase (DPMD), generating IPP (Rodríguez-Concepción et al. 2012) (Fig. 5.1). IPP isomerase (IDI) catalyzes a reversible isomerization phase in IPP that produces DMAPP, enabling the IPP/DMAPP condensation process to produce terpenoid molecules (Simkin et al. 2011).

## 5.4 Methylerythritol 4-Phosphate/Deoxyxylulose 5-Phosphate (MEP) Pathway

Following the discovery of the MVA route, major uncertainty was brought about as a consequence of variations in the data provided by labeling studies. It was not possible to reconcile the observed labelling patterns with the biochemical knowledge of the MVA route that had been developed (Krause et al. 2023). The MEP route, sometimes referred to as the non-mevalonate or 1-deoxy-D-xylulose 5-phosp (DXP) pathway, was first discovered in the gram-negative bacteria *E. coli* (Rohmer et al. 1993). In photosynthetic organisms such as plants and some protists, IPP and DMAPP biosynthesis occurs within the plastid cellular compartments (Perveen 2018). The first step of the MEP pathway is assisted by a protein called 1-deoxy-D-xylulose 5-phosphate synthase (DXS) (Fig. 5.1). This enzyme is responsible for converting the pyruvate and glyceraldehyde 3-phosphate precursors into DXP via a condensation process (Tian et al. 2022). This irreversible reaction dedicates carbon to the MEP pathway and results in the release of $CO_2$. In *E. coli*, DXP is a critical node in the metabolic pathway, as it is also utilized to synthesize pyridoxal phosphate and thiamine diphosphate (Wiesner and Jomaa 2013). In the next step, the enzyme DXP reductoisomerase (DXR) catalyzes the transformation of DXP into MEP in a process involving an intramolecular rearrangement and subsequent reduction, which represents the first committed step of the pathway (Noushahi et al. 2022). The key enzymes in the MEP pathway are DXR and DXS (Tian et al. 2022). Using a diphosphate bridge, 2-C-methyl-D-erythritol 4-phosphate cytidyl transferase (MCT) creates CDP-ME from a cytidyl moiety. The intermediate is phosphorylated by 4-(cytidine-5′-diphospho)-2-C methyl-D-erythritol kinase (CMK) and cyclizes once the cytidyl group is removed to generate CDP-MEP. The enzyme 2-C-methyl-D-erythritol 2,4-cyclodiphosphate synthases this process. The enzyme 4-hydroxy-3-methylbut-2-enyl diphosphate synthase (HDS) produces HMB-PP in the next step. In the final stage of the pathway, the enzyme HMBPP reductase converts HMBPP into a mixture of IPP and DMAPP in a 5:1 ratio (Wolff et al. 2002). In accordance with the findings of Uchida et al. (2018), it is generally acknowledged that HDR and DXS, as well as DXR to a certain extent, are the essential enzymes responsible for regulating metabolic flow via the MEP. Since the HDR enzyme is responsible for the production of both IPP and DMAPP simultaneously, the isomerase is a step in the MEP route that is unnecessary. Isoprenoid end products are known to increase when there is an increase in the overexpression of IDI, despite this being the case. A hypothesis has been put out that the MEP pathway is responsible for producing the precursors necessary for synthesizing gibberellins, diterpenoids, carotenoids, and chlorophylls.

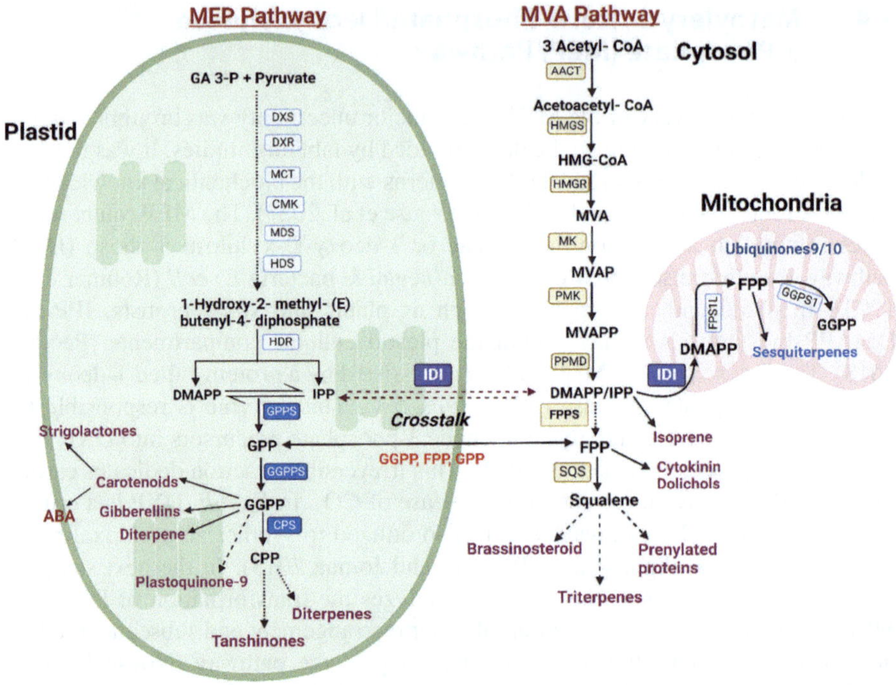


**Fig. 5.1** Subcellular compartmentalization of two types of isoprenoid biosynthesis pathways in plants. (1) The MVA pathway occurs in cytosol (2) The MEP pathway occur in the Plastid. Abbreviations: GA-3-P (glyceraldehyde 3-phosphate); DXS (1-deoxy-D-xylulose 5-phosphate synthase); DXR (1-deoxy-D-xylulose 5-phosphate reductoisomerase); MCT (2C-methyl-D-erythritol 4-phosphate cytidyl transferase); CMK (4-diphosphocytidyl-2C-methyl-D-erythritol kinase); MDS (2C-methyl-D-erythritol 2,4-cyclodiphosphate synthase); HDS (1-hydroxy-2-methyl-2-(E)-butenyl 4-diphosphate synthase); HDR (1-hydroxy-2-methyl-2-(E)-butenyl 4-diphosphate reductase); DMAPP (dimethylallyl diphosphate); IPP (isopentenyl diphosphate); GPPS (GPP synthase); GPP (geranyl diphosphate); GGPP (geranylgeranyl diphosphate); CPP (copalyl diphosphate); CPS (CPP synthases); FPP (farnesyl diphosphate); AACT (acetoacetyl-CoA thiolase); HMGS (3-hydroxy-3-methylglutaryl-CoA synthase); HMGR (3-hydroxy-3-methylglutaryl-CoA reductase); IDI (isopentenyl diphosphate isomerase); HMG-CoA (3-hydroxy-3-methylglutaryl-coenzyme A); MVA (mevalonate); MK (mevalonate kinase); MVAP (mevalonate 5-phosphate); PMK (phosphomevalonate kinase); MVAPP (mevalonate 5-diphosphate); PPMD (diphospho-mevalonate decarboxylase); FPPS (FPP synthase); SQS (squalene synthase); ABA (Abscisic acid)

## 5.5 Mineral Nutrient and Isoprenoid-Derived Responses Under Heat Stress

Plant scientists are increasingly concerned about global warming due to the significant contribution of gaseous emissions and other human activities to the buildup of greenhouse gases. Multiple global circulation models have forecasted a progressive rise in the global ambient temperature due to the accumulation of greenhouse gases, namely $CO_2$, methane, nitrous oxide, and chlorofluorocarbons (Imran et al. 2021).

According to the Sixth Assessment of the Intergovernmental Panel on Climate Change (IPCC), the Earth is experiencing its warmest period in the past 125,000 years, with global temperatures increasing by 1.1 °C above pre-industrial levels. While it is true that warming has been shown to have positive benefits on agricultural productivity in some high-latitude locations, the growing danger of climate change has had adverse consequences on crop yields across various countries. These harmful effects will worsen in the next century (Porter et al. 2016). Every plant species has an inherent temperature range ideal for its development and reproduction. Therefore, significant deviations from this range may quickly impact the thermodynamics of intracellular macromolecules (such as nucleic acids and proteins) and substructures inside plant cells (Hatfield and Prueger 2015; Niu and Xiang 2018). Direct Injuries caused by elevated temperatures can result in the denaturation and clumping together of proteins, along with an increase in the flexibility of membrane lipids. Slower heat-related injuries may occur indirectly through deactivating enzymes in chloroplasts and mitochondria, leading to reduced protein synthesis, breakdown, and compromised membrane stability (Hemantaranjan et al. 2014).

Plants have a challenging task in dealing with high temperatures, which is further complicated by environmental variables such as soil and air temperature, which may impact crop productivity (Lobell and Gourdji 2012). The respiration rate increases as temperature rises until a certain threshold, beyond which it begins to decrease. Similarly, photorespiration, which is less sensitive to temperature, likewise declines once the temperature exceeds a critical level (Posch et al. 2019). Elevated temperatures adversely impact plant development, crop duration, respiration, carbon assimilation, and other growth processes (Maestri et al. 2002). A 10 °C rise in temperature doubles enzymatic reactions, yet the optimal temperature range for enzymatic activity is 30–40 °C. Temperatures exceeding 45 °C result in enzymatic deactivation or denaturation (Bokszczanin and Fragkostefanakis 2013; Mathur et al. 2014).

Heat-induced damage may affect both shoots and roots. Root development is hindered by high-temperature stress (Huang et al. 2012). The impairment in the root system, in turn, results in detrimental changes in above-ground tissue as it reduces mineral and water absorption, phytohormone synthesis, and photosynthate transfer from source to sink (Hao et al. 2012; Giri et al. 2017). The decline in nutrient uptake per unit root can be attributed to the depletion of labile carbon and the subsequent decrease in energy within the roots (Giri et al. 2017). Alternately, this might result from an increase in root respiration or a decrease in the amount of carbon transported from the shoots to the roots. Heat stress may also interfere with enzymes responsible for nutrient metabolism, such as the absorption of nitrate and ammonium, by checking the synthesis and function of proteins responsible for nutrient absorption (Hungria and Kaschuk 2014; Cochavi et al. 2020). The damage caused by heat stress is fairly dependent upon the types of species and developmental stage of the plant. For example, Chenopodium, when subjected to 30–35 °C for 11 days during the flowering phase, decreased certain elements (arsenic (As), cadmium (Cd), rubidium (Rb), and strontium (Sr)) concentrations in the primary inflorescences (Tovar et al. 2022). Likewise, tomato plants exhibited decreased root concentrations of carbon, N, and iron (Fe) during acute heat stress and reduced shoot N content. On the contrary, the uptake rate of B increased under moderate heat stress, suggesting that B uptake was not impaired by heat stress (Giri et al. 2017). Moreover, when the lentil (*Lens culinaris*) was exposed to a temperature of 37 °C, Zn and Fe concentrations were reduced compared to the surrounding conditions (El Haddad et al. 2021). Heat stress experienced during the grain filling stage of cereal crops can also lead to changes in nutrient content in specific areas. For example, when bread wheat and durum wheat genotypes were subjected to heat stress, the amounts of Fe in the culm and leaves of bread wheat fell, but the levels of Fe in durum wheat rose significantly. On the other hand, it was observed that the amount of Fe in the spike increased (Dias et al. 2009).

The insufficient and uneven distribution of essential mineral elements and compromised soil fertility pose significant challenges, resulting in a decline in worldwide food production, particularly in emerging nations. The provision of adequate nourishment is of utmost importance to preserve the structural integrity of plants and to provide support for essential physiological activities. N and magnesium (Mg) are necessary for chlorophyll to function correctly. Phosphorous (P), on the other hand, being a structural component of nucleic acids, is essential in creating and storing energy and K regulates the osmotic equilibrium and enzyme activation (Waraich et al. 2011). Several research studies have shown that nutrients positively impact mitigating different abiotic stressors. For example, Tahir et al. (2011) found that Si has sound effects on enhancing salt tolerance in wheat. Similarly, Munns and Tester (2008) observed that K increases salinity tolerance. In the future, research focused on plant nutrition will be paramount, as it plays a crucial role in enhancing crop yield and maintaining soil fertility. When subjected to environmental challenges, agricultural plants' survival and productivity rely on their capacity to generate adaptive mechanisms to either evade or endure stress (Waraich et al. 2012). Nutrients are crucial in alleviating plant heat stress and facilitating critical physiological processes. These include regulating stomatal opening, preserving cell membrane integrity, activating stress response enzymes, and enhancing antioxidant defense mechanisms (Mishra et al. 2023). Heat stress diminishes the root hair mass, resulting in impaired nutrient absorption by plants. Applying external nutrients can help mitigate root damage and enhance the passive transfer of nutrients through various transporters, such as CDPKs, NRT-1 and 2, IRT-1, PHT-1, BOR-1, and SULTR-1 and 2 (Angon et al. 2024). Nutrient concentration in cereal crops is also influenced by heat stress during grain-filling. For example, bread wheat genotypes exhibited lower Fe contents in the culm and leaves under heat stress than durum wheat genotypes. Conversely, Fe content increased in the spike (Dias et al. 2009). The same study revealed a notable rise in shoot manganese (Mn) concentrations during grain loading. Cabral et al. (2016) reported that heat-stressed wheat plants preferentially allocate N to grains rather than tillers, although heat stress reduces phosphorus concentration in these plants. Heat stress also increases the K/Ca ratio, affecting plant ion mobility. Heat stress elevated the carbon-to-nitrogen (C:N) ratio and reduced the mineral nutrients in maize (*Zea mays*) leaves (Zhang et al. 2013).

Isoprenoids, however, are found in all living organisms, be it eukaryotes or prokaryotes. Many $C_5$ isoprenoid units can combine to form linear monoterpenes ($C_{10}$), diterpenes($C_{20}$), triterpenes ($C_{30}$) and polyterpenes (>$C_{40}$) (Connolly and Hill 1991). They undergo rearrangement reactions, cyclization oxidation and extensive intracellular modifications, which enable them to serve various biological roles like attractants, repellents, hormones, growth inhibitors, pigments and quinones (Tarkowská and Strnad 2018). Moreover, plant species release isoprene which is believed to enhance thermotolerance by supporting carbon assimilation to cope with high temperatures (Sharkey et al. 2001). Under high-temperature stress, isoprenoid responses are mediated through alteration of ROS, indirect modification of ROS signal cascade and lipid membrane stabilization and confer plant to resist oxidative stress inside the cell (Jalil and Ansari 2020). ABA, a non-volatile $C_{15}$ terpenoid carboxylic acid, is widely recognized as a chemical messenger mediating cell-cell communication. Apart from seed development regulation, shoot elongation, and stomatal movements, it mediates several stress-related responses in plants. ABA pretreatment to rice seedlings proffers heat tolerance through decreased leaf withering, cell injury, and reduced expression of apoptosis-related genes OsKOD1, OsCP1 and OsNAC4, along with improved antioxidant defense capacity (Liu et al. 2022). In another study, CK and its antagonistic action with ABA signify the hormonal interaction to regulate water relation, photosynthesis and stomatal movement in *Cichorium endivia* plants under stress condition (Romero-Muñoz et al. 2022). It has been suggested that gene related to DNA-replication, cell wall organization and microtubule-based movement are highly responsive to brassinolide treatments and optimal BR content is required for seedling development for both 24-epitestosterone and BR biosynthesis inhibitor reduced total P and K levels in shoots and affected several micronutrients like Mg, Fe, Mn and molybdenum (Mo) (Cheng et al. 2021). In wheat plants, the combined application of GA and sodium nitroprusside (SNP-NO source) proved beneficial in combating heat stress (Zhang et al. 2023). There are other instances where GA treatment improved germination rate, germination index, nutrient uptake, lengths and weight of coleoptile and radicle, regulation of endogenous salicylic acid and ABA accumulation and antioxidant defense responses and in different plant species under high-temperature conditions (Cavusoglu and Kabar 2007; Khan et al. 2020).

## 5.6 Interaction of Isoprenoid Derivatives with Mineral Nutrients Under Heat Stress

### 5.6.1 Gibberellic Acid

GA, often known as $GA_3$, is a plant hormone that plays an important role in various physiological processes. These activities include plant height, leaf expansion, accumulation of dry matter, differentiation of tissues, cell division, blooming, photosynthesis, and transpiration rate (Ullah et al. 2017). Moreover, $GA_3$ is a diterpenoid compound that significantly contributes to stress tolerance in several crops by affecting physiological, morphological, and enzymatic processes (Ji et al. 2015; Uzal and Yasar 2017; Morris et al. 2024). According to the findings of the study that was carried out by Guo et al. (2022), a foliar application of GA was used to reduce heat stress in tomato plants. Compared to other $GA_3$ treatments and the control, the plants treated with 75 mg $L^{-1}$ $GA_3$ collected the highest biomass. This indicates that $GA_3$ is beneficial in enhancing plant growth and mitigating the impacts of heat stress. Heat stress frequently compromises root systems or alters soil properties, hindering nutrient uptake. By maintaining root activity and nutrient absorption efficiency, GA mitigates these effects. GA enhances root surface area and the capacity to absorb essential nutrients, including N, P, K, and micronutrients, promoting root growth and development. Priming radish (*Raphanus sativus*) seeds with $GA_3$ enhances germination, increases coleoptile and radicle length, and boosts seedling fresh weight, improving resistance to high-temperature stress (Cavusoglu and Kabar 2007). Moreover, $GA_3$ treatment maintained redox balance in gladiolus (*Gladiolus hortulanus*) under varying temperatures by bolstering membrane stability, elevating antioxidant enzyme actions, and reducing membrane leakage (Saeed et al. 2014).

### 5.6.2 Abscisic Acid

ABA governs important physiological changes in plants under heat stress. Appropriate agricultural practices can modulate the endogenous production of ABA, leading to enhanced crop production. Elevated temperature stimulates ABA production, followed by stomatal closure and supporting plant growth. ABA contributes to the induction of many biological pathways essential to combat heat stress in different plant species (Maestri et al. 2002; Ding et al. 2010; Talanova et al. 2006). ABA-mediated reduction in malondialdehyde and $H_2O_2$ content and upregulation of different HSPs significantly helped in thermotolerance responses in plants (Sezgin Muslu and Kadıoğlu 2021; Wang et al. 2023). ABA also works synergistically with nicotinamide adenine dinucleotide (NAD) biosynthesis, contributing to thermal tolerance. Additionally, ABA helps maintain effective soil-water relationships, facilitating nutrient absorption and sustaining cell turgor. Heat stress impairs the uptake of micronutrients like Zn and Fe, which are crucial for enzymatic activities. ABA can mediate these responses by modulating the transpiration rate, helping conserve water and improve nutrient availability (Tao et al. 2022). Under abiotic stress conditions, ABA application helps promote deeper root development, increasing plant-water-nutrient balance under unfavorable conditions. The synergistic approach of nutrients and ABA showed promising results in improving plant tolerance. For instance, the application of ABA and calcium nanoparticles (Ca-NPs) regulated plant growth and drought tolerance in rapeseed in terms of increased photosynthetic pigmentations and nutritional profile. The treatment also enhanced the expression of ROS homeostasis genes, photosynthetic proteins, and related transcription factors (Ayyaz et al. 2024). Moreover, Hosseini et al. (2021) reported positive interaction among melatonin, ABA content, sugar, and phenolic compound abundance of lemon verbena (*Lippia citriodora*) under drought stress. Interaction of

ABA with NO has also been proved using c-PTIO (2–4 carboxyphenyl-4,4,5,5-tetramethylimidazoline-1-oxyl-3-oxide) where reduction in ABA-induced osmolytes and antioxidant metabolism suggested NO-dependent ABA responses under heat stress (Iqbal et al. 2022). Ca pretreatment enhanced antioxidant enzyme activity and lowered lipid peroxidation in ABA-treated seedlings in maize when subjected to heat stress (Gong et al. 1998). The application of triacontanol, a ubiquitous saturated primary alcohol, reversed the deleterious effect of heat stress by showing an increase in ABA and jasmonic acid levels, leading to better plant performance (Waqas et al. 2016). ABA signaling pathways activated at high temperatures increase ABA production and accumulation (Rai et al. 2024). This hormonal response affects significant mineral nutrient absorption, transport, and assimilation. ABA increases ion transporter gene expression, which helps cells absorb K and Ca under heat stress (Mostofa et al. 2022). ABA-mediated stomatal closure conserves water and regulates transpiration stream nutrient absorption. ABA-mediated antioxidant defense system requires Zn and Mn to repair oxidative damage under heat stress (Hsu et al. 2021). ABA regulates plant nutrition and stress tolerance with other phytohormones and signaling molecules. Understanding the complicated relationships between ABA and mineral nutrients under heat stress is essential for creating climate-resilient agricultural strategies.

### 5.6.3 Brassinosteroids

BRs are steroid hormones that regulate various plant functions, from plant growth to nutrient uptake and transport. The transcriptomic report of differentially expressed genes in soybeans states that BR and its biosynthetic inhibitor affected K and P levels along with other minor nutrients. BR regulates nutrient concentration through transcriptional regulation of mineral transporter genes like *GmPHT1*s, *GmKT*s, Gm*VIT2*, *GmZIP*s, and *GmMOT1* (Cheng et al. 2021). It has previously been reported that BR regulates Fe contents, showing distinct trends for the rice shoot and root, respectively (Wang et al. 2015), while BR increases the concentration of Fe in the shoot of $Ca(NO_3)_2$ stressed cucumber (*Cucumis sativus*) plants (Yuan et al. 2015). This implies that the environmental conditions in different species may impact BR's regulation of mineral concentration. Likewise, synthetic BRs, when applied in mustard (*Brassica juncea*) and tomato under temperature stress, generate thermotolerance (Dhaubhadel et al. 1999; Singh and Shono 2005). Strategies like maintenance of membrane stability, ROS scavenging enzymes, encoding HPSs, and induction of molecular chaperon signaling support plants to overcome heat stress. BRs regulate various aspects of plant development and the above-mentioned strategic responses under heat stress. For instance, a transcription factor and a positive regulator of BR signal, Brassinazole resistant 1 (BZR1), inhibits (*ERF49)* expression by binding with *ERF49*, which otherwise showed higher sensitivity to heat stress. *HSFA2*, DREB2A, and HPS transcription levels were significantly reduced under heat stress in *ERF49*-overexpressed transgenic plants. BZR1 repressed *ERF49* expression in *Arabidopsis* under heat stress by targeting downstream genes

(Chen et al. 2022). 24-epibrassinolide (EBR) induces heat tolerance through the higher synthesis of HSP at transcriptional and translational levels in mustard seedlings (Dhaubhadel et al. 2002).

### 5.6.4 Cytokinin

Increasing temperature beyond the threshold limits plant reproductive stage development, mainly kernel development, grain growth and yield stability. Cheikh and Jones (1994) recorded the peak in zeatin riboside and zeatin levels under controlled conditions, unlike the kernel subjected to 4d of heat stress. Application of benzyl adenine under heat stress resulted in reduced kernel abortion and hormonal balance thus increasing thermotolerance and providing high yield and stability in maize (Cheikh and Jones 1994). CK pool modification through inhibitor of cytokinin degradation (INCYDE-F) turned out to be effective in heat acclimation responses by increasing CK content. INCYDE-F inhibited CK degradation improving plant recovery under heat stress (Prerostova et al. 2020). The decrease in grain yield in wheat under heat stress is due to lower zeatin riboside content and higher $GA_3$ content, resulting in a smaller kernel number and a lower grain filling rate. The application of 6-BA significantly increased the grain filling rate and mitosis in endosperm and healthy grain settings compared to plants exposed to 6-BA without heat stress conditions (Yang et al. 2016). Moreover, during combined stress exposure (drought and heat), downregulation of CK receptor gene *SIHK2* enhanced abiotic tolerance with increased osmoprotectant accumulation, membrane stability, photosynthesis, and antioxidant defense. Therefore, the regulation of CK signaling is crucial in abiotic tolerance responses (Mushtaq et al. 2022). Degradation of CK through manipulating cytokine dehydrogenase (CKX) enzymes could hamper crop yield, stress tolerance, root growth, and nutrient absorption. There are several instances where higher CKX activity corresponds to higher accumulation of Zn, sulfur, Fe, and Mn and a stronger root system helped scavenge water and nutrients to the deeper end of the soil (Prasad 2022). The proximity between N and CK signaling has been established through regulation of the first step of CK synthesis and gene encoding *cytokinin responsive type-A response regulators (ARRs)* and *Cytokinin Responses Factors (CRFs)* in Arabidopsis. CK is also involved in nitrate foraging, enhancing nitrate acquisition (Menz et al. 2016; Liu et al. 2017).

### 5.6.5 Strigolactones

SLs are carotenoid-derived molecules primarily involved in regulating plant architecture, and other processes like root and shoot patterning, leaf senescence, and responses to nutrient deficiencies. Their role in mediating plant responses to abiotic stress is still emerging (Bhoi et al. 2021; Alvi et al. 2022). Interaction of SL with phytohormones, such as auxin, CK, ABA, and gibberellins in response to abiotic stress suggests that SL mediate stress adaptation responses. Workers reported SL-mediated heat and cold tolerance in tomatoes through induction of ABA, hsp70 accumulation, *C-REPEAT BINDING FATOR 1(CBF1)* transcription and antioxidant enzyme activity (Chi et al. 2021). Application of synthetic SLs (GR24) promoted root elongation in tall fescue under heat stress by maintaining active cell division and transcriptional regulation of gene associated with meristematic zone cell elongation followed by cell division and root elongation (Hu et al. 2018).

Undoubtedly, SL is closely associated with plant mineral absorption. In rice, SL mediates the balance between P and N by downregulating genes essential for N assimilation while upregulating genes associated with P uptake (Kun Yuan et al. 2023). In several studies, SL helped strengthen mycorrhization, plant biomass, N uptake, secondary root growth and, overcome secondary growth development. The work so far suggested that SL can be used to target crop optimization for less productive soils (Chesterfield et al. 2020; Pokluda et al. 2018; Kun Yuan et al. 2023). Not much about SL-mediated defense involving mineral nutrients under heat stress has been discovered. Under heat stress conditions, SL regulates thermotolerance during seed germination and seedling development in *Lupine angustifolius*. The thermotolerance response also includes enhanced proline content, lower lipid peroxidation, higher antioxidant enzyme activity and glyoxalase system (Omoarelojie et al. 2020). Some representative studies on the interaction of isoprenoid derivatives with mineral nutrients under heat stress have been shown in Table 5.1.

## 5.7 Conclusion

The increasing population and constant climate change significantly threaten food security. Isoprenoid derivatives like isoprene, carotenoid, and tocopherols play significant role in scavenging ROS, thereby minimizing oxidative damage during heat stress. They also stabilize lipid membranes, protecting them from heat-induced fluidity and preserving their structural integrity. The isoprenoid-derived phytohormones act as signaling molecules and optimize water usage and photosynthetic performance under high temperatures. On the other hand, mineral nutrients are crucial for osmotic adjustments, phospholipid synthesis, and providing cofactors required for antioxidant enzymes and HSPs.

The interconnected role of isoprenoid derivatives and mineral nutrients suggests the existence of an effective signaling cascade that enables plants to respond to heat stress. The combined action of two components ensures plants withstand high temperatures more effectively than individual contributions. While isoprenoids protect cellular components and stress signaling, mineral nutrients ensure efficient enzymatic activity, structural integrity, and metabolic adjustments. To promote thermotolerance in plants, more interactive research is required on the interplay between isoprenoid derivatives and mineral nutrients, focusing on unravelling the signaling cascade and key pathways involved.

**Table 5.1** Representative studies on the interaction of isoprenoid derivatives with mineral nutrients under heat stress

| Isoprenoid derivatives | Plant species | Mineral element | Outcome of interaction | References |
|---|---|---|---|---|
| Gibberellic acid | *Phoenix deactylifera* | Si | Heat tolerance response through ABA signaling and upregulation of genes associated with antioxidant defense and *HsfA3* | Khan et al. (2020) |
| | *Raphanus sativus* | N, P, K | Enhancement of root and shoot growth, fresh weight and thermotolerance | Cavusoglu and Kabar (2007) |
| Abscisic acid | *Triticum aestivum* and *Zea mays* | Zn and Fe | Maintenance of plant-water-nutrient balance and enzymatic activity | Tao et al. (2022) |
| | *Triticum aestivum* | N | Osmolyte and ROS regulation | Iqbal et al. (2022) |
| | – | K and Ca | Up-regulation of ion transport, membrane stability, prevention of protein denaturation | Mostofa et al. (2022) |
| | *Zea mays* | Ca | High antioxidant activity, decreased lipid peroxidation and improved growth | Gong et al. (1998) |
| Brassinosteroid | *Brassica* and *Lycopersicon eslulantus* | Mineral nutrients | ROS scavenging, induction of chaperon, increased growth and development | Dhaubhadel et al. (1999) |
| | *Zea mays* | $N/P_2O_5/$ $K_2O$ | The administration of BR enhanced P accumulation while having no effect on N or K accumulation. The volume and proportion of nutrients translocated and their contribution to grain yield increased | Xu et al. (2022) |
| Strigolactone | *Lupine angustifolius* | N and P | Enhanced proline content, seed germination and growth antioxidant defence and glyoxalase system and lower lipid peroxidation | Omoarelojie et al. (2020) |

## References

Alvi AF, Sehar Z, Fatma M, Masood A, Khan NA (2022) Strigolactone: an emerging growth regulator for developing resilience in plants. Plants 11:2604

Alvi AF, Khan S, Khan NA (2024) Hydrogen sulfide and ethylene regulate sulfur-mediated stomatal and photosynthetic responses and heat stress acclimation in rice. Plant Physiol Biochem 207:108437

Angon PB, Das A, Roy AR, Khan JJ, Ahmad I, Biswas A, Yeasmin ST (2024) Plant development and heat stress: role of exogenous nutrients and phytohormones in thermotolerance. Discover Plants 1:17

Ayyaz A, Zhou Y, Batool I, Hannan F, Huang Q, Zhang K, Shahzad K, Sun Y, Farooq MA, Zhou W (2024) Calcium nanoparticles and Abscisic acid improve drought tolerance, mineral nutrients uptake and inhibitor-mediated photosystem II performance in Brassica napus. J Plant Growth Regul 43:516–537

Bhoi A, Yadu B, Chandra J, Keshavkant S (2021) Contribution of strigolactone in plant physiology, hormonal interaction and abiotic stresses. Planta 254:1–21

Bokszczanin KL, Fragkostefanakis S (2013) Perspectives on deciphering mechanisms underlying plant heat stress response and thermotolerance. Front Plant Sci 4:315

Cabral C, Ravnskov S, Tringovska I, Wollenweber B (2016) Arbuscular mycorrhizal fungi modify nutrient allocation and composition in wheat (*Triticum aestivum* L.) subjected to heat-stress. Plant Soil 408:385–399

Cavusoglu K, Kabar K (2007) Comparative effects of some plant growth regulators on the germination of barley and radish seeds under high temperature stress. EurAsian J BioSci 1:1–10

Cheikh N, Jones RJ (1994) Disruption of maize kernel growth and development by heat stress (role of cytokinin/abscisic acid balance). Plant Physiol 106:45–51

Chen X, Xue H, Zhu L (2022) ERF49 mediates brassinosteroid regulation of heat stress tolerance in *Arabidopsis thaliana*. BMC Biol 20:254

Cheng L, Li M, Min W, Wang M, Chen R, Wang W (2021) Optimal brassinosteroid levels are required for soybean growth and mineral nutrient homeostasis. Int J Mol Sci 22:8400

Chesterfield R, Vickers C, Beveridge CA (2020) Translation of strigolactones from plant hormone to agriculture: achievements, future perspectives, and challenges. Trends Plant Sci 25:1087–1106

Chi C, Xu X, Wang M, Zhang H, Fang P, Zhou J, Xia X, Shi K, Zhou Y, Yu J (2021) Strigolactones positively regulate abscisic acid-dependent heat and cold tolerance in tomato. Horticult Res 8:237

Cochavi A, Cohen IH, Rachmilevitch S (2020) The role of different root orders in nutrient uptake. Environ Exp Bot 179:104212

Connolly JD, Hill RA (1991) Dictionary of terpenoids. https://doi.org/10.1007/978-1-4899-4513-6

Das A, Rushton PJ, Rohila JS (2017) Metabolomic profiling of soybeans (*Glycine max* L.) reveals the importance of sugar and nitrogen metabolism under drought and heat stress. Plants 6:21

Dhaubhadel S, Chaudhary S, Dobinson KF, Krishna P (1999) Treatment with 24-epibrassinolide, a brassinosteroid, increases the basic thermotolerance of Brassica napus and tomato seedlings. Plant Mol Biol 40:333–342

Dhaubhadel S, Browning KS, Gallie DR, Krishna P (2002) Brassinosteroid functions to protect the translational machinery and heat-shock protein synthesis following thermal stress. Plant J 29:681–691

Dias AS, Lidon FC, Ramalho JC (2009) IV. Heat stress in *Triticum*: kinetics of Fe and Mn accumulation. Braz J Plant Physiol 21:153–164

Ding W, Song L, Wang X, Bi Y (2010) Effect of abscisic acid on heat stress tolerance in the calli from two ecotypes of Phragmites communis. Biol Plant 54:607–613

El Haddad N, Sanchez-Garcia M, Visioni A, Jilal A, El Amil R, Sall AT, Bassi FM (2021) Crop wild relatives crosses: multi-location assessment in durum wheat, barley, and lentil. Agronomy 11:2283

Giri A, Heckathorn S, Mishra S, Krause C (2017) Heat stress decreases levels of nutrient-uptake and-assimilation proteins in tomato roots. Plants 6:6

Gong M, Li YJ, Chen SZ (1998) Abscisic acid-induced thermotolerance in maize seedlings is mediated by calcium and associated with antioxidant systems. J Plant Physiol 153:488–496

Guo T, Gull S, Ali MM, Yousef AF, Ercisli S, Kalaji HM, Telesiński A, Auriga A, Wróbel J, Radwan NS, Ghareeb RY (2022) Heat stress mitigation in tomato (*Solanum lycopersicum* L.) through foliar application of gibberellic acid. Sci Rep 12:1–13

Hao HP, Jiang CD, Zhang SR, Tang YD, Shi L (2012) Enhanced thermal-tolerance of photosystem II by elevating root zone temperature in Prunus mira Koehne seedlings. Plant Soil 353:367–378

Hatfield JL, Prueger JH (2015) Temperature extremes: effect on plant growth and development. Weather Clim Extremes 10:4–10

Hemantaranjan A, Bhanu AN, Singh MN, Yadav DK, Patel PK, Singh R, Djapar K (2014) Heat stress responses and thermotolerance. Adv Plants Agric Res 1:1–10
Holstein SA, Hohl RJ (2004) Isoprenoids: remarkable diversity of form and function. Lipids 39:293–309
Hong SW, Vierling E (2000) Mutants of Arabidopsis thaliana defective in the acquisition of tolerance to high temperature stress. Proc Natl Acad Sci 97:4392–4397
Hosseini MS, Samsampour D, Zahedi SM, Zamanian K, Rahman MM, Mostofa MG, Tran LSP (2021) Melatonin alleviates drought impact on growth and essential oil yield of lemon verbena by enhancing antioxidant responses, mineral balance, and abscisic acid content. Physiol Plant 172:1363–1375
Hsu PK, Dubeaux G, Takahashi Y, Schroeder JI (2021) Signaling mechanisms in abscisic acid-mediated stomatal closure. Plant J 105:307–321
Hu Q, Zhang S, Huang B (2018) Strigolactones and interaction with auxin regulating root elongation in tall fescue under different temperature regimes. Plant Sci 271:34–39
Huang B, Rachmilevitch S, Xu J (2012) Root carbon and protein metabolism associated with heat tolerance. J Exp Bot 63:3455–3465
Hungria M, Kaschuk G (2014) Regulation of N2 fixation and $NO^{3-}/NH^{4+}$ assimilation in nodulated and N-fertilized Phaseolus vulgaris L. exposed to high temperature stress. Environ Exp Bot 98:32–39
Imran QM, Falak N, Hussain A, Mun BG, Yun BW (2021) Abiotic stress in plants; stress perception to molecular response and role of biotechnological tools in stress resistance. Agronomy 11:1579
Iqbal N, Fatma M, Gautam H, Umar S, Sofo A, D'ippolito I, Khan NA (2021) The crosstalk of melatonin and hydrogen sulfide determines photosynthetic performance by regulation of carbohydrate metabolism in wheat under heat stress. Plants 10:1778
Iqbal N, Sehar Z, Fatma M, Umar S, Sofo A, Khan NA (2022) Nitric oxide and abscisic acid mediate heat stress tolerance through regulation of osmolytes and antioxidants to protect photosynthesis and growth in wheat plants. Antioxidants 11:372
Jalil SU, Ansari MI (2020) Isoprenoids in plant protection against abiotic stress. In: Protective chemical agents in the amelioration of plant abiotic stress, pp 424–436
Jha UC, Nayyar H, Siddique KHM (2021) Role of phytohormones in regulating heat stress acclimation in agricultural crops. J Plant Growth Regul 41:1041–1064
Ji P, Tang X, Jiang Y, Tong YA, Gao P, Han W (2015) Potential of gibberellic acid 3 ($GA_3$) for enhancing the phytoremediation efficiency of *Solanum nigrum* L. Bull Environ Contam Toxicol 95:810–814
Khan A, Bilal S, Khan AL, Imran M, Shahzad R, Al-Harrasi A, Al-Rawahi A, Al-Azhri M, Mohanta TK, Lee IJ (2020) Silicon and gibberellins: synergistic function in harnessing ABA signaling and heat stress tolerance in date palm (*Phoenix dactylifera* L.). Plants 9:620–629
Khan F, Jeong GJ, Tabassum N, Kim YM (2023) Functional diversity of c-di-GMP receptors in prokaryotic and eukaryotic systems. Cell Communication and Signaling, 21(1):259
Korek M, Marzec M (2023) Strigolactones and abscisic acid interactions affect plant development and response to abiotic stresses. BMC Plant Biol 23:1–18
Krause T, Wiesinger P, González-Cabanelas D, Lackus N, Köllner TG, Klüpfel T, Schmidt A (2023) HDR, the last enzyme in the MEP pathway, differently regulates isoprenoid biosynthesis in two woody plants. Plant Physiol 192:767–788
Liao P, Hemmerlin A, Bach TJ, Chye ML (2016) The potential of the mevalonate pathway for enhanced isoprenoid production. Biotechnol Adv 34:697–713
Liu KH, Niu Y, Konishi M, Wu Y, Du H, Sun Chung H, Li L, Boudsocq M, McCormack M, Maekawa S, Ishida T (2017) Discovery of nitrate–CPK–NLP signalling in central nutrient–growth networks. Nature 7654:311–316
Liu X, Ji P, Yang H, Jiang C, Liang Z, Chen Q, Lu F, Chen X, Yang Y, Zhang X (2022) Priming effect of exogenous ABA on heat stress tolerance in rice seedlings is associated with the upregulation of antioxidative defense capability and heat shock-related genes. Plant Growth Regul 98:23–38

Lobell DB, Gourdji SM (2012) The influence of climate change on global crop productivity. Plant Physiol 160:1686–1697
Maestri E, Klueva N, Perrotta C, Gulli M, Nguyen HT, Marmiroli N (2002) Molecular genetics of heat tolerance and heat shock proteins in cereals. Plant Mol Biol 48:667–681
Mathur S, Agrawal D, Jajoo A (2014) Photosynthesis: response to high temperature stress. J Photochem Photobiol B Biol 137:116–126
Menz J, Li Z, Schulze WX, Ludewig U (2016) Early nitrogen-deprivation responses in Arabidopsis roots reveal distinct differences on transcriptome and (phospho-) proteome levels between nitrate and ammonium nutrition. Plant J 88:717–734
Mishra S, Spaccarotella K, Gido J, Samanta I, Chowdhary G (2023) Effects of heat stress on plant-nutrient relations: an update on nutrient uptake, transport, and assimilation. Int J Mol Sci 24:15670
Moore CE, Meacham-Hensold K, Lemonnier P, Slattery RA, Benjamin C, Bernacchi CJ, Lawson T, Cavanagh AP (2021) The effect of increasing temperature on crop photosynthesis: from enzymes to ecosystems. J Exp Bot 72:2822–2844
Morris B, Khan S, Iqbal N, Al-Hashimi A, Khan NA (2024) Gibberellic acid reverses glucose-inhibited photosynthesis and growth via reduced glutathione and hormonal crosstalk in chromium-treated mustard. Plant Stress 14:100694
Mostofa MG, Rahman MM, Ghosh TK, Kabir AH, Abdelrahman M, Khan MAR, Tran LSP (2022) Potassium in plant physiological adaptation to abiotic stresses. Plant Physiol Biochem 186:279–289
Munns R, Tester M (2008) Mechanisms of salinity tolerance. Annu Rev Plant Biol 59:651–681
Mushtaq N, Wang Y, Fan J, Li Y, Ding J (2022) Down-regulation of cytokinin receptor gene SlHK2 improves plant tolerance to drought, heat, and combined stresses in tomato. Plants 11:154
Nagegowda DA, Gupta P (2020) Advances in biosynthesis, regulation, and metabolic engineering of plant specialized terpenoids. Plant Sci 294:110457
Niu Y, Xiang Y (2018) An overview of biomembrane functions in plant responses to high-temperature stress. Front Plant Sci 9:366067
Noushahi HA, Khan AH, Noushahi UF, Hussain M, Javed T, Zafar M, Shu S (2022) Biosynthetic pathways of triterpenoids and strategies to improve their biosynthetic efficiency. Plant Growth Regul 97:439–454
Omoarelojie LO, Kulkarni MG, Finnie JF, Pospíšil T, Strnad M, Van Staden J (2020) Synthetic strigolactone (rac-GR24) alleviates the adverse effects of heat stress on seed germination and photosystem II function in lupine seedlings. Plant Physiol Biochem 155:965–979
Pantoja-Benavides AD, Garces-Varon G, Restrepo-Díaz H (2022) Foliar cytokinins or brassinosteroids applications influence the rice plant acclimatization to combined heat stress. Front Plant Sci 13:983276
Perveen S (2018) Introductory chapter: terpenes and terpenoids. In: Perveen S, Al-Taweel A (eds) Terpens and terpenoids. Intech Open, London
Pokluda R, Shehata SM, Kopta T (2018) Vegetative, chemical status and productivity of zucchini squash (*Cucurbita pepo* L.) plants in responses to foliar application of pentakeep and strigolactones under NPK rates. Gesunde Pflanzen 70:21–29
Pollastri S, Baccelli I, Loreto F (2021) Isoprene: an antioxidant itself or a molecule with multiple regulatory functions in plants? Antioxidants 10:684–694
Porter SD, Reay DS, Higgins P, Bomberg E (2016) A half-century of production-phase greenhouse gas emissions from food loss & waste in the global food supply chain. Science of the Total Environment 571:721–729
Posch BC, Kariyawasam BC, Bramley H, Coast O, Richards RA, Reynolds MP, Atkin OK (2019) Exploring high temperature responses of photosynthesis and respiration to improve heat tolerance in wheat. J Exp Bot 70:5051–5069
Prasad R (2022) Cytokinin and its key role to enrich the plant nutrients and growth under adverse conditions-an update. Front Genet 13:883924

Prerostova S, Dobrev PI, Kramna B, Gaudinova A, Knirsch V, Spichal L, Zatloukal M, Vankova R (2020) Heat acclimation and inhibition of cytokinin degradation positively affect heat stress tolerance of Arabidopsis. Front Plant Sci 11:503541

Pu X, Dong X, Li Q, Chen Z, Liu L (2021) An update on the function and regulation of methylerythritol phosphate and mevalonate pathways and their evolutionary dynamics. J Integr Plant Biol 63:1211–1226

Rai GK, Khanday DM, Choudhary SM, Kumar P, Kumari S, Martínez-Andújar C, Pérez-Alfocea F (2024) Unlocking nature's stress buster: Abscisic acid's crucial role in defending plants against abiotic stress. Plant Stress 11:100359

Rohmer M, Knani M, Simonin P, Sutter B, Sahm H (1993) Isoprenoid biosynthesis in bacteria: a novel pathway for the early steps leading to isopentenyl diphosphate. Biochem J 295:517–524

Rodríguez-Concepción M, Campos N, Ferrer A, Boronat A (2012) Biosynthesis of Isoprenoid. In: Bach T, Rohmer M (Eds.). Isoprenoid Synthesis in Plants and Microorganisms: New Concepts and Experimental Approaches. Springer New York 439

Romero-Muñoz M, Albacete A, Gálvez A, Piñero MC, Del Amor FM, López-Marín J (2022) The use of ecological hydromulching improves growth in Escarole (*Cichorium endivia* L.) plants subjected to drought stress by fine-tuning cytokinins and abscisic acid balance. Agronomy 12:459–459

Saeed T, Hassan I, Abbasi NA, Jilani G (2014) Effect of gibberellic acid on the vase life and oxidative activities in senescing cut gladiolus flowers. Plant Growth Regul 72:89–95

Sarwar M, Saleem MF, Ullah N, Ali S, Rizwan M, Shahid MR, Alyemeni MN, Alamri SA, Ahmad P (2019) Role of mineral nutrition in alleviation of heat stress in cotton plants grown in glasshouse and field conditions. Sci Rep 9:1–17

Sezgin Muslu A, Kadıoğlu A (2021) Role of abscisic acid, osmolytes and heat shock factors in high temperature thermotolerance of *Heliotropium thermophilum*. Physiol Mol Biol Plants 27:861–871

Sharkey TD, Chen X, Yeh S (2001) Isoprene increases thermotolerance of fosmidomycin-fed leaves. Plant Physiol 125:2001–2006

Shi L, Ren A, Mu D, Zhao M (2010) Current progress in the study on biosynthesis and regulation of ganoderic acids. Appl Microbiol Biotechnol 88:1243–1251

Simkin AJ, Guirimand G, Papon N, Courdavault V, Thabet I, Ginis O, Clastre M (2011) Peroxisomal localisation of the final steps of the mevalonic acid pathway in planta. Planta 234:903–914

Singh I, Shono M (2005) Physiological and molecular effects of 24-epibrassinolide, a brassinosteroid on thermotolerance of tomato. Plant Growth Regul 47:111–119

Sun AZ, Guo FQ (2016) Chloroplast retrograde regulation of heat stress responses in plants. Front Plant Sci 7:183947

Tahir MA, Aziz T, Rahmatullah. (2011) Silicon-induced growth and yield enhancement in two wheat genotypes differing in salinity tolerance. Commun Soil Sci Plant Anal 42:395–407

Talanova VV, Topchieva LV, Titov AF (2006) Effect of abscisic acid on the resistance of cucumber seedlings to combined exposure to high temperature and chloride. Biol Bull 33:619–622

Tao Z, Yan P, Zhang X, Wang D, Wang Y, Ma X, Yang Y, Liu X, Chang X, Sui P, Chen Y (2022) Physiological mechanism of abscisic acid-induced heat-tolerance responses to cultivation techniques in wheat and maize. Agronomy 12:1579

Tarkowská D, Strnad M (2018) Isoprenoid-derived plant signaling molecules: biosynthesis and biological importance. Planta 247:1051–1066

Tian S, Wang D, Yang L, Zhang Z, Liu Y (2022) A systematic review of 1-Deoxy-D-xylulose-5-phosphate synthase in terpenoid biosynthesis in plants. Plant Growth Regul 96:221–235

Tovar JC, Berry JC, Quillatupa C, Castillo SE, Acosta-Gamboa L, Fahlgren N, Gehan MA (2022) Heat stress changes mineral nutrient concentrations in Chenopodium quinoa seed. Plant Direct 6:e384

Uchida H, Sumimoto K, Oki T, Nishii I, Mizohata E, Matsunaga S, Okada S (2018) Isolation and characterization of 4-hydroxy-3-methylbut-2-enyl diphosphate reductase gene from *Botryococcus braunii*, race B. J Plant Res 131:839–848

Ullah S, Anwar S, Rehman M, Khan S, Zafar S, Liu L, Peng D (2017) Interactive effect of gibberellic acid and NPK fertilizer combinations on ramie yield and bast fibre quality. Sci Rep 7:10647

Uzal O, Yasar F (2017) Effect of $GA_3$ hormone treatments on ion uptake and growth of pepper plant under cadmium stress. Appl Ecol Environ Res 15:4–25

Vranová E, Coman D, Gruissem W (2013) Network analysis of the MVA and MEP pathways for isoprenoid synthesis. Annu Rev Plant Biol 64:665–700

Wahid A, Gelani S, Ashraf M, Foolad MR (2007) Heat tolerance in plants: an overview. Environ Exp Bot 61:199–223

Wang B, Li G, Zhang WH (2015) Brassinosteroids are involved in Fe homeostasis in rice (*Oryza sativa* L.). J Exp Bot 66:2749–2761

Wang JQ, Xiang RH, Li ZG (2023) The essential role of $H_2S$-ABA crosstalk in maize thermotolerance through the ROS-scavenging system. Int J Mol Sci 24:12264

Waqas M, Shahzad R, Khan AL, Asaf S, Kim YH, Kang SM, Bilal S, Hamayun M, Lee IJ (2016) Salvaging effect of triacontanol on plant growth, thermotolerance, macro-nutrient content, amino acid concentration and modulation of defense hormonal levels under heat stress. Plant Physiol Biochem 99:118–125

Waraich EA, Ahmad R, Ashraf MY, Saifullah AM (2011) Improving agricultural water use efficiency by nutrient management in crop plants. Acta Agric Scand Sect B Soil Plant Sci 61:291–304

Waraich EA, Ahmad R, Halim A, Aziz T (2012) Alleviation of temperature stress by nutrient management in crop plants: a review. J Soil Sci Plant Nutr 12:221–244

Wiesner J, Jomaa H (2013) 1-Deoxy-D-Xylulose 5-phosphate pathway. In: Hommel M, Kremsner P (eds) Encyclopedia of malaria. Springer, New York. https://doi.org/10.1007/978-1-4614-8757-9_10-1

Wolff M, Seemann M, Grosdemange-Billiard C, Tritsch D, Campos N, Rodríguez-Concepción M, Rohmer M (2002) Isoprenoid biosynthesis via the methylerythritol phosphate pathway. (E)-4-Hydroxy-3-methylbut-2-enyl diphosphate: chemical synthesis and formation from methylerythritol cyclodiphosphate by a cell-free system from Escherichia coli. Tetrahedron Lett 43:2555–2559

Xiao Y, Savchenko T, Baidoo EE, Chehab WE, Hayden DM, Tolstikov V, Corwin JA, Kliebenstein DJ, Keasling JD, Dehesh K (2012) Retrograde signaling by the plastidial metabolite MEcPP regulates expression of nuclear stress-response genes. Cell 149:1525–1535

Xu Y, Zhan X, Yang H, Lu D (2022) Effects of exogenous brassinolide application at the silking stage on nutrient accumulation, translocation and remobilization of waxy maize under post-silking heat stress. Agriculture 12:572

Yang JY, Sun Y, Sun AQ, Yi SY, Qin J, Li MH, Liu J (2006) The involvement of chloroplast HSP100/ClpB in the acquired thermotolerance in tomato. Plant Mol Biol 62:385–395

Yang D, Li Y, Shi Y, Cui Z, Luo Y, Zheng M, Chen J, Li Y, Yin Y, Wang Z (2016) Exogenous Cytokinins increase grain yield of winter wheat cultivars by improving stay-green characteristics under heat stress. PLoS One 11:e0155437

Yeh YS, Jheng HF, Iwase M, Kim M, Mohri S, Kwon J, Goto T (2018) The mevalonate pathway is indispensable for adipocyte survival. IScience 9:175–191

Yuan L, Zhu S, Shu S, Sun J, Guo S (2015) Regulation of 2,4-epibrassinolide on mineral nutrient uptake and ion distribution in $Ca(NO_3)_2$ stressed cucumber plants. J Plant Physiol 188:29–36

Yuan K, Zhang H, Yu C, Luo N, Yan J, Zheng S, Hu Q, Zhang D, Kou L, Meng X, Jing Y (2023) Low phosphorus promotes NSP1–NSP2 heterodimerization to enhance strigolactone biosynthesis and regulate shoot and root architecture in rice. Mol Plant 16:1811–1831

Zhang M, Fang Y, Ji Y, Jiang Z, Wang L (2013) Effects of salt stress on ion content, antioxidant enzymes and protein profile in different tissues of Broussonetia papyrifera. S Afr J Bot 85:1–9

Zhang X, Ma M, Wu C, Huang S, Danish S (2023) Mitigation of heat stress in wheat (*Triticum aestivum* L.) via regulation of physiological attributes using sodium nitroprusside and gibberellic acid. BMC Plant Biol 23:1–13

# PGPR: Decoding the Secrets of Plant Resilience Under Abiotic Stress

# 6

Gholamreza Abdi, Anilava Chatterjee,
Md Sabir Ahmed Mondol, Ubaida Akbar, Aritra Mukherjee,
and Tadela Susmitha

**Abstract**

Abiotic stress causes plants to create and store ROS in subcellular compartments. Oxidative stress from high ROS levels damages cells, disrupts membranes, and kills them. Cellular activity depends on antioxidant enzymes and detoxification processes to balance ROS generation and removal. By increasing antioxidant defenses and lowering oxidative damage, plant growth-promoting rhizobacteria (PGPRs) strengthen crops under stress. Their regulation of enzymatic and non-enzymatic antioxidants, including catalase, superoxide dismutase, ascorbate peroxidase, glutathione reductase, α-tocopherol, flavonoids, and phenolics, protects plants even under normal circumstances. Recent research on microorganisms in antioxidant defense and plant responses to drought, salt, temperature changes, and heavy metal exposure is discussed in this chapter.

G. Abdi (✉)
Department of Biotechnology, Persian Gulf Research Institute, Persian Gulf University, Bushehr, Iran
e-mail: abdi@pgu.ac.ir

A. Chatterjee · M. S. A. Mondol
Department of Agricultural Biochemistry, Bidhan Chandra Krishi Viswavidyalaya, Mohanpur, Nadia, West Bengal, India

U. Akbar
Department of Food Technology and Nutrition, Lovely Professional University, Phagwara, Punjab, India

A. Mukherjee
Department of Seed Science and Technology, Bidhan Chandra Krishi Viswavidyalaya, Mohanpur, Nadia, West Bengal, India

T. Susmitha
Department of Agricultural Biochemistry, School of Agricultural Sciences, Mallareddy University, Hyderabad, Telangana, India



N. Iqbal et al. (eds.), *Plant Signaling Molecules in Regulation of ROS-Scavenging System*, https://doi.org/10.1007/978-981-95-3377-0_6

**Keywords**

PGPR · ROS · Oxidative stress · Abiotic stress

## 6.1 Introduction

Heavy metal toxicity, heat stress, salt, and drought reduce plant growth and output. Stressors impair metabolic processes, inhibiting growth, yield, and plant death. Salinity affects ion balance and promotes ionic toxicity, whereas drought reduces turgor pressure. Lead, cadmium, and arsenic cause cellular toxicity, whereas temperature changes denatured proteins and disrupt metabolism. ROS communicate plant stress responses but cause oxidative stress when overproduced. Excess ROS damage lipids, proteins, and nucleic acids, threatening cell integrity. An antioxidant defense system of enzymatic (SOD, CAT, APX, GR) and non-enzymatic (ascorbate, glutathione, flavonoids, phenolics) antioxidants helps plants maintain ROS equilibrium and stress tolerance. Plant Growth-Promoting Rhizobacteria (PGPRs) improve plant resilience by facilitating mineral solubilization, hormone production, and stress adaptation. PGPRs enhance stress tolerance by regulating biochemical and physiological processes, activating antioxidant defense mechanisms, and reducing oxidative stress, enabling plants to thrive in harsh environmental conditions.

## 6.2 ROS and Plant Stress Response

### 6.2.1 Generation of ROS in Subcellular Compartments

ROS, oxygen-derived chemicals produced during oxygen metabolism, signal and harm plant cells. ROS result from chloroplast, mitochondria, peroxisome, and plasma membrane metabolism.

Most ROS are produced by chloroplasts during photosynthesis. Disrupted light absorption under drought, excessive light, or high temperatures can cause electron transport chain over-reduction, leading to increased $O_2^-$ production. Superoxide dismutase (SOD) transforms radicals into $H_2O_2$, a signaling or oxidative stress agent.

ROS are also produced by electron leakage from the electron transport chain in mitochondria, especially under stress. Excess ROS in mitochondria destroy DNA, lipids, and proteins, reducing energy generation. Peroxisomes produce ROS by photorespiration, fatty acid β-oxidation, glycolate oxidation, and fatty acid catabolism, resulting in $H_2O_2$ ROS overproduction in peroxisomes overwhelms the antioxidant defense system, causing oxidative damage.

Additionally, plasma membrane NADPH oxidases produce $O_2^-$ by transferring electrons from NADPH to oxygen. While ROS are second messengers in defense and stress signaling, excessive ROS production disturbs cellular homeostasis and causes oxidative damage.

### 6.2.2 Effects of ROS Accumulation on Cellular Components

Critical biological components including lipids, proteins, and nucleic acids are severely damaged when ROS levels rise over the cell's detoxifying capability and cause oxidative stress. One typical consequence of ROS buildup is lipid peroxidation. ROS, especially $OH^-$, target lipids in the membrane and induce chain reactions resulting in disruption of lipid bilayer. This changes membrane fluidity, enhances permeability, and results in leakage of electrolytes. Such changes compromise cell integrity and induce ion imbalances, which compromise cell function and viability. Membrane damage is extremely cytotoxic in chloroplasts and mitochondria, where it stops energy production and metabolic processes. Oxidation of proteins is another ROS accumulation non-specific effect. ROS oxidize amino acid residues of protein structure, stopping enzyme function and structural stability. Oxidative modification causes protein misfolding, aggregation, and degradation, stopping cellular metabolism. Enzymes of photosynthetic, respiratory, and stress transduction pathways are extremely ROS-mediated oxidative modification sensitive, which affects plant growth and development under stress to a large extent. Nucleic acids are also targeted by oxidative damage. ROS, particularly $OH^-$, preferentially cause strand breakage, base modification, and crosslinking of DNA. Such damage disrupts genetic integrity, promoting mutation and stopping replication and transcription processes. ROS-mediated DNA damage can potentially trigger DNA repair processes, but such excess damage has a tendency to overload repair processes, leading to genomic instability and cell death. This is of utmost importance during plant development and stress adaptation, where genomic integrity is a survival and reproductive determinant.

### 6.2.3 Mechanisms of Oxidative Damage and Its Consequences

The overproduction of ROS leads to oxidative damage, activating a cascade of pathological responses within the plant cell, ultimately inhibiting development and stress resistance capacity. Lipid peroxidation destroys the membrane structural integrity, thus impairing the regulation of ion transport and nutrient distribution. Such effects impede cellular osmoregulation, thus inhibiting vital physiological processes, including photosynthesis, respiration, and signal transduction. Furthermore, destabilization of the cell membrane inhibits intercellular communication, thus inhibiting the ability of the plant to cope with environmental stress. Protein oxidation is vital to maintain metabolic activities and stress resistance. Inhibition of vital enzymes involved in protein synthesis, photosynthesis, respiration, and antioxidant mechanisms inhibits energy production and the ability to counteract ROS in the plant. Furthermore, protein oxidation is able to activate cellular dysfunction and decrease the ability of the plant to resist stress. Thus, the plant stimulates proteolytic pathways to dispose of misfolded proteins and recycle amino acids for protein synthesis. ROS-dependent DNA damage activates long-term impacts on gene expression and cell cycle regulation. Oxidative DNA structure modifications impede transcriptional processes, thus leading to the expression of aberrant gene expression patterns. Programmed cell death (PCD) pathways are triggered in response to extreme DNA damage as a protection mechanism for removal of irreversibly damaged cells. Although PCD is essential to prevent the spread of damage, its uncontrolled activation may initiate tissue necrosis and eventually lead to the death of the plant.

## 6.3 Antioxidant Defense Mechanisms in Plants

Plants have complex ROS scavenging mechanisms to combat oxidative stress and preserve homeostasis. They neutralize ROS and reduce oxidative damage with enzymatic and non-enzymatic antioxidants. Enzymatic antioxidants convert ROS into harmless molecules, while non-enzymatic antioxidants stabilize free radicals and inhibit chain reactions.

### 6.3.1 Enzymatic Antioxidants

I. *Catalase (CAT)*—The antioxidant enzyme CAT breaks down $H_2O_2$ to oxygen and water and prevents its accumulation in cells. CAT is found in peroxisomes and prevents extracellular stress-induced oxidative damage.
II. *Superoxide Dismutase (SOD)*—$O_2$- is dismutated into oxygen and $H_2O_2$ by SOD. Many SOD isoforms are found in the cytosol, mitochondria, and chloroplasts. SOD is essential for ROS detoxification and stress tolerance.
III. *Ascorbate Peroxidase (APX)*—APX reduces $H_2O_2$ to water using ascorbate as an electron source. APX is essential to the ascorbate-glutathione cycle and metabolic processes in chloroplasts, mitochondria, and cytosols.
IV. *Glutathione Reductase (GR)*—Glutathione reductase, abbreviated as GR, is a crucial enzyme in the glutathione cycle that catalyses the conversion of oxidised glutathione (GSSG) to reduced glutathione (GSH). GR maintains GSH in a constant level of availability, a mechanism crucial for detoxifying reactive oxygen species and enhancing stress tolerance.

### 6.3.2 Non-enzymatic Antioxidants

I. *Ascorbate (Vitamin C)*—Not only is ascorbate a part of the ascorbate-glutathione cycle, but it is also a very potent free radical scavenger. Not only that, it is an antioxidant enzyme cofactor.
II. *Glutathione (GSH)*—Protecting cells from oxidative damage, GSH is a tripeptide redox buffer. Both the detoxification and signal transduction pathways rely on it.

III. *a-Tocopherol (Vitamin E)*—Tocopherols prevent lipid peroxidation in cell membranes because they are lipid-soluble antioxidants. In situations of stress, they are necessary for the stabilisation of membranes.
IV. *Flavonoids and Phenolic Compounds*—These secondary metabolites shield cells from oxidative stress and fight reactive oxygen species (ROS). Plants use them for defense and stress signaling as well.

## 6.4 The Role of Plant Growth-Promoting Rhizobacteria (PGPRs) in Alleviating Oxidative Stress

Plant growth-promoting rhizobacteria (PGPRs) regulate oxidative stress defenses. They boost antioxidant activity, gene expression, and antioxidant enzyme function to detoxify ROS. PGPRs also increase stress protein and metabolite synthesis to eliminate ROS. Induced systemic resistance (ISR) from PGPRs boosts plant resilience to environmental stresses.

### 6.4.1 PGPR Regulation of the Enzymatic Antioxidant System

Oxidative stress occurs when ROS production exceeds the plant's detoxification capacity, causing cellular damage. PGPRs enhance antioxidative activity by regulating key enzymatic antioxidants, including SOD, CAT, APX, and GR, aiding in ROS detoxification.

#### 6.4.1.1 Superoxide Dismutase (SOD) Supplementation

Superoxide dismutase (SOD) is a crucial enzyme in antioxidant pathways because it oxidises ozone to hydrogen peroxide and oxygen molecules. To protect plant cells from oxidative stress, this biochemical activity is crucial. By increasing SOD activity and regulating oxidative stress more effectively, plant growth-promoting rhizobacteria (PGPRs) are a boon to plant health.

#### 6.4.1.2 Catalase (CAT) Regulation

Catalase (CAT), an important antioxidant enzyme, catalyses the conversion of hydrogen peroxide ($H_2O_2$) into oxygen and water, hence mitigating detrimental peroxide accumulation. Numerous studies indicate that plants infected with PGPR have elevated CAT activity, hence enhancing stress tolerance and reducing oxidative damage.

#### 6.4.1.3 Activation of Ascorbate Peroxidase (APX) and Glutathione Reductase (GR)

The ascorbate-glutathione pathway mitigates $H_2O_2$ to maintain redox balance, with APX and GR playing key roles. GR reduces oxidized glutathione (GSSG) to GSH, aiding oxidative stress management. PGPR enhances plant antioxidant defense by boosting APX and GR activity, leading to increased antioxidant enzyme expression. This reduces oxidative stress, improving plant resilience to drought, salinity, and heavy metal toxicity.

### 6.4.2 PGPRs' Role in Non-enzymatic Antioxidant Accumulation

Besides enzymatic antioxidant control, PGPRs also constitute the non-enzymatic antioxidants group. The molecules act as central control in cells' redox homeostasis and against oxidative stress. The most important non-enzymatic antioxidants regulated by PGPRs include glutathione, ascorbate, phenolic acids, and flavonoids.

#### 6.4.2.1 Biosynthesis of Glutathione and Ascorbate

Ascorbate (vitamin C) and glutathione (GSH) are some of the major antioxidants involved in neutralizing ROS and recycling ROS. PGPR biosynthetic pathways are regulated through the regulation of their biosynthesis enzymes. Improved activity of ascorbate and glutathione leads to increased stress tolerance and improved plant growth in stressful conditions of the environment.

#### 6.4.2.2 Flavonoid and Phenolic Compounds Synthesis

Phenolic acids and flavonoids are high-antioxidant-activity secondary metabolites. Both participate in ROS scavenging, inhibition of lipid peroxidation, and general stress alleviation. PGPRs trigger phenolic acid and flavonoid biosynthesis and hence trigger plant resistance to oxidative stress.

### 6.4.3 Reducing Oxidative Stress in Plants by PGPR

In stressed plants, PGPRs' enzymatic and non-enzymatic antioxidant defense mechanisms minimize oxidative damage. Abiotic stresses including dryness, salt, heavy metals, and severe temperatures increase ROS formation and harm cells. PGPR-inoculated plants detoxify ROS better, reducing oxidative damage and preserving physiological activities.

PGPRs also possess some mechanisms by which they can render the plant resistant to oxidative stress. These are:

I. *Induction of Antioxidant Enzymes*—Additionally, PGPRs increase ROS detoxification by inducing the genes that code for antioxidant enzymes, which in turn biosynthesise well-known antioxidants including SOD, CAT, APX, and GR.
II. *Compatible Solutes Biosynthesis*—Protecting cell structures and proteins from oxidative stress, PGPRs initiate the biosynthetic production of suitable solutes such as proline and trehalose.
III. *Phytohormone Synthesis*—Certain PGPRs produce phytohormones like gibberellins, cytokinins, and indole-3-acetic acid (IAA), which regulate plant defense mechanisms and growth, improving stress tolerance.

IV. *Stress-Responsive Gene Regulation*—Plant growth regulators (PGPRs) modulate the expression of genes that react to stress, enhancing plants' resistance to oxidative stress.
V. *Siderophore and ACC Deaminase Production*—Antioxidant enzyme action requires iron chelation, which siderophores facilitate. The deaminase enzyme activity of 1-aminocyclopropane-1-carboxylate (ACC) decreases ethylene levels, relieving stress-induced growth inhibition.

## 6.5 Influence of Microorganisms on Plants Under Specific Abiotic Stresses

### 6.5.1 Drought Stress

#### 6.5.1.1 Effects on ROS Generation and Antioxidant Response

Dryland conditions significantly hinder crop productivity due to limited soil moisture, with climate variability being a major source of uncertainty. To maintain steady crop yields, it is essential to manage plant stress efficiently and mitigate its adverse consequences. Plants inherently manage cellular stress by regulating the production of reactive oxygen species (ROS) and mitigating their effects through an antioxidant system.

Enzymes like superoxide dismutase generate reactive oxygen species (ROS). Fukai and indicate that the next step in its conversion is the transfer of $O_2-$ into $H_2O_2$. Low $H_2O_2$ levels can operate as secondary messengers, but high levels can cause programmed cell death. Moreover, $H_2O_2$ oxidizes thiol groups, inactivating several enzymes. It is essential for stomatal regulation and xylem differentiation, found that the OH− radical interacts with cellular macromolecules and plays a crucial role in plant development, particularly during vegetative growth, reproduction, and stomatal closure responses to abiotic stress. However, it is extremely reactive and difficult to remove without an enzymatic mechanism, resulting in cell death from excess accumulation (Das and Roychoudhury 2014).

#### 6.5.1.2 Contribution of PGPRs to Drought Tolerance

Numerous studies indicate that drought stress can significantly inhibit plant development and agricultural productivity. An environmentally sustainable answer to this issue is plant growth-promoting rhizobacteria, commonly referred to as PGPR. PGPR enhance drought resistance through many mechanisms (Fig. 6.1).

### 6.5.2 Salinity Stress

#### 6.5.2.1 ROS Dynamics Under Salt Stress

Under salt stress, ROS generation and scavenging imbalances harm lipids, proteins, and DNA (Hossain and Dietz 2016). ROS production occurs mostly in chloroplasts due to stomatal closure limiting $CO_2$ for photosynthesis, leading to electron

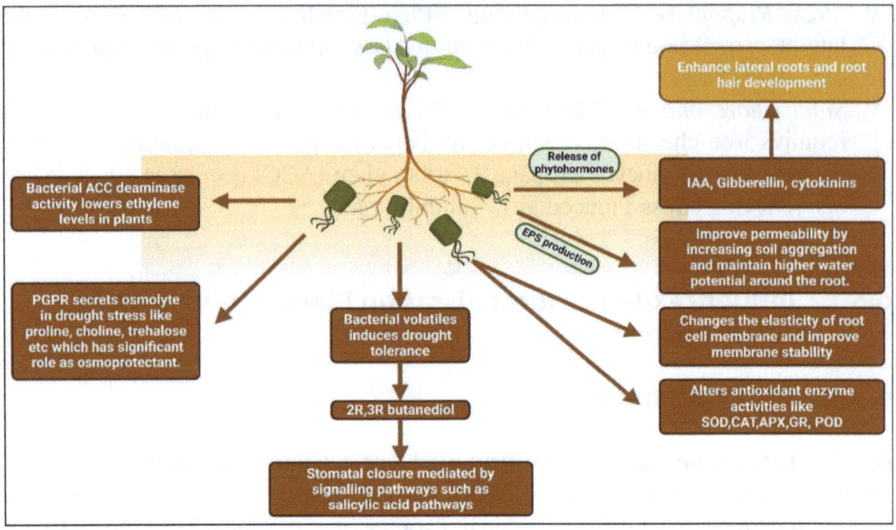


**Fig. 6.1** Contribution of PGPRs to stress tolerance

transport chain over-reduction and creation of $^3O_2$ and $O_2^-$. By converting $O_2^-$ into $H_2O_2$, superoxide dismutase (SOD) can produce hydroxyl radicals ($OH^-$), the most damaging ROS species (Akyol et al. 2020) Salt stress in C3 plants leads to enhanced photorespiration in peroxisomes, a significant source of $H_2O_2$.

Electron leakage in mitochondrial complexes I and III and oxidative protein folding in the endoplasmic reticulum increase ROS generation during salt stress. Plants fight oxidative stress using SOD, CAT, APX, and glutathione peroxidase, and non-enzymatic antioxidants include ascorbate, glutathione, carotenoids, and tocopherols. Antioxidant-rich halophytes survive salt stress better than glycophytes. ROS signals stress reactions. According to, NADPH oxidase (RBOH) in the apoplast converts $O_2^-$ to $H_2O_2$, a crucial signaling molecule, very quickly. Under salt stress, RBOH increases ROS, activates $Ca^{2+}$ signaling pathways, enhances stress responses, and regulates ion homeostasis and osmotic balance.

#### 6.5.2.2 Role of PGPRs in Salinity Stress Mitigation

Crop yields decline when plants experience salt stress, which impedes growth through ionic imbalance, osmotic stress, and oxidative damage. PGPR enhance plant growth and yield in saline conditions through several ways.

1. *Maintenance of Ionic Homeostasis*: Under salinity stress, cells accumulate harmful $Na^+$ and $Cl^-$ ions, disrupting normal functions. Plant growth-promoting rhizobacteria (PGPRs) help regulate ion absorption, transpiration, and translocation to maintain ionic balance. They enhance the expression of HKT and SOS1, key proteins involved in potassium and sodium transport.
2. *Production of Exopolysaccharides (EPS)*: Exopolysaccharides (EPS) produced by PGPR help mitigate salt stress in plants by forming a protective biofilm

around roots, trapping $Na^+$ ions, and reducing ionic toxicity. Inoculation with *Halomonas variabilis* (HT1) and *Planomicrobium chinense* (RT4) enhanced soil aggregation and plant growth in chickpeas under saline conditions through EPS rhizosheath formation. Similarly, EPS-producing *Bacillus* and *Pseudomonas* strains improve soil structure and water retention, promoting plant growth in saline environments.

3. *Enhancement of Antioxidant Defense*: Overproduction of ROS under saline stress harms cells. Plant defense relies on antioxidant enzymes like SOD, CAT, and APX, which PGPR boosts. CAT activity in basil rose with Pseudomonas fluorescens inoculation, while APX and GPX activities increased with Pseudomonas sp., Bacillus lentus, and Azospirillum brasilense. Bacillus amyloliquefaciens SQR9 increased glutathione, peroxidase (POD), and CAT in maize to enhance salt tolerance.
4. *Phytohormone Modulation*: Important phytohormones, including auxins, cytokinins, gibberellins, and abscisic acid (ABA) are produced by PGPR and are required for development and subsequent stress control. One example of an endophyte that can promote root development in cotton plants subjected to salt stress is Pseudomonas putida. Plants can increase their gibberellin levels and thrive in salty environments when these two are added: Two species of Promicromonospora, MJ-1 and Bacillus cereus SE188. Moreover, these rhizobacteria provide a product called ACC deaminase which lowers the effect of ethylene through degradation of its precursor; ACC so that ethylene induced stress response diminishes (Glick et al. 2007). Enterobacter cloacae HSNJ4 has proven to lower levels of ethylene by wheat for greater tolerance to salt.

### 6.5.3 Temperature Stress

#### 6.5.3.1 Impact of High and Low Temperatures on ROS Levels

The extremes in temperature-high or low-greatly affect the levels of ROS in an organism and subsequently all physiological processes associated with them. While ROS have their role as critical signaling molecules during stress responses, they must not be present at a damaging concentration level.

##### Impact of High Temperatures on ROS Levels

- At high temperatures, plants produce more reactive oxygen species (ROS), which play a key role in the signalling pathways that respond to stress. found that this causes cysteine-mediated oxidative protein modifications and shifts in cellular redox states.
- Negative Effects of Excess ROS: The effects of heat on the production of ROS in excess and subsequent impairment to plant growth and stress acclimatization are really telling. But the plants possess ROS-scavenging systems that help in acclimatization to such conditions.

- Impact on Fertility: Male sterility, caused by slowed meiotic progression and delayed pollen formation, results when rice plants are subjected to extremely high temperatures, which disrupt equilibrium ROS. The reason behind this is that there is an excess of reactive oxygen species (ROS) and not enough scavengers to deal with them.

**Impact of Low Temperatures on ROS Levels**

- Restoration of ROS Homeostasis: Exposure to lower temperatures could assist the overcoming of the ROS homeostasis in rice that leads to improved fertility through such action through developing antioxidant systems and reduced ROS.
- Regulation of Pollen Fertility: In rice lines that are sensitive to temperature, low temperatures induce the normal development of anthers and the fertility of their pollen by triggering a normal buildup of ROS, which is necessary for programmed cell death in tapetal cells at the right time.

#### 6.5.3.2 Mechanisms of ROS Regulation

Scavenging systems for ROS are essential to regulate levels of ROS and prevent oxidative stress to promote stress tolerance. Hormonal interactions with ROS also modulate stress responses as transcriptomic and biochemical states toward improved acclimation and survival. Further, antioxidants like ascorbic acid dramatically help maintain an optimal ROS balance, especially under different temperatures.

#### 6.5.3.3 PGPR-Mediated Temperature Stress Tolerance

Plant growth-promoting rhizobacteria are well-known for aiding plant nutrient intake and hormonal balance. By increasing these two variables, plants can grow fully at high temperatures. found that PGPR improves osmotic and ionic balance, increasing plant tolerance to abiotic temperature shocks. PGPR controls plant defense and growth, improving temperature tolerance, through complex signaling cascades that link several phytohormones. emphasize phytohormones and PGPR signals in stress management. PGPR synthesizes antioxidant enzymes, including CAT and POD to defend against high-temperature oxidative damage. link this mechanism to a complex stress signaling network that regulates stress-related gene production.

### 6.5.4 Heavy Metal Stress

#### 6.5.4.1 ROS-Induced Damage from Heavy Metals

Biological systems are adversely affected by the generation of ROS due to the presence of toxins like mercury, lead, chromium, cadmium, and arsenic. Under such oxidative stress, cellular structures undergo oxidative damage and exhibit alterations in cellular activities. Cadmium exposure heightens the production of ROS that leads to oxidative injury and cell death in neurons and cells of the liver. This disturbance primarily disrupts the Nrf2-Keap1 pathway, which is the most important regulatory mechanism in cellular antioxidant responses. The formation of ROS from chromium is directly responsible for inducing oxidative stress, damaging the DNA, and breaking down the cell membranes. It also inhibits photosynthesis as well as other plant metabolic pathways.

#### 6.5.4.2 Mechanisms of ROS-Induced Damage by Heavy Metals

1. *ROS Generation and Oxidative Stress*: Heavy metals produce ROS, which alter the cell's oxidant-antioxidant equilibrium and cause oxidative stress. Oxidative stress may damage lipids, proteins, and nucleic acids and disrupt biological processes, causing cell death. The penetrability of the mitochondrial membrane is impaired by cadmium and chromium, which in turn lead to increased ROS generation and oxidative stress.
2. *Cellular and Molecular Damage*: Owing to the development of oxidative stress, heavy metals disintegrate DNA molecules, enzymes, and lipids. These damages induce apoptosis, necrosis, and different types of cell death. Cadmium generates mitochondrial ROS, which inactivates the XIAP pathway and induces neuronal death.
3. *Impact on Antioxidant Defense Systems*: Heavy metals diminish the cellular antioxidant defenses, aggravating oxidative stress. These inhibit antioxidant enzymes such as SOD and CAT and damage the ability of the cell to scavenge the ROS molecules. Chromium has cytotoxic and genotoxic effects on plants by inhibiting the enzymatic antioxidant system of plants.

#### 6.5.4.3 PGPR Strategies for Heavy Metal Detoxification

1. *Nitrogen Fixation*: There are nitrogen-fixing PGPR like Rhizobium spp. that can convert atmospheric nitrogen to forms accessible to plants. Because of their symbiotic relationship with these bacteria in their root nodules, legumes find this to be very significant. In addition to bolstering structures subjected to heavy metal stress, PGPR increases nitrogen availability, an essential component for plant development.
2. *Phosphorus Solubilization*: PGPR solubilizes soil insoluble phosphorus for plant absorption. It generates stable compounds that immobilize heavy metals and supply the required nutrients for plants. Such organic acids and enzymes anteriorly produced by PGPR help absorb phosphorus and reduce heavy metal bioavailability for plants.
3. *Siderophore Production*: Siderophores from PGPR serve to bind with iron and chelate heavy metals. The metals sequestered by the siderophores are less toxic to tissues of the plants. It thus detoxifies the metals and supplies plants with iron.
4. *Phytohormone Production*: The phytohormones produced by PGPR like auxins (e.g., IAA), cytokinins, and gibberellins are important for the growth of plants. These hormones enhance root development and the ability of the plants to absorb nutrients in times of heavy metal stress. IAA increases the root surface area so as to enhance nutrient absorption and heavy metal sequestration.

**Heavy Metal Transformation and Immobilization** Some PGPR strains reduce heavy metal toxicity by chelating and immobilizing metals in soil, limiting plant uptake. These bacteria produce chelating compounds, exopolysaccharides (EPS), and siderophores to stimulate plant development and stress tolerance.

PGPR improves crop resilience, according to several research. EPS and siderophore production by Bacillus cereus MG257494.1 and Alcaligenes faecalis strains improve Sorghum bicolor growth Myroides odoratinimus TCR22 and Bacillus cereus TCR17 boost plant protein and antioxidant enzymes. Serratia sp. ZTB increases antioxidant activity and solubilizes phosphate and potassium, helping Zea mays flourish.

PGPR also aid in phytoremediation. *Burkholderia contaminans* improves *Glycine max* growth under cadmium stress by increasing nitrogen accumulation and producing ACC deaminase. *Pseudomonas fluorescens* enhances lateral root development in *Sedum alfredii* for cadmium phytoremediation. *Arachis hypogaea* benefits from *Halobacillus* sp. SB2 and *Bacillus* sp. SB1 for nitrogen fixation and phosphate solubilization.

Further, *Ochrobactrum cytisi* and *Bradyrhizobium* sp. 750 improve nitrogen fixation in *Lupinus luteus*, while *Bacillus cereus* and *Bacillus thuringiensis* promote *Broussonetia papyrifera* biomass yield via phosphate solubilization and IAA synthesis. *Bacillus vietnamensis* AB403 and *Kocuria flava* AB402 enhance *Oryza sativa* growth in heavy metal-contaminated soils, and *Rhodobacter sphaeroides* aids *Triticum aestivum* development under Cd and Zn stress.

## 6.6 Applications and Future Perspectives

### 6.6.1 Potential Use of PGPRs in Sustainable Agriculture

Sustainable agriculture relies on soil microbiota to boost plant growth, reduce synthetic fertilizers and pesticides, and increase stress tolerance. Plant development-Promoting Rhizobacteria (PGPR) improve nutrient absorption, nitrogen fixation, and phosphorus solubilization, improving plant health, root development, and disease resistance. Nutritional acquisition is PGPR's main function. These bacteria fix nitrogen and solubilize phosphorus, lowering synthetic fertilizer use. Insoluble soil phosphorus is converted to absorbable by PGPR secreting organic acids and enzymes. Nitrogen-fixing bacteria like Rhizobium, Azospirillum, and Azotobacter transform atmospheric nitrogen into ammonia, reducing the need for expensive nitrogen fertilizers.

PGPR boosts plant resistance to drought, salinity, and harsh temperatures. They increase stress protein and antioxidant synthesis, giving plants systemic resistance. PGPR also aids plant development in tough situations by retaining water and absorbing nutrients. PGPR improve root architecture, allowing plants to acquire water and nutrients from deeper soil layers and respond to environmental stress.

In addition to their capacity for abiotic stress tolerance, plant growth-promoting rhizobacteria (PGPR) have the significant function of alleviating biotic stresses from insect pests and infections by pathogen microorganisms. Empirical evidence depicts that PGPR secrete antimicrobial metabolites, prevent plant pathogen access to nutrients necessary for them to grow, and trigger the growth of systemic defense mechanisms in the plant, and collectively lead to suppression of disease incidence. In particular, species of PGPR that are exemplified by members in the Bacillus and Pseudomonas genera are of unusually efficient biocontrol activities, and consequently, protect crops from damage by soil pathogens such as Phytophthora and Fusarium. By controlling disease incidence, PGPR further support minimization of chemical pesticide use, thus increasing agroecosystem sustainability. In addition to their promotion of plant growth, PGPR further contribute to the promotion of environmental sustainability through their alleviation of dependence on chemical inputs for farming. Overuse of synthetic fertilizers results in acidification of soil, nutrient loss, and water pollution in aquatic ecosystems by leaching and runoff. By the natural increase in the fertility of soil, PGPR enable conservation of soil diversity, structural integrity, and microbial balance. Further, the use of PGPR equates to lower production of greenhouse gas emissions for the production and application of nitrogen fertilizers. Further, their use in agriculture practice conforms to principles of sustainable agriculture, thus enhancing food security without compromising environmental integrity. Overall, PGPR are significant tools for sustainable agriculture through enhanced nutrient acquisition, enhancing stress resistance, promotion of root elongation, and resistance against disease. Their capacity to lower the reliance on synthetic fertilizers and pesticides qualifies them as deserving elements in areas of soil fertility, biodiversity, and groundwater. The application of PGPR technology in modern agriculture allows practitioners to utilize environmentally friendly agricultural practices with smaller ecological footprints, thus maintaining ecosystem long-term productivity and stability (Fig. 6.2).

### 6.6.2 Challenges and Future Directions of Research

Plant growth-promoting rhizobacteria (PGPR) might replace agrochemicals by increasing crop yield and resistance sustainably. PGPR biostimulants have various barriers to broad use. Selecting environmentally stress-tolerant bacterial strains is crucial for PGPR in plant cultivation. The kind of soil, climatic circumstances, and compatibility with certain plant species affect the microbes' efficacy, requiring substantial study to increase their use. Genomic, transcriptomic, and proteomic methods have improved PGPR-host plant interaction knowledge. These findings show how PGPR promotes plant growth and abiotic stress tolerance. By "priming" plants to resist abiotic challenges like temperature variations, drought, and soil salt, scientists may now construct tailored tactics. By activating stress response pathways, these priming tactics increase plant resilience and reduce synthetic input use, boosting sustainable agricultural output.

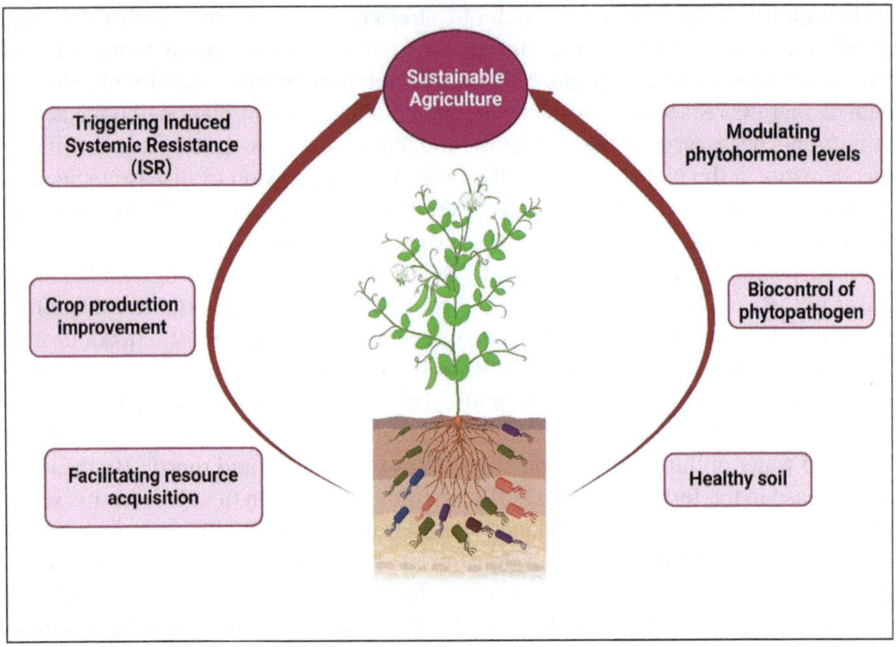


**Fig. 6.2** Potential use of PGPRs in sustainable agriculture

Despite their promise for optimization, plant growth-promoting rhizobacteria (PGPR) treatments must be used cautiously to minimize environmental impacts. Complex interactions of PGPR, plant-microbe, and native soil microbiota require substantial investigation before being used in agriculture models. Demystifying complicated interaction processes is essential along with the launch of eco-friendly and effective microbial products. To determine the long-term impacts of PGPR on soil fertility, biodiversity, and agricultural product, current regulations must include monitoring and regulating frameworks. In addition to scientific research and regulation, farmers and agrifoodstuffs stakeholders must be educated to employ PGPR-based systems. To integrate PGPR into agricultural practice, farmers must be properly taught and given accurate knowledge on its benefits and implementation. Governments, agricultural extension agencies, and research institutes must collaborate to promote PGPR as a sustainable alternative to chemical fertilizers and pesticides through awareness campaigns, practical demonstrations, and incentive programs. In short, PGPR has huge potential to improve sustainable agriculture, but commercialization and implementation are difficult. Omics technology allowed intense PGPR mechanism research, but further studies and field tests are needed. Regulatory management, environmental risk evaluation, and farmer education will drive safe and efficient PGPR use in modern agriculture. Farmers, politicians, and scientists may apply PGPR at scale to protect the environment, improve soil fertility, and secure global food security against climate change.

## 6.7 Conclusion

The use of PGPR has been recognized as an effective method for assisting plants in managing abiotic stress and as a cost-effective and more sustainable alternative to pesticides in agriculture. Regarding climate-smart agriculture, PGPR is at the forefront because of its capacity to alter antioxidant defense mechanisms, reduce oxidative stress, and facilitate plant response to environmental stressors. To assist plants in overcoming abiotic stressors such as drought, salinity, elevated temperatures, and heavy metals, PGPR modulate antioxidant enzymes and non-enzymatic antioxidants to maintain regulated levels of reactive oxygen species (ROS). PGPR offers significant potential to enhance crop yield while mitigating environmental effects in contemporary agricultural systems. Strain selection, field compatibility, and regulation via regulatory mechanisms are the primary constraints to their implementation. Biotechnological and molecular biology advancements provide new directions to investigate the interaction of PGPR and plant defense response with room for targeted use with enhanced effectiveness. For effective utilization of PGPR-based technologies, farmers, policymakers, and scientists will have to join hands. Policy support, research grants, and awareness generation can popularize PGPR in large numbers and thus make agriculture strong and healthy. Future research will have to optimize PGPR formulations, characterize microbial interactions at the molecular level, and formulate new ideas towards their effective application. With consistent innovation and judicious application, PGPR can be a game-changer towards global food security in the context of impending climatic changes (Table 6.1).

**Author Contributions Statement** The authors contributed to this book chapter as follows:

**Gholamreza Abdi**: Led the conceptualization and design of the chapter, provided overall supervision, and spearheaded critical revisions of the intellectual content, and writing—original draft.

**Anilava Chatterjee**: Developed the theoretical framework, contributed to methodological design, and participated in substantive editing and refinement.

**Md Sabir Ahmed Mondol**: Conducted data collection and analysis, contributed to results interpretation, and co-drafted significant portions of the manuscript.

**Ubaida Akbar**: Designed and prepared technical elements, created visual representations of data, and contributed specialized technical content, and writing—original draft.

**Aritra Mukherjee**: Orchestrated chapter organization, synthesized interdisciplinary perspectives, and led the editorial coordination and integration process.

**Tadela Susmitha**: Assisted in analytical validation, contributed to content refinement, and participated in quality assurance of the final manuscript.

**Table 6.1** Representative PGPR strains and consortia enhancing crop resilience under salinity stress through diverse physiological and molecular mechanisms

| PGPR strain/consortium | Target crop | Mechanisms of action | References |
|---|---|---|---|
| *Achromobacter piechaudii ARV8* | Pepper, Tomato | Induces systemic tolerance, controls the level of ethylene, and increases the resistance of the crop | Mayak et al. (2004) |
| *Arthrobacter nitroguajacolicus* | Wheat | Enhances iron absorption, optimal nutrition assimilation, and salt tolerance | Safdarian et al. (2019) |
| *Azospirillum brasilense* | Maize | Inhibits sodium and enhances potassium and calcium uptake; enhances nitrogenase activity to produce enhanced stress resistance | Hamdia et al. (2004) |
| *Azospirillum lipoferum FK1* | Chickpea | Enhances antioxidant mechanisms, optimizes nutrient uptake, and enhances growth performance under salinity | El-Esawi et al. (2019) |
| *Bacillus amyloliquefaciens SQR9* | Maize | Induces the activation of antioxidant enzyme activities and stress-inducible gene expressions for enhanced salt tolerance | |
| *Bacillus aryabhattai M53* | Rice | Solubilizes phosphate, IAA, and siderophores and stimulates growth in saline environments | Sultana et al. (2020) |
| *Bacillus firmus SW5* | Soybean | Triggers antioxidant defense system and root growth in salt stress | El-Esawi et al. (2018) |
| *Bacillus megaterium NRCB001, Bacillus subtilis NRCB002, Bacillus subtilis NRCB003* | Alfalfa | Enhances seed germination, vegetative growth, and suppresses oxidative damage | Zhu et al. (2020) |
| *Bacillus pumilus* | Rice | Represses sodium uptake, increases antioxidant enzyme activity, and increases stress tolerance | |
| *Bacillus subtilis OKB105* | Rice | Induces nutrient metabolism and transport genes that help alleviate salt stress | Xie et al. (2015) |
| *Bacillus subtilis Rhizo SF 48* | Tomato | Mitigates oxidative stress through ACC deaminase activity and thus improves salinity and drought stress tolerance | |
| *Burkholderia phytofirmans PsJN* | Arabidopsis | Regulates ion homeostasis genes (HKT1, SOS1, NHX2) and increases salt tolerance mechanisms | |

(continued)

**Table 6.1** (continued)

| PGPR strain/consortium | Target crop | Mechanisms of action | References |
|---|---|---|---|
| *Chryseobacterium gleum* sp. *SUK* | Wheat | Production of ACC deaminase, promotion of stress tolerance, and promotion of plant health in salinity | Bhise et al. (2017) |
| *Dietzia natronolimnaea* | Wheat | Controls stress-related gene networks, controls ion homeostasis, and enhances ABA signaling | Bharti et al. (2016) |
| *Enterobacter cloacae HSNJ4* | Wheat | Regulates ethylene production via ACC deaminase to increase salt stress tolerance | |
| *Enterobacter* sp. *EJ01* | Tomato, Arabidopsis | Causes salt stress-responsive gene expression (DREB2b, RD29A, RAB18), causes systemic resistance, and proline accumulation | Kim et al. (2014) |
| *Enterobacter* sp. *UPMR18* | Okra | Activates the activity of ROS scavenging enzymes (SOD, CAT, POD) and also inhibits oxidative damage | Habib et al. (2016) |
| *Halomonas variabilis (HT1), Planomicrobium chinense (RT4)* | Chickpea | Production of exopolysaccharide (EPS), increases soil aggregation, increases water holding capacity, and restricts sodium intake | |
| *Kosakonia sacchari MSK1* | Mung Bean | Sustains ion uptake, improves antioxidant defense, and increases yield under salinity | Shahid et al. (2021) |
| *Microbacterium oleivorans KNUC7074, Brevibacterium iodinum KNUC7183, Rhizobium massiliae KNUC7586* | Pepper | Reduces salt stress by facilitating enhanced nutrient transportation and growth | Hahm et al. (2017) |
| *Microbacterium oleivorans KNUC7074, Brevibacterium iodinum KNUC7183, Rhizobium massiliae KNUC7586* | Pepper | Reduces salt stress by facilitating enhanced nutrient transportation and growth | Hahm et al. (2017) |
| *Pseudomonas aeruginosa T15, Pseudomonas fluorescens NT1, Pseudomonas stutzeri C4* | Tomato | Reduces ethylene-induced stress, induces root and shoot growth, and salt tolerance | Tank and Saraf (2010) |
| *Pseudomonas argentinensis, Pseudomonas azotoformans* | Mustard | Controls ethylene metabolism, promotes plant growth, and enhances salt tolerance | Phour and Sindhu (2020) |

(continued)

**Table 6.1** (continued)

| PGPR strain/consortium | Target crop | Mechanisms of action | References |
|---|---|---|---|
| *Pseudomonas fluorescens, Pseudomonas migulae* | Tomato | Inhibits ethylene production by ACC deaminase activity, improves salt stress tolerance, and improves growth of the plant | |
| *Pseudomonas mendocina* | Lettuce | Facilitates nutrient uptake and stress alleviation through ACC deaminase activity | Alguacil et al. (2009) |
| *Pseudomonas putida* | Cotton | Enhances the uptake of magnesium, potassium, and calcium and inhibits sodium uptake | |
| *Pseudomonas putida UVA* | Canola | Possesses ACC deaminase production, regulation of ethylene production, and improving nutrient uptake to facilitate stress alleviation. | Cheng et al. (2012) |
| *Pseudomonas simiae AU* | Soybean | Enhances chlorophyll content, phosphorus solubilization, IAA and siderophore production, and sodium uptake inhibition | |
| *Pseudomonas syringae, Pseudomonas fluorescens, Enterobacter aerogenes* | Maize | Increases the chlorophyll level, promotes growth, and increases yield in salinity | |
| *Rhizobium leguminosarum + Azotobacter chroococcum + Melatonin* | Faba bean | Increases photosynthetic pigments, decreases Na+/K+ imbalance, and stress resistance | Abd El-Ghany and Attia (2020) |
| *Rhizobium, Pseudomonas* | Maize | Decreases electrolyte leakage, enhances proline accumulation, and selectively accumulates potassium to render it more tolerant to stress | Bano and Fatima (2009) |
| *Stenotrophomonas maltophilia SBP-9* | Wheat | Triggers the plant defense for biotic and abiotic stress with salt tolerance induction | Singh and Jha (2017) |
| *Streptomyces* sp. | Tomato | Regulates the growth of the plant by releasing IAA, ACC deaminase, and phosphate solubilization | Palaniyandi et al. (2014) |

## References

Abd El-Ghany MF, Attia M (2020) Effect of exopolysaccharide-producing bacteria and melatonin on faba bean production in saline and non-saline soil. Agronomy 10(3):316

Akyol TY, Yilmaz O, UZİLDAY B, Uzilday RÖ, Türkan I (2020) Plant response to salinity: an analysis of ROS formation, signaling, and antioxidant defense. Turk J Bot 44(1):1–13

Alguacil MDM, Kohler J, Caravaca F, Roldán A (2009) Differential effects of Pseudomonas mendocina and Glomus intraradices on lettuce plants physiological response and aquaporin

PIP2 gene expression under elevated atmospheric CO2 and drought. Microbial ecology 58(4):942–951
Bano A, Fatima M (2009) Salt tolerance in Zea mays (L). following inoculation with Rhizobium and Pseudomonas. Biology and fertility of soils, 45(4):405–413
Bharti N, Pandey SS, Barnawal D, Patel VK, Kalra A (2016) Plant growth promoting rhizobacteria Dietzia natronolimnaea modulates the expression of stress responsive genes providing protection of wheat from salinity stress. Scientific reports 6(1):34768
Bhise KK, Bhagwat PK, Dandge PB (2017) Synergistic effect of Chryseobacterium gleum sp. SUK with ACC deaminase activity in alleviation of salt stress and plant growth promotion in Triticum aestivum L. 3 Biotech 7(2):105
Cheng Z, Woody OZ, McConkey BJ, Glick BR (2012) Combined effects of the plant growth promoting bacterium Pseudomonas putida UW4 and salinity stress on the Brassica napus proteome. Applied Soil Ecology 61:255–263
Das K, Roychoudhury A (2014) Reactive oxygen species (ROS) and response of antioxidants as ROS scavengers during environmental stress in plants. Front Environ Sci 2:53
El-Esawi MA, Alaraidh IA, Alsahli AA, Alamri SA, Ali HM, Alayafi AA (2018) Bacillus firmus (SW5) augments salt tolerance in soybean (Glycine max L.) by modulating root system architecture, antioxidant defense systems and stress-responsive genes expression. Plant Physiology and Biochemistry 132:375–384
El-Esawi MA, Al-Ghamdi AA, Ali HM, Alayafi AA (2019) Azospirillum lipoferum FK1 confers improved salt tolerance in chickpea (Cicer arietinum L.) by modulating osmolytes, antioxidant machinery and stress-related genes expression. Environ Exp Bot 159:55–65
Glick BR, Todorovic B, Czarny J, Cheng Z, Duan J, McConkey B (2007) Promotion of plant growth by bacterial ACC deaminase. CritAl Rev Plant Sci 26(5-6):227–242
Habib SH, Kausar H, Saud HM (2016) Plant growth-promoting rhizobacteria enhance salinity stress tolerance in okra through ROS-scavenging enzymes. BioMed Research International 2016(1):6284547
Hahm MS, Son JS, Hwang YJ, Kwon DK, Ghim SY (2017) Alleviation of salt stress in pepper (Capsicum annum L.) plants by plant growth-promoting rhizobacteria. Journal of microbiology and biotechnology 27(10):1790–1797
Hamdia MAES, Shaddad MAK, Doaa MM (2004) Mechanisms of salt tolerance and interactive effects of Azospirillum brasilense inoculation on maize cultivars grown under salt stress conditions. Plant Growth Regul 44(2):165–174
Hossain MS, Dietz KJ (2016) Tuning of redox regulatory mechanisms, reactive oxygen species and redox homeostasis under salinity stress. Front Plant Sci 7:548
Kim K, Jang YJ, Lee SM, Oh BT, Chae JC, Lee KJ (2014) Alleviation of salt stress by Enterobacter sp. EJ01 in tomato and Arabidopsis is accompanied by up-regulation of conserved salinity responsive factors in plants. Molecules and cells 37(2):109–117
Mayak S, Tirosh T, Glick BR (2004) Plant growth-promoting bacteria confer resistance in tomato plants to salt stress. Plant Physiol Biochem 42(6):565–572
Palaniyandi SA, Damodharan K, Yang SH, Suh JW (2014) Streptomyces sp. strain PGPA39 alleviates salt stress and promotes growth of 'Micro Tom'tomato plants. Journal of applied microbiology 117(3):766–773
Phour M, Sindhu SS (2020) Amelioration of salinity stress and growth stimulation of mustard (Brassica juncea L.) by salt-tolerant Pseudomonas species. Applied Soil Ecology 149:103518
Safdarian M, Askari H, Shariati JV, Nematzadeh G (2019) Transcriptional responses of wheat roots inoculated with Arthrobacter nitroguajacolicus to salt stress. Sci Rep 9(1):1792
Shahid M, Ameen F, Maheshwari HS, Ahmed B, AlNadhari S, Khan MS (2021) Colonization of Vigna radiata by a halotolerant bacterium Kosakonia sacchari improves the ionic balance, stressor metabolites, antioxidant status and yield under NaCl stress. Applied Soil Ecology 158:103809
Singh RP, Jha PN (2017) The PGPR Stenotrophomonas maltophilia SBP-9 augments resistance against biotic and abiotic stress in wheat plants. Frontiers in microbiology 8:1945

Sultana S, Paul SC, Parveen S, Alam S, Rahman N, Jannat B, Karim MM (2020) Isolation and identification of salt-tolerant plant-growth-promoting rhizobacteria and their application for rice cultivation under salt stress. Can J Microbiol 66(2):144–160

Tank N, Saraf M (2010) Salinity-resistant plant growth promoting rhizobacteria ameliorates sodium chloride stress on tomato plants. Journal of Plant Interactions 5(1):51–58

Xie S, Wu H, Chen L, Zang H, Xie Y, Gao X (2015) Transcriptome profiling of Bacillus subtilis OKB105 in response to rice seedlings. BMC microbiology 15(1):21

Zhu Z, Zhang H, Leng J, Niu H, Chen X, Liu D, Ying H (2020) Isolation and characterization of plant growth-promoting rhizobacteria and their effects on the growth of Medicago sativa L. under salinity conditions. Antonie Van Leeuwenhoek 113(9):1263–1278

# Proteomics Studies to Investigate Signaling Processes for ROS Reduction in Fruits and Vegetables

7

Giulia Franzoni and Antonio Ferrante

**Abstract**

Fruit vegetables, such as tomato, pepper, and melon, are essential components of the human diet, providing a wide array of micronutrients and bioactive compounds. However, their productivity and quality are highly susceptible to abiotic stresses, which trigger a cascade of physiological and molecular changes, including the overproduction of reactive oxygen species (ROS). While ROS can lead to cellular damage at excessive levels, they also play a pivotal role as signaling mediators in adaptive responses at controlled levels. Proteomics has emerged as a powerful approach to identify stress-responsive proteins and elucidate the molecular networks key to stress adaptation in fruit vegetables. This review focuses on the application of proteomics to investigate ROS signaling and its modulation in response to abiotic stresses during fruit development and ripening. Proteomic studies have revealed the involvement of antioxidant enzymes, ASC-GSH cycle components, and carbonylated proteins in maintaining redox homeostasis during climacteric (e.g., tomato) and non-climacteric (e.g., pepper) fruit ripening. Additionally, proteomic analyses have shed light on the mechanisms of action of compounds such as melatonin and benzothiadiazole (BTH) in delaying fruit senescence and enhancing stress tolerance. Future research should focus on integrating proteomics with other omics approaches to gain a more comprehensive understanding of ROS signaling networks in fruit vegetables under abiotic stress conditions.

G. Franzoni
Department of Agricultural and Environmental Science, University of Milan, Milan, Italy

A. Ferrante (✉)
Institute of Crop Science, Sant'Anna School of Advanced Studies of Pisa, Pisa, Italy
e-mail: antonio.ferrante@santannapisa.it

 
N. Iqbal et al. (eds.), *Plant Signaling Molecules in Regulation of ROS-Scavenging System*, https://doi.org/10.1007/978-981-95-3377-0_7

**Keywocrds**

Fruit vegetables · Abiotic stress · Proteomics · Antioxidant enzymes · ASC-GSH cycle · Ripening

## 7.1 Introduction

Fruit and vegetables constitute a fundamental component of the human diet, providing a diverse array of micronutrients, including essential minerals and vitamins, as well as a wide spectrum of secondary metabolites such as phenolics, flavonoids, and carotenoids. These bioactive compounds play key roles in maintaining physiological functions, preventing chronic diseases, and supporting overall nutritional status and human health. Despite representing less than 1% of global plant species, vegetables exhibit extraordinary genetic, anatomical, and morphological diversity as a taxonomic and agronomic group (Radovich 2018). Several classification systems have been used to group vegetables, and one of the most widely adopted, especially by growers, is based on the edible part of a plant. According to this criterion, which is informed with specific cultural techniques; vegetables are grouped into classes such as leafy, root, tuber, and fruit vegetables (Tsao and Lo 2003). The latter group includes those species in which the edible part is the botanical fruit. This category includes a wide range of botanical families and crop species, such as tomato, pepper, eggplant, zucchini, watermelon, melon, and cucumber. These are among the most important horticultural crops worldwide due to their high economic value and significant contribution to human nutrition.

### 7.1.1 ROS Production and Signaling Under Abiotic Stress in Fruit

The productivity and quality of fruit vegetable crops are highly exposed to environmental challenges, also due to their particularly high sensitivity to *abiotic stress* compared to other crop types. Indeed, organs such as flowers and developing fruits are among the most stress-sensitive tissues in plants, and even short-term exposure to abiotic stress conditions during critical developmental stages such as flowering or fruit set can result in yield or quality losses (Albertos et al. 2019; Kumari et al. 2024; Zinn et al. 2010). Moreover, vegetables crops have high water and nutrient requirements, especially when cultivated in open fields, and have a relatively narrow optimal range for temperature and soil salinity, which limits their adaptability under variable environmental conditions. At the same time, the adaptive mechanisms activated by plants in response to abiotic stress impose a significant energetic cost, often leading to trade-offs that negatively affect growth and productivity (Munns and Millar 2023; Heinemann and Hildebrandt 2021).

While abiotic stresses manifest at the macroscopic level through reductions in yield and quality, they also trigger a cascade of physiological and molecular changes within the plant. These changes include the modulation of numerous genes and metabolic pathways, many of which remain only partially understood. With the growing impact of climate change, gaining comprehension into how crops cope with abiotic stress has become a key focus in plant biology and agriculture. The overproduction of *reactive oxygen species (ROS)* is a common response in plants exposed to various abiotic stresses, such as drought, salinity, excessive light, extreme temperatures, and nutrient deficiencies. Although excessive ROS accumulation can lead to cellular damage, at controlled/physiological levels these molecules play a pivotal role as signaling mediators in the activation of adaptive responses. Their role in signal transduction implies the existence of tightly coordinated regulatory networks that maintain ROS at nontoxic levels. This requires a delicate balance between ROS production and the activity of ROS-scavenging pathways (You and Chan 2015). While ROS have been extensively studied in model plants such as *Arabidopsis*, particularly in leaf tissues under various stress conditions, the roles of *ROS in fruit* development and physiology remain comparatively underexplored, representing a notable gap in the current literature. ROS are continuously produced by oxygen metabolism in almost all plant cell compartments, especially from the electron transport chains of photosynthesis in different organelles (Decros et al. 2019). Mitochondria represent the main source of energy and of ROS in non-photosynthetic tissues, such as fruit. Different organelles also possess antioxidant systems involved in the regulation of ROS accumulation and determining their concentration and lifetime (Meitha et al. 2020). Among them, antioxidant enzymes as superoxide dismutase (SOD, EC 1.15.1.1), catalase (CAT; EC 1.11.1.6), and GPX are also present in fruit, where they regulate ROS levels under stressful conditions and are involved in fruit ripening and development (Hu et al. 2011). Proteomic analyses on *bell pepper* fruit revealed that *chilling injuries* during postharvest mainly affect the redox homeostasis and carbohydrate metabolism. In particular, the protein connected with ASC-GSH cycle (Fig. 7.1) is altered, and CAT is downregulated. Moreover, Enolase and GAPDH have been identified as potential key regulators in the onset and progression of chilling injury.

### 7.1.2 Proteomic Approaches to Study ROS Dynamics in Fruit

*Proteins* are major targets of ROS, accounting for up to 68% of oxidized cellular molecules. *Proteomics* has emerged as a powerful approach for identifying stress-responsive proteins with differential expression patterns. Traditional proteomic analysis of fruit tissues has relied on two-dimensional gel electrophoresis (2DE), which involves protein separation, spot excision, enzymatic digestion, and mass spectrometry (MS) identification. However, 2DE is limited in throughput and compatibility with advanced MS techniques. As a result, gel-free approaches have gained importance, offering higher sensitivity and scalability. These include MudPIT, iTRAQ, SILAC, ICAT, LOPIT, and 2D-DIGE, which combine protease digestion, liquid chromatography (LC), and MS-based quantification. Among these, 2D-DIGE enables comparative analysis of samples labelled with distinct

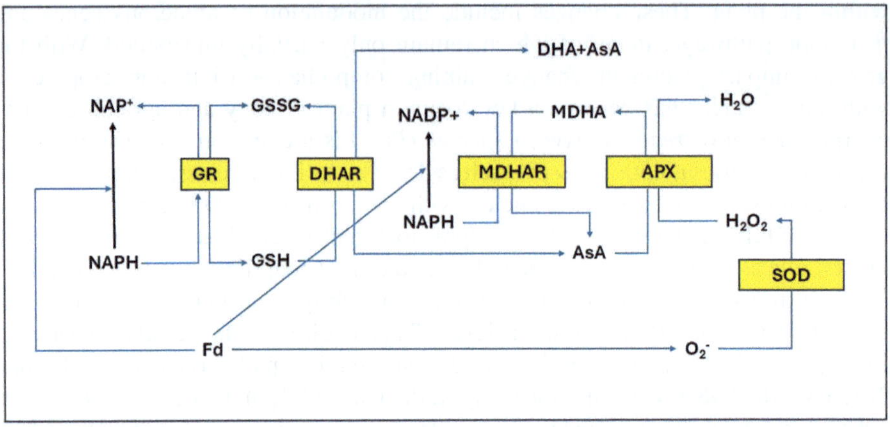


**Fig. 7.1** Ascorbate–glutathione cycle. The SOD—superoxide dismutase converts the superoxide anion in hydrogen peroxide and the ascorbic acid (AsA) into monodehydroascorbate (MDHA). The APX—ascorbate peroxidase converts the hydrogen peroxide in water. The MDHA is reconverted into AsA by the monodehydroascorbate reductase (MDHAR). AsA can also be derived from dehydroascorbate (DHA) by action of dehydroascorbate reductase (DHAR) that also converts the reduced glutathione (GSH) to oxidized glutathione (GSSG). The glutathione reductase (GR), using the NADPH converts the GSSG to GSH, forming NADP+ with the involvement of ferrodoxin (Fd)

fluorescent dyes. These gel-based and gel-free techniques are complementary, and their integration enhances proteomic resolution and coverage in fruit research (Chan 2013). Redox proteomics, a branch of functional proteomics, has been applied by integrating 1DE or 2DE with MS to specifically identify and analyse carbonylated proteins (Bykova and Rampitsch 2013). The identification and quantification of these proteins represent a useful strategy to better understand the mitochondrial redox status in fruit, especially during *ripening*. Proteomic analysis of *non-model fruits* has always been challenging due to the lack of optimized protocols for such species, the inherently low protein abundance in fruit tissues, and the difficulty in distinguishing proteins from different cellular compartments. Recently, iQTRAP and bioinformatic tools have facilitated the proteome analyses in non-model fruits also at the subcellular level. Even though the ripening process is particularly investigated, a proteomic approach based on matrix-assisted laser desorption/ionisation-time of flight and peptide mass fingerprinting analysis has been used on tomato fruit for the identification of proteins involved in *blossom-end rot (BER)*, a physiopathy causing disorganisation, cell breakage, and darkening of the tissues. An increased accumulation of oxidative stress-associated proteins (ASC-GSH cycle and the pentose phosphate pathway) was observed in BER-affected fruits (Casado-Vela et al. 2005). Table 7.1 shows the proteins and enzymes involved in ROS detoxification.

**Table 7.1** Protein or enzymes that are associated with ROS accumulation and their role

| Protein/enzyme | Function | Role in ROS regulation | Fruits |
|---|---|---|---|
| Superoxide Dismutase (SOD) | Converts superoxide radicals to hydrogen peroxide | First line of defense against ROS accumulation | Apple, banana, grape, tomato, mango, kiwi, blueberry |
| Catalase (CAT) | Breaks down hydrogen peroxide into water and oxygen | Detoxifies $H_2O_2$, preventing oxidative damage | Apple, citrus, banana, mango, tomato, strawberry |
| Ascorbate Peroxidase (APX) | Uses ascorbate to reduce hydrogen peroxide | Key enzyme in the ascorbate-glutathione cycle | Tomato, grape, kiwi, mango, strawberry, blueberry |
| Glutathione Reductase (GR) | Regenerates reduced glutathione (GSH) from oxidized form | Maintains redox balance and supports ROS scavenging | Apple, grape, citrus, mango, kiwi |
| Peroxiredoxins (Prx) | Reduce peroxides using thioredoxin | Involved in fine-tuning ROS signaling | Grape, citrus, mango |
| Alternative Oxidase (AOX) | Bypasses mitochondrial electron transport chain | Reduces ROS production in mitochondria | Banana, tomato, apple |
| Respiratory Burst Oxidase Homologs (RBOH) | NADPH oxidases producing ROS as signaling molecules | Involved in defense signaling and development | Tomato, grape, strawberry, apple |
| Metallothioneins (MTs) | Metal-binding proteins with antioxidant properties | Scavenge ROS and regulate metal homeostasis | Apple, grape, mango |
| Glutaredoxins (GRXs) | Redox enzymes using glutathione as a cofactor | Regulate redox-sensitive proteins and ROS signaling | Tomato, citrus, kiwi |
| Thioredoxins (TRXs) | Maintain protein thiol redox state | Modulate ROS signaling and antioxidant defense | Apple, grape, mango, kiwi |

### 7.1.3 Fruit Ripening and ROS in Climacteric and Non-climacteric Fruit

The ripening of climacteric fruits involves oxidative metabolism and alteration of redox homeostasis, leading to the production of ROS such as superoxide ($O_2^-$), hydrogen peroxide ($H_2O_2$), and hydroxyl radical ($O\dot{H}$). Tomato (*Solanum lycopersicum* L.) and melon (*Cucumis melo* L.) are among the most representative *climacteric fruit vegetables* exhibiting this oxidative ripening behaviour. Understanding how ROS shapes plant responses to stress requires a holistic view, something proteomics is well-suited to provide (Momo et al. 2022a). By examining the global protein changes and interactions, it reveals the molecular networks key to stress adaptation. Several studies have reported a strong association between ROS levels, particularly hydrogen peroxide ($H_2O_2$), and tomato fruit ripening and senescence.

Specifically, an experiment with *rin* mutant showed that low ROS concentrations are associated with delayed senescence, whereas elevated ROS levels accelerate this process (Kumar et al. 2016; Cai et al. 2018). A critical moment in tomato is the breaker stage, when chloroplasts transform into chromoplasts, the photosynthetic apparatus is disassembled, and there is a huge transformation in fruit structure (Faurobert et al. 2007). Carbonylation alters protein content, structure, and function, and is widely used as a marker of oxidative stress and protein damage. Proteomic profiling during ripening revealed that a significant proportion of differentially expressed proteins is associated with stress response pathways and senescence, providing further evidence that fruit ripening is tightly linked to oxidative processes (Palma et al. 2011). A proteomic analysis of carbonylated proteins, together with the study of the ASC–GSH cycle components, was carried out in mitochondria from *tomato* fruit at two ripening stages. The accumulation of $H_2O_2$ in mitochondria was similar, regardless of the ripening stage, probably due to the activity of antioxidant enzymes and transporters. Moreover, Mn-SOD and APX activity did not change in tomato fruit from beaker to light red stage. In contrast, enzymes involved in the ASC–GSH cycle, such as MDHAR, DHAR, and GR, showed a significant increase during tomato fruit ripening, indicating the essential role of this enzymatic pathway in maintaining mitochondrial redox homeostasis (Fig. 7.1). Additionally, a significant increase in GSH and GSSH levels in mitochondria was observed during tomato fruit ripening. The analysis of posttranslational modification showed an increase in carbonylated proteins of 55% in tomato fruit from the breaker to the light red stage, suggesting that oxidative stress caused by ROS may modulate the ripening process. (López-Vidal et al. 2016). An iodoacetyl tandem mass tag (iodoTMT)-based redox proteomic approach was used to identify redox-sensitive proteins in tomato fruit ripening. The authors suggested that ROS may regulate fruit ripening by targeting different critical enzymes, including polygalacturonase 2A (PG2A) and 1-aminocy clopropane-1-carboxylate oxidase-like protein (E8), are associated with cell wall degradation and ethylene biosynthesis (Wang et al. 2021). A proteomic approach was used to investigate the proteome of two *tomatoes ecotypes*. The results showed that, although most proteins overlapped between the two ecotypes, specific proteins could still be distinguished during fruit maturation. For instance, the peptide corresponding to methionine sulphoxide reductase, also known as the fruit-ripening protein E4 and involved in key defence mechanisms against ROS, was found to be induced in the regional ecotype during maturation (Rocco et al. 2006). Although *pepper* belongs to the same botanical family as tomato, it is classified as a *non-climacteric fruit* and therefore undergoes a ripening process different from that of tomato. Moreover, proteomic studies on the *Capsicum* genus are relatively recent, beginning in the late 1990s with the identification of enzymes involved in capsaicinoid biosynthesis. Since then, progress in elucidating the proteomic basis of trait development and stress responses in *Capsicum* has remained limited (Camejo et al. 2015; Momo et al. 2022b). However, a recent study focused on the proteomic identification of mitochondrial carbonylated proteins in two maturation stages of pepper fruits. The experiment reported a distinct pattern compared to that observed in tomato, characterised by a significant decrease in carbonylated proteins, an increase in APX activity, accompanied by a decrease in MDHAR, DHAR, and GR activities throughout maturation. The most abundant categories of protein found in *sweet pepper* during *ripening* include secondary and antioxidant metabolisms. At the same time, the analyses of mitochondrial proteins showed an abundance of proteins involved in oxidation-reduction events. Mitochondria supply cellular energy via the TCA cycle and oxidative phosphorylation; energy deficits can impair membrane function and elevate ROS production, which, in combination with mitochondrial nitric oxide NO, may form reactive species like peroxynitrite that alter protein function and affect fruit quality. Moreover, a central role in the modulation of cytochrome c oxidase (COX) by ROS and reactive nitrogen species (RNS) in *pepper* fruit during *ripening* was reported (González-Gordo et al. 2022). While the role of NO in fruit ripening is recognized, the underlying metabolic networks involving NO and other RNS remain to be fully elucidated. RNS are known to influence cellular signaling by inducing posttranslational protein modifications. The analyses of RNS profile in sweet pepper revealed increased levels of S-nitrosothiols and protein tyrosine nitration in *ripening* fruit. Nitrated proteins were associated with key functions, including photosynthesis, NADPH generation, proteolysis, amino acid biosynthesis, and oxidative metabolism. Additionally, reduced CAT levels in red fruit suggest diminished $H_2O_2$ scavenging capacity, potentially enhancing lipid peroxidation during late ripening stages (Sánchez-Bel et al. 2012; Chaki et al. 2015). Antioxidant metabolites, naturally present in fruits at ripening stage-dependent levels, help counteract ROS-induced damage. A targeted quantitative proteomic approach (MRM) was used to investigate multiple proteins and isomers simultaneously, and provides direct quantitative evidence of the dynamic changes in these enzymes during strawberry fruit ripening. The results showed significant changes in antioxidant enzyme levels, including increased abundance of AKR, SOD, and GST, and decreased levels of CAT, APX, and peroxiredoxins, indicating activation of a specific redox system centred on glutathione and superoxide dismutase metabolism (Song et al. 2020).

### 7.1.4 Effects of Exogenous Treatments on ROS in Fruit

Proteomic analysis has proven useful in elucidating the *mechanisms of action* of certain compounds on fruit ripening. Exogenous application of *melatonin* has been observed to exert a positive effect in delaying fruit senescence by modulating redox balance. This is achieved through its ability to enhance the activity of antioxidant enzymes, which in turn reduces the accumulation of ROS during fruit ripening (Manzoor et al. 2023). Proteomic analysis proved essential in revealing that melatonin influences a wide range of proteins involved in tomato ripening, including those related to carbohydrate and fatty acid biosynthesis, flavonoid metabolism, and cell wall modification. Moreover, a differential quantitative proteomics label-free technique was carried out to investigate the effect of melatonin in tomato fruit during postharvest, and results showed numerous changes in the expression levels of proteins involved in senescence and the antioxidant system, proving the role of melatonin as a powerful free radical scavenger (Sun et al. 2016). *Arsenic (As) treatment* of tomato plants, with subsequent As accumulation in fruits, provided a model to study antioxidant metabolism under combined endogenous (ripening) and exogenous (As) oxidative stress. Proteomic analysis of tomato fruit treated with As and As+Si revealed a broad reprogramming of stress-related proteins, including those linked to abiotic stress (e.g., HSPs, calmodulins, growth-regulating factors, and ubiquitination proteins), as well as proteins specifically involved in oxidative stress responses, such as sulfotransferases, enzymes of lipid metabolism, and those involved in secondary metabolite biosynthesis (Marmiroli et al. 2017). Proteomic analyses have enabled the identification of proteins associated with dehydration stress, likely reflecting responses to environmental fluctuations during fruit development (Faurobert et al. 2007).

Another example is provided by the proteomic study of muskmelon fruit in response to *BTH* application. BTH-induced resistance is mediated through the activation of ROS metabolism, stimulation of the phenylpropanoid pathway, and accumulation of pathogenesis-related proteins in infected fruit. Notably, BTH treatment led to a marked induction of proteins associated with oxidative stress responses and the ubiquitin proteasome system, suggesting their key roles in regulating ripening and senescence in melon. Additionally, BTH was shown to delay the depletion of ascorbic acid (AsA), further supporting its positive effect on the fruit's antioxidant defence system (Li et al. 2015).

The proteins that are associated with ROS are mainly associated with the detoxification action. These proteins have the function to reduce the ROS accumulation or maintain them under specific thresholds that do not compromise the cell functionality. Low ROS concentrations can have a positive effect on increasing the secondary metabolism and increasing the antioxidant compounds of fruit.

## 7.2 Conclusion

The role of proteins in the regulation of the ROS concentration has been studied, but there are many aspects that should be further investigated. Most of the studies have been focused on protein accumulation under stress or senescence. It will be interesting to study the functional analysis using specific protein inhibitors to understand the mode of action of the single enzyme or their simultaneous action.

## References

Albertos P, Wagner K, Poppenberger B (2019) Cold stress signalling in female reproductive tissues. Plant Cell Environ 42(3):846–853

Bykova NV, Rampitsch C (2013) Modulating protein function through reversible oxidation: redox-mediated processes in plants revealed through proteomics. Proteomics 13(3–4):579–596

Cai J, Wang P, Tian S, Qin G (2018) Quantitative proteomic analysis reveals the involvement of mitochondrial proteins in tomato fruit ripening. Postharvest Biol Technol 145:213–221

Camejo D, Jiménez A, Palma JM, Sevilla F (2015) Proteomic identification of mitochondrial carbonylated proteins in two maturation stages of pepper fruits. Proteomics 15(15):2634–2642

Casado-Vela J, Sellés S, Bru Martínez R (2005) Proteomic approach to blossom-end rot in tomato fruits (Lycopersicon esculentum M.): antioxidant enzymes and the pentose phosphate pathway. Proteomics 5(10):2488–2496

Chaki M, Alvarez de Morales P, Ruiz C, Begara-Morales JC, Barroso JB, Corpas FJ, Palma JM (2015) Ripening of pepper (Capsicum annuum) fruit is characterized by an enhancement of protein tyrosine nitration. Ann Bot 116(4):637–647

Chan Z (2013) Proteomic responses of fruits to environmental stresses. Front Plant Sci 3:311

Decros G, Baldet P, Beauvoit B, Stevens R, Flandin A, Colombié S et al (2019) Get the balance right: ROS homeostasis and redox signalling in fruit. Front Plant Sci 10:1091

Faurobert M, Mihr C, Bertin N, Pawlowski T, Negroni L, Sommerer N, Causse M (2007) Major proteome variations associated with cherry tomato pericarp development and ripening. Plant Physiol 143(3):1327–1346

González-Gordo S, Rodríguez-Ruiz M, Paradela A, Ramos-Fernández A, Corpas FJ, Palma JM (2022) Mitochondrial protein expression during sweet pepper (Capsicum annuum L.) fruit ripening: iTRAQ-based proteomic analysis and role of cytochrome c oxidase. J Plant Physiol 274:153734

Heinemann B, Hildebrandt TM (2021) The role of amino acid metabolism in signaling and metabolic adaptation to stress-induced energy deficiency in plants. J Exp Bot 72(13):4634–4645

Hu H, Liu Y, Shi GL, Liu YP, Wu RJ, Yang AZ et al (2011) Proteomic analysis of peach endocarp and mesocarp during early fruit development. Physiol Plant 142(4):390–406

Kumar V, Irfan M, Ghosh S, Chakraborty N, Chakraborty S, Datta A (2016) Fruit ripening mutants reveal cell metabolism and redox state during ripening. Protoplasma 253:581–594

Kumari S, Basu S, Kumar G (2024) A systematic review on the implications of concurrent heat and drought stress in modulating floral development in plants. Plant Sci 349:112248

Li X, Bi Y, Wang J, Dong B, Li H, Gong D et al (2015) BTH treatment caused physiological, biochemical and proteomic changes of muskmelon (Cucumis melo L.) fruit during ripening. J Proteomics 120:179–193

López-Vidal O, Camejo D, Rivera-Cabrera F, Konigsberg M, Villa-Hernández JM, Mendoza-Espinoza JA et al (2016) Mitochondrial ascorbate–glutathione cycle and proteomic analysis of carbonylated proteins during tomato (Solanum lycopersicum) fruit ripening. Food Chem 194:1064–1072

Manzoor MA, Xu Y, Xu J, Wang Y, Sun W, Liu X et al (2023) Melatonin: a multi-functional regulator of fruit crop development and abiotic stress response. Sci Hortic 321:112282

Marmiroli M, Mussi F, Imperiale D, Lencioni G, Marmiroli N (2017) Abiotic stress response to As and As+ Si, composite reprogramming of fruit metabolites in tomato cultivars. Front Plant Sci 8:2201

Meitha K, Pramesti Y, Suhandono S (2020) Reactive oxygen species and antioxidants in postharvest vegetables and fruits. Int J Food Sci 2020(1):8817778

Momo J, Rawoof A, Kumar A, Islam K, Ahmad I, Ramchiary N (2022a) Proteomics of reproductive development, fruit ripening, and stress responses in tomato. J Agric Food Chem 71(1):65–95

Momo J, Kumar A, Islam K, Ahmad I, Rawoof A, Ramchiary N (2022b) A comprehensive update on Capsicum proteomics: advances and future prospects. J Proteome 261:104578

Munns R, Millar AH (2023) Seven plant capacities to adapt to abiotic stress. J Exp Bot 74(15):4308–4323

Palma JM, Corpas FJ, del Río LA (2011) Proteomics as an approach to the understanding of the molecular physiology of fruit development and ripening. J Proteome 74(8):1230–1243

Radovich TJ (2018) Biology and classification of vegetables. In: Handbook of vegetables and vegetable processing, pp 1–23

Rocco M, d'Ambrosio C, Arena S, Faurobert M, Scaloni A, Marra M (2006) Proteomic analysis of tomato fruits from two ecotypes during ripening. Proteomics 6(13):3781–3791

Sánchez-Bel P, Egea I, Sánchez-Ballesta MT, Martinez-Madrid C, Fernandez-Garcia N, Romojaro F et al (2012) Understanding the mechanisms of chilling injury in bell pepper fruits using the proteomic approach. J Proteome 75(17):5463
Song J, CampbellPalmer L, Vinqvist-Tymchuk M, Fillmore S, Forney C, Luo H, Zhang Z (2020) Proteomic changes in antioxidant system in strawberry during ripening. Front Plant Sci 11:594156
Sun Q, Zhang N, Wang J, Cao Y, Li X, Zhang H et al (2016) A label-free differential proteomics analysis reveals the effect of melatonin on promoting fruit ripening and anthocyanin accumulation upon postharvest in tomato. J Pineal Res 61(2):138–153
Tsao SJJ, Lo HF (2003) I. Definition of a vegetable. In: Handbook of vegetable preservation and processing, p 1
Wang P, Li X, Wang Y, Wang W, Tian S, Qin G (2021) Redox proteomic analysis reveals the involvement of oxidative post-translational modification in tomato fruit ripening. Postharvest Biol Technol 178:111556
You J, Chan Z (2015) ROS regulation during abiotic stress responses in crop plants. Front Plant Sci 6:1092
Zinn KE, Tunc-Ozdemir M, Harper JF (2010) Temperature stress and plant sexual reproduction: uncovering the weakest links. J Exp Bot 61(7):1959–1968

# Studying Underlying Mechanisms of PGPR in Reducing Abiotic Stress-Induced Oxidative Stress

8

Nitika Kapoor, Manu Sharma, Amandeep Singh, Nahida Rehman Mir, Jaspreet Kour, and Renu Bhardwaj

**Abstract**

Abiotic stresses, including drought, salinity, extreme temperatures, and heavy metal contamination, significantly impair plant growth and productivity. One of the major consequences of abiotic stress is excessive production of ROS, which leads to oxidative stress. Oxidative stress damages various cellular components, such as lipids, proteins, and nucleic acids, thereby affecting plant morphology, physiology, and overall plant growth. Plants have evolved enzymatic and non-enzymatic antioxidant defense mechanisms to counteract oxidative stress. Microbes, especially plant growth-promoting rhizobacteria (PGPR), play a pivotal role in mitigating abiotic stress in plants through various processes and underlying mechanisms that include biological nitrogen fixation, phosphate solubilization, phytohormone production, and the synthesis of stress-related proteins and other metabolites. Besides, PGPR enhances plant tolerance to abiotic stress by maintaining ion homeostasis, enhancing water uptake and retention, and activating stress-responsive genes and detoxification mechanisms. They also reduce oxidative damage/stress by modulating the ROS scavenging system, including the activities of antioxidants and antioxidant enzymes. The present chapter explores the underlying mechanisms of PGPR mediated alleviation abiotic stress-induced oxidative stress, highlighting their role in enhancing plant resilience. The potential applications in sustainable agriculture are also discussed, emphasizing their importance in developing crops resistance to stress to ensure food security in the face of increasing environmental challenges.

N. Kapoor · M. Sharma · A. Singh · N. R. Mir · J. Kour
Department of Botanical and Environmental Sciences, Guru Nanak Dev University, Amritsar, Punjab, India

R. Bhardwaj (✉)
University Centre for Research and Development, Chandigarh University, Mohali, Punjab, India



N. Iqbal et al. (eds.), *Plant Signaling Molecules in Regulation of ROS-Scavenging System*, https://doi.org/10.1007/978-981-95-3377-0_8

**Keywords**

Abiotic stress · Defense · Plant growth promoting rhizobacteria · Rhizosphere

## 8.1 Introduction

Stress in plants can be understood as any unfavorable condition or substance that disrupts their normal physiological functioning (Álvarez and Acosta-Motos 2023). This stress can arise from both natural and anthropogenic factors, significantly impacting plant growth and productivity, which in turn poses challenges to global food security. Notably, it is a leading cause of crop loss around the globe, often reducing average yields by over 50%. Factors contributing to this stress include extreme temperature variability, high light intensity, including UV light, osmotic stresses, water stress, and heavy metal stress (Ciarmiello et al. 2011). The negative effects of these stressors are primarily linked to the substantial production of reactive oxygen species (ROS). An excess of ROS leads to oxidative stress in plants, characterized by a disturbance in the critical balance between the generation of ROS and their elimination. This imbalance can result in oxidative damage to crucial cellular components, ultimately impairing plant health and productivity (Bartosz 1997). ROS are highly reactive molecules that can interact with various cellular components, potentially resulting in severe metabolic dysfunction and cell death. They inflict damage on critical cellular structures, including lipids, proteins, and DNA (Qamer et al. 2021; Nadarajah 2020). Additionally, oxidative stress induced by ROS modifies signaling pathways and gene expression with the disruption of key cellular processes, leading to diminished metabolic functions, changes in biochemical composition, morphological alterations in plants, and overall agricultural productivity (Khozeeva et al. 2020; Zhang and Lu 2024).

To survive such damage caused by oxidative stress, plants have evolved antioxidative defense systems, including enzymatic and nonenzymatic processes, to maintain ROS homeostasis and safeguard against oxidative damage (Sharma et al. 2019; Nadarajah 2020). Also, various strategies have been proposed, including the breeding of stress-tolerant hybridized crops, optimized cultural practices, resource management, and the use of ameliorators and/or amendments like plant growth promoters, biochar, compost, nanoparticles, and PGPRs to encourage plant growth (Kaur et al. 2024; Khozeeva et al. 2020). These approaches aim to enhance crop resilience and maintain sustainable agriculture in the midst of growing environmental problems that pose a threat to agricultural production in the current climate change context. PGPR offers a sustainable solution to mitigate abiotic stresses in agriculture. PGPR comes forward as an economical, less time-consuming and promising approach for enhancing plant performance and stress tolerance. These beneficial microorganisms colonize the rhizosphere and enhance plant growth through various mechanisms. PGPR induces phytohormone production, nutrient acquisition improvement, production of antioxidant enzymes and osmolytes, modulating physiological responses, and improves soil properties. Using PGPR as bioinoculants presents an eco-friendly alternative to synthetic agrochemicals, promoting sustainable agriculture and food security (Sylia et al. 2021; Chandran et al. 2021).

## 8.2 Abiotic Stress-Mediated Oxidative Stress in Plants

The generation of ROS is a major consequence of environmental stresses (biotic/abiotic) on plants (Ravi et al. 2023; Wang et al. 2024). Mostly abiotic stress, including drought, heavy metal, salinity, thermal, and pesticide stress etc., is responsible for the excessive production of ROS (Mishra et al. 2023; Thiruvengadam et al. 2024). For instance, drought stress can lead to reduced water availability and impaired photosynthesis, resulting in the production of excessive ROS (Ahluwalia et al. 2021; Kumar et al. 2024a, b). Metal toxicity, such as cadmium (Cd), copper (Cu) mercury (Hg), lead (Pb), nickel (Ni), aluminum (Al), chromium (Cr), arsenic (As), etc., in the soil, can disrupt cellular processes, leading to the overproduction of ROS (Goyal et al. 2020; Bhardwaj et al. 2024). Similarly, salinity stress, caused by high levels of salt in the soil, can impair water uptake and cause oxidative burst in plants, resulting in the generation of ROS and further, thermal stress, arising from extreme temperatures, can also increase ROS production by disrupting cellular membranes and leading to oxidative damage (Kumar et al. 2020; Khan et al. 2024a, b). Pesticide stress, especially those that have oxidative properties or inhibit antioxidant defense mechanisms, can induce ROS production in plants (Liu et al. 2021).

ROS are highly reactive molecules, including superoxide radicals ($O_2^{\bullet -}$), hydrogen peroxide ($H_2O_2$), hydroxyl radicals ($HO^-$), peroxide ion ($O_2^{\bullet 2-}$), perhydroxyl radical ($HO^{\bullet}{}_2$), alkoxyl radical ($RO^{\bullet}$), peroxyl radical ($ROO^{\bullet}$), Singlet oxygen ($^1O_2$), and hypochlorous acid (HOCL) (Villalpando-Rodriguez and Gibson 2021; Singh et al. 2024a, b). ROS are produced as byproducts of various metabolic processes in plants, including photosynthesis, respiration, and cell wall metabolism (Hasanuzzaman et al. 2020). Under normal conditions, plants can cope with low levels of ROS through their antioxidant defense system. However, under unfavorable abiotic stress conditions, the balance between ROS production and the antioxidant defense system is disrupted, leading to an accumulation of ROS in plant cells, damaging cellular structures, such as lipids, proteins, and DNA, leading to oxidative stress (Sachdev et al. 2021; Iqbal et al. 2024). A multitude of studies observed that abiotic stress causes excessive production of ROS, which leads to oxidative stress. According to Khan et al. (2024a, b), rice plants cultivated under salinity and drought stress had reduced water balance, relative water content, and increased electrolyte leakage that impacted shoot and root growth and overall physiology of plants. Likewise, Cd toxicity increased hydrogen peroxide ($H_2O_2$) and malondialdehyde (MDA) content, lipid peroxidation, and accumulation of Cd in plant tissue and reduced chlorophyll content, overall growth, and development of rice plant (Iqbal et al. 2024). In *Beta vulgaris L.* plant, Pb and Cd toxicity enhanced metal accumulation and oxidative stress by suppressing morphological and biochemical attributes (Badawy et al. 2024). Likewise, in another research conducted on rice plants, the phytotoxic effect of Hg increased oxidative damage, lipid peroxidation in leaves, and Hg accumulation in leaves (Huang et al. 2024). Cd toxicity declined leaf exchange parameters like photosynthetic rate, stomatal conductance, transpiration rate, and intercellular $CO_2$ concentration by increasing oxidative damage in *Lonicera japonica* plant (Liu et al. 2023). In another study conducted on cotton plants cultivated under excessive amounts of salt have shown decreased plant height, chlorophyll content, and biochemical activities (Choudhary et al. 2024). Nawaz et al. (2024) also reported that Cd stress leads to decreased pigment concentration of chlorophyll a, chlorophyll b, chlorophyll a/b ratio, and carotenoid content in wheat plants. Another research conducted on *Brassica napus L.* plant grown under salt stress conditions has shown decreased shoot and root length, and chlorophyll a and b content by increased $Na^+$ accumulation, leading to oxidative damage in canola plants (Ilyas et al. 2024). The cumulative effect of oxidative damage caused by ROS accumulation under abiotic stress conditions suppresses plant growth and development. Various studies related to the effect of abiotic stress on plants are summarized in Table 8.1. Stress factors can lead to stunted growth, reduced biomass accumulation, impaired root development, inhibition of flower and seed formation, and even plant death under severe stress conditions (Ali et al. 2024; Asad et al. 2024). Furthermore, the impaired growth and development caused by abiotic stress-induced oxidative damage can have negative consequences on crop productivity and yield (Fig. 8.1).

## 8.3 Defense Mechanisms of Plants Against Oxidative Stress

Plants undergo various morphological and physiological adaptations to cope with oxidative stress. For instance, under drought stress, plants may close their stomata to reduce water loss and limit oxidative damage. They may also develop specialized structures, such as trichomes or cuticles, which act as physical barriers against stressors. Plants adapt biochemically by detoxifying harmful substances, creating new proteins, and adjusting hormone levels to survive these stresses (Kumar et al. 2024a, b). Plants have also developed enzymatic and non-enzymatic antioxidant defense systems to counteract ROS (Wani et al. 2021; Yaqoob et al. 2022). Under stress, there is an increase in activities of antioxidative enzymes such as superoxide dismutase (SOD), catalase (CAT), peroxidase (POD), ascorbate peroxidase (APOX), Glutathione reductase (GR), monodehydroascorbate reductase (MDHAR), dehydroascorbate reductase (DHAR), glutathione peroxidase (GPOX), etc. to neutralize the overproduced ROS (Kopecká et al. 2023; Khan et al. 2024a, b). They also produce non-enzymatic antioxidants like ascorbic acid and glutathione to scavenge the excessive ROS and maintain cellular homeostasis. The production of antioxidant enzymes and nonenzymatic antioxidants helps plants mitigate oxidative stress and minimize damage to cellular components. Additionally, plants can activate stress signaling pathways in response to ROS. These pathways lead to changes in gene expression and the synthesis of stress-responsive proteins, which play important roles in stress tolerance (Feki et al. 2021). For example, certain stress-responsive

**Table 8.1** Effect of abiotic stress-induced excessive ROS/oxidative stress on plants

| Stress | Plant species | Effect of stress on plant | Reference |
|---|---|---|---|
| As | *Zea mays* | Reduced shoot length and root length under As stress | Du et al. (2017) |
| High temperature (35/25 °C) | *Picea crassifolia and P. wilsonii* | Lowered photosynthetic rate in response to high temperature stress | Zhang et al. (2018) |
| Chlorpyrifos (Pesticide) | *Brassica juncea* L. | Lowered Germination potential, vigor index and root and shoot length | Bakshi et al. (2021) |
| Salinity | *Portulaca oleracea L.* | Reduced stomatal conductance, water potential, and increased $CO_2$ assimilation, proline and MDA content | Hnilickova et al. (2021) |
| High temperature (42 °C) | *Solanum lycopersicum* L. | Reduced pigment content, gas exchanged parameters, and inhibition of chlorophyll fluorescence | Jahan et al. (2021) |
| UV-B radiations | *Pinus nigra* Arnold. | Decrease in germination speed and germination percentage, root and shoot dry weight, root collar diameter and rooting percentage | Ozel et al. (2021) |
| Salinity | *Oryza sativa L.* | Dry weight, relative water content, pigment content, gas exchange parameters, starch and Sucrose content reduced under salt conditions | Yan et al. (2021) |
| High temperature (32 °C) | *Lens culinaris* Medikus L. | Lowered grain yield, pods per plant, protein, Fe, and Zn content | Choukri et al. (2022) |
| Salinity | *Lycopersicon esculentum Mill.* | Reduced pigment content, seedling vigour, plant height, biomass accumulation, osmotic potential, and water potential | Raziq et al. (2022) |
| Cd | *Hordeum vulgare* | Increased MDA, SOD, APX, CAT and POD content | He et al. (2023) |
| Cd | *Lonicera japonica* | Declined in leaf exchange parameters like photosynthetic rate, stomatal conductance, transpiration rate and intercellular $CO_2$ concentration | Liu et al. (2023) |
| As | *Oryza sativa L.* | Reduced lateral root growth | Piacentini et al. (2023) |
| Pb and Cd | *Beta vulgaris L.* | Enhanced metal accumulation and oxidative stress by suppressed morphological, and biochemical attributes | Badawy et al. (2024) |
| Salinity | *Gossypium hirusutam* | Decreased plant height, chlorophyll content, biochemical activities | Chaudhary et al. (2024) |
| Cd, cu, Pb, and Ni | *Capsicum annuum L.* | Increased MDA, $H_2O_2$ content and electrolyte leakage under heavy metal stress | El-Saadony et al. (2024) |

(continued)

**Table 8.1** (continued)

| Stress | Plant species | Effect of stress on plant | Reference |
|---|---|---|---|
| Hg | Bamboo | Reduced plant growth and biomass, root dry weight, shoot length, and plant tolerance as compared to the control | Emamverdian et al. (2024) |
| Hg | *Oryza sativa L.* | Increased oxidative damage, lipid peroxidation in leaves and Hg accumulation in leaves | Huang et al. (2024) |
| Cd | *Cajanus cajan* | Reduced moisture content, fresh weight, dry weight, lateral roots, root length, and shoot length | Kaushik et al. (2024) |
| As | *Vigna radiata L.* | Increased SOD, POD, CAT, and APX content | Kumar et al. (2024a, b) |
| Cd | *Triticum aestivum* | Cd stress leads to decreased pigment concentration, chlorophyll a, chlorophyll b, chlorophyll a/b ratio, and carotenoid content. | Nawaz et al. (2024) |
| UV-B radiations | *Sorghum bicolor* | Lowered plant height, hemicellulose content in *bmr* and *dm* mutants under UV-B stress | Olson and Ruhland (2024) |
| High temperature (35 °C, 40 °C and 45 °C) | *Picea pungens, Picea abies, and Picea omorika* | Parched and chlorotic needles, chlorosis, and lowered relative water content | Ouyang et al. (2024) |
| Al | *Lens culinaris Medikus* | Increased accumulation of Al in the roots, low root tolerance index and malate secretion under Al stress was observed | Singh et al. (2024a, b) |
| Al | *Vigna radiata L.* | Upregulation of *VrMATE19, VrMATE30,* and *VrALMT13* genes under the Al stress provides tolerance to plants against stress | Singh et al. (2024a, b) |
| Cd | *Matthiola incana* | Reduction in shoot and root dry weight | Zulfiqar et al. (2024) |

proteins act as ROS scavengers, while others are involved in repairing damaged biomolecules or maintaining cellular integrity (Kour et al. 2021). Antioxidant genes like *APX* and *SOD* are responsible for ROS balancing, and their expression was found to be upregulated when plants were given salinity stress (Ali et al. 2022). Expression of ionic homeostasis genes *SOS1* (plasma membrane $Na^+/H^+$ transporter) and *NHX1* (tonoplast $Na^+/H^+$ antiporter), and osmosis-related proteins, *AQP* (Aquaporins), and *OSM-34* (Osmotin) were also enhanced under salt stress, which maintained ionic and osmotic balance under stress conditions (Alnusairi et al. 2021). Various studies indicated that enzymatic and nonenzymatic antioxidant machinery play a crucial role in the alleviation of toxic effects of environmental stress in plants and promote growth and development of plants, further enhancing crop productivity. Their activity is further enhanced by the addition of several plant

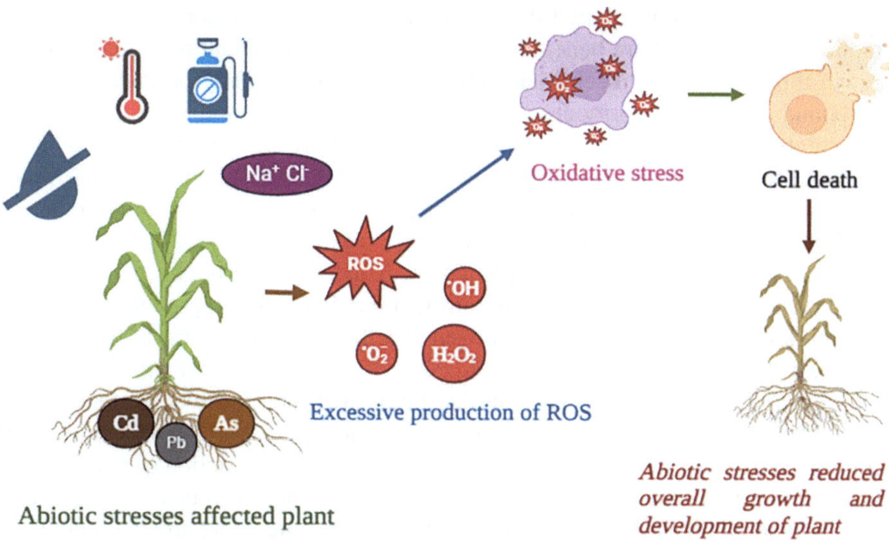


**Fig. 8.1** It is represented that abiotic stress is accountable for excessive ROS production, which leads to oxidative damage and promotes cell death, resulting in reduced overall growth and development of plants

growth promoters (PGRs). According to Choudhary et al. (2024), salicylic acids (SAs) application in fenugreek seedlings enhanced SOD, POD, CAT, and GR antioxidant enzyme content upon heat stress as compared to stressed plant. Likewise, in another study conducted on barley plants cultivated under excessive salt, SA application enhanced Chl a, b, carotenoids content and antioxidant enzyme activities as compared to control plants (Hanif et al. 2024). Application of synthetic strigolactones (GR 24) in sunflower plants reduced salinity effect by decreasing $H_2O_2$ and MDA content and by increasing antioxidant activities as compared to the control plant (Zulfiqar et al. 2024). Overall, plants have evolved a complex network of antioxidant defense systems, regulatory mechanisms, and adaptations to counteract oxidative stress and survive under adverse conditions. Understanding these mechanisms can help in developing crop varieties with enhanced stress tolerance and improve agricultural productivity in challenging environments.

## 8.4 PGPR and Their Role in Mitigating Abiotic Stress in Plants

Various environmental factors like temperature, salt, drought, and heavy metals affect crop plants worldwide, leading them to develop defense mechanisms to cope with these challenges (Kumar and Ohri 2023; Gull et al. 2019). Stress causes changes in plant structure and function, affecting their cells and leading to issues like ion imbalance and protein damage (Ali and Ohri 2023; Samynathan et al. 2021).

As an eco-friendly strategy, PGPR has also made a notable impact in reducing these abiotic stresses. PGPR are beneficial bacteria that colonize plant roots and enhance growth through various mechanisms (Hasan et al. 2024). They produce bioactive compounds like phytohormones, antimicrobial substances and siderophores that help plants to absorb nutrients, suppress pathogen activity, degrade toxic contaminants and serve as bioremediation agents for cleaning contaminated soil and water (Fig. 8.2). Root exudates from plants also influence the microbial community in the rhizosphere, promoting beneficial interactions (Ranjan et al. 2024; Kumar et al. 2024a, b). Rhizobacteria enhances plant growth and health through various mechanisms such as nutrient solubilization, nitrogen fixation, and the production of phytohormones, siderophores, and pathogen-repelling biomolecules (Ahemad and Kibret 2014).

To alleviate plant stress, PGPR utilize both direct and indirect strategies. The direct mechanisms include complex actions like improved mineral uptake, increased nutrient availability, enhanced water absorption, and the production of exopolysaccharides, biofilms, and various organic solutes such as sugars, organic acids, amino acids, and polyamines (Kazerooni et al. 2021). On the other hand, the indirect methods by which PGPR mitigate plant stress involve processes like chemotaxis, phytohormone synthesis and modulation, changes in phytohormone levels, activation of antioxidant defense mechanisms, ACC (1-Aminocyclopropane-1-carboxylate) deaminase activity, and the regulation of stress-responsive genes in plants. These combined activities influence a variety of biological processes in plants, promoting both local and systemic tolerance to abiotic stress induced by rhizobacteria (Rosier

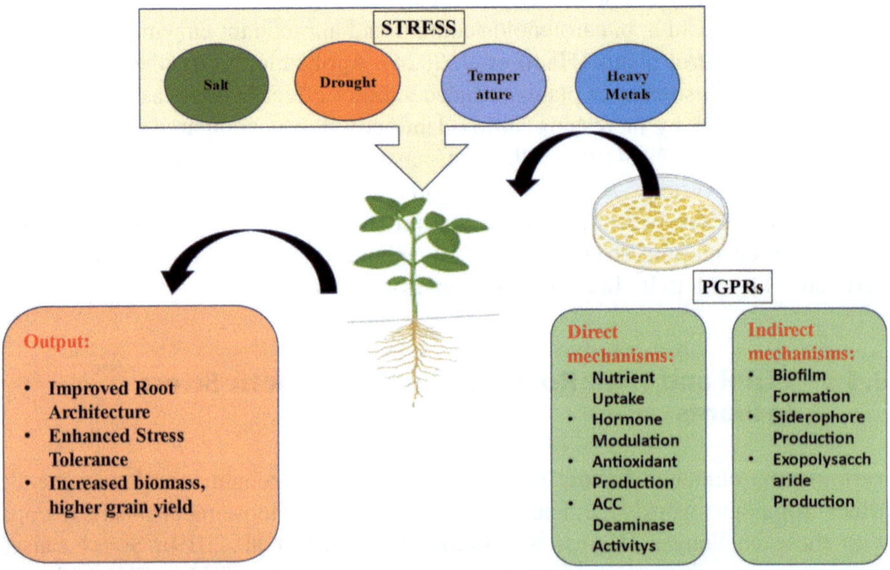


**Fig. 8.2** Illustration of PGPRs role in ameliorating stress and improving the growth and development in stressed plants via various direct and indirect mechanisms

et al. 2018). In their study, Ali and Khan (2021) highlighted the significant role of PGPR in alleviating drought stress. They found that these beneficial microbes can alter root structure, regulate genes associated with stress responses, and produce a range of substances such as phytohormones, osmolytes, siderophores, volatile organic compounds, and exopolysaccharides. Additionally, they enhance ACC deaminase activity, which collectively helps plants cope with drought conditions (Ahmad et al. 2022). For instance, a study was conducted to understand the drought tolerance potential of PGPRs (*PGPR-FS2 and PGPR-VHH4*) isolated from a hot spring in wheat. The study found that these bacteria upregulated wheat genes related to drought and growth, especially *PGPR-VHH4*, which improved root structure to handle dry conditions more effectively (Ali et al. 2024). Similarly, Ferioun et al. (2024) evaluate the effectiveness of two PGPR-based consortia in aiding barley cultivars, both drought-sensitive and tolerant, to recover from drought stress. Inoculation with these consortia improves drought tolerance by enhancing relative water content, SPAD index, and Fv/Fm, while reducing electrolyte leakage. Biochemically, the consortia increased proline, soluble sugars, $H_2O_2$, and MDA levels, and boosted CAT and APOX activities. Agro-morphologically, they contributed to higher root and shoot dry weights and better grain yield. The most significant recovery was observed in the tolerant cultivar treated with Consortium 1, showing comparable traits to unstressed plants.

Another research focuses on two new strains of PGPR, *Bacillus subtilis*-FAB1 and *Pseudomonas azotoformans*-FAP3, which form biofilms and help plants grow better, especially during drought conditions. The study showed that these strains can survive drought and improve wheat plant health by enhancing root growth and soil quality, which helps the plants cope with stress (Ansari et al. 2023). Like drought, PGPRs can also help plants grow well in heat, heavy metal, and salt-stressed conditions. As highlighted by Lu et al. (2024), their study identifies a novel PGPR strain, *Streptomyces lasalocidi* JCM 3373T, which improves salt tolerance in soybeans. Among 36 *Streptomycete* strains tested, *S. lasalocidi* JCM 3373T was found to alleviate salt stress without colonizing soybean roots. The strain secretes indole-3-carboxaldehyde, which helps modulate soybean root architecture by regulating stress-responsive (*GmVSP, GmPHD2*, and *GmWRKY54*) and root growth-related (*GmPIN1a, GmPIN2a, GmYUCCA5*, and *GmYUCCA6*) genes. Likewise, Ansari et al. (2024) used a novel halotolerant PGPR strain, *Pantoea agglomerans* FAP10, to protect wheat from salinity stress by forming biofilms and enhancing plant growth. FAP10 improves salinity tolerance by regulating photosynthesis, antioxidant systems, sulfur metabolism, and sustaining soil health. In a study conducted on bacterial strains isolated from rhizospheric samples in arid regions of Tunisia, *Bacillus halotolerans* Gb67, *Bacillus subtilis* All3, and *Bacillus mojavensis* Gb7 were found to significantly enhance salinity tolerance in both tomato and *Arabidopsis thaliana*. Among the 24 isolates tested, Gb67 showed the most notable improvements in tomato growth parameters, including fresh weight, plant height, root length, and chlorophyll content. These strains exhibited various beneficial traits such as nitrogen fixation, phosphate solubilization, and ACC deaminase production, highlighting their potential as PGPR for use in saline agricultural systems (Abdelkefi et al. 2024).

El-Saadony et al. (2024) carried out research leveraging PGPR to mitigate heavy metal toxicity in pepper plants. Their study identified four heavy metal-tolerant PGPR strains, *Pseudomonas azotoformans*, *Serratia rubidaea*, *Paenibacillus pabuli*, and *Bacillus velezensis,* which, particularly in combination, improved growth and antioxidant defense in heavy metal-stressed pepper plants and reduced heavy metal accumulation. Similarly, Malik et al. (2024) isolated Cr resistant bacteria *Bacillus cereus* NM8 and *Bacillus subtilis* NM28 to investigate their effect on *Capsicum annum* (L.). The study showed that the strains enhanced shoot and root growth, dry weight, relative water content, and other growth metrics, while reducing Cr uptake. Morphological and biochemical analyses identified the strains as *Bacillus cereus* NM8 and *Bacillus subtilis* NM28, with NM28 being more effective in promoting growth under Cr stress. Likewise, in the study conducted by Zhao et al. (2024), three PGPRs (P-solubilizing bacteria G1, Si-solubilizing bacteria YS1, and K-solubilizing bacteria HL18) were used to remediate the soil contaminated with Cd, Pb, and Zn (Zinc). The study found that inter-root inoculation of PGPR improved soil quality and ryegrass growth, increasing soil redox potential, enzyme activities, and nutrient content. Simultaneous use of three PGPR strains significantly enhanced the ryegrass's ability to extract Cd, Pb, and Zn, while reducing their soil concentrations. The research carried out by Oubohssaine et al. (2024) assessed *Rhodococcus qingshengii* strain LMR356's potential under Pb stress. Inoculation with LMR356 improved *Sulla* spinosissima growth and Pb tolerance, enhancing biomass, antioxidant activity, and pigment content. The strain also promoted plant growth in non-contaminated sand and polluted soil, making it a promising biofertilizer for soil restoration and stabilization in polluted environments. Majhi et al. (2024) found that *Micrococcus yunnanensis* GKSM13, a Cu-tolerant PGPR, improved rice seedling growth under Cu stress by producing growth factors, reducing Cu uptake, and alleviating oxidative stress through antioxidant enzyme activation. The study conducted by El-Ballat et al. (2023) demonstrated that *Azospirillum brasilense* EMCC1454 effectively enhances chickpea growth and Cr stress tolerance. This PGPR strain, which can tolerate up to 260 μM Cr, promotes plant growth by increasing antioxidant activity, osmolyte production, and nutrient acquisition, while reducing oxidative stress and modulating stress-related gene expression. Few other studies related to PGPRs' role in coping with abiotic stress are listed in Table 8.2.

## 8.5 Mechanism of PGPR-Mediated Diminution of Oxidative Stress

Due to growing industrialization and increasing population at a high pace, plants face abiotic stress at a great ordeal, which results in the production of ROS, including hydrogen peroxide, superoxide anion, singlet oxygen, and hydroxy radical that are responsible for oxidative stress in plants (Shanker et al. 2005; Ma et al. 2017; Saleem et al. 2018). Due to reactive nature of these ROS molecules generated, they

**Table 8.2** PGPRs role in ameliorating stress in plants

| S. no. | PGPR species | Plant species | Role in stress alleviation | References |
|---|---|---|---|---|
| 1. | *R. leguminosarum*, *Pseudomonas fluorescens*, *Luteibacter* sp. and *Variovorax* sp. | *Lathyrus sativus* L. | Boosted non-protein thiol synthesis, mitigated oxidative stress, reduced membrane damage, and improved antioxidant enzyme activities | Abdelkrim et al. (2023) |
| 2. | *Azospirillum brasilense* EMCC1454 | *Cicer arietinum* L. | Promoted plant growth by enhancing antioxidant activity, osmolyte production, and nutrient acquisition, while reducing oxidative stress and modulating stress-related genes | El-Ballat et al. (2023) |
| 3. | *Acinetobacter beijerinckii* and *Raoultella planticola* | *Glycine max* L. | Enhanced phosphate solubilization, plant's antioxidant response, reduced oxidative damage, and supported better growth under metal stress | Hussain et al. (2023) |
| 4. | *Trichosporon* sp. CA1, *Rhizobium* sp. Zn1–1, *Shinella* sp. Zn5–6, and *Pseudomonas* sp. CHA1–4 | *Miscanthus* × *giganteus* | Improved its growth, increasing biomass and chlorophyll levels while enhancing its ability to tolerate toxic elements | Nurzhanova et al. (2023) |
| 5. | Bacterial strains (JS1 and JS2) | *Spinacia oleracea* | Improved chlorophyll content, root and shoot length and their dry weight | Sarwar et al. (2023) |
| 6. | *Pseudomonas fluorescens* SBW25 and *Pseudomonas putida* KT2440 | *Hordeum vulgare* L. | Improved nutrient uptake, helping barley plants grow better and resist drought and salinity stress | Zaib et al. (2023) |
| 7. | *Pantoeaagglomerans*-FAP10 | *Triticum aestivum*; var. 343 | Beneficial in managing photosynthetic traits, the antioxidant system, and plant metabolism | Ansari et al. (2024) |
| 8. | *B. cereus* | *Triticum aestivum* | Alleviated ROS stress by enhancing SOD activity | Direk et al. (2024) |
| 9. | *Bacillus* and *Enterobacter* | *Zea mays* L. | A notable enhancement in the growth of the stem, leaf, and root was observed | ELjiati et al. (2024) |
| 10. | *Bacillussubtilis*CNBG-PGPR-1 | *Solanum lycopersicum* | Cellular homeostasis and photosynthetic efficiency improved, while ion toxicity and osmotic stress were reduced | Feng et al. (2024) |

(continued)

**Table 8.2** (continued)

| S. no. | PGPR species | Plant species | Role in stress alleviation | References |
|---|---|---|---|---|
| 11. | *Bacillus aerius* and *Bacillus cereus* | *Spinacia oleracea* L. | Heavy metal-induced oxidative stress was significantly reduced by boosting protein content, seed germination percentage, seedling growth, and nitrogen levels | Jamil et al. (2024) |
| 12. | *Micrococcus yunnanensis* GKSM13 | *Oryza sativa* | Enhanced seedling growth by increasing growth factors, lowering Cu uptake, and reducing oxidative stress with antioxidant enzymes | Majhi et al. (2024) |
| 13. | *Bacillus velezensis* strain UTB96 | *Triticum aestivum* | Grain yield and harvest index increased, along with improved APX enzyme activity | Mehrabi et al. (2024) |
| 14. | *Bacillus aryabhattai* | Soybean | Enhanced root and shoot length, increased fresh and dry weight, and improved chlorophyll content | Mun et al. (2024) |
| 15. | *M. morganii* strains ABT3 and ABT9 | *Arabidopsis* | Increased the activity of antioxidant enzymes, growth, yield, and chlorophyll content | Naqqash et al. (2024) |
| 16. | *Rhodococcus qingshengii* strain LMR356 | | Enhanced biomass, antioxidant activity, and pigment content | Oubohssaine et al. (2024) |
| 17. | *Klebsiella* sp. SBP-8 | *B. juncea* | Enhances POD enzyme activity, boosts photosynthetic pigment content, and increases both shoot and root fresh and dry weights | Shah et al. (2024) |
| 18. | *Bacillus mycoides* | *Triticum aestivum* | Boosted protein and sugar content, as well as shoot length and leaf width | Shahzad et al. (2024) |
| 19. | *Ralstonia* sp. YDR | *Oryza sativa* | Boosts rice growth and elevates the activity of CAT, glutathione, peroxidase, and SOD | Yin et al. (2024) |

are responsible for damage to cells and ultimately cell death (Gupta et al. 2020). It has been observed that Respiratory burst oxidase homolog (RBOH), which is a plasma bound reduced NADPH oxidase, is involved in the production of ROS in plants under stress (Sharma et al. 2021). In many studies, it has been observed that application of PGPR diminishes the ROS content by influencing downregulation of the RBOH gene. PGPR (*Pseudomonas aeruginosa* and *Burkholderia gladioli*) application has shown decreased RBOH1 expression in *Brassica juncea* L. seedlings grown under chlorpyrifos (CP) toxicity (Bakshi et al. 2021). It has been observed that MDA accumulation (end product of lipid peroxidation) in plants under stress results in damage to chloroplasts and mitochondria. Moreover, the accumulation of MDA and electrolyte leakage is also responsible for affecting the stability and permeability of the cell membrane (Abdelaal et al. 2021). However, application of PGPRs in the stressed plants has been observed to show improved activities of both enzymatic and non-enzymatic antioxidants, reducing electrolyte leakage and accumulation of MDA. Moreover, damage to the cell membrane was also decreased, thereby improving the stability of the cell membrane (Asghari et al. 2020).

In order to cope with the oxidative stress caused by abiotic stress and maintain the level of ROS generated, plants have developed a complex antioxidative defense system (Ashraf et al. 2021). There are several enzymatic antioxidants, such as CAT, GPX, SOD, APOX, GST, DHAR, GPOX, and MDHAR, which are responsible for protecting plants against oxidative stress caused by the generation of ROS under abiotic stress (Gill and Tuteja 2010). Several non-enzymatic antioxidants, such as ascorbic acid, glutathione, and tocopherol, are also responsible for protecting plants against several abiotic stresses in a well-coordinated manner. It has been observed that these antioxidants in the plants are responsible for reducing the oxidative stress caused due to overproduction of ROS by behaving as quenchers of lipid radicals and ROS (Holländer-Czytko et al. 2005; Ulhassan et al. 2019). In many studies, it has been reported that the application of several rhizobacteria improved the growth of the plant by reducing the toxicity caused by abiotic stresses. PGPRs do so by modulating the activity of enzymatic antioxidants. Moreover, PGPRs are also responsible for enhancing the synthesis of several non-enzymatic antioxidants (Islam et al. 2014). From the reports of Akhtar et al. (2021), it has been observed that due to the application of *Bacillus cereus* in plants of *Brassica nigra* that was grown under Cr toxicity, membrane integrity was maintained along with lowering the levels of MDA. Similarly, it was observed that PGPRs were responsible for reducing the oxidative stress in tomato plants that were grown under metal toxicity by increasing the levels of both enzymatic and non-enzymatic antioxidants (Gupta et al. 2020). Further, it was observed by Batool et al. (2020) that the plants of *Solanum tuberosum* grown in drought stress were able to combat the stress due to application of *Bacillus subtilis*. This *B. subtilis* was responsible for increasing the activity of enzymatic antioxidants like POD, SOD, and CAT by 51%, 63% and 68% respectively. Similarly, in the case of nonenzymatic antioxidants, it was observed that when the plants of *Hyoscyamus niger* grown under drought conditions were treated with the strains of *Pseudomonas putida* and *Pseudomonas fluorescens*, the concentration of tropane alkaloid was enhanced, which helped plants to combat the stress conditions (Ghorbanpour et al. 2013). Similarly, in another study by Chiappero et al. (2019), it was reported that when drought-stressed plants were inoculated with PGPRs, the content of total phenols was increased as compared to plants that were not treated with the PGPRs. Due to inoculation of PGPRs in plants of *Cucumis sativus* grown under drought stress, the activity and generation of CAT and SOD, along with proline was increased in the leaves of plants under stress (Wang et al. 2012a). Similarly, in case of *Lycopersicon esculentum* under drought stress, treatment of *Bacillus cereus* AR156 caused improvement in the activity of POD, CAT, and SOD (Wang et al. 2012b). In *Zea mays* grown under drought stress, the content of proline was enhanced due to inoculation of *Pseudomonas fluorescens* (Ansary et al. 2012). In addition to the biochemical studies conducted on the role of PGPRs in reducing the oxidative stress caused by different abiotic factors, observations on the expression of several genes related to stress response have also been made. It has been reported by Tiwari et al. (2016) that due to inoculation of PGPRs in stressed plants, ROS scavenging was enhanced due to overexpression of antioxidative genes, including *GST, CAT* and *APX.* Similarly, it was observed in plants of *Oryza sativa* grown under drought stress that due to application of *Bacillus megaterium*, *Rhizobium* sp., and *Pseudomonas azotoformans*, the expression of stress-related genes, including *GST* was increased (Omar et al. 2021). Upregulation of SOD, APOX, DHAR, and GR genes was observed when consortium treatment of *Pseudomonas aeruginosa* and *Burkholderia gladioli*) was applied to *Brassica juncea* under pesticide stress (Bakshi et al. 2021).

## 8.6 Conclusion

In conclusion, the impact of environmental concerns on global crop production is undeniable. To effectively address these challenges, we must adopt sustainable solutions that can mitigate stress, which poses significant risks to plant growth and productivity. Among the various strategies developed, the application of PGPR has emerged as a highly effective approach in alleviating abiotic stress conditions. By enhancing plant performance through both direct and indirect benefits, PGPR offers a promising avenue for significantly improving crop yields, especially in areas vulnerable to environmental stressors. PGPR stands out not only for its effectiveness but also for its sustainable and eco-friendly nature. A deeper understanding of the mechanisms by which PGPR mediates oxidative stress responses is essential for maximizing agricultural productivity and addressing the pressing issue of global food security. It is important that the development and commercialization of efficient PGPR formulations must be prioritized making them accessible to farmers worldwide. By doing so, agricultural communities can be empowered, crop yields can be enhanced, and this contributes to a more sustainable and secure food future for all.

## References

Abdelaal K, AlKahtani M, Attia K, Hafez Y, Király L, Künstler A (2021) The role of plant growth-promoting bacteria in alleviating the adverse effects of drought on plants. Biology 10(6):520

Abdelkefi N, Louati I, Mechichi HZ, Sayahi N, El-Sayed WS, El Nayal A et al (2024) Enhanced salt stress tolerance in tomato plants following inoculation with newly isolated plant growth-promoting rhizobacteria. Sci Hortic 328:112921

Abdelkrim S, Abid G, Chaieb O, Taamalli W, Mannai K, Louati F et al (2023) Plant growth promoting rhizobacteria modulates the antioxidant defense and the expression of stress-respon-

sive genes providing Pb accumulation and tolerance of grass pea. Environ Sci Pollut Res 30(4):10789–10802
Ahemad M, Kibret M (2014) Mechanisms and applications of plant growth promoting rhizobacteria: current perspective. J King Saud Univ Sci 26(1):1–20
Ahluwalia O, Singh PC, Bhatia R (2021) A review on drought stress in plants: implications, mitigation and the role of plant growth promoting rhizobacteria. Resour Environm Sustainability 5:100032
Ahmad HM, Fiaz S, Hafeez S, Zahra S, Shah AN, Gul B et al (2022) Plant growth-promoting rhizobacteria eliminate the effect of drought stress in plants: a review. Front Plant Sci 13:875774
Akhtar N, Ilyas N, Yasmin H, Sayyed RZ, Hasnain ZA, Elsayed E, El Enshasy HA (2021) Role of Bacillus cereus in improving the growth and phytoextractability of Brassica nigra (L.) K. Koch in chromium contaminated soil. Molecules 26(6):1569
Ali S, Khan N (2021) Delineation of mechanistic approaches employed by plant growth promoting microorganisms for improving drought stress tolerance in plants. Microbiol Res 249:126771
Ali M, Ohri P (2023) Deciphering the synergistic effect of Jasmonic acid and Spermine in mitigating root-knot nematode stress in tomato plants through enhancing growth and activity of antioxidant enzymes. S Afr J Bot 161:21–35
Ali B, Hafeez A, Ahmad S, Javed MA, Sumaira, Afridi MS et al (2022) Bacillus thuringiensis PM25 ameliorates oxidative damage of salinity stress in maize via regulating growth, leaf pigments, antioxidant defense system, and stress responsive gene expression. Front Plant Sci 13:921668
Ali Q, Sami A, Haider MZ, Ashfaq M, Javed MA (2024) Antioxidant production promotes defense mechanism and different gene expression level in Zea mays under abiotic stress. Sci Rep 14(1):7114
Alnusairi GS, Mazrou YS, Qari SH, Elkelish AA, Soliman MH, Eweis M et al (2021) Exogenous nitric oxide reinforces photosynthetic efficiency, osmolyte, mineral uptake, antioxidant, expression of stress-responsive genes and ameliorates the effects of salinity stress in wheat. Plants 10(8):1693
Álvarez, S., & Acosta-Motos, J. R. (Eds.). (2023). Advances in plant physiology of abiotic stresses. MDPI-Multidisciplinary Digital Publishing Institute, 260.
Ansari FA, Ahmad I, Pichtel J (2023) Synergistic effects of biofilm-producing PGPR strains on wheat plant colonization, growth and soil resilience under drought stress. Saudi J Biol Sci 30(6):103664
Ansari FA, Ahmad I, Pichtel J, Husain FM (2024) Pantoea agglomerans FAP10: A novel biofilm-producing PGPR strain improves wheat growth and soil resilience under salinity stress. Environ Exp Bot 222:105759
Ansary MH, Rahmani HA, Ardakani MR, Paknejad F, Habibi D, Mafakheri S (2012) Effect of Pseudomonas fluorescens on proline and phytohormonal status of maize (Zea mays L.) under water deficit stress. Ann Biol Res 3:1054–1062
Asad MAU, Zhang Y, Zhou L, Guan X, Cheng F (2024) How abiotic stresses trigger sugar signaling to modulate leaf senescence? Plant Physiol Biochem 210:108650
Asghari B, Khademian R, Sedaghati B (2020) Plant growth promoting rhizobacteria (PGPR) confer drought resistance and stimulate biosynthesis of secondary metabolites in pennyroyal (Mentha pulegium L.) under water shortage condition. Sci Hortic 263:109132
Ashraf MA, Rasheed R, Zafar S, Iqbal M, Saqib ZA (2021) Menadione sodium bisulfite neutralizes chromium phytotoxic effects in okra by regulating cytosolutes, lipid peroxidation, antioxidant system and metal uptake. Int J Phytoremediation 23(7):736–746
Badawy AA, Alamri AA, Hussein HAA, Salem NF, Mashlawi AM, Kenawy SK, El-Shabasy A (2024) Glycine betaine mitigates heavy metal toxicity in beta vulgaris (L.): an antioxidant-driven approach. Agronomy 14(4):797
Bakshi P, Chouhan R, Sharma P, Mir BA, Gandhi SG, Landi M et al (2021) Amelioration of chlorpyrifos-induced toxicity in Brassica juncea L. by combination of 24-epibrassinolide and plant-growth-promoting rhizobacteria. Biomolecules 11(6):877

Batool T, Ali S, Seleiman MF, Naveed NH, Ali A, Ahmed K, Abid M, Rizwan M, Shahid MR, Alotaibi M, Al-Ashkar I (2020) Plant growth promoting rhizobacteria alleviates drought stress in potato in response to suppressive oxidative stress and antioxidant enzymes activities. Sci Rep 10(1):16975

Bartosz G (1997) Oxidative stress in plants. Acta Physiol Plant 19:47–64

Bhardwaj T, Kour J, Chouhan R, Devi K, Singh H, Gandhi SG et al (2024) Integrated transcriptomic and physio-molecular studies unveil the melatonin and PGPR induced protection to photosynthetic attributes in Brassica juncea L. under cadmium toxicity. J Hazard Mater 476:134875

Chandran H, Meena M, Swapnil P (2021) Plant growth-promoting rhizobacteria as a green alternative for sustainable agriculture. Sustainability 13(19):10986

Chaudhary MT, Majeed S, Rana IA, Ali Z, Jia Y, Du X et al (2024) Impact of salinity stress on cotton and opportunities for improvement through conventional and biotechnological approaches. BMC Plant Biol 24(1):20

Chiappero J, del Rosario Cappellari L, Alderete LGS, Palermo TB, Banchio E (2019) Plant growth promoting rhizobacteria improve the antioxidant status in Mentha piperita grown under drought stress leading to an enhancement of plant growth and total phenolic content. Ind Crops Prod 139:111553

Choudhary S, Bhat TM, Alwutayd KM, Abd El-Moneim D, Naaz N (2024) Salicylic acid enhances thermotolerance and antioxidant defense in Trigonella foenum graecum L. under heat stress. Heliyon 10(6):e27227

Choukri H, El Haddad N, Aloui K, Hejjaoui K, El-Baouchi A, Smouni A et al (2022) Effect of high temperature stress during the reproductive stage on grain yield and nutritional quality of lentil (Lens culinaris Medikus). Front Nutr 9:857469

Ciarmiello LF, Woodrow P, Fuggi A, Pontecorvo G, Carillo P (2011) Plant genes for abiotic stress. In: Abiotic stress in plants—mechanisms and adaptations, pp 283–308

Direk A, Arikan-Abdulveli B, Ozfidan-Konakci C, Yildiztugay E, Uysal A (2024) Effects of Bacillus cereus on physiological and biochemical characteristics of wheat under arsenic and cadmium stress: a biological agent to reduce heavy metal stress. Plant Stress 12:100458

Du L, Xia X, Lan X, Liu M, Zhao L, Zhang P, Wu Y (2017) Influence of arsenic stress on physiological, biochemical, and morphological characteristics in seedlings of two cultivars of maize (Zea mays L.). Water Air Soil Pollut 228:1–14

El-Ballat EM, Elsilk SE, Ali HM, Ali HE, Hano C, El-Esawi MA (2023) Metal-resistant PGPR strain Azospirillum brasilense EMCC1454 enhances growth and chromium stress tolerance of chickpea (Cicer arietinum L.) by modulating redox potential, osmolytes, antioxidants, and stress-related gene expression. Plants 12(11):2110

ElJiati A, Elmaati Y, Ouchaou H (2024) Plant growth promoting Rhizobacteria (PGPR) isolated from an arid soil in Saudi Arabia improve maize growth. Atlas J Biol:830–838. https://doi.org/10.5147/ajb.vi.245

El-Saadony MT, Desoky ESM, El-Tarabily KA, AbuQamar SF, Saad AM (2024) Exploiting the role of plant growth promoting rhizobacteria in reducing heavy metal toxicity of pepper (Capsicum annuum L.). Environ Sci Pollut Res 31(18):27465–27484

Emamverdian A, Khalofah A, Pehlivan N, Li Y, Chen M, Liu G (2024) Iron nanoparticles in combination with other conventional Fe sources remediate mercury toxicity-affected plants and soils by nutrient accumulation in bamboo species. Ecotoxicol Environ Saf 278:116431

Feki K, Tounsi S, Mrabet M, Mhadhbi H, Brini F (2021) Recent advances in physiological and molecular mechanisms of heavy metal accumulation in plants. Environ Sci Pollut Res 28:1–20

Feng L, Li Q, Zhou D, Jia M, Liu Z, Hou Z et al (2024) B. subtilis CNBG-PGPR-1 induces methionine to regulate ethylene pathway and ROS scavenging for improving salt tolerance of tomato. Plant J 117(1):193–211

Ferioun M, Bouhraoua S, Belahcen D, Zouitane I, Srhiouar N, Louahlia S, El Ghachtouli N (2024) PGPR consortia enhance growth and yield in barley cultivars subjected to severe drought stress and subsequent recovery. Rhizosphere 31:100926

Ghorbanpour M, Hatami M, Khavazi K (2013) Role of plant growth promoting rhizobacteria on antioxidant enzyme activities and tropane alkaloid production of Hyoscyamus Niger under water deficit stress. Turk J Biol 37(3):350–360

Gill SS, Tuteja N (2010) Reactive oxygen species and antioxidant machinery in abiotic stress tolerance in crop plants. Plant Physiol Biochem 48(12):909–930

Goyal D, Yadav A, Prasad M, Singh TB, Shrivastav P, Ali A et al (2020) Effect of heavy metals on plant growth: an overview. In: Contaminants in agriculture: sources, impacts and management, pp 79–101

Gull A, Lone AA, Wani NUI (2019) Biotic and abiotic stresses in plants. In: Abiotic and biotic stress in plants, vol 7, pp 1–9

Gupta P, Kumar V, Usmani Z, Rani R, Chandra A, Gupta VK (2020) Implications of plant growth promoting Klebsiella sp. CPSB4 and Enterobacter sp. CPSB49 in luxuriant growth of tomato plants under chromium stress. Chemosphere 240:124944

Hanif S, Mahmood A, Javed T, Bibi S, Zia MA, Asghar S et al (2024) Exogenous application of salicylic acid ameliorates salinity stress in barley (Hordeum vulgare L.). BMC Plant Biol 24(1):1–16

Hasan A, Tabassum B, Hashim M, Khan N (2024) Role of plant growth promoting rhizobacteria (PGPR) as a plant growth enhancer for sustainable agriculture: a review. Bacteria 3(2):59–75

Hasanuzzaman M, Bhuyan MB, Zulfiqar F, Raza A, Mohsin SM, Mahmud JA et al (2020) Reactive oxygen species and antioxidant defense in plants under abiotic stress: revisiting the crucial role of a universal defense regulator. Antioxidants 9(8):681

He S, Lian X, Zhang B, Liu X, Yu J, Gao Y et al (2023) Nano silicon dioxide reduces cadmium uptake, regulates nutritional homeostasis and antioxidative enzyme system in barley seedlings (Hordeum vulgare L.) under cadmium stress. Environ Sci Pollut Res 30(25):67552–67564

Hnilickova H, Kraus K, Vachova P, Hnilicka F (2021) Salinity stress affects photosynthesis, malondialdehyde formation, and proline content in Portulaca oleracea L. Plants 10(5):845

Holländer-Czytko H, Grabowski J, Sandorf I, Weckermann K, Weiler EW (2005) Tocopherol content and activities of tyrosine aminotransferase and cystine lyase in Arabidopsis under stress conditions. J Plant Physiol 162(7):767–770

Huang Y, Yi J, Li X, Li F (2024) Transcriptomics and physiological analyses reveal that sulfur alleviates mercury toxicity in rice (Oryza sativa L.). J Environ Sci 135:10–25

Hussain A, Shah M, Hamayun M, Iqbal A, Qadir M, Alataway A et al (2023) Phytohormones producing rhizobacteria alleviate heavy metals stress in soybean through multilayered response. Microbiol Res 266:127237

Ilyas M, Maqsood MF, Shahbaz M, Zulfiqar U, Ahmad K, Naz N et al (2024) Alleviating salinity stress in canola (Brassica napus L.) through exogenous application of salicylic acid. BMC Plant Biol 24(1):611

Iqbal A, Khan R, Hussain Q, Imran M, Mo Z, Hua T et al (2024) Vermicompost application enhances soil health and plant physiological and antioxidant defense to conferring heavy metals tolerance in fragrant rice. Front Sustain Food Syst 8:1418554

Islam F, Yasmeen T, Ali Q, Ali S, Arif MS, Hussain S, Rizvi H (2014) Influence of Pseudomonas aeruginosa as PGPR on oxidative stress tolerance in wheat under Zn stress. Ecotoxicol Environ Saf 104:285–293

Jahan MS, Guo S, Sun J, Shu S, Wang Y, Abou El-Yazied A et al (2021) Melatonin-mediated photosynthetic performance of tomato seedlings under high-temperature stress. Plant Physiol Biochem 167:309–320

Jamil M, Malook I, Rehman SU, Aslam MM, Fayyaz M, Shah G et al (2024) Inoculation of heavy metal resistant bacteria alleviated heavy metal-induced oxidative stress biomarkers in spinach (Spinacia oleracea L.). BMC Plant Biol 24(1):221

Kaur S, Kalia A, Sharma S (2024) Bioformulation of Azotobacter and Streptomyces for improved growth and yield of wheat (Triticum aestivum L.): a field study. J Plant Growth Regul:1–17

Kaushik S, Ranjan A, Singh AK, Sirhindi G (2024) Methyl jasmonate reduces cadmium toxicity by enhancing phenol and flavonoid metabolism and activating the antioxidant defense system in pigeon pea (Cajanus cajan). Chemosphere 346:140681

Kazerooni EA, Maharachchikumbura SS, Adhikari A, Al-Sadi AM, Kang SM, Kim LR, Lee IJ (2021) Rhizospheric Bacillus amyloliquefaciens protects Capsicum annuum cv. Geumsugangsan from multiple abiotic stresses via multifarious plant growth-promoting attributes. Front Plant Sci 12:669693

Khan Z, Jan R, Asif S, Farooq M, Jang YH, Kim EG et al (2024a) Exogenous melatonin induces salt and drought stress tolerance in rice by promoting plant growth and defense system. Sci Rep 14(1):1214

Khan Z, Thounaojam TC, Rajkumari JD, Deka BK, Rahman R, Upadhyaya H (2024b) Arsenic induced chromosomal aberrations, biochemical and morphological changes in Vigna radiata L. (Mung bean) seedlings and its amelioration by Curcumin. Vegetos 37(3):1185–1194

Khozeeva EV, Zimina YA, Sroslova GA (2020) Oxidative stress of plants: chemistry, physiology, methods of protection. Vestnik Volgogradskogo Gosudarstvennogo Universiteta. Seriâ 11, Estestvennye Nauki 10(4)

Kopecká R, Kameniarová M, Černý M, Brzobohatý B, Novák J (2023) Abiotic stress in crop production. Int J Mol Sci 24(7):6603

Kour J, Kohli SK, Khanna K, Bakshi P, Sharma P, Singh AD et al (2021) Brassinosteroid signaling, crosstalk and, physiological functions in plants under heavy metal stress. Front Plant Sci 12:608061

Kumar D, Ohri P (2023) Say "NO" to plant stresses: unravelling the role of nitric oxide under abiotic and biotic stress. Nitric Oxide 130:36–57

Kumar A, Singh S, Gaurav AK, Srivastava S, Verma JP (2020) Plant growth-promoting bacteria: biological tools for the mitigation of salinity stress in plants. Front Microbiol 11:1216

Kumar D, Ali M, Sharma N, Sharma R, Manhas RK, Ohri P (2024a) Unboxing PGPR-mediated management of abiotic stress and environmental cleanup: what lies inside? Environ Sci Pollut Res 31:1–38

Kumar A, Bhattacharya T, Shaikh WA, Roy A (2024b) Sustainable soil management under drought stress through biochar application: immobilizing arsenic, ameliorating soil quality, and augmenting plant growth. Environ Res 259:119531

Liu T, Li T, Zhang L, Li H, Liu S, Yang S et al (2021) Exogenous salicylic acid alleviates the accumulation of pesticides and mitigates pesticide-induced oxidative stress in cucumber plants (Cucumis sativus L.). Ecotoxicol Environ Saf 208:111654

Liu Z, Lu Q, Zhao Y, Wei J, Liu M, Duan X, Lin M (2023) Ameliorating effects of graphene oxide on cadmium accumulation and eco-physiological characteristics in a greening Hyperaccumulator (Lonicera japonica Thunb.). Plants 13(1):19

Lu L, Liu N, Fan Z, Liu M, Zhang X, Tian J et al (2024) A novel PGPR strain, Streptomyces lasalocidi JCM 3373T, alleviates salt stress and shapes root architecture in soybean by secreting indole-3-carboxaldehyde. Plant Cell Environ 47(6):1941–1956

Ma Q, Cao X, Ma J, Tan X, Xie Y, Xiao H, Wu L (2017) Hexavalent chromium stress enhances the uptake of nitrate but reduces the uptake of ammonium and glycine in pak choi (Brassica chinensis L.). Ecotoxicol Environ Saf 139:384–393

Majhi K, Let M, Bandopadhyay R (2024) Efficacious use of Micrococcus yunnanensis GKSM13 for the growth of rice seedlings under copper stress with elucidation into genomic traits. Curr Plant Biol 37:100318

Malik N, Ahmad M, Malik Z, Hussain A, Waseem M, Ali A, Rizwan M (2024) Isolation and characterization of chromium-resistant bacteria and their effects on germination, growth, and Cr accumulation in Capsicum annum (L.) under Cr stress. Plant Physiol Biochem 214:108955

Mehrabi SS, Sabokdast M, Bihamta MR, Dedičová B (2024) The coupling effects of PGPR inoculation and foliar spraying of Strigolactone in mitigating the negative effect of salt stress in wheat plants: insights from phytochemical, growth, and yield attributes. Agriculture 14(5):732

Mishra N, Jiang C, Chen L, Paul A, Chatterjee A, Shen G (2023) Achieving abiotic stress tolerance in plants through antioxidative defense mechanisms. Front Plant Sci 14:1110622

Mun BG, Hussain A, Park YG, Kang SM, Lee IJ, Yun BW (2024) The PGPR Bacillus aryabhattai promotes soybean growth via nutrient and chlorophyll maintenance and the production of butanoic acid. Front Plant Sci 15:1341993

Nadarajah KK (2020) ROS homeostasis in abiotic stress tolerance in plants. Int J Mol Sci 21(15):5208
Naqqash T, Aziz A, Gohar M, Khan J, Ali S, Radicetti E et al (2024) Heavy metal-resistant rhizobacteria fosters to alleviate the cadmium toxicity in Arabidopsis by upregulating the plant physiological responses. Int J Phytoremediation 26(4):557–568
Nawaz M, Saleem MH, Khalid MR, Ali B, Fahad S (2024) Nitric oxide reduces cadmium uptake in wheat (Triticum aestivum L.) by modulating growth, mineral uptake, yield attributes, and antioxidant profile. Environ Sci Pollut Res 31(6):9844–9856
Nurzhanova AA, Pidlisnyuk V, Berzhanova R, Nurmagambetova AS, Terletskaya N, Omirbekova N et al (2023) PGPR-driven phytoremediation and physiobiochemical response of Miscanthus× giganteus to stress induced by the trace elements. Environ Sci Pollut Res Int 30(42):96098–96113
Olson DR, Ruhland CT (2024) Ultraviolet-B stress increases epidermal UV-screening effectiveness and alters growth and Cell-Wall constituents of the Brown midrib bmr6 and bmr12 mutants of Sorghum bicolor. J Agron Crop Sci 210(4):e12723
Omar SA, Fetyan NA, Eldenary ME, Abdelfattah MH, Abd-Elhalim HM, Wrobel J, Kalaji HM (2021) Alteration in expression level of some growth and stress-related genes after rhizobacteria inoculation to alleviate drought tolerance in sensitive rice genotype. Chem Biol Technol Agric 8:1–19
Oubohssaine M, Sbabou L, Aurag J (2024) Potential of the plant growth-promoting rhizobacterium Rhodococcus qingshengii LMR356 in mitigating lead stress impact on Sulla spinosissima L. Environ Sci Pollut Res 31(33):46002–46022
Ouyang F, Sun M, Cui X, Tigabu M, Zhang H, Deng J et al (2024) Picea pungens exhibits greatest tolerance to short-time thermal stress compared to Picea abies, and Picea omorika. New Forests 55(4):769–784
Ozel HB, Abo Aisha AES, Cetin M, Sevik H, Zeren Cetin I (2021) The effects of increased exposure time to UV-B radiation on germination and seedling development of Anatolian black pine seeds. Environ Monit Assess 193(7):388
Piacentini D, Della Rovere F, Lanni F, Cittadini M, Palombi M, Fattorini L et al (2023) Brassinosteroids interact with nitric oxide in the response of rice root systems to arsenic stress. Environ Exp Bot 209:105287
Qamer Z, Chaudhary MT, Du X, Hinze L, Azhar MT (2021) Review of oxidative stress and antioxidative defense mechanisms in Gossypium hirsutum L. in response to extreme abiotic conditions. J Cotton Res 4(1):9
Ranjan A, Rajput VD, Prazdnova EV, Gurnani M, Sharma S, Bhardwaj P et al (2024) Augmenting abiotic stress tolerance and root architecture: the function of phytohormone-producing PGPR and their interaction with nanoparticles. S Afr J Bot 167:612–629
Ravi B, Foyer CH, Pandey GK (2023) The integration of reactive oxygen species (ROS) and calcium signalling in abiotic stress responses. Plant Cell Environ 46(7):1985–2006
Raziq A, Wang Y, Mohi Ud Din A, Sun J, Shu S, Guo S (2022) A comprehensive evaluation of salt tolerance in tomato (Var. ailsa craig): responses of physiological and transcriptional changes in RBOH's and ABA biosynthesis and signalling genes. Int J Mol Sci 23(3):1603
Rosier A, Medeiros FH, Bais HP (2018) Defining plant growth promoting rhizobacteria molecular and biochemical networks in beneficial plant-microbe interactions. Plant Soil 428:35–55
Sachdev S, Ansari SA, Ansari MI, Fujita M, Hasanuzzaman M (2021) Abiotic stress and reactive oxygen species: generation, signaling, and defense mechanisms. Antioxidants 10(2):277
Saleem M, Asghar HN, Zahir ZA, Shahid M (2018) Impact of lead tolerant plant growth promoting rhizobacteria on growth, physiology, antioxidant activities, yield and lead content in sunflower in lead contaminated soil. Chemosphere 195:606–614
Samynathan R, Shanmugam K, Nagarajan C, Murugasamy H, Ilango RVJ, Shanmugam A et al (2021) The effect of abiotic and biotic stresses on the production of bioactive compounds in tea (Camellia sinensis (L.) O. Kuntze). Plant Gene 27:100316

Sarwar MJ, Shabaan M, Asghar HN, Ayyub M, Ali Q, Zulfiqar U et al (2023) Interaction of chromium (Cr) resistant plant growth promoting rhizobacteria with compost to phytostabilize Cr in spinach rhizosphere. Plant Stress 10:100261

Shah AA, Zafar S, Usman S, Javad S, Aslam M, Noreen Z et al (2024) Zinc oxide nanoparticles and Klebsiella sp. SBP-8 alleviates chromium toxicity in Brassica juncea by regulation of antioxidant capacity, osmolyte production, nutritional content and reduction in chromium adsorption. Plant Physiol Biochem 210:108624

Shahzad A, Aslam U, Ferdous S, Qin M, Siddique A, Billah M, Naeem M, Mahmood Z, Kayani S (2024) Combined effect of endophytic Bacillus mycoides and rock phosphate on the amelioration of heavy metal stress in wheat plants. BMC Plant Biol 24(1):125

Shanker AK, Cervantes C, Loza-Tavera H, Avudainayagam S (2005) Chromium toxicity in plants. Environ Int 31(5):739–753

Sharma P, Jha AB, Dubey RS (2019) Oxidative stress and Antioxidative defense system in plants growing under abiotic stresses. In: Pessarakli M (ed) Handbook of plant and crop stress, 4th edn. CRC Press, Boca Raton, p 44

Sharma A, Sharma P, Kumar R, Sharma V, Bhardwaj R, Sharma I (2021) Role of reactive oxygen species in the regulation of abiotic stress tolerance in legumes. In: Singh VP, Singh S, Tripathi DK, Prasad SM, Bhardwaj R, Chauhan DK (eds) Abiotic stress and legumes. Academic/Elsevier, Cambridge, pp 217–243

Singh D, Maithreyi S, Taunk J, Singh MP (2024a) Physiological and proteomic characterization revealed the response mechanisms underlying aluminium tolerance in lentil (Lens culinaris Medikus). Physiol Plant 176(3):e14298

Singh D, Tripathi A, Mitra R, Bhati J, Rani V, Taunk J et al (2024b) Genome-wide identification of MATE and ALMT genes and their expression profiling in mungbean (Vigna radiata L.) under aluminium stress. Ecotoxicol Environ Saf 280:116558

Sylia AB, Corrêa A, Cruz C, Yadav AN, Nabti E (2021) Plant growth promoting microbes as biofertilizers: promising solutions for sustainable agriculture under climate change associated abiotic stresses. Plant Sci Today 8(sp1):60–76

Thiruvengadam R, Venkidasamy B, Easwaran M, Chi HY, Thiruvengadam M, Kim SH (2024) Dynamic interplay of reactive oxygen and nitrogen species (ROS and RNS) in plant resilience: unveiling the signaling pathways and metabolic responses to biotic and abiotic stresses. Plant Cell Rep 43(8):1–24

Tiwari S, Lata C, Chauhan PS, Nautiyal CS (2016) Pseudomonas putida attunes morphophysiological, biochemical and molecular responses in Cicer arietinum L. during drought stress and recovery. Plant Physiol Biochem 99:108–117

Ulhassan Z, Gill RA, Huang H, Ali S, Mwamba TM, Ali B et al (2019) Selenium mitigates the chromium toxicity in Brassicca napus L. by ameliorating nutrients uptake, amino acids metabolism and antioxidant defense system. Plant Physiol Biochem 145:142–152

Villalpando-Rodriguez GE, Gibson SB (2021) Reactive oxygen species (ROS) regulates different types of cell death by acting as a rheostat. Oxidative Med Cell Longev 2021(1):9912436

Wang CJ, Yang W, Wang C, Gu C, Niu DD, Liu HX et al (2012a) Induction of drought tolerance in cucumber plants by a consortium of three plant growth-promoting rhizobacterium strains. PLoS One 7(12):e52565

Wang C, Guo Y, Wang C, Liu H, Niu D, Wang Y, Guo J (2012b) Enhancement of tomato (Lycopersicon esculentum) tolerance to drought stress by plant-growth-promoting rhizobacterium (PGPR) Bacillus cereus AR156. J Agric Biotechnol 20:1097–1105

Wang P, Liu WC, Han C, Wang S, Bai MY, Song CP (2024) Reactive oxygen species: multidimensional regulators of plant adaptation to abiotic stress and development. J Integr Plant Biol 66(3):330–367

Wani KI, Naeem M, Castroverde CDM, Kalaji HM, Albaqami M, Aftab T (2021) Molecular mechanisms of nitric oxide (NO) signaling and reactive oxygen species (ROS) homeostasis during abiotic stresses in plants. Int J Mol Sci 22(17):9656

Yan F, Zhang J, Li W, Ding Y, Zhong Q, Xu X et al (2021) Exogenous melatonin alleviates salt stress by improving leaf photosynthesis in rice seedlings. Plant Physiol Biochem 163:367–375

Yaqoob U, Jan N, Raman PV, Siddique KH, John R (2022) Crosstalk between brassinosteroid signaling, ROS signaling and phenylpropanoid pathway during abiotic stress in plants: does it exist? Plant Stress 4:100075

Yin DX, Niu LL, Liu J, Yang R, Han B, Liu ZY et al (2024) Cadmium-resistant bacterium Ralstonia sp. YDR alleviated Cd toxicity in rice seedlings by enhancing antioxidant defense and inhibiting Cd2+ influx and H+ efflux. Environ Technol Innov 34:103614

Zaib S, Zubair A, Abbas S, Hussain J, Ahmad I, Shakeel SN (2023) Plant growth-promoting Rhizobacteria (PGPR) reduce adverse effects of salinity and drought stresses by regulating nutritional profile of barley. Appl Environ Soil Sci 2023(1):7261784

Zhang H, Lu L (2024) Transcription factors involved in plant responses to cadmium-induced oxidative stress. Front Plant Sci 15:1397289

Zhang X, Chen L, Wang J, Wang M, Yang S, Zhao C (2018) Photosynthetic acclimation to long-term high temperature and soil drought stress in two spruce species (Picea crassifolia and P. Wilsonii) used for afforestation. J For Res 29(2):363–372

Zhao Y, Yao J, Li H, Sunahara G, Li M, Tang C et al (2024) Effects of three plant growth-promoting bacterial symbiosis with ryegrass for remediation of Cd, Pb, and Zn soil in a mining area. J Environ Manag 353:120167

Zulfiqar F, Moosa A, Ali HM, Hancock JT, Yong JWH (2024) Synergistic interplay between melatonin and hydrogen sulfide enhances cadmium-induced oxidative stress resistance in stock (Matthiola incana L.). Plant Signal Behav 19(1):2331357

# Nutrient Dynamics and ROS Homeostasis: A Critical Link in Plant Stress Tolerance

9

Gholamreza Abdi, Aritra Mukherjee,
Md Sabir Ahmed Mondol, Mero Yangfo, Tadela Susmitha,
Anilava Chatterjee, Ubaida Akbar, and Amisha Rani

**Abstract**

An integral part of plant development, growth, and stress tolerance is a process that involves the interplay between nutritional dynamics and the homeostasis of reactive oxygen species (ROS). This chapter describes how crucial the relationship of nutrient availability concerning ROS balance could be with respect to nutrient deficiency or excess, causing oxidative stress and how oxidative stress

G. Abdi (✉)
Department of Biotechnology, Persian Gulf Research Institute, Persian Gulf University, Bushehr, Iran
e-mail: abdi@pgu.ac.ir

A. Mukherjee
Department of Seed Science and Technology, Bidhan Chandra Krishi Viswavidyalaya, Mohanpur, Nadia, West Bengal, India

M. S. A. Mondol · A. Chatterjee
Department of Agricultural Biochemistry, Bidhan Chandra Krishi Viswavidyalaya, Mohanpur, Nadia, West Bengal, India

M. Yangfo
North Eastern Regional Institute of Science and Technology, Nirjuli, Arunachal Pradesh, India

T. Susmitha
Department of Agricultural Biochemistry, School of Agricultural Sciences, Mallareddy University, Hyderabad, Telangana, India

U. Akbar
Department of Food Technology and Nutrition, Lovely Professional University, Phagwara, Punjab, India

A. Rani
Department of Forest Products, Dr. Yashwant Singh Parmar University of Horticulture and Forestry, Nauni, Solan, India



N. Iqbal et al. (eds.), *Plant Signaling Molecules in Regulation of ROS-Scavenging System*, https://doi.org/10.1007/978-981-95-3377-0_9

influences nutrient uptake and signaling pathways. The interplay of macronutrients, including nitrogen, phosphorus, potassium, sulfur, iron, zinc, manganese, as well as other similar major and minor elements, with ROS metabolism, synergistic, and antagonistic with one another, and is focused on maintaining cellular redox homeostasis. Excessive reactive oxygen species generation and oxidative damage to proteins, lipids, and nucleic acids are common outcomes of dietary deficits, especially macronutrient shortages. In plants, nutrient-ROS interactions are complicated, and ROS's dual role as harmful agents and signaling molecules reflects that. It is now much more important to modify nutrition to prevent ROS damage when subjected to abiotic conditions like drought, salt, and high temperature. Nutrient supplementation and priming strategies for increasing stress tolerance through ROS modulation were explored. This has also highlighted genetic engineering approaches through the manipulation of antioxidant genes and nutrient transporters as promising avenues in the development of stress-resilient crops. Roles of beneficial microorganisms were also noted in improving nutrient availability and ROS balance, which can be enhanced in the future towards sustainable agriculture.

**Keywords**

Abiotic stress · Microorganisms · Nutrients · Reactive oxygen species

## 9.1 Introduction

### 9.1.1 The Significance of Plant Nutrient Dynamics

Nutrient dynamics is a resourceful process, where macro and micronutrients are absorbed, translocated, utilized, and consequently recycled in plants; and is therefore one of the indispensable components of growth, development, and yield, and is affected by plant genetics, ecology, and agricultural practices (Toor et al. 2022). The nutritional requirements of the plant change along its life cycle. With plant growth, new nutrient sinks become established and are used over time, while others are depleted in time. Post-anthesis, maturing grains come to be regarded as the major nutrient depository. Effective nutrient management requires internal nutrient cycling and recycling. Roots receive nutrients, which can be used in several ways, such as canopy construction. Over time, nutrients are reallocated from older tissues to developing seeds to provide nutrient storage for the next generation and reduce nutrient losses due to litterfall decomposition. Several factors like plant genes, environment, and farming practices affect the nutrient absorbtion. Root structure, soil condition, and climate can influence the nutrient availability and absorption. Agriculture systems, for example, tilling and fertilizer, have great effects (Toor et al. 2021). Nutrients have synergistic or antagonist relations among them and therefore intervene in absorption, distribution, and utilization such in synergistic or antagonistic way (Toor et al. 2021; Wang et al. 2021a, b, c, d, e; Hanikenne et al. 2021; Kumar et al. 2021a, b).

Reactive oxygen species (ROS) are a group of chemicals that are produced during the metabolic processes of aerobic organisms, including plants. They are crucial for many biological functions, including signaling throughout development. Excess reactive oxygen species (ROS) can cause oxidative stress and the breakdown of proteins, lipids, and nucleic acids; yet, ROS are necessary for several biological functions (Tavanti et al. 2021; Mansoor et al. 2022; Zandi and Schnug 2022; Qamer et al. 2021). The development of a climate change-resistant cultivar, improved crop stress tolerance, and reduced fertilizer application costs might result from a deeper understanding of the complexities of reactive oxygen species (ROS) in plants.

### 9.1.2 Nutritional Status and Reactive Oxygen Species Homeostasis Relationship

Damage to electron transport in chloroplasts and mitochondria, leading to leaky electron flow and excess ROS, is a prominent effect of macronutrient deficiencies on plant metabolic activities. For instance, essential elements like sulphur (S), nitrogen (N), and phosphorus (P) play a key role in ensuring electron transport efficiency. Due to poor antioxidant defences, sulphur shortage causes glutathione production to be restricted, which in turn increases ROS buildup. Alternatively, apoplastic enhanced ROS generation is stimulated by nitrogen and phosphorus deficiency-induced upregulation of respiratory burst oxidase homologs (RBOHs). While potassium (K) serves as a signaling molecule to regulate potassium transport, potassium shortage causes ROS buildup. Elevated levels of hydrogen peroxide ($H_2O_2$) caused by iron (Fe) deficiency over time alter gene expression related to iron (Fe) transport and absorption, reinforcing the feedback loop between reactive oxygen species (ROS) and nutritional homeostasis.

Despite their usual reputation as harmful metabolic byproducts, reactive oxygen species (ROS) are now being recognized as signaling molecules that control food intake and transport. Deprivation of nutrients often stimulates the increase of cytosolic calcium ion ($Ca^{2+}$) concentrations in plants, which activate the downstream signaling pathways such as calcium-calmodulin kinase (CaM-K) and mitogen-activated protein kinases (MAPKs). Signaling cascades phosphorylate transcription factors responsible for modulating nutrient absorption and transport, hence enabling plants to adapt to nutrient imbalance. Nutrient homeostasis is also subject to regulation by this complex network, whose central node is PHR1 as far as transcription factors are concerned. Under phosphorus deficiency conditions, PHR1 regulates the sulfate transporter genes for the uptake of sulfur. It is also the reason behind the regulation of iron-associated genes (FER1) and zinc-associated genes (ZIP2, ZIP4), and thus contributes to metal ion homeostasis. It, in return, ties in nitrogen availability to that of phosphorus through important regulators such as PHR1 and PHO2 (Table 9.1).

**Table 9.1** Macro and Micro nutrients and their functions

| Nutrient type | Nutrient | Primary functions | Role in physiological/ metabolic processes | Deficiency effects | References |
|---|---|---|---|---|---|
| Macronutrients | Nitrogen (N) | Essential for amino acids, proteins, nucleic acids, chlorophyll, enzymes | Electron transport system (ETS) in chloroplasts and mitochondria; growth, development, productivity | Stunted growth, reduced productivity | Wang et al. (2021a, b, c, d, e), Tewari et al. (2021), Fan et al. (2021), and Kumar et al. (2021a, b) |
| | Phosphorus (P) | Integral to nucleic acids, membrane lipids, ATP, signaling molecules | Photosynthesis, respiration, root development, metabolic processes | Poor root development, reduced growth | Wang et al. (2021a, b, c, d, e), Tewari et al. (2021), Fan et al. (2021) and Kumar et al. (2021a, b) |
| | Potassium (K) | Maintains turgor pressure, regulates stomatal movement, activates enzymes | Water use efficiency, disease resistance, metabolic pathways | Reduced water efficiency, increased disease susceptibility | Wang et al. (2021a, b, c, d, e), Tewari et al. (2021), Toor et al. (2021) |
| | Calcium (ca) | Forms and maintains cell wall, secondary messenger in signal transduction | Stress responses, cell division, physiological regulation | Malfunctioning cell division, stunted growth, developmental abnormalities | Fan et al. (2021), Toor et al. (2021) and Li et al. (2021a, b, c) |
| | Magnesium (Mg) | Central element of chlorophyll, activates enzymes, stabilizes ribosomes | Photosynthesis, nutrient transport | Chlorosis, yellowing of leaves | Tewari et al. (2021), Fan et al. (2021) and Li et al. (2021a, b, c) |
| | Sulfur (S) | Part of amino acids (cysteine, methionine), proteins, glutathione | Coenzyme production, secondary metabolites, antioxidant defense, growth, development | Reduced growth, poor antioxidant defense | Tewari et al. (2021), Fan et al. (2021), Kumar et al. (2021a, b) and Li et al. (2021a, b, c) |

| | | | | | |
|---|---|---|---|---|---|
| Micronutrients | Iron (Fe) | Essential for chlorophyll synthesis | Photosynthesis, enzyme cofactor | Chlorosis, reduced photosynthesis | Kumar et al. (2021a, b) |
| | Copper (Cu) | Cofactor for enzymes, antioxidant defense | Metabolic pathways, oxidative damage protection | Reduced enzyme activity, oxidative damage | Basu et al. (2023), Toor et al. (2021) |
| | Zinc (Zn) | Enzyme activation, osmoregulation, cellular development | Cell division, elongation, water balance, stress tolerance | Stunted growth, poor cellular development | Basu et al. (2023), Toor et al. (2021) |
| | Manganese (Mn) | Enzyme cofactor, chlorophyll synthesis | Photosynthesis, antioxidant defense | Reduced photosynthesis, oxidative damage | Basu et al. (2023), Toor et al. (2021) |
| | Boron (B) | Regulates water balance, cell wall formation | Osmoregulation, drought stress tolerance | Poor water balance, reduced stress tolerance | Basu et al. (2023) |
| | Silicon (Si) | Reinforces cell walls, improves stress resistance | Drought stress tolerance, structural integrity | Reduced stress resistance | Basu et al. (2023) |

## 9.2 Principles of Plant Nutrient Dynamics

### 9.2.1 Absorption, Translocation, and Distribution of Nutrients in Plants

Absorption of nutrients in plants is a complicated process that requires specialized proteins and processes. Nutrient absorption is mainly conducted through the root system. The effectiveness of nutrient uptake highly depends on plant genetics, soil conditions, and environmental factors.

**Nutrient Transport Mechanisms** Transporting nutrients in plants is mediated by the functioning of special transporter proteins that help in transporting them across cellular membranes. These transporters are classified into families based on structural and functional references. Channels and transporters from families such as Shaker, TPK, TPC, KT/HAK/KUP, HKT, and CPA are known to play an important role in potassium (K+) uptake (Wang et al. 2021a, b, c, d, e). NPF and NRT2 transporters for nitrate (NO3-) and AMT1 transporters for ammonium (NH4+) play critical roles in nitrogen (N) uptake, and some aquaporins are also involved in the transport of urea as well (Muratore et al. 2021; Barzana et al. 2021). Phosphorus (P) is taken up as inorganic phosphate (Pi) through either high-affinity PHT transporters or low-affinity paths involving aquaporins (Barzana et al. 2021). Apart from reduction-based or chelation-based strategies of acquisition, iron (Fe) acquisition occurs through strategy I and strategy II, while zinc (Zn) uptake occurs through ZIP proteins and is often modified by cation interactions with other metals like iron (Hanikenne et al. 2021). Silicon is absorbed in the form of its polymer Si(OH)4 and later appends itself to cell walls, forms structures like the endoplasmic reticulum, and intercellular spaccs (Singhal et al. 2023). NSCC and HKT transporters are Na + uptake ones whose function is significant, especially among halophytes and some C4 plants (Singhal et al. 2023). All of these mechanisms work together to ensure the efficient uptake and use of nutrients in plants.

**Interactions Among Nutrients** Interactions, both positive and negative, are common during the absorption of nutrients by plants. For example, mugineic acids increase iron and zinc absorption because of their presence or activity, which is an example of a synergistic interaction (Hanikenne et al. 2021). On the contrary, antagonistic interactions prevent the absorption of some nutrients as with excess zinc which inhibits iron uptake (Hanikenne et al. 2021). Thus, they are essential for nutrient balancing optimizing growth.

### 9.2.2 Nutrient Detection and Signaling Pathways

Plants have developed intricate processes to detect and react to the presence of vital nutrients in their surroundings. The nutrient-sensing pathways are essential for the plant's survival and its capacity to flourish under diverse nutritional situations. Presented below are several critical insights:

*Distinct Nutrient-Specific Sensing Systems*: Plants have unique systems for sensing various nutrients.

*Nitrate*: a primary nitrogen source for plants, is detected by NRT1.1, a dual-affinity nitrate transporter and transceptor (Li et al. 2021a, b, c). NRT1.1 is capable of detecting nitrate both extracellularly and intracellularly (Li et al. 2021a, b, c). In conditions of nitrogen deprivation, plants induce the expression of CLE1/3/4/7 peptides, which inhibit lateral root growth (Li et al. 2021a, b, c). They also generate CEPs that propagate from roots to shoots to indicate nitrogen need (Li et al. 2021a, b, c).

*Phosphorus*: Internal phosphate concentrations are detected by the interaction of SPX domain proteins with inositol polyphosphate (InsP8) (Li et al. 2021a, b, c). This interaction modulates the function of PHR1, a key transcription factor governing phosphate starvation responses (Li et al. 2021a, b, c).

*Potassium*: In calcium signaling pathways, plants detect potassium deprivation. In the case of low potassium contents, two independent calcium signals would be triggered and appropriate actions would be undertaken to cope with the deficit effects (Wang et al. 2021a, b, c, d, e).

## 9.3 Reactive Oxygen Species Biology in Plants

### 9.3.1 Categories of Reactive Oxygen Species in Plant Cells

The most prevalent ROS identified in plant cells include superoxide radicals ($O_2-$), hydrogen peroxide ($H_2O_2$), hydroxyl radicals (·OH), and singlet oxygen ($^1O_2$). These molecules play dual roles in plants, acting as both signaling molecules in essential physiological processes and as contributors to oxidative stress under unfavorable conditions.

The *superoxide radical* ($O_2-$) is primarily produced in chloroplasts by electron transport activities that occur during photosynthesis. These processes encompass the actions of Photosystem I and the photorespiratory cycle, as suggested by many studies (Tavanti et al. 2021; Mansoor et al. 2022; Mangal et al. 2023; Dvořák et al. 2021). According to Tavanti et al. (2021) and Tewari et al. (2021), it can also be produced in mitochondria when the electron transport chain mechanism is disrupted. Because of their extreme reactivity, superoxide radicals jeopardize biological integrity by damaging DNA, lipids, and proteins (Tavanti et al. 2021). $H_2O_2$ is produced when superoxide radicals are dismutated by the enzyme superoxide dismutase (Mansoor et al. 2022; Tewari et al. 2021). Hydrogen peroxide is both stable and able to pass across cell membranes, in contrast to superoxide, which is neither. Stress responses, plant growth, and programmed cell death are all impacted by this crucial signaling molecule (Tavanti et al. 2021; Mansoor et al. 2022).

The *hydroxyl radical* (OH) is thought to be the ROS form that is more reactive and harmful. By destroying DNA, lipids, and proteins, hydroxyl radicals may severely impair regular biological processes. A series of studies conducted by Tavanti et al. (2021), Mansoor et al. (2022), Qamer et al. (2021), and Tomar et al. (2021) have demonstrated that the Fenton reaction, which involves the interaction of hydrogen peroxide with transition metals like iron or copper as catalysts, may yield hydroxyl radicals.

*Singlet oxygen* ($^1O_2$) is an oxygen state that is created during photosynthesis, primarily in chloroplasts, by being activated by light. Tavanti et al. (2021), Mansoor et al. (2022), Dvořák et al. (2021), Qamer et al. (2021), and Tomar et al. (2021) all found that photosynthetic pigments, lipids, and other biological components can be damaged by singlet oxygen due to its high reactivity. Reduced photosynthetic efficiency and increased oxidative stress are two detrimental outcomes of singlet oxygen buildup in stressful environments.

### 9.3.2 Origins of Reactive Oxygen Species (ROS) Production

Essential for the activity of plant cells are reactive oxygen species (ROS), a class of chemicals produced from oxygen. Numerous metabolic pathways result in the production of reactive oxygen species (ROS), with the majority of this production occurring in specialized organelles, including chloroplasts, peroxisomes, and mitochondria. Overproduction of reactive oxygen species (ROS) causes oxidative stress, which damages cellular organelles and components, despite ROS's critical function in cell signaling and proliferation.

**Origin of ROS Production in Plant Cells**

Peroxisomes contribute significantly to the production of reactive oxygen species (ROS), whether they are engaged in photorespiration, fatty acid oxidation, or the detoxifying function. Oxidative stress is caused and signalled by the generation of hydrogen peroxide during various metabolic activities (Mangal et al. 2023; Tavanti et al. 2021; Mansoor et al. 2022; Wani et al. 2021).

Superoxide radical generation occurs from the mitochondrial ETC, including the combined activities of normal respiration and stress conditions, with said mitochondria being another major site of ROS production during such physiological and environmental situations for ROS formation (Mangal et al. 2023; Tavanti et al. 2021; Mansoor et al. 2022; Redza-Dutordoir and Averill-Bates 2021).

Plasma membranes also contribute to superoxide radical synthesis via NADPH oxidases (NOXs), and the apoplast contributes to ROS production through cell wall-associated peroxidases, generating hydrogen peroxide to harden and fortify the plant's cell walls and plant defensive mechanisms (Mansoor et al. 2022; Dvořák et al. 2021).

Superoxide highlights yet another important source of ROS. It includes the aforementioned generation of superoxide anions from mitochondrial ETC during normal respiration and stress conditions. Therefore, making these organelles key players in the creation of ROS under various physiological and environmental conditions (Mangal et al. 2023; Tavanti et al. 2021; Mansoor et al. 2022; Redza-Dutordoir and Averill-Bates 2021).

Above these organelles came some other cellular structures, such as the endoplasmic reticulum and membranes, which come as sources of ROS under specific conditions (Mansoor et al. 2022). Moreover, external environmental conditions such as UV radiation, air pollutants, and heavy metals intensify the level of ROS, making the oxidative stress worse (Tavanti et al. 2021; Redza-Dutordoir and Averill-Bates 2021).

### 9.3.3 Reactive Oxygen Species Scavenging Systems

Plants rely on a highly sophisticated ROS scavenging system to deal with oxidative stresses due to metabolic byproducts, especially under adverse environmental conditions. This system consists of both groupings of enzymatic and non-enzymatic antioxidants, which collectively synergize to detoxify ROS and maintain a cellular redox equilibrium.

One of the most important enzymatic antioxidants, superoxide dismutase (SOD), breaks down sulfur-containing radicals into oxygen and hydrogen peroxide. This is how it prevents the production of the extremely dangerous hydroxyl radical. The multiple SOD isoforms work in distinct parts of the cell to detoxify ROS, and their metal cofactors make them unique (Tewari et al. 2021; Mansoor et al. 2022). Catalase is an additional important enzyme that is mostly found in peroxisomes and helps break down the metabolic byproduct hydrogen peroxide into oxygen and water (Tavanti et al. 2021). Some also include peroxidases (POD) that use other electron donors such as ascorbate and glutathione, to decompose hydrogen peroxide. These enzymes help with ROS detoxification as well as with lignification and pathogen defense processes (Tewari et al. 2021). Ascorbate peroxidase (APX) and glutathione reductase (GR) eliminate hydrogen peroxide and recycle antioxidants such as ascorbate and glutathione in the ascorbate-glutathione cycle, which is another important defence mechanism (Tewari et al. 2021). In order to avoid lipid peroxidation and guarantee cellular integrity, glutathione peroxidase (GPX) reduces organic hydroperoxides and hydrogen peroxide utilizing glutathione (Tavanti et al. 2021).

When it comes to detoxifying reactive oxygen species (ROS), plants not only absorb enzymatic antioxidants but also non-enzymatic antioxidants. One of the best ROS scavengers and oxidatively damaging substances is ascorbate, which is a key component of the ascorbate-glutathione cycle (Mansoor et al. 2022). Similarly, glutathione, a water-soluble antioxidant, effectively detoxifies ROS and allows recycling of antioxidants in the cell-based defense system (Tomar et al. 2021). Tocopherols or Vitamin E are agents that act as lipid-soluble antioxidants that protect cellular membranes against lipid peroxidation caused by ROS (Tewari et al. 2021). Carotenoids are the only ones localized primarily in chloroplasts, and they help neutralize singlet oxygen while also safeguarding photosynthetic pigments from oxidative stress (Tomar et al. 2021). Also, phenolic compounds include flavonoids and phenolic acids, serving as ROS scavengers, metal ion chelators, and modifiers of antioxidant enzyme activity, further strengthening the plant's capability to withstand oxidative stress (Dvořák et al. 2021).

### 9.3.4 Reactive Oxygen Species as Secondary Messengers

**Stress Responses** Reactive oxygen species (ROS) mostly function as secondary messengers among the numerous biotic and abiotic stimuli that plants respond to. Some examples of these are exposure to salt, low or high temperatures, lack of nutrients, intense light, and diseases. The cellular redox equilibrium was disrupted by these stressors, leading to an increase in ROS generation and subsequent activation of signaling pathways further down the cell signal transduction cascade (Mangal et al. 2023; Wang et al. 2021a, b, c, d, e; Tewari et al. 2021; Rivero et al. 2022; Manna et al. 2021; Mansoor et al. 2022; Dvoőák et al. 2021; Wani et al. 2021; Tomar et al. 2021).

**Developmental Processes** They also regulate important developmental processes such as proliferation and differentiation of cells, apoptosis, formation of roots, and interactions with phytohormones (Mangal et al. 2023; Rivero et al. 2022; Mansoor et al. 2022; Dvořák et al. 2021; Wani et al. 2021).

**Long-Distance Signaling** Through a "ROS wave," a self-sustaining mechanism involving NADPH oxidases, calcium channels, and calcium fluxes, ROS can function as long-distance signaling molecules (Tewari et al. 2021; Rivero et al. 2022; Mansoor et al. 2022; Dvořák et al. 2021). Systemic acquired acclimatization is what the "ROS wave" facilitates, enabling plants to create memories of stress for resilience in the future.

### 9.3.5 Mechanisms of ROS Signaling

**Redox Modifications** ROS can directly change the structure and functioning of target proteins in target cells by oxidizing certain amino acids, namely cysteine and methionine (Zandi and Schnug 2022; Dvořák et al. 2021). Such modifications are responsible for changes in protein activity, stability, and interactions with downstream molecules, affecting downstream signaling pathways.

**Calcium Signaling** (Mangal et al. 2023; Wang et al. 2021a, b, c, d, e; Tewari et al. 2021) Calcium-dependent protein kinases and other calcium-binding proteins cause an influx of calcium into the cytosol, which in turn generates reactive oxygen species (ROS). The convergence of reactive oxygen species (ROS) and calcium enhances and fortifies every mechanism in the body that reacts to stress (Fig. 9.3).

**MAPK Pathway Signaling** The mitogen-activated protein kinase (MAPK) cascades, initiated by reactive oxygen species (ROS) (Mangal et al. 2023; Mansoor et al. 2022; Dvořák et al. 2021), are responsible for regulating gene expression and transmitting signals regarding abiotic stress. The phosphorylation of transcription factors regulates the expression of genes involved in stress adaptation and other activities; it is the last stage in these cascades (Fig 9.2).

**Phytohormonal Crosstalk** Reactive oxygen species (ROS) signaling crosses several pathways that regulate plant development, growth, and stress responses (Dvořák et al. 2021; Mangal et al. 2023; Mansoor et al. 2022). The pathways in question comprise auxin, cytokinin, abscisic acid, gibberellins, and ethylene. For example, it affects auxin transport and signaling, which in turn affects root development and stress adaptation.

### 9.3.6 Oxidative Stress and Its Ramifications

New research shows that plants experience oxidative stress when the levels of reactive oxygen species (ROS) exceed their antioxidant capacity (Tewari et al. 2021; Sousa et al. 2022). Consequently, the crop experiences physiological harm, which limits its maturation, growth, and ultimate downfall. Numerous studies have shown that this imbalance disrupts membrane function and leads to major cellular damage, such as lipid peroxidation (Basu et al. 2023; Mangal et al. 2023; Tewari et al. 2021; Tavanti et al. 2021; Bijalwan et al. 2022) Reactive oxygen species (ROS) have the potential to damage DNA, create mutations, and ultimately lead to cell death. A decrease in enzymatic activity can occur when reactive oxygen species (ROS) alter the structure and function of proteins (Mangal et al. 2023; Tavanti et al. 2021; Bijalwan et al. 2022).

Damage caused by free radicals changes important cellular functions. According to several studies, this has a detrimental effect on nutritional metabolism, mitochondrial function, and chloroplast photosynthesis. The works of the aforementioned writers and others like Tewari et al. (2021), Tomar et al. (2021), Sousa et al. (2022), and Mangal et al. (2023) are all part of this body of knowledge. In addition, it can trigger internal programmed cell death (Mangal et al. 2023; Tewari et al. 2021; Bijalwan et al. 2022). According to Sousa et al. (2022), Mangal et al. (2023), and Qamer et al. (2021), these elements can have a negative effect on plant health, which in turn can cause stunted development and growth, decreased agricultural yields, and an increased sensitivity to illnesses (Figs. 9.1, 9.2 and 9.3).

## 9.4 Interactions Between Nutrients and Reactive Oxygen Species

### 9.4.1 Impact of Nutritional Status on Reactive Oxygen Species Generation

Minerals are vital for cellular structure and metabolic processes, with macronutrients (N, P, K, Ca, Mg) serving important structural and regulatory functions (Tewari

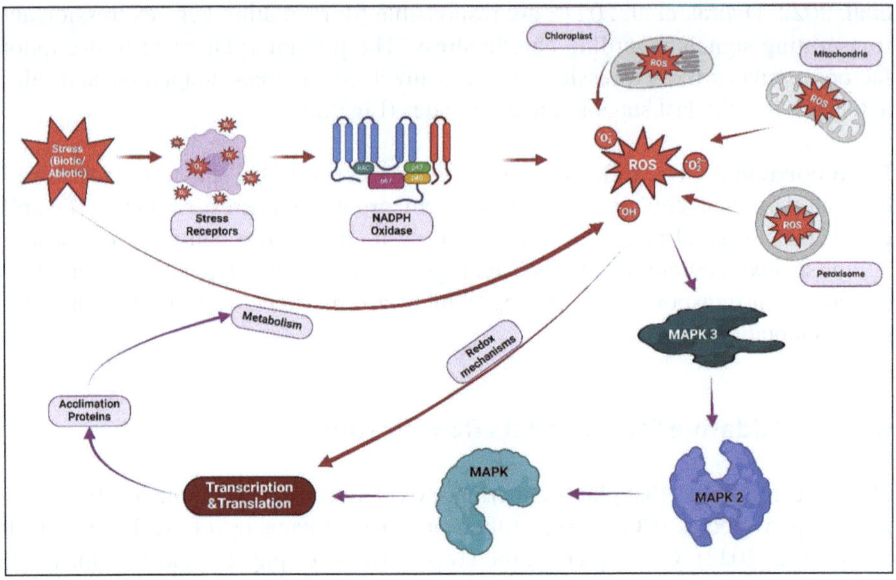


**Fig. 9.1** Signaling pathways associated with ROS, and the mitogen-activated protein kinase cascade in plant cells, particularly towards biotic and abiotic stresses. It involves a complex regulation network which is used by the plants in terms of the handling of the responses to stresses with the use of ROS and MAPK pathways

et al. 2021). Nutrient deficiencies have a considerable effect on the generation of ROS in plants. Nitrogen deficiency impedes growth, induces early senescence, and elevates secondary metabolites, suggesting oxidative stress (Tewari et al. 2021).

Phosphorus deficiency typically results in the improved production of reactive oxygen species in roots and also exhibits potential influence on the calcium signaling generated by stress conditions (Tewari et al. 2021). Potassium deficiency would cause induction of ROS in roots, which subsequently influences the transcription of AtHAK5 together with activation of AtGORK and AtSKOR channels (Wang et al. 2021a, b, c, d, e). Deficiency of magnesium, which leads to the disruption of phloem loading, would lead to sugar accumulation that would subsequently produce ROS, while the decrease in ETS activity and $CO_2$ fixation due to sulfur deficiency would further add to the amount of ROS through contribution from RUBISCO degradation (Tewari et al. 2021).

As a defence mechanism against oxidative stress, plants have both enzymatic and non-enzymatic mechanisms to neutralize processes based on reactive oxygen species (ROS). According to research conducted by Mansoor et al. (2022), damage and cell death can occur when there is an imbalance in the production and elimination of reactive oxygen species (ROS).

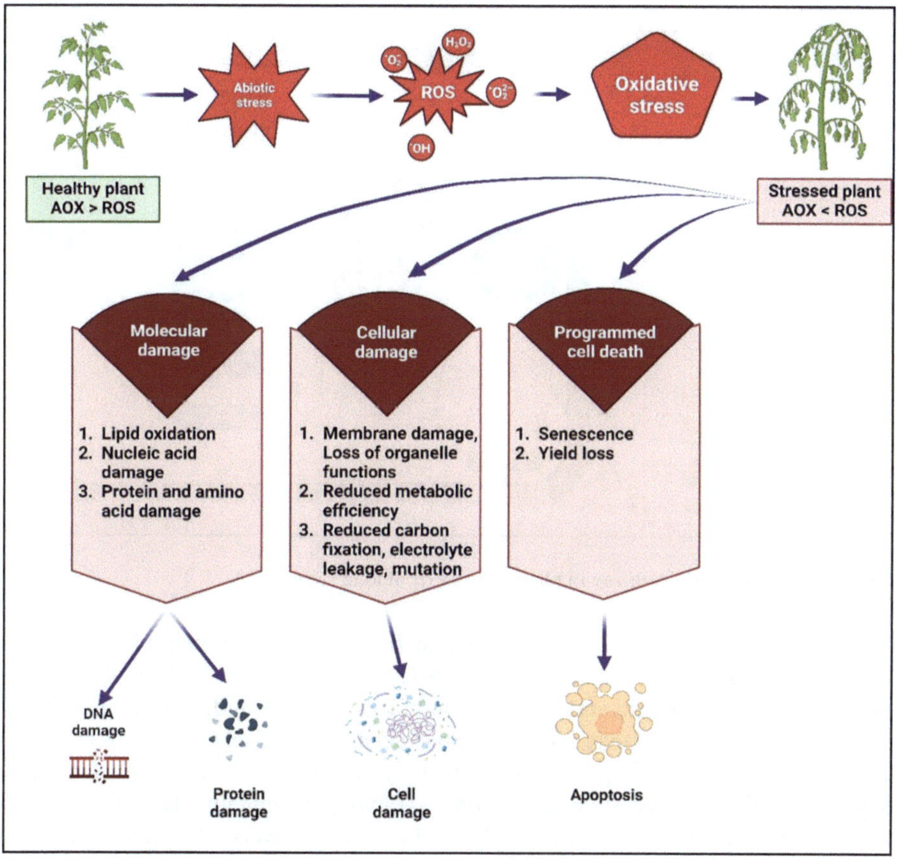


**Fig. 9.2** Influence of oxidative stress upon growth and cellular constancy in plants focusing at the roles played with reactive oxygen species (ROS) and antioxidants (AOX) that could cause molecular and cellular damage, resulting in reduced metabolic efficiency, growth, and yield

### 9.4.2 Influence of Reactive Oxygen Species on Nutrient Absorption and Assimilation

ROS play a crucial role in root architecture that is important for nutrient acquisition (Considine and Foyer 2021; Tarkowski et al. 2023; Foyer and Hanke 2022). Under stress conditions, ROS act on the cytoskeleton and tubulin polymers to disrupt or inhibit cell division and root development. Regulation of this process occurs through a transcription factor named UPBEAT1 (UPB1), which controls peroxidase genes and thus affects the ROS level and root development as well (Considine and Foyer 2021).

ROS gradients in root tips control the growth of roots and root hairs to regulate the capacity of nutrient absorption (Tarkowski et al. 2023). Nutrient deficiency, predominantly in nitrogen and phosphate, would heighten production of reactive

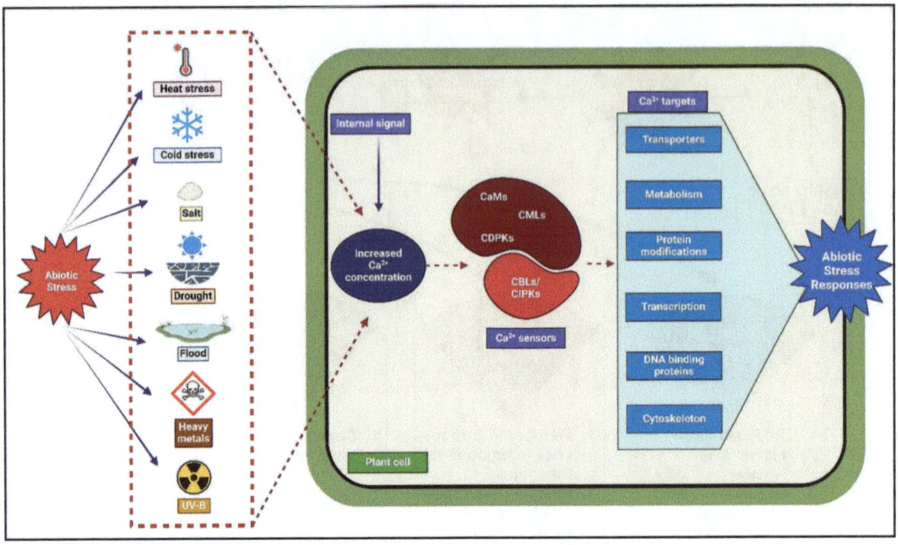


**Fig. 9.3** $Ca^{2+}$ signaling pathway in plant cells under abiotic stress

oxygen species in roots, thus activating signaling molecules to help plants through nutritional stress (Tarkowski et al. 2023). Such signals relayed by ROS direct gene expression and metabolic and developmental modulations under nutrient-deficient conditions (Foyer and Hanke 2022; Tarkowski et al. 2023; Sachdev et al. 2021). Iron sufficiency is crucial for phosphate deficiency responses and balance of reactive oxygen species. Under such conditions, Arabidopsis mutants lacking the iron absorption gene $FRO_2$ demonstrate elevated root ROS levels when grown without phosphate, supporting the interplay between iron and antioxidant strategies (Tarkowski et al. 2023).

### 9.4.3 Antioxidant Mechanisms and Nutrient Accessibility

The relationship between the availability of nutrition and the performance of antioxidant systems in plants is seemingly strongest. This contains the idea that reactive oxygen species and nutrient uptake and absorption are very similar to each other. The need for metal cofactors such as iron and manganese for important antioxidant enzymes like SOD and CAT further supports this probability (Wang et al. 2021a, b, c, d, e). Changes in nutrient availability may directly influence the ability of the plant to control levels of ROS, thus affecting the signaling pathways governing nutrient uptake and digestion.

### 9.4.4 Antioxidant Systems Reliant on Nutrients

Iron has key importance for plant responses under deficiency of phosphate. Arabidopsis mutants deficient in the gene $FRO_2$ have been characterized by a lack of phosphate, resulting in the induction of high ROS in the primary root due to poor uptake of iron (Tarkowski et al. 2023). Antioxidant enzymes require metal cofactors, as superoxide dismutase (SOD) and catalase (CAT) (Sachdev et al. 2021). The maintenance of optimal activity for these two antioxidant enzymes depends upon metals. SOD, which catalyzes the disproportionation of superoxide radicals to hydrogen peroxide, is present in a number of isoforms with iron as its cofactor or manganese as in the case of some. CAT, which converts hydrogen peroxide into water and oxygen, is also dependent on iron or manganese for its catalytic activity (Sachdev et al. 2021). Thus, it can be analyzed that the direct availability of these metals has a very significant effect on the efficiency of those antioxidant enzymes which are essential for the survival of the organism in a given environment.

## 9.5 Macronutrients and Reactive Oxygen Species Homeostasis

### 9.5.1 Balance of Nitrogen and Reactive Oxygen Species

Nitrogen is a vital element for energy homeostasis in plants, influencing photosynthesis and respiration. Absorption and translocation of nitrogen are also processes that consume a lot of energy. Any overuse of nitrogen or nutrient imbalances, especially between phosphorus and potassium, may reduce nitrogen use efficiency (Jiaying et al. 2022). It also regulates ROS inside the plant and helps reduce oxidative damage caused by increased levels of hydrogen peroxide ($H_2O_2$) and malondialdehyde (MDA) (Jiaying et al. 2022).

Without a balanced nutrient supply such as nitrogen and iron, plants cannot achieve optimal growth. Iron deficiency results in chlorosis as well as reduced photosynthesis and metabolic imbalances (Song et al. 2023). Controlled and uncontrolled production of ROS, which are produced during stress and normal condition development, acts as signaling molecules for maintaining normal cellular activities. However, their excessive production results in damage (Zhou et al. 2021). Another function of nitrogen-containing radicals, such as nitric oxide (NO), is to participate in various processes such as defense against invading organisms, metabolism, and even absorption or assimilation of nitrogen (Kumar et al. 2021a, b; Tewari et al. 2021).

Reactive oxygen species (ROS), which also comprise reactive sulphur species (RSS), reactive nitrogen species (RNS), and reactive carbonyl species (RCS), play a significant role in important physiological processes. This is how they work together. Hormone signaling and reactive oxygen species (ROS) production are regulated by lipid peroxidation regulatory complexes (RCS); protein glutathionylation is one way RSS maintains redox equilibrium (Zhou et al. 2021). Fertilization that is both balanced and optimized for nutrients is crucial for these interactions since it increases plant yield and resilience to stress.

### 9.5.2 Deficiency of Phosphorus and Oxidative Stress

Lack of phosphorus is a common cause of oxidative stress in plants (Tewari et al. 2021; Meng et al. 2021). Phosphorus deficiency worsens chloroplast and mitochondrial electron transport system (ETS) function and increases ROS generation, the same as nitrogen starvation does (Tewari et al. 2021). Let's have a look at the processes involved and the effects on plants.

Phosphorus is essential for energy production and regulation since it is a component of ATP, DNA, and phospholipids (Jiaying et al. 2022; Liu et al. 2022a, b). A reduction in orthophosphate levels in chloroplast stroma, which in turn decreases ATP synthase performance, is one result of photosynthetic phosphorus deficiency (Khan et al. 2023a, b; Tewari et al. 2021; Bechtaoui et al. 2021). Furthermore, the photosynthetic electron transport chain might be disrupted due to hydrogen ion buildup in the thylakoids, which would be another consequence of chloroplast energization (Tewari et al. 2021). Furthermore, it can possibly super saturate the electron transport system (ETS), which would lead to electron leakage and ROS development, such as superoxide radicals. Injury to the apparatus is brought about when excess energy remains undissipated (Tewari et al. 2021).

### 9.5.3 Plant Responses to Phosphorus Deficiency and Oxidative Stress

To cope with environmental stress, plants have evolved a variety of mechanisms, including morphological alterations, energy dissipation systems, and antioxidant modes. Antioxidant defence can increase the activity of many enzymes, as far as anybody knows. They are SOD, CAT, POD, APX, GR, and DHAR, among others. Higher concentrations of the non-enzymatic antioxidants glutathione (GSH) and ascorbic acid (AA) are also detected (Meng et al. 2021; Tewari et al. 2021). Multiple energy-dissipation mechanisms, including xanthophyll cycles, water-water cycles, and photorespiratory cycles, shield the entire photosynthetic apparatus from ROS damage at stressful conditions (Tewari et al. 2021). Improved lateral root development and cluster root formation are two potential results. Changes in physiology, including the insertion of parameters that permit problem-free operation (Tewari et al. 2021). Economical, conservational, and responseal shifts are conditional on how species and genotype interact with additional stressors such as heavy metals, drought, and salt (Liu et al. 2022a, b; Bechtaoui et al. 2021; Meng et al. 2021).

### 9.5.4 The Role of Potassium in the Regulation of Reactive Oxygen Species

Decreased photosynthesis and redox state disruption due to decreased photo-assimilate movements are consequences of low potassium. The NADPH oxidases' enhanced activity, ABA connections, and the main signaling component hydrogen peroxide are all discussed in recent studies (Johnson et al. 2022; Tewari et al. 2021). Potassium is crucial in protecting against reactive oxygen species (ROS) since it increases the activity of detoxifying enzymes such as DHAR, GPX, SOD, CAT, APX, and POX. Johnson et al. (2022) and Sardans and Peñuelas (2021) found that at sufficient doses, it inhibits NADPH oxidase activity while enhancing the generation of AsA and GSH. An increase in intracellular calcium and activation of HAK5 transcription are both brought about by a ROS-calcium ($Ca^{2+}$) signaling cascade that is set off by a deficiency of potassium. According to Wang et al. (2021a, b, c, d, e) and Tewari et al. (2021), the $Ca^2$ and calmodulin pathway regulates ROS homeostasis, whereas the AtRAP2.11 transcription factor controls boron transport. Consuming an adequate amount of potassium is crucial for cellular stability, stress tolerance, antioxidant system maintenance, and other related processes (Johnson et al. 2022; Tewari et al. 2021).

### 9.5.5 Metabolism of Sulfur and Synthesis of Antioxidants

Sulfur is essential for biomolecule synthesis, including amino acids (cysteine, methionine), vitamins (thiamine, biotin), glutathione, thioredoxins, lipoic acids, glucosinolates, and phytochelations, which together contribute to plant growth and stress tolerance (Tewari et al. 2021; Narayan et al. 2023; Shah et al. 2022).

The last step in the sulfate reduction, leading to cysteine biosynthesis, starts with the sulfate uptake. Sulfates are converted to adenosine 5′-phosphosulfate (APS), reduced to sulfite and sulfide, and then reacted with O-acetylserine, thereby forming a final molecule of cysteine at this pathway, which in turn links sulfur and nitrogen. It is useful to provide valuable precursors for antioxidants (Narayan et al. 2023; Shah et al. 2022; Andrés-Barrao et al. 2021).

Glutathione (GSH) performs some essential functions such as oxidative stress relief, scavenging ROS, the ascorbic acid-glutatione cycle, and protecting proteins through mixed disulfide formation, as well as detoxification of heavy metals. Sulfur deficiency places GSH harmony in a backward spin that induces further vulnerability to oxidative stress (Narayan et al. 2023; Zhou et al. 2021; Tewari et al. 2021; Zenda et al. 2021).

For enhanced sulfur assimilation and GSH production during stress, RSS energizes ion channels and merges with ROSs in this regard to perform functional assignments in the regulation of protein glutathionylation and induction of antioxidant systems. Under these conditions, ethylene further intensifies sulfur assimilation, which is thought to fortify the cellular defense mechanism (Zhou et al. 2021; Fatma et al. 2021).

### 9.5.6 Calcium Signaling in Reactive Oxygen Species-Mediated Stress Responses

**Calcium as a Secondary Messenger** Calcium ions (Ca2+) serve as essential secondary messengers, analogous to ROS, facilitating the transduction of stress signals into cellular responses (Tewari et al. 2021; Bhar et al. 2023; Lee and Seo 2021; Li et al. 2022a, b, c).

**Stress-Induced $Ca^{2+}$ Signatures** Abiotic stressors provoke swift elevations in cytosolic $Ca^{2+}$ ($[Ca^{2+}]$cyt) from both external and internal origins. The distinctive "calcium signatures" are recognized by $Ca^{2+}$ sensors to initiate downstream signaling (Tewari et al. 2021; Lee and Seo 2021; Li et al. 2022a, b, c; Ren et al. 2021; Bhar et al. 2023).

**ROS-Induced Calcium Influx** Calcium influx can be enhanced in phenomena such as ROS-induced calcium rise (RICR) when reactive oxygen species (ROS) directly activate calcium channels, such as annexins, cyclic nucleotide-gated channels (CNGCs), and SKOR (Bhar et al. 2023; Lee and Seo 2021; Li et al. 2022a, b, c).

**$Ca^{2+}$ Regulation of ROS Production** The formation of reactive oxygen species (ROS) is regulated by calcium ions. The activation of ROS-generating respiratory burst oxidase homologs (RBOHs) is dependent on calcium signaling. According to Marcec and Tanaka (2021), the generation of reactive oxygen species (ROS) is enhanced when respiratory burst oxidase homologs (RBOHs) are phosphorylated by calcium-dependent protein kinases.

**Reciprocal Calcium-ROS Interaction** The interaction between calcium and reactive oxygen species (ROS) entails positive feedback mechanisms, wherein ROS-induced $Ca^{2+}$ influx activates RBOHs, hence augmenting ROS production and amplifying systemic $Ca^{2+}$ waves for adaptive responses (Sardans and Peñuelas 2021; Lee and Seo 2021).

## 9.6 Micronutrients in Reactive Oxygen Species Metabolism

### 9.6.1 Iron and Fenton Reaction in Reactive Oxygen Species Formation

**The Role of Iron in Reactive Oxygen Species Production**
Overconsumption of iron, due to its redox characteristics, causes oxidative stress by upsetting the cellular redox balance and facilitating the creation of reactive oxygen species via the Fenton reaction (Lapaz et al. 2022). Free iron generates a pro-oxidant milieu, increasing the likelihood of oxidative damage (Lapaz et al. 2022).

**Fenton Reaction Mechanism**
Iron decomposes hydrogen peroxide ($H_2O_2$) to generate hydroxyl radicals (HO•), a highly reactive ROS. The reaction is represented as $Fe^{2+} + H_2O_2 \rightarrow Fe^{3+} + HO\bullet + OH^-$ (Lapaz et al. 2022; Regon et al. 2022). The Haber–Weiss cycle facilitates this process by converting $Fe^{3+}$ back to $Fe^{2+}$ by contact with superoxide anions ($O_2\bullet^-$) (Lapaz et al. 2022).

**Impact of pH and Metal Chelation**
The Fenton reaction is influenced by pH and iron concentration, with oxoiron (IV) species predominant at pH levels exceeding 3, while hydroxyl radicals (HO•) prevail in acidic environments (Rai et al. 2021). Chelation of iron and other metals, such as copper, assists organisms in regulating ROS production (Kumar et al. 2021a, b; Krämer and Clemens 2006).

**Cellular Damage Induced by Reactive Oxygen Species and Plant Defenses**
According to Lapaz et al. (2022), reactive oxygen species (ROS) produced by Fenton have the potential to harm proteins, lipids, and nucleic acids, which in turn can cause a disruption of biological processes, lipid peroxidation, and tissue degeneration. Enzymatic defences (e.g., superoxide dismutase, catalase, and peroxidase) and non-enzymatic antioxidants (e.g., ascorbate and glutathione) help plants reduce these impacts (Lapaz et al. 2022).

### 9.6.2 The Structural Function of Zinc in Antioxidant Enzymes

**The Structural Function of Zinc in Antioxidant Enzymes** Zinc has an important structural role in copper-zinc superoxide dismutase (Cu-Zn SOD). The tetrahedral complexation of zinc makes it quite stable and suitable for sensitive biomolecules such as DNA (Rai et al. 2021). Zinc gives structural support to Cu-Zn superoxide dismutase, and copper acts as the catalytic center (Zewide and Sherefu 2021).

**Zinc and the Defense Against Oxidative Stress** According to Zewide and Sherefu (2021), deficiency of zinc lowers Cu-Zn SOD activity and induces the buildup of superoxide radicals lipid peroxidation. Hence, structural function involves the role of zinc in Cu-Zn superoxide dismutase in the plant defense mechanism against oxidative stress (Krämer and Clemens 2006).

**Significance of Zinc** Zinc serves in making the cofactors in enzymes, which, in turn, help in the DNA transcription, protein synthesis, and enzyme activity in metabolism, that is metabolism of carbohydrates, nucleic acids, and lipids (Rai et al. 2021; Bashir et al. 2016). Zinc deficiency brings about poor growth, chlorosis, causing smaller leaf size, spikelet sterility, compromised protein activities, and increased susceptibility to infections (Rai et al. 2021). Zinc also has a structural role in antioxidant enzymes like Cu-Zn SOD that are necessary for mitigating oxidative stress, besides rendering a plethora of cofactor roles, which altogether make it non-negotiable for any general plant health or abilities to combat stresses.

### 9.6.3 Manganese and the Activity of Superoxide Dismutase

**Manganese and Manganese Superoxide Dismutase (Mn-SOD)** The enzyme manganese superoxide dismutase, or Mn-SOD, is an important component of mitochondria. Manganese superoxide dismutase (Mn-SOD) prevents oxidative stress in cells by transforming superoxide radicals into oxygen and hydrogen peroxide. When paired with other enzymes that degrade hydrogen peroxide, such as catalases and peroxidases, Mn-SOD has a more robust protective function, as stated by Dumanović et al. (2021) and Abedi and Shahpiri (2024).

**The Function of Mn-SOD in Stress Tolerance** It is indicated that the gene that encodes manganese superoxide dismutase (AmSOD2) in the Avicennia mangrove plant is upregulated due to various stress conditions like salinity, hydrogen peroxide, and heavy metals. This has emphasized its importance in the antioxidant defense of the plant (Abedi and Shahpiri 2024).

**Transport of Manganese for Manganese Superoxide Dismutase (Mn-SOD)** Specialized mitochondrial transporters such as AtMTM1 and AtMTM2 enable the transfer of manganese to Mn-SOD in Arabidopsis. Alterations in these transporters can reduce Mn-SOD activity, underscoring the importance of manganese availability (Hu and Jinn 2022).

**The Specificity of Manganese for Mn-Superoxide Dismutase** Manganese therapy enhances Mn-SOD activity, however other metals such as Fe, Cu, and Zn are unable to sufficiently replace it. Manganese is essential for the proper folding and assembly of Mn-SOD into its active tetrameric configuration (Hu and Jinn 2022; Abedi and Shahpiri 2024).

**Manganese and Additional Antioxidants** There is evidence that manganese stress regulates the activities of other antioxidant enzymes, including peroxidase and catalase (Zhenggang et al. 2023; Mangal et al. 2023).

### 9.6.4 The Dual Function of Copper in the Generation and Scavenging of ROS

Copper serves a dual function in plants, functioning as an essential nutrient while also posing a possible toxic threat.

Excess copper can promote the generation of ROS through Fenton-like processes, resulting in photosynthetic impairment and chlorophyll degradation (Wairich et al. 2022; Krämer and Clemens 2006; Xu et al. 2024; Kumar et al. 2021a, b).

Cu-Zn superoxide dismutase and other antioxidant enzymes rely on copper as a cofactor (Rai et al. 2021; Krämer and Clemens 2006; Wang et al. 2022). Plants also play a role in the battle against ROS. When copper is unavailable in plants, other enzymes based on metals are utilized, as stated by Rai et al. (2021). According to Rai et al. (2021), Wairich et al. (2022), and Xu et al. (2024), copper homeostasis cannot be achieved without specific transport, chaperoning, and sequestration mechanisms. Copper is essential for homeostasis maintenance, which depends on many transport, chaperoning, and sequestration mechanisms (Rai et al. 2021; Wairich et al. 2022). Among the several transcription factors that control the copper response is SPL7. As stated by Wairich et al. (2022).

## 9.7 Interactions Between Nutrients and Reactive Oxygen Species in Tolerance to Abiotic Stress

### 9.7.1 Drought-Induced Stress

**Reactive Oxygen Species Production and Effects under Drought Stress** Lipid peroxygenation, reduced photosynthetic efficiency, and stunted development are all effects of drought stress-induced aberrant increases in the production of superoxide anion radicals and hydrogen peroxide ($H_2O_2$). Sun et al. 2020 found that food availability and reactive oxygen species are two factors that impact plant tolerance mechanisms.

**Accumulation of Osmolytes and Antioxidant Defense Mechanisms** When plants experience drought stress, they acquire soluble sugars and add proline, an osmolyte, to stabilize membrane proteins (Sun et al. 2020). Furthermore, they promote antioxidant defences by activating several enzymes such as SOD, POD, CAT, GR, and APX, which can eliminate excess ROS (Sun et al. 2020).

**The Function of Nutrients in Drought Resilience** Plant viability is significantly impacted by drought stress, which lowers nitrogen absorption. Phosphorus, potassium, and nitrogen are nutrients that help plants withstand drought by boosting their antioxidant response. Their absence lowers the ability to scavenge reactive oxygen species (Lee et al. 2021; Swain et al. 2023; Ihtisham et al. 2023).

Nitrogen is crucial for plant physiology as it helps form chlorophyll, amino acids, and proteins needed to process the stressful conditions (Swain et al. 2023). Further, potassium plays a role in osmosis and activation of multiple enzymes, thus contributing to the equilibrium of water in plant cells (Ihtisham et al. 2023). Besides, phosphorus is a fundamental part of storing energy, and during stressful phases when plants need a lot of energy to continue metabolic processes (Ihtisham et al. 2023).

**Indirect Nutrient Influences on Drought Resilience**
Modulation of phytohormones along with nutrients, like abscisic acid and cytokinins, leads to modification in drought responses (Li et al. 2022a, b, c; Swain et al. 2023). Other nutrients like nitrogen and phosphorus encourage root development, besides the microbiome improvement in the rhizosphere during drought conditions, which leads to augmented uptake of water and nutrients (Kang et al. 2022).

### 9.7.2 Salinity Stress

**Reactive Oxygen Species and Salinity Stress** According to Tarkowski et al. (2023) and Ihtisham et al. (2023), plants that are subjected to salt stress produce an excess of reactive oxygen species (ROS), which causes oxidative stress, cellular damage, and stunted growth. In order to build tolerance mechanisms and mitigate the negative effects of nutrients, plants rely on these nutrient–ROS interactions (Swain et al. 2023; Kang et al. 2022).

**The Role of Nutrients in Reactive Oxygen Species Balance** In response to salt stress, antioxidant enzymes stabilize ROS levels, shielding cells from oxidative damage (Ihtisham et al. 2023). The synthesis of chlorophyll and proteins/amino acids—which are critical in reacting to stress—both require nitrogen, as stated by Swain et al. (2023). According to Tarkowski et al. (2023), potassium helps control osmotic pressure, activates enzymes, and keeps the water balance inside cells. Producing energy and surviving under harsh conditions are hence dependent on phosphorus (Ihtisham et al. 2023).

**Instances of Nutrient-Reactive Oxygen Species Interactions**
Nitrogen fertilizers increase the salinity tolerance level of plants through increased antioxidant defense pathways, increased carbon allocation to plant tissue, and the storage of carbohydrates (Swain et al. 2023). It is potassium that properly regulates sodium absorption by modulating action at the NRT1.5 transporter, which is very important in maintaining the balance of ions and the resistance of plants against salt stress (Pantha et al. 2023). Whereas, zinc enhances tolerance to salt by increasing proline levels, which is an osmolyte that prevents cells from being damaged, while at the same time facilitating photosynthesis and chelating harmful metals (Mushtaq et al. 2023).

The plants thus try to maintain potassium and sodium levels in a ratio that causes immobility to these ions in case of salt stress for nutrient and ion homeostasis. However, excess potassium can be as harmful as high sodium, which may require proportional balancing methods (Pantha et al. 2023).

Soils with salts also hinder nutrient assimilation by accumulating salts obstructing the uptake of vital nutrients such as potassium, calcium, and magnesium, just making the effects of stress more miserable (Swain et al. 2023).

### 9.7.3 Temperature Extremes (High and Low)

**Temperature-Induced ROS Generation** Both hot and cold extreme temperatures increase intracellular ROS levels in plants and produce oxidative stress that impairs cellular activities, disrupts membranes, and retards growth. The plants respond to such stresses through mechanisms influenced by food availability that allows the activities of antioxidant enzymes vital for clearance of ROS and redox homeostasis.

**Interactions Between Heat Stress and Reactive Oxygen Species** Membrane fluidity is reduced by heat stress, increasing cytosolic calcium levels and then triggering calcium signaling pathways and ROS production by the action of Respiratory Burst Oxidase Homolog (RBOH) in the plasma membrane (Devireddy et al. 2021; Kumar et al. 2021a, b). The ROS produced during the heat stress will serve as signals to upregulate the protective mechanisms, such as heat shock proteins (HSPs) for the chaperoning of cellular components (Babbar et al. 2021).

**Essential Nutrient Functions in Thermal Resilience** Nitrogen is vital for the synthesis of chlorophyll, amino acids, and proteins necessary for regulating heat stress (Swain et al. 2023). Potassium facilitates osmotic control and enzyme activation, enhancing water balance and metabolic function under elevated temperatures (Ihtisham et al. 2023).

Phosphorus has a vital role in energy storage and synthesis, essential for managing heat stress (Ihtisham et al. 2023).

**Instances of Nutrient-Heat Stress Interactions**

*Potassium*: Modulates stomatal activity to sustain interior temperature (Swain et al. 2023).

*Boron, manganese, and selenium*: These micronutrients mitigate heat stress by enhancing physiological and metabolic functions; selenium combined with salicylic acid increases antioxidants, neutralizing reactive oxygen species, and stabilizing membranes (Swain et al. 2023).

*Mechanisms of Cold Stress and Reactive Oxygen Species*: Cold stress diminishes photosynthetic efficiency, resulting in electron leakage from the electron transport chain and consequent accumulation of reactive oxygen species in chloroplasts (Devireddy et al. 2021). It diminishes membrane fluidity, activates calcium channels, and initiates RBOH-mediated reactive oxygen species generation. The C-REPEAT BINDING FACTOR/COLD-REGULATED (CBF/COR) pathway is essential for cold tolerance, governing genes responsible for ROS detoxification and other protective mechanisms, whereas abscisic acid (ABA) controls CBF and COR genes to improve cold resistance (Devireddy et al. 2021).

*Nutritional Contributions to Cold Resilience*: Nitrogen and phosphorus enhance reactive oxygen species scavenging and membrane stability, hence conferring tolerance to cold (Ihtisham et al. 2023). Potassium facilitates growth, enzymatic activity, charge equilibrium, and osmotic control in cold environments (Ihtisham et al. 2023).

### 9.7.4 Heavy Metal Poisoning

Heavy metals, including lead, cadmium, copper, and zinc, cause oxidative stress in plants, which in turn reduces their growth and production. Metals, similar to heat, dryness, and salt, produce an excess of reactive oxygen species (ROS), which damage cells. Tarkowski et al. (2023), Pardo-Hernández et al. (2021), and Cha et al. (2022) are the sources that they reference. In low concentrations, reactive oxygen species (ROS) in the blood serve as signaling molecules that activate heavy metal defence systems. According to two studies (Wani et al. 2021; Li et al. 2022a, b, c), metal toxicity can be reduced by the activation of detoxification and antioxidant defence genes through the phosphorylation of transcription factors by a ROS-responsive MAPK cascade. The absorption of essential nutrients is hindered by heavy metal stress, particularly lead and cadmium (Pardo-Hernández et al. 2021). While imbalanced nutrition weakens plant defences, healthy eating boosts stress tolerance by increasing antioxidant responses and protective pathways mediated by reactive oxygen species (ROS).

### 9.7.5 Stress Due to Nutrient Insufficiency

**Reactive Oxygen Species and Nutrient Deficiency** Elevated levels of reactive oxygen species are characteristic of nutritional deficiency in plants, especially in tissues that are lacking in nutrients. According to studies conducted by Khan et al. (2023a, b) and Tarkowski et al. (2023), an activation signal and disturbance of cellular metabolism are the probable causes of an increase in reactive oxygen species (ROS). In response to nutritional stress, this signal activates pathways.

**Root Architecture and Robot Operating System Indication** The architecture of root systems serves nutrient acquisition, while ROS substantially dictates root development patterns under nutrient deprivation. Nitrogen deficiency in Arabidopsis thaliana causes local accumulation of ROS right at the tips of root hairs, thus promoting their elongation. This increases the surface area of the roots, hence improving the capacity of the plant to acquire limited nutrients from the soil (Tarkowski et al. 2023).

**Interaction of Reactive Oxygen Species with Hormones** Signaling through reactive oxygen species interacts with the hormonal pathways in influencing plant response to a shortage in food. Dysfunction of nutrients modifies ROS homeostasis and concentrations of hormones that finally impact the cell functions, such as cell division and growth. Such changes are visible in root growth alterations that enhance nutrient uptake during stress (Tarkowski et al. 2023; Rane et al. 2022).

**Dynamics of Spatial and Temporal Reactive Oxygen Species** To understand how plants react to nutritional stress, one must know when and where in their tissues reactive oxygen species are produced. Subsequent signaling processes may be dictated by certain ROS signals in particular cell types and tissue locations, impacting the adaptive response of the plant (Tarkowski et al. 2023).

## 9.8 Interaction Between Nutrients and Reactive Oxygen Species in Biotic Stress Responses

### 9.8.1 Defensive Systems Against Pathogens

Plants and pathogens are constantly engaged in a dynamic scenario of an "arms race" with regard to nutrient availability and ROS during infections. The nutrient deficiencies in nitrogen would lead to enhanced virulence of the pathogen by activation of several genes involved in swarming motility, type III secretion systems (T3SS), and other metabolic pathways such as GABA and polyketide synthesis. These are examples seen with *Pseudomonas syringae* (Tripathi et al. 2022). Furthermore, nitrogen starvation induces effector genes in *Magnaporthe oryzae*, which include hrp, avr, and MPG1, to inhibit host defenses and control infections (Tripathi et al. 2022).

Nitrogen plays an important role in plant defence mechanisms by enhancing the production of reactive oxygen species in diseased regions. They do more than just strengthen cell walls; they also cause oxidative damage to pathogens, indicate systemic resistance, and induce programmed cell death (PCD) (Prasad et al. 2023; Yang et al. 2022). On the other hand, these mutual mechanisms are counteracted by pathogens through the introduction of enzymes like catalases and peroxidases that detoxify ROS or by employing effectors that inhibit ROS production and immune responses (Mandon et al. 2021; Zhang et al. 2023). For instance, *Magnaporthe oryzae* alters host condition through MoOsm1 kinase phosphorylation and the organelle enzyme MoKat2 to reduce host ROS and increase susceptibility (Liu and Zhang 2021; Zhang et al. 2023). This situation demonstrates the fact that certain pathogens adaptively counteract plant defenses.

### 9.8.2 Responses of Herbivores to Attacks

ROS production, nutrient poring, and other mechanisms are among the several mechanisms that plants employ to counter herbivory. Mechanical damage or ingestion of herbivore saliva triggers the production of ROS, including hydrogen peroxide, which deters herbivores by damaging their digestive systems or disrupting their activity (Gandhi et al. 2021; Mostafa et al. 2022). It also seems that the ability of the plants to induce "herbivore-induced resource sequestration" or redirection of allocation of nutrients into source tissues to resource-locked areas, such as roots, is indirectly involved both in hormone signaling and in the production and metabolism of reactive oxygen species (Thompson et al. 2022). Availability of nutrients also determines herbivory and plant defense; for example, nitrogen deficiency in maize, for instance, affects the performance of fall armyworm larvae, as well as the mechanism of plant defense (Mason et al. 2022).

### 9.8.3 Reactive Oxygen Species in Symbiotic Interactions

Of the beneficial or detrimental roles that ROS play, plant–microbe interactions fall into this category. In symbiotic relationships, for example, with rhizobia or ectomycorrhizal fungi, ROS are vital for the onset and maintenance of the association. Their dual role is defined by the concentration: low concentrations promote signaling for symbiosis, whereas high concentrations inhibit it, pointing to the critical nature of achieving the right balance between reduction and oxidation (Khan et al. 2023a, b; Parra-Aguilar et al. 2023; Kumar et al. 2021a, b).

**The Symbiosis Between Rhizobia and Legumes**

In legumes during initial stages of infection by rhizobia, ROS act as main signaling molecules, which cause curling of root hairs followed by formation of infection threads (Minguillón et al. 2022). ROS continue to further regulate the proliferation and differentiation of cells, which are important to the development of nodules. Nitrogen-fixing nodules are formed under conditions created by ROS that lead to microoxic conditions essential for the survival and functioning of rhizobia, differentiating into bacteroids, which possess the ability to fix nitrogen. These conditions are properly regulated by ROS as well as antioxidant activity to ensure proper nitrogenase functioning (Minguillón et al. 2022).

**Mycorrhizal Symbiosis**

In mycorrhizal symbiosis, the generation of ROS becomes important in the early colonization stages of fungi and in plant-fungus recognition. They convey signaling roles as preconditions for establishing symbiosis and the further colonizing process of plant roots by mycorrhizal fungi. Additionally, they are involved in the formation and function of arbuscules, which are protrusions in plants that enable the exchange of nutrients between the two organisms. Normal ROS signaling would mean that arbuscules were correctly developed within root cells to ensure proper nutrient transfer (Parra-Aguilar et al. 2023; Kumar et al. 2021a, b).

**Reactive Oxygen Species as Two Edges of a Sword**

Among their many functions, the ROS act as important signaling molecules that activate defense genes and regulate the expression of other genes essential for the establishment and maintenance of the symbiosis (Khan et al. 2023a, b; Parra-Aguilar et al. 2023). But if this ROS gets too much, then it will destroy more amounts of benefits, rendering the delicate balance for beneficial plant-microbe interactions.

**Managing Oxidative Stress**

Enzymatic components (such as superoxide dismutase and catalase) and non-enzymatic compounds (such as ascorbate and glutathione) make up the antioxidant systems of plants and microorganisms, which are crucial for maintaining ROS homeostasis (Khan et al. 2023a, b; Parra-Aguilar et al. 2023; Minguillón et al. 2022). Such systems mitigate oxidative stress so as to preserve the stability of symbiotic interaction while preventing damage inside cells.

## 9.9 Molecular Mechanisms Connecting Nutrient Availability and Reactive Oxygen Species Homeostasis

### 9.9.1 Transcription Factors Implicated in Nutrient-ROS Interactions

**Nitrogen Deficiency and Reactive Oxygen Species Generation** Under conditions of nitrogen deficiency, there will be a generation of ROS. The role of glutaredoxins in the nuclear response to nitrate starvation has also been shown, where the elimination of ROS can also suppress this response (Ludewig et al. 2021). The NIGT1/HHO family of transcriptional repressors is involved in phosphate deficiency responses while concurrently inhibiting responses to nitrate scarcity, underscoring the complicated nature of the connection between nutritional signaling and reactive oxygen species (Ludewig et al. 2021).

**Phosphate Deficiency and Reactive Oxygen Species** Impacts of Nutrient Deprivation: Deficiencies in potassium, nitrogen, or phosphate can result in increased hydrogen peroxide ($H_2O_2$) concentrations in different areas of Arabidopsis roots, signifying a stress response associated with nutritional deficiency (Ludewig et al. 2021).

**Transcription Factors Modulating Reactive Oxygen Species Metabolism** Transcription factors are crucial in regulating ROS metabolism during abiotic stress, affecting antioxidant enzyme activity and gene expression (Table 9.2).

### 9.9.2 Posttranslational Alterations in Reactive Oxygen Species Signaling

Cellular responses to stresses such as biotic and abiotic stress largely depend on post-translational modifications (PTMs) associated with ROS signaling (Matamoros and Becana 2021; Rai and Kaushik 2023). There might be changes in protein function, stability, localization, and interaction, all of which facilitate plant response to stresses from the environment (Matamoros and Becana 2021).

**Principal Posttranslational Modifications Encompass**

- *Methionine sulfoxidation*: A reversible oxidative modification of methionine residues that affects protein structure and function. Oxidation of cysteine thiols to several forms, influencing protein functionality and redox signaling.
- *S-nitrosylation*: The reversible binding of nitric oxide to cysteine thiols, which plays a role in plant immunity and stress responses.
- *S-glutathionylation*: The conjugation of glutathione to cysteine thiols, which safeguards proteins against permanent oxidation and influences signal transduction.

**Table 9.2** Details of transcription factors

| Transcription factor | Organism/source | Function | Key effects | References |
|---|---|---|---|---|
| WRKY (e.g., AtWRKY30, ZmWRKY40) | Arabidopsis, Maize | The resistance to drought has been improved by raising the activity of antioxidant enzymes (POD, CAT, SOD) and upregulating the expression of stress-related genes (for example, RD29A, DREB2A) | Reduces ROS levels. | Wang et al. (2018), Wang et al. (2021a, b, c, d, e) |
| AhWRKY75 | Peanut | Enhances salt tolerance | Boosts ROS scavenging and optimizes photosynthetic efficiency | Zhu et al. (2021) |
| EjWRKY17 | Eriobotrya japonica | Enhances drought tolerance | Reduces ROS levels in Arabidopsis | Wang et al. (2021a, b, c, d, e) |
| FtWRKY46 | Tartary buckwheat | Improves stress tolerance | Lowers ROS levels | Lv et al. (2020) |
| NAC (e.g., JUNGBRUNNEN1, SlNAC6) | Tomato | Enhances drought tolerance by activating drought-response genes and increasing glutathione synthesis | Reduces ROS buildup | Thirumalaikumar et al. (2018), Jian et al. (2021) |
| BnaNAC55 | Brassica napus | Modulates genes for ROS generation and scavenging | Enhances drought tolerance | Niu et al. (2016) |
| AP2/ERF (e.g., NtERF172, JERF3) | Tobacco | Stimulates expression of catalase (CAT) and superoxide dismutase (SOD) | Modulates $H_2O_2$ homeostasis, enhancing drought tolerance | Liu et al. (2022a, b) |
| MYB (e.g., GbMYB5, TaMyb1D, PbrMYB21) | Cotton, Tobacco, Pyrus betulaefolia | Regulates antioxidant enzyme genes and polyamine accumulation | Assists hydrogen peroxide removal, improving drought tolerance | Chen et al. (2015), Wei et al. (2017) and Li et al. (2017) |
| bHLH (e.g., AhbHLH112, MdbHLH130) | Peanut, Apple | Enhances ROS scavenging. | Improves drought stress responses | Li et al. (2021a, b, c), Zhao et al. (2020) |
| AhbHLH112 | Peanut | Directly stimulates peroxidase (POD) gene | Enhances drought resistance in Arabidopsis | Li et al. (2021a, b, c) |
| bHLH123 | Tobacco | Enhances salt tolerance via NtbHLH123-NtRbohE signaling | Improves stress response mechanisms | Liu et al. (2021) |

- *Persulfidation*: The transformation of thiols derived from cysteine into persulfides can affect the structure and function of proteins and may also offer some protection against oxidative stress.

Other irreversible post-translational modifications, including carbonylation and glycation, have effects on the functionality and degradation of such proteins. Further studies should focus on understanding the mechanisms, quantification, and assessment of the impacts of PTMs on plant stress responses (Matamoros and Becana 2021).

### 9.9.3 Epigenetic Modulation of Stress Reactions

When faced with environmental stresses such as drought, salt, heat, or cold, or even nutritional deficits, it employs epigenetic processes that include DNA methylation and histone changes to activate or repress particular genes (Liu et al. 2022a, b; Yung et al. 2023; Nunez-Vazquez et al. 2022).

**DNA Methylation** Methylation of DNA is an epigenetic change that may activate or silence genes (Liu et al. 2022a, b; Yung et al. 2023). According to Liu et al. (2022a, b), the change in DNA methylation pattern is often a result of abiotic stress and controls the expression of genes associated with stress. One example is the adaptive response of plants to various stresses, which includes the repression of gene expression in response to drought by hypermethylation of the promoter region of ABA-inducible genes and a similar phenomenon in response to salt stress by hypomethylation of cytosine residues (Liu et al. 2022a, b).

**Histone Modifications** Modification of histones, acetylation, methylation, and phosphorylation, includes changes in the expression of genes by affecting their accessibility for chromatin (Liu et al. 2022a, b; Nunez-Vazquez et al. 2022). Heat stress causes acetylation of histones, activating the heat shock protein genes, whereas drought is responsible for deacetylation of histones, thus downregulating growth-related genes (Liu et al. 2022a, b). In this way, they act as dynamic regulators of gene expression during abiotic stress (Nunez-Vazquez et al. 2022).

**Interaction of DNA Methylation with Histone Modifications** In response to stress, many genes are regulated by a process that involves DNA methylation and histone changes (Halder et al. 2022; Kumar et al. 2021a, b). To illustrate the point, when iron levels are low, histones are acetylated and demethylated at the promoters of genes that respond to iron, which causes these genes to be expressed (Liu et al. 2022a, b).

**Epigenetic Memory** Plants develop resilience to future shocks through the formation of "epigenetic memories" when some alterations have long-lasting consequences even when the stressor is no longer present (Liu et al. 2022a, b; Miryeganeh 2021). One kind of memory is somatic memory, which allows for adaptation to changing environments. For instance, when people experience drought stress, it can lead to changes in DNA methylation that are passed down through generations, allowing for increased drought resistance (Miryeganeh 2021).

### 9.9.4 Hormonal Connections in Nutrient-ROS Signaling

Plant hormones regulate various stress responses by interacting with nutrient signaling and ROS signaling. Growth regulators (auxins, cytokinins, gibberellins, brassinosteroids) and stress hormones (ABA, JA, SA, ethylene) mediate adaptation to abiotic stress (Napieraj et al. 2023). ABA occupies a central position in the stress response: it upregulates ROS production via NADPH oxidases, while ROS feedback in stimulating ABA biosynthesis downregulates its own production (due to stimulation of metabolism) (Liu et al. 2022a, b; Li et al. 2021a, b, c). Solute disruption and electron leakage during stress serve to amplify oxidative stress. Deficiency of certain nutrients, such as nitrogen or phosphorous, brings ROS to accumulating in the cells. But ABA and JA can help cells maintain homeostasis even when they're under stress by reducing oxidative damage and improving nutrient absorption (Kim et al. 2021; According to Napieraj et al. (2023). A study conducted by Singh et al. (2024) found that the balance of reactive oxygen species (ROS) is impacted by the interplay between these hormones, which in turn affects the control of antioxidant defence mechanisms and the production of genes related to stress tolerance. An even larger increase in ROS levels may be necessary to activate stress response pathways after NADPH oxidase activation. Hormones and reactive oxygen species (ROS) interact to form a dynamic regulatory network, which changes during development in response to various stimuli (Considine and Foyer 2021; Ravi et al. 2023; Li et al. 2022a, b, c; Kolupaev et al. 2023).

## 9.10 Strategies for Augmenting Stress Tolerance via Nutrient and Reactive Oxygen Species Management

### 9.10.1 Nutrient Supplementation and Priming to Modulate ROS Levels

Plants experience abiotic constraints of various types that limit their growth and productivity. These stresses can be mitigated using different strategies carried out by plants to control ROS levels. Priming is the best pre-treating technique that increases stress tolerance, activating plant defense systems via control of ROS (Zulfiqar et al. 2022; Islam et al. 2021; Kumar et al. 2021a, b). For example, priming agents are those natural or synthetic chemicals, plant hormones, or beneficial microbes.

These nutrients support cellular activities and antioxidant defences. Potassium, calcium, magnesium, nitrogen, or micronutrients are necessary for the production of antioxidant enzymes (SOD, CAT, POD, and APX) and nonenzymatic antioxidants (ascorbic acid and glutathione) (Kumari et al. 2021; Wei et al. 2021).

**Seed Priming with Phytohormones** Pre-sowing treatments using phytohormones like ABA, SA, ethylene, or BRs improve seedling establishment and stress tolerance. For instance, ABA priming in indica rice enhances salt tolerance through proline biosynthesis, whereas salicylic-acid priming improves growth and photosynthesis under cadmium stress in menthol mint (Wei et al. 2021; Zaid et al. 2022).

**Foliar Nutrient Application** Spraying nutrients boosts nutrition and stress tolerance. For example, silicon foliar application improved salt tolerance of *Lathyrus odoratus* by enhancing antioxidant capacity and potassium uptake while reducing sodium accumulation (El-Scrafy ct al. 2021).

**Nutritional Supplementation with Priming Agents** Supplementation of priming treatment with nutrients will yield positive results. The effect of ethylene and sulfur application on mustard under salt stress using proline metabolism modulation and antioxidant increase is the enhancement of tolerance. These influences vary according to the plant species and type of stress, priming agent, nutrient combination and concentration, as well as the application method and timing (Zulfiqar et al. 2022; Hernándiz et al. 2022).

### 9.10.2 Genetic Engineering Strategies to Regulate Reactive Oxygen Species and Nutritional Dynamics

By modifying the dynamics of reactive oxygen and nutrients inside the plant, genetic engineering provides a solid foundation for increasing plants' stress tolerance (Kaur et al. 2022a, b; Niazian et al. 2021; Esmaeili et al. 2022). Despite ROS's importance as plant signaling molecules, an excess of them can cause oxidative damage (Huchzermeyer et al. 2022). Consequently, it is crucial to alter plants in order to keep the reactive oxygen species balance. This may occur as a result of regulating ROS degradation by enzymatic processes or enhanced scavenging of ROS by antioxidant production. (Huchzermeyer et al. 2022).

The cycle of ascorbate-glutathione is mostly useful for the detoxification of ROS. Modified triplication in genes like OsDHAR1 might also be helpful for augmenting rice salt tolerance due to increased ascorbate accumulation and improvement in redox equilibrium (Kim et al. 2022).

Apart from reactive oxygen species, plant resilience, rather than nutrient dynamics, is very important. For persistent agriculture, improvement of nutrient-use efficiency (NtUE) is necessary. Through CRISPR-Cas, an understanding and modification of NtUE can be achieved (Sathee et al. 2022; Mangal et al. 2023). Altering ion transporters, such as overexpression of Na+/H+ antiporters, can improve salt tolerance (Niazian et al. 2021; Esmaeili et al. 2022).

According to Zhao et al.(2024), stress tolerance could be enhanced by combining many beneficial genes. This approach can modify the multilineal natures of stress responses and integrate various advantages of specific genes for better performance overall (Zhao et al. 2024; Niazian et al. 2021; Esmaeili et al. 2022). Such co-expression of genes, for instance, AVP1, OsSIZ1, and Fld could synergistically boost plant development and tolerance toward different stresses (Zhao et al. 2024).

Thorough investigation of gene interactions and crosstalk prevents unwanted results from effective gene pyramiding (Zhao et al. 2024; Esmaeili et al. 2022). Thus, "stress training" can carry out the prep against future adversity through ROS-generating nanoparticles like AgNPs (Chen et al. 2023). This nanobiostimulant method enables metabolic and transcriptional reprogramming, preparing plants for a quicker and more resilient response to future stressors (Chen et al. 2023).

### 9.10.3 Bio Stimulant Interventions to Alleviate ROS Stress

One environmentally friendly approach to helping plants cope with abiotic stressors, including drought, heat, and salt is to use biostimulants (Hasanuzzaman et al. 2021; Ma et al. 2022; Kumar et al. 2021a, b). Reactive oxygen species (ROS) damage reduction is their primary focus. To illustrate this point, consider the impact of Ascophyllum nodosum extracts on rice. These extracts boost pigment content and growth maintenance of ionic equilibrium, two factors that lead to salt tolerance. In low-light situations, these biostimulants improve gas exchange and PS II efficiency, allowing photosynthesis to proceed. Additionally, they influence the antioxidant system by boosting metabolite accumulation for greater salt tolerance, improving the activity of enzymes like catalase (CAT) and superoxide dismutase (SOD), and so on, in addition to reducing ROS levels (Hasanuzzaman et al. 2021).

By controlling the quantity of hydrogen peroxide ($H_2O_2$), copper chlorophyllins (Cu-chl) were able to lessen the damage caused by oxidative stress, as demonstrated by Islam et al. (2021). To prime seeds with phytohormones that improve responses to osmotic and salt stress, encourage development, and impart stress tolerance, polyamines (PAs) such as putrescine and 1,3-diaminopropane are crucial (Hernándiz et al. 2022). Researchers have found that polysaccharide extracts from Lobularia maritima (PSLm) can reduce membrane leakage, boost antioxidant content and chlorophyll fluorescence, and regulate ion homeostasis. These effects help durum wheat to withstand salinity stress (Bouteraa et al. 2022; Kumar et al. 2021a, b). One potential change might be the use of biostimulants in sustainable agriculture to protect crops from abiotic stress.

### 9.10.4 The Function of Beneficial Microorganisms in Enhancing Nutrient Accessibility and ROS Equilibrium

Essential rhizosphere beneficial bacteria, like as Pseudomonas, Rhizobium, and Bacillus control reactive oxygen species (ROS) and nutrient availability in response to biotic and abiotic stress (Morcillo et al. 2022; Munir et al. 2022). These microorganisms provide root growth stimulation, for example, through microbial phytohormone production, nutrient solubilization, nitrogen fixation, and association with mycorrhiza for improved phosphorus uptake (Munir et al. 2022).

They counteract stress through antioxidative actions such as catalase and superoxide dismutase that scavenge ROS (Lopes et al. 2021). They also synthesize osmoprotectants such as proline for the normal regulation of water balance and ACC deaminase to reduce levels of ethylene, which can inhibit plant growth (Munir et al. 2022; Lopes et al. 2021). Beneficial microbes also enhance soil structural and water retention through exopolysaccharides and biofilm formation, which ultimately promotes nutrient and water uptake (Bhagat et al. 2021). In addition, both plants and microorganisms secrete or excrete secondary metabolites that provide infection protection mechanisms (Koza et al. 2022).

### 9.10.5 Key Areas of Concentration in Plant Stress Research

**Heat Stress and Plant Responses** A primary topic of investigation centers on comprehending plant reactions to thermal stress. It is essential to differentiate among many categories of heat stress tests: heat shock, heat wave, and warming trials (Jagadish et al. 2021).

Future study should prioritize plant tissue temperatures instead of exclusively depending on air temperature measurements (Jagadish et al. 2021). Leaf size, shape, and transpirational cooling are factors that affect leaf temperature and must be considered in heat stress research.

**Reproductive Effects** Plant reproduction is significantly impacted by heat stress (Jagadish et al. 2021; Rivero et al. 2022). Early stages, such as the beginning and growth of floral meristems, need further investigation, even though pollen, seed set, and grain filling have already been studied. Breeding programs must prioritize heat-stress recovery (Jagadish et al. 2021). If we are just focused on increasing heat tolerance, we risk ignoring post-stress recovery, an essential process for developing robust phenotypes.

**Seasonal Variations in Plant Stress Tolerance** Plant stress tolerance exhibits seasonal fluctuations (Grossman 2023). Comprehending adaptation and deacclimation to thermal, frigid, and hydric stress, associated with plant phenology, is crucial for predicting and managing reactions to climate change.

Plants frequently encounter simultaneous stressors, resulting in intricate interaction and trade-offs among signaling pathways (Rivero et al. 2022; Leisner et al. 2023). Comprehending these connections is essential for cultivating stress-resistant crops.

**Multi-omics and Advanced Technologies** Genome editing, high-throughput phenotyping, and integrated omic technologies have been the subject of several recent investigations on their ability to enhance crop tolerance to both biotic and abiotic challenges. Tinte et al. (2021), Sandrini et al. (2022), Kumar et al. (2021a, b), and Galieni et al. (2021) are among these. These cutting-edge tools show a lot of potential for understanding metabolic processes, which might lead to answers on stress tolerance in the future.

There is consensus among Trivedi et al. (2022), Khan (2023), and Gonzalez Guzman et al. (2022) that microbiomes are crucial to plant defence and health. In order to create effective stress reduction measures, it is essential to understand how microbiomes are changing as a result of climate change.

### 9.10.6 The Advancement of Research on Plant Stress

Indeed, interdisciplinary research is critical to meet the complex threats imposed by climate change, as was strongly emphasized in the words of Rivero et al. (2022), Bäurle et al. (2023), Kumar et al. (2021a, b), and Leisner et al. (2023). Understanding fully the plant stress response mechanisms and developing climate-resilient crops will require some integration of plant physiology, genetics, molecular biology, ecology, and microbiology, as suggested by Rivero et al. (2022), Leisner et al. (2023). Research activities should also aim to do so within the context of realistic future climate scenarios and precision-agriculture tools that can offer sustainable pathways to food security (Rivero et al. 2022; Leisner et al. 2023). Given the increase in global population, one can expect a growing demand for food, urging the necessity of efficient development into highly climate-resilient crops, which becomes a priority (Galieni et al. 2021). This area remains open and critical to addressing this issue and ensuring long-term food security.

## 9.11 Conclusion

Nutrient dynamics in plants are highly interdependent on cellular signaling pathways involved in development and growth. Nitrogen, phosphorus, and potassium are examples of macronutrients, and plants have unique systems for absorbing these and other micronutrients, which are then transported to various parts of the plant for use in growth and storage.

The connections of these nutrients are significantly related to signaling via ROS. Once thought to be harmful waste, ROS are now recognized as having vital functions in the regulation of development and responses to stress. To ensure homeostasis, the plant creates a defense mechanism, resulting in a very delicate balance within the plant itself concerning overall vitality. During stressful conditions, the relationship between nutrients and ROS becomes of heightened importance, as depletion of certain nutrients results in excess ROS accumulation, which in turn initiates signaling pathways affecting root architecture and nutrient uptake. This signaling is regulated within complex molecular pathways, comprising transcription factors, post-translational modifications, and epigenetic regulation, all of which are finely modulated by phytohormones.

It is important to understand these relationships for addressing the threats brought about by climate change. Plant genetic engineering, improved nutrient management, and beneficial microbes are all promising avenues for the culture of stress-resistant crops and better plant resilience. More research is needed to extend the comprehension of these relationships and build sustainable farming practices for our changing environment.

**Authors Contribution**

**Gholamreza Abdi:** Conceptualization, overall supervision, and critical revision of the manuscript, literature synthesis, and writing—original draft.
**Aritra Mukherjee:** Chapter structuring, literature synthesis, and writing—original draft.
**Md Sabir Ahmed Mondol:** Data curation, analysis, and drafting key sections
**Mero Yangfo:** Research assistance, visualization (figures/tables), and editing.
**Tadela Susmitha:** Methodology support, validation, and manuscript refinement.
**Anilava Chatterjee:** Theoretical framework development, writing—review & editing.
**Ubaida Akbar:** Technical inputs, data interpretation, and specialized content drafting.
**Amisha Rani:** Proofreading, formatting, and editorial oversight.

All authors reviewed, revised, and approved the final version of the chapter.

## References

Abedi H, Shahpiri A (2024) Functional characterization of a manganese superoxide dismutase from Avicennia marina: insights into its role in salt, hydrogen peroxide, and heavy metal tolerance. Sci Rep 14(1):406

Ali S, Tyagi A, Mir RA, Rather IA, Anwar Y, Mahmoudi H (2023) Plant beneficial microbiome a boon for improving multiple stress tolerance in plants. Front Plant Sci 14:1266182

Altaf MA, Shahid R, Ren MX, Naz S, Altaf MM, Khan LU et al (2022) Melatonin improves drought stress tolerance of tomato by modulating plant growth, root architecture, photosynthesis, and antioxidant defense system. Antioxidants 11(2):309

Andrés-Barrao C, Alzubaidy H, Jalal R, Mariappan KG, De Zelicourt A, Bokhari A et al (2021) Coordinated bacterial and plant sulfur metabolism in Enterobacter sp. SA187–induced plant salt stress tolerance. Proc Natl Acad Sci 118(46):e2107417118

Babbar R, Karpinska B, Grover A, Foyer CH (2021) Heat-induced oxidation of the nuclei and cytosol. Front Plant Sci 11:617779

Barzana G, Rios JJ, Lopez-Zaplana A, Nicolas-Espinosa J, Yepes-Molina L, Garcia-Ibañez P, Carvajal M (2021) Interrelations of nutrient and water transporters in plants under abiotic stress. Physiol Plant 171(4):595–619

Bashir K, Rasheed S, Kobayashi T, Seki M, Nishizawa NK (2016) Regulating subcellular metal homeostasis: the key to crop improvement. Front Plant Sci 7:1192

Basu S, Prabhakar AA, Kumari S, Aabha, Kumar RR, Shekhar S et al (2023) Micronutrient and redox homeostasis contribute to Moringa oleifera-regulated drought tolerance in wheat. Plant Growth Regul 100(2):467–478

Bäurle I, Laplaze L, Martin A (2023) Preparing for an uncertain future: molecular responses of plants facing climate change. J Exp Bot 74(5):1297–1302

Bechtaoui N, Rabiu MK, Raklami A, Oufdou K, Hafidi M, Jemo M (2021) Phosphate-dependent regulation of growth and stresses management in plants. Front Plant Sci 12:679916

Bhagat N, Raghav M, Dubey S, Bedi N (2021) Bacterial exopolysaccharides: insight into their role in plant abiotic stress tolerance. J Microbiol Biotechnol 31(8):1045

Bhar A, Chakraborty A, Roy A (2023) The captivating role of calcium in plant-microbe interaction. Front Plant Sci 14:1138252

Bijalwan P, Sharma M, Kaushik P (2022) Review of the effects of drought stress on plants: a systematic approach. Bioscience 9(1):1–6. https://doi.org/10.18782/2582-2845.8527

Bouteraa MT, Mishra A, Romdhane WB, Hsouna AB, Siddique KH, Saad RB (2022) Biostimulating effect of natural polysaccharides from Lobularia maritima on durum wheat seedlings: improved plant growth, salt stress tolerance by modulating biochemical responses and ion homeostasis. Plants 11(15):1991

Cha JY, Jeong SY, Ahn G, Shin GI, Ji MG, Lee SC et al (2022) The thiol-reductase activity of YUCCA6 enhances nickel heavy metal stress tolerance in Arabidopsis. Front Plant Sci 13:1007542

Chen T, Li W, Hu X, Guo J, Liu A, Zhang B (2015) A cotton MYB transcription factor, GbMYB5, is positively involved in plant adaptive response to drought stress. Plant Cell Physiol 56(5):917–929

Chen S, Pan Z, Zhao W, Zhou Y, Rui Y, Jiang C et al (2023) Engineering climate-resilient rice using a nanobiostimulant-based "stress training" strategy. ACS Nano 17(11):10760–10773

Considine MJ, Foyer CH (2021) Stress effects on the reactive oxygen species-dependent regulation of plant growth and development. J Exp Bot 72(16):5795–5806

Devireddy AR, Tschaplinski TJ, Tuskan GA, Muchero W, Chen JG (2021) Role of reactive oxygen species and hormones in plant responses to temperature changes. Int J Mol Sci 22(16):8843

Dumanović J, Nepovimova E, Natić M, Kuča K, Jaćević V (2021) The significance of reactive oxygen species and antioxidant defense system in plants: a concise overview. Front Plant Sci 11:552969

Dvořák P, Krasylenko Y, Zeiner A, Šamaj J, Takáč T (2021) Signaling toward reactive oxygen species-scavenging enzymes in plants. Front Plant Sci 11:618835

El-Serafy RS, El-Sheshtawy ANA, Atteya AK, Al-Hashimi A, Abbasi AM, Al-Ashkar I (2021) Seed priming with silicon as a potential to increase salt stress tolerance in Lathyrus odoratus. Plants 10(10):2140

Esmaeili N, Shen G, Zhang H (2022) Genetic manipulation for abiotic stress resistance traits in crops. Front Plant Sci 13:1011985

Fan X, Zhou X, Chen H, Tang M, Xie X (2021) Cross-talks between macro-and micronutrient uptake and signaling in plants. Front Plant Sci 12:663477

Fatma M, Iqbal N, Gautam H, Sehar Z, Sofo A, D'Ippolito I, Khan NA (2021) Ethylene and sulfur coordinately modulate the antioxidant system and ABA accumulation in mustard plants under salt stress. Plants 10(1):180

Feng Z, Lu G, Sun M, Jin Y, Xu Y, Liu X et al (2023) Comparative study of the priming effect of abscisic acid on tolerance to saline and alkaline stresses in rice seedlings. Agronomy 13(11):2698

Foyer CH, Hanke G (2022) ROS production and signalling in chloroplasts: cornerstones and evolving concepts. Plant J 111(3):642–661

Galieni A, D'Ascenzo N, Stagnari F, Pagnani G, Xie Q, Pisante M (2021) Past and future of plant stress detection: an overview from remote sensing to positron emission tomography. Front Plant Sci 11:609155

Gandhi A, Kariyat R, Harikishore A, Ayati M, Bhunia A, Sahoo N (2021) Deciphering the role of ion channels in early defense signaling against herbivorous insects. Cells 10(9):2219

Giannakoula A, Therios I, Chatzissavvidis C (2021) Effect of lead and copper on photosynthetic apparatus in citrus (Citrus aurantium L.) plants. The role of antioxidants in oxidative damage as a response to heavy metal stress. Plants 10(1):155

Gonzalez Guzman M, Cellini F, Fotopoulos V, Balestrini R, Arbona V (2022) New approaches to improve crop tolerance to biotic and abiotic stresses. Physiol Plant 174(1):e13547

Grossman JJ (2023) Phenological physiology: seasonal patterns of plant stress tolerance in a changing climate. New Phytol 237(5):1508–1524

Halder K, Chaudhuri A, Abdin MZ, Majee M, Datta A (2022) Chromatin-based transcriptional reprogramming in plants under abiotic stresses. Plants 11(11):1449
Hanikenne M, Esteves SM, Fanara S, Rouached H (2021) Coordinated homeostasis of essential mineral nutrients: a focus on iron. J Exp Bot 72(6):2136–2153
Hasanuzzaman M, Parvin K, Bardhan K, Nahar K, Anee TI, Masud AAC, Fotopoulos V (2021) Biostimulants for the regulation of reactive oxygen species metabolism in plants under abiotic stress. Cells 10(10):2537
Hawkesford MJ, Buchner PH, Riche AB (2018) Nutrient dynamics in wheat. Ann Plant Rev 3(November):1–14
Hernándiz AE, Aucique-Perez CE, Ćavar Zeljković S, Štefelová N, Salcedo Sarmiento S, Spíchal L, De Diego N (2022) Priming with small molecule-based biostimulants to improve abiotic stress tolerance in Arabidopsis thaliana. Plants 11(10):1287
Hirakawa T, Tanno S, Ohara K (2023) N-acetylglutamic acid alleviates oxidative stress based on histone acetylation in plants. Front Plant Sci 14:1165646
Hu SH, Jinn TL (2022) Impacts of Mn, fe, and oxidative stressors on MnSOD activation by AtMTM1 and AtMTM2 in arabidopsis. Plants 11(5):619
Huchzermeyer B, Menghani E, Khardia P, Shilu A (2022) Metabolic pathway of natural antioxidants, antioxidant enzymes and ROS providence. Antioxidants 11(4):761
Ihtisham M, Hasanuzzaman M, El-Sappah AH, Zaman F, Khan N, Raza A et al (2023) Primary plant nutrients modulate the reactive oxygen species metabolism and mitigate the impact of cold stress in overseeded perennial ryegrass. Front Plant Sci 14:1149832
Iqbal S, Iqbal MA, Li C, Iqbal A, Abbas RN (2023) Overviewing drought and heat stress amelioration—from plant responses to microbe-mediated mitigation. Sustainability 15(2):1671
Irshad K, Shaheed Siddiqui Z, Chen J, Rao Y, Hamna Ansari H, Wajid D et al (2023) Bio-priming with salt tolerant endophytes improved crop tolerance to salt stress via modulating photosystem II and antioxidant activities in a sub-optimal environment. Front Plant Sci 14:1082480
Islam MT, Ckurshumova W, Fefer M, Liu J, Uddin W, Rosa C (2021) A plant based modified biostimulant (copper chlorophyllin), mediates defense response in Arabidopsis thaliana under salinity stress. Plants 10(4):625
Jagadish SK, Way DA, Sharkey TD (2021) Plant heat stress: concepts directing future research. Plant Cell Environ 44(7):1992–2005
Jia Z, Giehl RF, von Wirén N (2022) Nutrient–hormone relations: driving root plasticity in plants. Mol Plant 15(1):86–103
Jian W, Zheng Y, Yu T, Cao H, Chen Y, Cui Q et al (2021) SlNAC6, A NAC transcription factor, is involved in drought stress response and reproductive process in tomato. J Plant Physiol 264:153483
Jiaying MA, Tingting C, Jie L, Weimeng F, Baohua F, Guangyan L et al (2022) Functions of nitrogen, phosphorus and potassium in energy status and their influences on rice growth and development. Rice Sci 29(2):166–178
Johnson R, Vishwakarma K, Hossen MS, Kumar V, Shackira AM, Puthur JT et al (2022) Potassium in plants: growth regulation, signaling, and environmental stress tolerance. Plant Physiol Biochem 172:56–69
Kang J, Peng Y, Xu W (2022) Crop root responses to drought stress: molecular mechanisms, nutrient regulations, and interactions with microorganisms in the rhizosphere. Int J Mol Sci 23(16):9310
Kaur C, Gupta M, Garai S, Mishra SK, Chauhan PS, Sopory S et al (2022a) Microbial methylglyoxal metabolism contributes towards growth promotion and stress tolerance in plants. Environ Microbiol 24(6):2817–2836
Kaur N, Sharma S, Hasanuzzaman M, Pati PK (2022b) Genome editing: a promising approach for achieving abiotic stress tolerance in plants. Int J Genomics 2022(1):5547231
Kesawat MS, Satheesh N, Kherawat BS, Kumar A, Kim HU, Chung SM, Kumar M (2023) Regulation of reactive oxygen species during salt stress in plants and their crosstalk with other signaling molecules—current perspectives and future directions. Plants 12(4):864

Khan AL (2023) The phytomicrobiome: solving plant stress tolerance under climate change. Front Plant Sci 14:1219366

Khan MN, Li Y, Fu C, Hu J, Chen L, Yan J et al (2022) CeO2 nanoparticles seed priming increases salicylic acid level and ROS scavenging ability to improve rapeseed salt tolerance. Global Chall 6(7):2200025

Khan F, Siddique AB, Shabala S, Zhou M, Zhao C (2023a) Phosphorus plays key roles in regulating plants' physiological responses to abiotic stresses. Plants 12(15):2861

Khan M, Ali S, Al Azzawi TNI, Saqib S, Ullah F, Ayaz A, Zaman W (2023b) The key roles of ROS and RNS as a signaling molecule in plant–microbe interactions. Antioxidants 12(2):268

Kim JS, Jeon BW, Kim J (2021) Signaling peptides regulating abiotic stress responses in plants. Front Plant Sci 12:704490

Kim YS, Park SI, Kim JJ, Shin SY, Kwak SS, Lee CH et al (2022) Over-expression of dehydroascorbate reductase improves salt tolerance, environmental adaptability and productivity in Oryza sativa. Antioxidants 11(6):1077

Kolupaev YE, Yastreb TO, Dmitriev AP (2023) Signal mediators in the implementation of jasmonic acid's protective effect on plants under abiotic stresses. Plants 12(14):2631

Kong Z, Liu H (2022) Modification of rhizosphere microbial communities: a possible mechanism of plant growth promoting rhizobacteria enhancing plant growth and fitness. Front Plant Sci 13:920813

Koza NA, Adedayo AA, Babalola OO, Kappo AP (2022) Microorganisms in plant growth and development: roles in abiotic stress tolerance and secondary metabolites secretion. Microorganisms 10(8):1528

Krämer U, Clemens S (2006) Functions and homeostasis of zinc, copper, and nickel in plants. In: Molecular biology of metal homeostasis and detoxification: from microbes to man, pp 215–271

Kumar SJ, Chintagunta AD, Reddy YM, Rajjou L, Garlapati VK, Agarwal DK et al (2021a) Implications of reactive oxygen and nitrogen species in seed physiology for sustainable crop productivity under changing climate conditions. Curr Plant Biol 26:100197

Kumar S, Kumar S, Mohapatra T (2021b) Interaction between macro-and micro-nutrients in plants. Front Plant Sci 12:665583

Kumari A, Bhinda MS, Sharma S, Chitara MK, Debnath A, Maharana C et al (2021) ROS regulation mechanism for mitigation of abiotic stress in plants. In: Reactive oxygen species, vol 28, pp 197–238

Kumawat KC, Sharma P, Nagpal S, Gupta RK, Sirari A, Nair RM et al (2021) Dual microbial inoculation, a game changer?–bacterial biostimulants with multifunctional growth promoting traits to mitigate salinity stress in spring mungbean. Front Microbiol 11:600576

Lapaz ADM, Yoshida CHP, Gorni PH, Freitas-Silva LD, Araújo TDO, Ribeiro C (2022) Iron toxicity: effects on the plants and detoxification strategies. Acta Bot Bras 36:e2021abb0131

Lee HJ, Seo PJ (2021) Ca2+ talyzing initial responses to environmental stresses. Trends Plant Sci 26(8):849–870

Lee YJ, Lee WJ, Le QT, Hong SW, Lee H (2021) Growth performance can be increased under high nitrate and high salt stress through enhanced nitrate reductase activity in arabidopsis anthocyanin over-producing mutant plants. Front Plant Sci 12:644455

Leisner CP, Potnis N, Sanz-Saez A (2023) Crosstalk and trade-offs: plant responses to climate change-associated abiotic and biotic stresses. Plant Cell Environ 46(10):2946–2963

Li K, Xing C, Yao Z, Huang X (2017) Pbr MYB 21, a novel MYB protein of Pyrus betulaefolia, functions in drought tolerance and modulates polyamine levels by regulating arginine decarboxylase gene. Plant Biotechnol J 15(9):1186–1203

Li C, Yan C, Sun Q, Wang J, Yuan C, Mou Y et al (2021a) The bHLH transcription factor AhbHLH112 improves the drought tolerance of peanut. BMC Plant Biol 21:1–12

Li L, Liu KH, Sheen J (2021b) Dynamic nutrient signaling networks in plants. Annu Rev Cell Dev Biol 37(1):341–367

Li Z, Tian P, Huang T, Huang J (2021c) Noncoding-RNA-mediated regulation in response to macronutrient stress in plants. Int J Mol Sci 22(20):11205

Li S, Han X, Lu Z, Qiu W, Yu M, Li H et al (2022a) MAPK cascades and transcriptional factors: regulation of heavy metal tolerance in plants. Int J Mol Sci 23(8):4463

Li S, Liu S, Zhang Q, Cui M, Zhao M, Li N et al (2022b) The interaction of ABA and ROS in plant growth and stress resistances. Front Plant Sci 13:1050132
Li Y, Liu Y, Jin L, Peng R (2022c) Crosstalk between Ca2+ and other regulators assists plants in responding to abiotic stress. Plants 11(10):1351
Liu Y, Zhang H (2021) Reactive oxygen species and nitric oxide as mediators in plant hypersensitive response and stomatal closure. Plant Signal Behav 16(12):1985860
Liu D, Li YY, Zhou ZC, Xiang X, Liu X, Wang J et al (2021) Tobacco transcription factor bHLH123 improves salt tolerance by activating NADPH oxidase NtRbohE expression. Plant Physiol 186(3):1706–1720
Liu P, Wu X, Gong B, Lü G, Li J, Gao H (2022a) Review of the mechanisms by which transcription factors and exogenous substances regulate ROS metabolism under abiotic stress. Antioxidants 11(11):2106
Liu Y, Wang J, Liu B, Xu ZY (2022b) Dynamic regulation of DNA methylation and histone modifications in response to abiotic stresses in plants. J Integr Plant Biol 64(12):2252–2274
Lopes MJDS, Dias-Filho MB, Gurgel ESC (2021) Successful plant growth-promoting microbes: inoculation methods and abiotic factors. Front Sustain Food Syst 5:606454
Ludewig U, Vatov E, Hedderich D, Neuhäuser B (2021) Adjusting plant nutrient acquisition to fluctuating availability: transcriptional co-regulation of the nitrate and phosphate deprivation responses in roots. J Exp Bot 72(10):3500–3503
Lv B, Wu Q, Wang A, Li Q, Dong Q, Yang J et al (2020) A WRKY transcription factor, FtWRKY46, from Tartary buckwheat improves salt tolerance in transgenic Arabidopsis thaliana. Plant Physiol Biochem 147:43–53
Ma Y, Freitas H, Dias MC (2022) Strategies and prospects for biostimulants to alleviate abiotic stress in plants. Front Plant Sci 13:1024243
Malook SU, Maqbool S, Hafeez M, Karunarathna SC, Suwannarach N (2022) Molecular and biochemical mechanisms of elicitors in pest resistance. Life 12(6):844
Mandon K, Nazaret F, Farajzadeh D, Alloing G, Frendo P (2021) Redox regulation in diazotrophic bacteria in interaction with plants. Antioxidants 10(6):880
Mangal V, Lal MK, Tiwari RK, Altaf MA, Sood S, Kumar D et al (2023) Molecular insights into the role of reactive oxygen, nitrogen and sulphur species in conferring salinity stress tolerance in plants. J Plant Growth Regul 42(2):554–574
Manna M, Thakur T, Chirom O, Mandlik R, Deshmukh R, Salvi P (2021) Transcription factors as key molecular target to strengthen the drought stress tolerance in plants. Physiol Plant 172(2):847–868
Mansoor S, Ali Wani O, Lone JK, Manhas S, Kour N, Alam P et al (2022) Reactive oxygen species in plants: from source to sink. Antioxidants 11(2):225
Mansoor S, Ali A, Kour N, Bornhorst J, AlHarbi K, Rinklebe J et al (2023) Heavy metal induced oxidative stress mitigation and ROS scavenging in plants. Plants 12(16):3003
Manzoor N, Ali L, Ahmed T, Noman M, Adrees M, Shahid MS et al (2022) Recent advancements and development in nano-enabled agriculture for improving abiotic stress tolerance in plants. Front Plant Sci 13:951752
Marcec MJ, Tanaka K (2021) Crosstalk between calcium and ROS signaling during flg22-triggered immune response in Arabidopsis leaves. Plants 11(1):14
Mason CJ, Ray S, Davidson-Lowe E, Ali J, Luthe DS, Felton G (2022) Plant nutrition influences resistant maize defense responses to the fall armyworm (Spodoptera frugiperda). Front Ecol Evol 10:844274
Matamoros MA, Becana M (2021) Molecular responses of legumes to abiotic stress: post-translational modifications of proteins and redox signaling. J Exp Bot 72(16):5876–5892
Meng X, Chen WW, Wang YY, Huang ZR, Ye X, Chen LS, Yang LT (2021) Effects of phosphorus deficiency on the absorption of mineral nutrients, photosynthetic system performance and antioxidant metabolism in Citrus grandis. PLoS One 16(2):e0246944
Minguillón S, Matamoros MA, Duanmu D, Becana M (2022) Signaling by reactive molecules and antioxidants in legume nodules. New Phytol 236(3):815–832
Miryeganeh M (2021) Plants' epigenetic mechanisms and abiotic stress. Genes 12(8):1106

Morcillo RJ, Manzanera M (2021) The effects of plant-associated bacterial exopolysaccharides on plant abiotic stress tolerance. Metabolites 11(6):337

Morcillo RJL, Baroja-Fernández E, López-Serrano L, Leal-López J, Muñoz FJ, Bahaji A et al (2022) Cell-free microbial culture filtrates as candidate biostimulants to enhance plant growth and yield and activate soil-and plant-associated beneficial microbiota. Front Plant Sci 13:1040515

Mostafa S, Wang Y, Zeng W, Jin B (2022) Plant responses to herbivory, wounding, and infection. Int J Mol Sci 23(13):7031

Munir N, Hanif M, Abideen Z, Sohail M, El-Keblawy A, Radicetti E et al (2022) Mechanisms and strategies of plant microbiome interactions to mitigate abiotic stresses. Agronomy 12(9):2069

Muratore C, Espen L, Prinsi B (2021) Nitrogen uptake in plants: the plasma membrane root transport systems from a physiological and proteomic perspective. Plants 10(4):681

Mushtaq NU, Alghamdi KM, Saleem S, Tahir I, Bahieldin A, Henrissat B et al (2023) Exogenous zinc mitigates salinity stress by stimulating proline metabolism in proso millet (Panicum miliaceum L.). Front Plant Sci 14:1053869

Nandal V, Solanki M (2021) The Zn as a vital micronutrient in plants. J Microbiol Biotechnol Food Sci 11(3):e4026

Napieraj N, Janicka M, Reda M (2023) Interactions of polyamines and phytohormones in plant response to abiotic stress. Plants 12(5):1159

Narayan OP, Kumar P, Yadav B, Dua M, Johri AK (2023) Sulfur nutrition and its role in plant growth and development. Plant Signal Behav 18(1):2030082

Niazian M, Sadat-Noori SA, Tohidfar M, Mortazavian SMM, Sabbatini P (2021) Betaine aldehyde dehydrogenase (BADH) vs. flavodoxin (Fld): two important genes for enhancing plants stress tolerance and productivity. Front Plant Sci 12:650215

Niu F, Wang C, Yan J, Guo X, Wu F, Yang B et al (2016) Functional characterization of NAC55 transcription factor from oilseed rape (Brassica napus L.) as a novel transcriptional activator modulating reactive oxygen species accumulation and cell death. Plant Mol Biol 92:89–104

Nunez-Vazquez R, Desvoyes B, Gutierrez C (2022) Histone variants and modifications during abiotic stress response. Front Plant Sci 13:984702

Pantha P, Oh DH, Longstreth D, Dassanayake M (2023) Living with high potassium: balance between nutrient acquisition and K-induced salt stress signaling. Plant Physiol 191(2):1102–1121

Pardo-Hernández M, López-Delacalle M, Martí-Guillen JM, Martínez-Lorente SE, Rivero RM (2021) ROS and NO phytomelatonin-induced signaling mechanisms under metal toxicity in plants: A review. Antioxidants 10(5):775

Parra-Aguilar TJ, Sarmiento-López LG, Santana O, Olivares JE, Pascual-Morales E, Jiménez-Jiménez S et al (2023) TETRASPANIN 8-1 from Phaseolus vulgaris plays a key role during mutualistic interactions. Front Plant Sci 14:1152493

Poudel M, Mendes R, Costa LA, Bueno CG, Meng Y, Folimonova SY et al (2021) The role of plant-associated bacteria, fungi, and viruses in drought stress mitigation. Front Microbiol 12:743512

Prasad A, Sharma S, Prasad M (2023) Suppressing plant defense: scavenge the ROS! Physiol Plant 175(3):e13942

Puig S, Andrés-Colás NURIA, García-Molina ANTONI, Penarrubia L (2007) Copper and iron homeostasis in Arabidopsis: responses to metal deficiencies, interactions and biotechnological applications. Plant Cell Environ 30(3):271–290

Qamer Z, Chaudhary MT, Du X, Hinze L, Azhar MT (2021) Review of oxidative stress and antioxidative defense mechanisms in Gossypium hirsutum L. in response to extreme abiotic conditions. J Cotton Res 4(1):9

Rai KK, Kaushik P (2023) Free radicals mediated redox signaling in plant stress tolerance. Life 13(1):204

Rai S, Singh PK, Mankotia S, Swain J, Satbhai SB (2021) Iron homeostasis in plants and its crosstalk with copper, zinc, and manganese. Plant Stress 1:100008

Rane J, Singh AK, Tiwari M, Prasad PV, Jagadish SK (2022) Effective use of water in crop plants in dryland agriculture: implications of reactive oxygen species and antioxidative system. Front Plant Sci 12:778270
Ravi B, Foyer CH, Pandey GK (2023) The integration of reactive oxygen species (ROS) and calcium signalling in abiotic stress responses. Plant Cell Environ 46(7):1985–2006
Raza A, Salehi H, Rahman MA, Zahid Z, Madadkar Haghjou M, Najafi-Kakavand S et al (2022) Plant hormones and neurotransmitter interactions mediate antioxidant defenses under induced oxidative stress in plants. Front Plant Sci 13:961872
Redza-Dutordoir M, Averill-Bates DA (2021) Interactions between reactive oxygen species and autophagy: special issue: death mechanisms in cellular homeostasis. Biochim Biophys Acta (BBA)-Mol Cell Res 1868(8):119041
Regon P, Dey S, Rehman M, Pradhan AK, Chowra U, Tanti B et al (2022) Transcriptomic analysis revealed reactive oxygen species scavenging mechanisms associated with ferrous iron toxicity in aromatic Keteki Joha rice. Front Plant Sci 13:798580
Ren H, Zhao X, Li W, Hussain J, Qi G, Liu S (2021) Calcium signaling in plant programmed cell death. Cells 10(5):1089
Rivero RM, Mittler R, Blumwald E, Zandalinas SI (2022) Developing climate-resilient crops: improving plant tolerance to stress combination. Plant J 109(2):373–389
Sachdev S, Ansari SA, Ansari MI, Fujita M, Hasanuzzaman M (2021) Abiotic stress and reactive oxygen species: generation, signaling, and defense mechanisms. Antioxidants 10(2):277
Sako K, Van Ha C, Matsui A, Tanaka M, Sato A, Seki M (2021) Transcriptome analysis of Arabidopsis thaliana plants treated with a new compound Natolen128, enhancing salt stress tolerance. Plants 10(5):978
Sandrini M, Nerva L, Sillo F, Balestrini R, Chitarra W, Zampieri E (2022) Abiotic stress and belowground microbiome: the potential of omics approaches. Int J Mol Sci 23(3):1091
Sardans J, Peñuelas J (2021) Potassium control of plant functions: ecological and agricultural implications. Plants 10(2):419
Sathee L, Jagadhesan B, Pandesha PH, Barman D, Adavi B S, Nagar S et al (2022) Genome editing targets for improving nutrient use efficiency and nutrient stress adaptation. Front Genet 13:900897
Shabir M, Amjad Z, Jalal AZ, Fatima I, Yousaf K, Nabi MMU, Khalil A (2023) New insights for plant genome optimization, biotic and abiotic stresses and agricultural applications. Haya Saudi J Life Sci 8(8):139–144
Shah SH, Islam S, Mohammad F (2022) Sulphur as a dynamic mineral element for plants: a review. J Soil Sci Plant Nutr 22(2):2118–2143
Shahzad R, Harlina PW, Gallego PP, Flexas J, Ewas M, Leiwen X, Karuniawan A (2023) The seaweed Ascophyllum nodosum-based biostimulant enhances salt stress tolerance in rice (Oryza sativa L.) by remodeling physiological, biochemical, and metabolic responses. J Plant Interact 18(1):2266514
Shome S, Tewari S, Bhattacharya MK, Panda SK, Upadhyaya H (2022) Phytofunctionalized ZnO nanoparticles ameliorate water stress and its recovery in Oryza sativa L. Acta Physiol Plant 44(12):137
Singh A, Mehta S, Yadav S, Nagar G, Ghosh R, Roy A et al (2022) How to cope with the challenges of environmental stresses in the era of global climate change: an update on ROS stave off in plants. Int J Mol Sci 23(4):1995
Singh M, Nara U, Kumar A, Jaswal C (2024) Signaling for abiotic stress tolerance in plants: a review. Agric Rev 45(2):239
Singhal RK, Fahad S, Kumar P, Choyal P, Javed T, Jinger D et al (2023) Beneficial elements: new players in improving nutrient use efficiency and abiotic stress tolerance. Plant Growth Regul 100(2):237–265
Snoeck S, Guayazán-Palacios N, Steinbrenner AD (2022) Molecular tug-of-war: plant immune recognition of herbivory. Plant Cell 34(5):1497–1513

Song Y, Wan GY, Wang JX, Zhang ZS, Xia JQ, Sun LQ et al (2023) Balanced nitrogen–iron sufficiency boosts grain yield and nitrogen use efficiency by promoting tillering. Mol Plant 16(10):1661–1677

Sousa B, Rodrigues F, Soares C, Martins M, Azenha M, Lino-Neto T et al (2022) Impact of combined heat and salt stresses on tomato plants—insights into nutrient uptake and redox homeostasis. Antioxidants 11(3):478

Sun Y, Wang C, Chen HY, Ruan H (2020) Response of plants to water stress: a meta-analysis. Front Plant Sci 11:978

Swain R, Sahoo S, Behera M, Rout GR (2023) Instigating prevalent abiotic stress resilience in crop by exogenous application of phytohormones and nutrient. Front Plant Sci 14:1104874

Tan S, Cao J, Xia X, Li Z (2022) Advances in 5-aminolevulinic acid priming to enhance plant tolerance to abiotic stress. Int J Mol Sci 23(2):702

Tarkowski ŁP, Signorelli S, Considine MJ, Montrichard F (2023) Integration of reactive oxygen species and nutrient signalling to shape root system architecture. Plant Cell Environ 46(2):379–390

Tavanti TR, de Melo AAR, Moreira LDK, Sanchez DEJ, dos Santos Silva R, da Silva RM, Dos Reis AR (2021) Micronutrient fertilization enhances ROS scavenging system for alleviation of abiotic stresses in plants. Plant Physiol Biochem 160:386–396

Tewari RK, Yadav N, Gupta R, Kumar P (2021) Oxidative stress under macronutrient deficiency in plants. J Soil Sci Plant Nutr 21(1):832–859

Thirumalaikumar VP, Devkar V, Mehterov N, Ali S, Ozgur R, Turkan I et al (2018) NAC transcription factor JUNGBRUNNEN 1 enhances drought tolerance in tomato. Plant Biotechnol J 16(2):354–366

Thompson MN, Grunseich JM, Marmolejo LO, Aguirre NM, Bradicich PA, Behmer ST et al (2022) Undercover operation: belowground insect herbivory modifies systemic plant defense and repels aboveground foraging insect herbivores. Front Ecol Evol 10:1033730

Tinte MM, Chele KH, van Der Hooft JJ, Tugizimana F (2021) Metabolomics-guided elucidation of plant abiotic stress responses in the 4IR era: an overview. Meta 11(7):445

Tomar RS, Kataria S, Jajoo A (2021) Behind the scene: critical role of reactive oxygen species and reactive nitrogen species in salt stress tolerance. J Agron Crop Sci 207(4):577–588

Toor MD, Adnan M, ur Rehman F, Tahir R, Saeed MS, Khan AU, Pareek V (2021) Nutrients and their importance in agriculture crop production: a review. Ind J Pure App. https://doi.org/10.18782/2582-2845.8527

Toor MD, Adnan M, ur Rehman F, Tahir R, Saeed MS, Khan AU, Pareek, Tripathi R, Tewari R, Singh KP, Keswani C, Minkina T, Srivastava AK et al (2022) Plant mineral nutrition and disease resistance: A significant linkage for sustainable crop protection. Front Plant Sci 13:883970

Trivedi P, Batista BD, Bazany KE, Singh BK (2022) Plant–microbiome interactions under a changing world: responses, consequences and perspectives. New Phytol 234(6):1951–1959

Tripathi R, Tewari R, Singh KP, Keswani C, Minkina T, Srivastava AK, Sansinenea E (2022) Plant mineral nutrition and disease resistance: A significant linkage for sustainable crop protection. Front Plant Sci 13:883970

Wairich A, De Conti L, Lamb TI, Keil R, Neves LO, Brunetto G et al (2022) Throwing copper around: how plants control uptake, distribution, and accumulation of copper. Agronomy 12(5):994

Wang CT, Ru JN, Liu YW, Yang JF, Li M, Xu ZS, Fu JD (2018) The maize WRKY transcription factor ZmWRKY40 confers drought resistance in transgenic Arabidopsis. Int J Mol Sci 19(9):2580

Wang D, Chen Q, Chen W, Liu X, Xia Y, Guo Q et al (2021a) A WRKY transcription factor, EjWRKY17, from Eriobotrya japonica enhances drought tolerance in transgenic Arabidopsis. Int J Mol Sci 22(11):5593

Wang L, Zhu B, Deng Y, Li T, Tian Q, Yuan Z et al (2021b) Biocatalytic and antioxidant nanostructures for ROS scavenging and biotherapeutics. Adv Funct Mater 31(31):2101804

Wang TJ, Huang S, Zhang A, Guo P, Liu Y, Xu C et al (2021c) JMJ17–WRKY40 and HY5–ABI5 modules regulate the expression of ABA-responsive genes in Arabidopsis. New Phytol 230(2):567–584
Wang Y, Chen YF, Wu WH (2021d) Potassium and phosphorus transport and signaling in plants. J Integr Plant Biol 63(1):34–52
Wang Y, Dai X, Xu G, Dai Z, Chen P, Zhang T, Zhang H (2021e) The Ca2+-CaM signaling pathway mediates potassium uptake by regulating reactive oxygen species homeostasis in tobacco roots under low-K+ stress. Front Plant Sci 12:658609
Wang J, Moeen-Ud-Din M, Yin R, Yang S (2022) ROS homeostasis involved in dose-dependent responses of Arabidopsis seedlings to copper toxicity. Genes 14(1):11
Wani KI, Naeem M, Castroverde CDM, Kalaji HM, Albaqami M, Aftab T (2021) Molecular mechanisms of nitric oxide (NO) signaling and reactive oxygen species (ROS) homeostasis during abiotic stresses in plants. Int J Mol Sci 22(17):9656
Wei Q, Zhang F, Sun F, Luo Q, Wang R, Hu R et al (2017) A wheat MYB transcriptional repressor TaMyb1D regulates phenylpropanoid metabolism and enhances tolerance to drought and oxidative stresses in transgenic tobacco plants. Plant Sci 265:112–123
Wei TJ, Wang MM, Jin YY, Zhang GH, Liu M, Yang HY et al (2021) Abscisic acid priming creates alkaline tolerance in alfalfa seedlings (Medicago sativa L.). Agriculture 11(7):608
Xu E, Liu Y, Gu D, Zhan X, Li J, Zhou K et al (2024) Molecular mechanisms of plant responses to copper: from deficiency to excess. Int J Mol Sci 25(13):6993
Yang C, Dolatabadian A, Fernando WD (2022) The wonderful world of intrinsic and intricate immunity responses in plants against pathogens. Can J Plant Pathol 44(1):1–20
Yung WS, Huang C, Li MW, Lam HM (2023) Changes in epigenetic features in legumes under abiotic stresses. Plant Genome 16(4):e20237
Zaid A, Mohammad F, Siddique KH (2022) Salicylic acid priming regulates stomatal conductance, trichome density and improves cadmium stress tolerance in Mentha arvensis L. Front Plant Sci 13:895427
Zandi P, Schnug E (2022) Reactive oxygen species, antioxidant responses and implications from a microbial modulation perspective. Biology 11(2):155
Zenda T, Liu S, Dong A, Duan H (2021) Revisiting Sulphur—the once neglected nutrient: it's roles in plant growth, metabolism, stress tolerance and crop production. Agriculture 11(7):626
Zewide I, Sherefu A (2021) Review paper on effect of micronutrients for crop production. J Nutr Food Process 4(7):1–8
Zhang J, Li H, Gu W, Zhang K, Liu X, Liu M et al (2023) Peroxisome dynamics determines host-derived ROS accumulation and infectious growth of the rice blast fungus. MBio 14(6):e02381–e02323
Zhao Q, Fan Z, Qiu L, Che Q, Wang T, Li Y, Wang Y (2020) MdbHLH130, an apple bHLH transcription factor, confers water stress resistance by regulating stomatal closure and ROS homeostasis in transgenic tobacco. Front Plant Sci 11:543696
Zhao G, Liu Y, Li L, Che R, Douglass M, Benza K et al (2024) Gene pyramiding for boosted plant growth and broad abiotic stress tolerance. Plant Biotechnol J 22(3):678–697
Zhenggang X, Li F, Mengxi Z, Yunlin Z, Huimin H, Guiyan Y (2023) Physiological dynamics as indicators of plant response to manganese binary effect. Front Plant Sci 14:1145427
Zhou X, Joshi S, Patil S, Khare T, Kumar V (2021) Reactive oxygen, nitrogen, carbonyl and sulfur species and their roles in plant abiotic stress responses and tolerance. J Plant Growth Regul 41:1–24
Zhu H, Jiang Y, Guo Y, Huang J, Zhou M, Tang Y et al (2021) A novel salt inducible WRKY transcription factor gene, AhWRKY75, confers salt tolerance in transgenic peanut. Plant Physiol Biochem 160:175–183
Zulfiqar F, Nafees M, Chen J, Darras A, Ferrante A, Hancock JT et al (2022) Chemical priming enhances plant tolerance to salt stress. Front Plant Sci 13:946922



# Sources of ROS Production in Plants Under Normal and Stressed Conditions

10

Hamzeh Amiri

**Abstract**

Reactive oxygen species (ROS) are essential yet potentially harmful molecules generated as byproducts of various metabolic processes in plants. Under normal physiological conditions, ROS are primarily produced in chloroplasts, mitochondria, and peroxisomes through photosynthetic and respiratory electron transport, photorespiration, and enzymatic reactions. These ROS function as key signaling molecules, regulating growth, development, and stress responses. However, under abiotic (e.g., drought, salinity, extreme temperatures, heavy metals) and biotic (e.g., pathogen attack) stress conditions, ROS production escalates due to enhanced activity of oxidases (e.g., NADPH oxidases, peroxidases) and disruptions in metabolic pathways. While controlled ROS levels mediate defense signaling, excessive accumulation leads to oxidative damage, necessitating a robust antioxidant system for detoxification. This chapter explores the major sources of ROS in plants, their roles in cellular signaling under normal growth conditions, and their overproduction during stress. Understanding these mechanisms is critical for improving plant stress tolerance and developing strategies to enhance crop resilience in a changing climate.

**Keywords**

Reactive oxygen species (ROS) · Oxidative stress · Photosynthesis · Mitochondria · Peroxisomes · NADPH oxidase · Abiotic stress

H. Amiri (✉)
Department of Biology, Lorestan University, Khorramabad, Iran



N. Iqbal et al. (eds.), *Plant Signaling Molecules in Regulation of ROS-Scavenging System*, https://doi.org/10.1007/978-981-95-3377-0_10

## 10.1 Introduction

Reactive oxygen species (ROS), which include superoxide radicals ($O2^{\bullet -}$), hydrogen peroxide ($H_2O_2$), and hydroxyl radicals ($^{\bullet}OH$), are intrinsic byproducts of biological processes in plants that play pivotal roles in signaling and metabolism (Apel and Hirt 2004). These molecules are generated through various enzymatic and non-enzymatic pathways primarily within key organelles such as chloroplasts, mitochondria, and peroxisomes. Although often associated with cell damage and oxidative stress, ROS also function as critical signaling molecules that regulate plant growth, development, and stress responses (Foyer and Noctor 2005).

Under normal physiological conditions, ROS are produced during photosynthesis, particularly through the photochemical reactions occurring in chloroplasts. The process of light capture and electron transfer can lead to the formation of excess ROS, especially under high light intensity or unfavorable conditions, resulting in photoinhibition (Baker 2008). Similarly, during mitochondrial respiration, electron leakage from the electron transport chain generates superoxide radicals, further contributing to the cellular pool of ROS (Møller 2001). Peroxisomes also play a significant role in ROS metabolism, producing hydrogen peroxide during fatty acid β-oxidation and metabolism of reactive intermediates (Noctor et al. 2018).

However, environmental stressors can drastically alter ROS dynamics. Drought, salinity, high temperatures, and pathogen infections are known to enhance ROS production, leading to a state of oxidative stress that can damage cellular structures and impair metabolic functions (Suzuki et al. 2012). For instance, water deficit conditions can promote the over-reduction of the electron transport chain in chloroplasts, resulting in elevated ROS levels (Fahad et al. 2017). Likewise, salt stress can disrupt cellular ionic balance and enhance ROS accumulation, ultimately impacting photosynthesis and growth (Wang et al. 2025).

The dual role of ROS—both as deleterious agents and as signaling molecules—underscores the complexity of plant responses to environmental challenges. Plants possess an array of antioxidant defense mechanisms, including enzymes such as superoxide dismutase, catalase, and peroxidases, which counterbalance ROS accumulation and mitigate oxidative damage (Halliwell and Gutteridge 2015). Understanding these sources and mechanisms of ROS production within the context of plant stress responses is essential for developing strategies to enhance plant resilience.

This chapter aims to provide a thorough examination of the sources of ROS production in plants, both under normal and stressed conditions. By exploring the pathways of ROS generation and their implications for plant health and adaptation, we seek to advance our understanding of the complex interplay between ROS and plant physiology. Insights gleaned from this exploration can inform agricultural strategies aimed at improving crop resilience in the face of environmental stresses and contribute to the broader goal of achieving sustainable food security.

## 10.2 Definition and Types of Reactive Oxygen Species

Reactive oxygen species are chemically reactive molecules derived from molecular oxygen ($O_2$) that play dual roles as both toxic byproducts of aerobic metabolism and crucial signaling molecules in biological systems (Halliwell and Gutteridge 2015). These oxygen-containing molecules exhibit higher reactivity than ground-state $O_2$ due to the presence of unpaired electrons in their outer orbitals (Birben et al. 2012). ROS can be broadly categorized into two groups: free radicals containing one or more unpaired electrons (e.g., superoxide anion [$O_2^{\bullet-}$], hydroxyl radical [$^{\bullet}OH$], and alkoxyl radicals [$RO^{\bullet}$]), and non-radical oxidizing agents (e.g., hydrogen peroxide [$H_2O_2$], singlet oxygen [$^1O_2$], and hypochlorous acid [HOCl]) (Schieber and Chandel 2014). The superoxide anion ($O_2^{\bullet-}$), produced primarily through electron leakage from mitochondrial electron transport chains (ETC), photosystems, and NADPH oxidase (NOX) enzymes, serves as the primary ROS from which most other species are derived (Murphy 2009). Hydrogen peroxide ($H_2O_2$), formed via superoxide dismutase (SOD)-catalyzed dismutation of $O_2^{\bullet-}$, is relatively stable and membrane-permeable, making it an important signaling molecule (Bienert and Chaumont 2014). The hydroxyl radical ($^{\bullet}OH$), generated through Fenton and Haber–Weiss reactions involving $H_2O_2$ and transition metals (particularly $Fe^{2+}$ and $Cu^{+}$), is the most reactive ROS with an extremely short half-life that causes severe oxidative damage to all biomolecules (Imlay 2013). Singlet oxygen ($^1O_2$), an excited state of molecular oxygen produced in chloroplasts during photosynthesis and by immune cells during oxidative bursts, preferentially oxidizes unsaturated lipids and aromatic amino acids (Triantaphylidès and Havaux 2009). Additional ROS include organic peroxides (ROOH), peroxynitrite ($ONOO^-$) formed from the reaction of $O_2^{\bullet-}$ with nitric oxide ($^{\bullet}NO$), and ozone ($O_3$) produced in certain physiological contexts (Lushchak 2014). The cellular concentration and biological impact of these species depend on their chemical properties, half-lives, and reaction specificities: while $H_2O_2$ can diffuse across membranes and act as a second messenger in redox signaling pathways, $^{\bullet}OH$ reacts indiscriminately with nearby molecules within nanometers of its generation site (Sies et al. 2017). In plants, ROS are generated constitutively in chloroplasts, mitochondria, peroxisomes, apoplast, and endoplasmic reticulum, with production rates increasing significantly under environmental stresses such as drought, salinity, extreme temperatures, and pathogen attack (Mittler et al. 2022). The evolutionary conservation of ROS-generating systems (e.g., NOX enzymes) and antioxidant networks across kingdoms underscores their fundamental importance in cellular physiology, where they regulate processes ranging from growth and development to immune responses and programmed cell death (Waszczak et al. 2018). Modern redox biology recognizes ROS not merely as harmful metabolic byproducts but as essential components of cellular communication networks, with their specific chemical properties determining their roles in oxidative damage versus redox signaling (Sies and Jones 2020).

## 10.3 Sources of ROS Production in Plants

Reactive oxygen species (ROS) are generated in various parts of plant cells, including chloroplasts, mitochondria, peroxisomes, plasma membranes, the endoplasmic reticulum, and the cell wall, under both normal and stressful conditions. In light conditions, chloroplasts and peroxisomes are the primary sources of ROS, whereas mitochondria are the main producers in darkness (Choudhury et al. 2013).

### 10.3.1 Photosynthesis

**Chloroplast**

The production of reactive oxygen species (ROS) in chloroplasts is a crucial aspect of photosynthesis and plant metabolism. The production of reactive oxygen species (ROS) in chloroplasts is a crucial aspect of photosynthesis and plant metabolism. During the light-dependent reactions of photosynthesis, absorbed light energy excites electrons in chlorophyll. These high-energy electrons are transferred through a series of proteins embedded in the thylakoid membrane, known as the electron transport chain (ETC). The chloroplast features a highly organized structure of thylakoid membranes that contains the photosynthetic apparatus and the necessary components for effective light absorption (Pfannschmidt 2003). The photosystems PSI and PSII, which constitute the core of this light-harvesting system, are the main contributors to ROS production. Environmental stressors such as drought, high salinity, and temperature extremes, which can induce water stress and limit $CO_2$ availability, along with excessive light, drive the generation of superoxide ($O_2^{\bullet-}$) at the photosystems through the Mehler reaction:

$$2O_2 + 2Fd_{red} \rightarrow 2O_2^{\bullet-} + 2Fd_{ox}$$

Afterward, a membrane-associated Cu/Zn superoxide dismutase (SOD) at PSI converts $O_2^{\bullet-}$ into hydrogen peroxide ($H_2O_2$) (Miller et al. 2010). Additionally, the 2Fe-2S and 4Fe-4S clusters that leak electrons from PSI's electron transport chain (ETC) contribute to ROS production. In PSII, electron leakage occurs via electron acceptors QA and QB, resulting in $O_2^{\bullet-}$ generation. This superoxide radical can transform into more harmful ROS like hydroxyl radicals ($OH^{\bullet}$) through the Fenton reaction involving $H_2O_2$ at iron–sulfur centers. PSII also produces singlet oxygen ($^1O_2$) through two mechanisms: first, environmental stress can disrupt the balance between light acquisition and energy consumption, resulting in the formation of triplet chlorophyll ($^3Chl^*$), which reacts with molecular oxygen ($^3O_2$) to produce $^1O_2$ (Karuppanapandian et al. 2011). Secondly, over-reduction of the electron transport chain can lead to the generation of $^1O_2$ at the light-harvesting complex (LHC) in PSII (Asada 2006). Accumulation of $^1O_2$ within the chloroplast can cause lipid peroxidation of membrane lipids, particularly polyunsaturated fatty acids (PUFAs), damage membrane proteins, and jeopardize the P680 reaction center in PSII, potentially leading to cell death (Møller et al. 2007; Triantaphylidès et al. 2008). The role of the chloroplast in oxidative stress-induced programmed cell death was highlighted when the anti-apoptotic protein Bcl-2 was expressed in transgenic tobacco (Chen and Dickman, 2004). Furthermore, $^1O_2$ can activate genetic pathways, specifically the EXECUTOR1 and EXECUTOR2 pathways, leading to growth inhibition in plants (Lee et al. 2007). Therefore, the chloroplast is a crucial site of ROS generation in plants, and effective management of ROS levels is essential for plant survival under stress, as evidenced by transgenic plants and stress-tolerant cultivars (Tseng et al. 2007).

### 10.3.2 Cellular Respiration

**Mitochondria**

Mitochondria are also capable of producing harmful reactive oxygen species (ROS) such as hydrogen peroxide ($H_2O_2$) and superoxide ($O_2^{\bullet-}$) (Navrot et al. 2007), albeit at lower levels compared to other sources. Plant mitochondria differ from those in animals by existing in environments rich in oxygen and carbohydrates (Rhoads et al.2006) and playing a role in photorespiration. The mitochondrial electron transport chain (mtETC) is primarily responsible for ROS production, as it contains energized electrons that can reduce $O_2$ to create ROS. The two main components of the mtETC involved in this process are Complex I and Complex III (Møller et al. 2007; Noctor et al. 2007). Complex I, also known as NADH Dehydrogenase, directly reduces $O_2$ to $O_2^{\bullet-}$in its flavoprotein region. The production of ROS at Complex I is further amplified by reverse electron flow from Complex III to Complex I, which occurs in the absence of $NAD^+$-linked substrates, a process regulated by ATP hydrolysis (Turrens 2003). In Complex III, the fully reduced form of ubiquinone donates an electron to Cytochrome c1, leaving behind an unstable ubisemiquinone semiradical that encourages electron leakage to $O_2$, resulting in the formation of $O_2^{\bullet-}$ (Murphy 2009). Additional ROS sources in the mitochondria include various enzymes present in the matrix. For instance, aconitase directly produces ROS, while enzymes like 1-Galactono-γ-lactone dehydrogenase (GAL) indirectly contribute to ROS production by supplying electrons to the ETC (Rasmusson et al. 2008). While $O_2^{\bullet-}$ is the predominant ROS in mitochondria, it is eventually converted into $H_2O_2$ by manganese superoxide dismutase (Mn-SOD) and ascorbate peroxidase (APX) (Sharma et al. 2012). It is estimated that about 1–5% of the total $O_2$ consumption in mitochondria is utilized for $H_2O_2$ production. Under normal conditions, the mitochondria produce ROS, but this production significantly increases during periods of abiotic stress (Pastore et al. 2007). Stressful conditions can disrupt the tight coupling between the electron transport chain and ATP synthesis, leading to an over-reduction of electron carriers such as the ubiquinone (UQ) pool, which in turn generates more ROS (Rhoads et al. 2006; Blokhina and Fagerstedt 2010). As the respiratory rate rises during drought conditions, mitochondrial ATP production increases to compensate for a reduced rate of ATP synthesis in chloroplasts, subsequently boosting ROS output in mitochondria (Atkin and Macherel, 2009). To mitigate oxidative stress in mitochondria, two key enzymes—Mitochondrial Alternative

Oxidase (AOX) and Mitochondrial SOD (Mn-SOD)—are essential. AOX helps maintain the reduced state of the UQ pool, thereby reducing ROS generation. This importance is underscored by the fact that Arabidopsis plants lacking functional AOX are more sensitive to drought stress and exhibit altered expression of various antioxidant machinery components (Ho et al. 2008). Conversely, enhanced activity of Mn-SOD has been shown to differentiate between salt-tolerant and salt-sensitive tomato cultivars under salinity stress (Mittova et al. 2003).

### 10.3.3 Glycolate and Fatty Acid Metabolism

**Peroxisomes**

Peroxisomes are spherical microbodies enclosed by a single membrane and serve as the primary sites for intracellular hydrogen peroxide ($H_2O_2$) production due to their essential oxidative metabolism (Luis et al. 2006; Palma et al.2009). They also produce superoxide ($O_2^{\bullet-}$), similar to chloroplasts and mitochondria, during various metabolic processes. $O_2^{\bullet-}$ is generated in two distinct locations within the peroxisome. First, the enzyme xanthine oxidase (E.C.1.17.3.2), located in the peroxisomal matrix, converts xanthine and hypoxanthine into uric acid while producing $O_2^{\bullet-}$ as a by-product. Second, a small NADPH-dependent electron transport chain (ETC) composed of NADH and cytochrome b, found in the peroxisomal membrane, uses O2 as the electron acceptor and results in the release of $O_2^{\bullet-}$ into the cytosol. Additionally, three integral membrane proteins, known as peroxisomal membrane polypeptides (PMPs) with molecular weights of 18, 29, and 132 kDa, are responsible for $O_2^{\bullet-}$ production. The 18 kDa and 32 kDa PMPs use NADH as the electron donor, while the 29 kDa PMP utilizes NADPH to reduce cytochrome c. Under stress conditions, where water availability is limited and stomata are closed, the ratio of $CO_2$ to $O_2$ decreases significantly, prompting increased photorespiration and the formation of glycolate. Glycolate is subsequently oxidized by glycolate oxidase in the peroxisome, producing $H_2O_2$, which makes peroxisomes a major source of $H_2O_2$ during photorespiration (Noctor et al. 2002). Furthermore, additional metabolic processes, such as the β-oxidation of fatty acids, the flavin oxidase pathway, and the disproportionation of $O_2^{\bullet-}$ radicals, contribute to ROS production in peroxisomes.

### 10.3.4 Lignification and Cell Wall Remodeling

**Apoplast**

The apoplast, which refers to the space surrounding the plant cell membrane, facilitates the conversion of incoming $CO_2$ into a soluble, diffusible form that enters the cytosol for photosynthesis. During adverse environmental conditions, stress signals combined with abscisic acid (ABA) elevate $H_2O_2$ production in the apoplast (Hu et al. 2006). NADPH oxidases, specifically AtRbohD and AtRbohF, expressed in guard cells and mesophyll cells of Arabidopsis, generate apoplastic ROS necessary for ABA-induced stomatal closure (Kwak et al. 2003). In addition to these enzymes, other ROS-generating enzymes, including pH-dependent peroxidases (POXs), cell wall-associated oxidases, germin-like oxalate oxidases, and polyamine oxidases, primarily produce $H_2O_2$.

### 10.3.5 Plasma Membranes

The plasma membrane surrounding each plant cell plays a crucial role in responding to changing environmental conditions, providing essential information for the cell's ongoing survival. NADPH-dependent oxidases located in the plasma membrane have garnered attention due to their gene expression and the presence of various homologs under different stress conditions (Apel and Hirt 2004). These NADPH oxidases produce $O_2^{\bullet-}$ by transferring electrons from cytosolic NADPH to $O_2$, which can either spontaneously dismutate to $H_2O_2$ or be catalyzed by superoxide dismutase (SOD). The significant role of NADPH oxidase in plant defense against pathogens and abiotic stress is well-documented (Kwak et al. 2003).

### 10.3.6 Cell Walls

In response to stress, the lipoxygenase (LOX) enzyme localized in the cell wall catalyzes the hydroperoxidation of polyunsaturated fatty acids (PUFAs), resulting in the production of reactive oxygen species (ROS) such as hydroxyl radicals ($OH^{\bullet}$), superoxide ($O_2^{\bullet-}$), hydrogen peroxide ($H_2O_2$), and singlet oxygen ($^1O_2$). Additionally, cell wall-localized diamine oxidases utilize diamines or polyamines to generate ROS within the cell wall. When pathogens attack, the precursors of lignin undergo significant cross-linking through $H_2O_2$-mediated pathways, which strengthens the cell wall with lignin (Higuchi 2006).

### 10.3.7 Endoplasmic Reticulum (ER)

In the endoplasmic reticulum (ER), NADPH-mediated electron transport involving cytochrome P450 (CytP450) results in the generation of superoxide ($O_2^{\bullet-}$) (Mittler, 2002). An organic substrate, designated as RH, interacts with CytP450 and is subsequently reduced by a flavoprotein, producing a free radical intermediate (Cyt P450 R−). This intermediate rapidly reacts with triplet oxygen ($^3O_2$) to create an oxygenated complex (Cyt P450-ROO−). Occasionally, this complex decomposes back to Cyt P450-RH, releasing $O_2^{\bullet-}$ as a by-product.

## 10.4 Sources of ROS Production in Plants Under Stressed Conditions

### 10.4.1 Reactive Oxygen Species (ROS) Production in Chloroplasts Under Stress Conditions: Photoinhibition and Excess Light Energy

Chloroplasts are major sites of ROS production in plant cells, particularly under stress conditions that disrupt the balance between light absorption and energy utilization. During normal photosynthesis, the photosynthetic electron transport chain (PETC) efficiently transfers electrons from water to $NADP^+$ through photosystems II (PSI) and I (PSII). However, when plants experience environmental stresses such as high light intensity, drought, or extreme temperatures, this balance is disturbed, leading to over-reduction of the PETC and enhanced ROS production (Asada 2006). The primary ROS generated in chloroplasts under stress conditions include superoxide radical ($O_2^-$), hydrogen peroxide ($H_2O_2$), and singlet oxygen ($^1O_2$). At PSI, over-reduction of the electron transport chain causes leakage of electrons to molecular oxygen, forming $O_2^-$ through the Mehler reaction (Foyer and Noctor 2003). This $O_2^-$ is rapidly converted to $H_2O_2$ by superoxide dismutase (SOD) in the stroma. While $H_2O_2$ can function as a signaling molecule at low concentrations, its accumulation leads to oxidative damage of cellular components (Mittler et al. 2004).
At PSII, excess light energy promotes the formation of chlorophyll triplet states ($^3Chl$), which react with ground-state oxygen ($^3O_2$) to generate highly reactive $^1O_2$ (Krieger-Liszkay 2005). Unlike other ROS, $^1O_2$ has an extremely short half-life but causes severe damage to the D1 protein of PSII reaction centers, leading to photoinhibition—the light-induced reduction in photosynthetic efficiency (Nishiyama et al. 2006). The damage to D1 protein requires constant repair through de novo synthesis, a process that becomes limiting under severe stress conditions (Murata et al. 2007).
Plants have evolved multiple protective mechanisms to cope with chloroplastic ROS. The water–water cycle involving ascorbate peroxidase (APX) and other antioxidants helps scavenge $H_2O_2$ (Asada 1999). Non-photochemical quenching (NPQ) mechanisms dissipate excess excitation energy as heat through the xanthophyll cycle (Müller et al. 2001). Additionally, alternative electron pathways such as cyclic electron flow around PSI help prevent over-reduction of the PETC (Miyake 2010). However, when these protective systems are overwhelmed, ROS accumulation leads to oxidative damage of lipids, proteins, and nucleic acids, ultimately affecting plant growth and productivity (Gill and Tuteja 2010).

Understanding the mechanisms of ROS production and scavenging in chloroplasts is crucial for developing strategies to enhance plant stress tolerance. Recent approaches include genetic engineering of antioxidant enzymes and manipulation of photoprotective mechanisms to improve crop performance under adverse environmental conditions (Foyer and Shigeoka 2011).

### 10.4.2 Reactive Oxygen Species (ROS) Production in Mitochondria under Stress Conditions: Energy Imbalance and Oxidative Stress

While chloroplasts are major sites of ROS production during light stress, mitochondria also generate significant amounts of ROS under various stress conditions when energy homeostasis is disrupted (Møller et al. 2007). Under normal conditions, mitochondrial electron transport chain (ETC) complexes transfer electrons to oxygen with high efficiency, with only 1–2% of electrons leaking to form superoxide ($O_2^-$) (Murphy 2009). However, during abiotic stresses like drought, salinity, extreme temperatures, or pathogen attack, mitochondrial ROS production increases dramatically due to impaired ETC function and altered cellular energy demands (Noctor et al. 2007).

The primary sites of mitochondrial ROS generation are Complex I (NADH dehydrogenase) and Complex III (ubiquinol-cytochrome c reductase) of the ETC (Murphy 2009). At Complex I, reverse electron transport from ubiquinol can generate substantial $O_2^-$ when the proton motive force is high (Liu et al. 2002). At Complex III, electron leakage at the ubiquinone-binding site (Qo site) produces $O_2^-$ that is released to both the matrix and intermembrane space (Chen et al. 2003). The generated $O_2^-$ is rapidly converted to $H_2O_2$ by manganese superoxide dismutase (Mn-SOD) in the matrix and copper/zinc SOD in the intermembrane space (Mittler et al. 2004).

Multiple environmental stressors enhance mitochondrial ROS generation through distinct mechanisms, including hypoxia/reoxygenation-induced ETC disruption due to oxygen fluctuations (Blokhina and Fagerstedt 2010), drought and salinity-mediated alterations in cellular ATP demand (Miller et al. 2010), cold stress-impaired membrane fluidity reducing ETC efficiency (Yoshida and Noguchi 2009), and heavy metal-induced structural damage to ETC components (Heyno et al. 2008). The resulting ROS overproduction triggers oxidative damage to mitochondrial macromolecules (Møller and Kristensen 2004), activates programmed cell death pathways (Van Breusegem and Dat 2006), initiates retrograde signaling to modulate nuclear gene expression (Rhoads and Subbaiah 2007), and disrupts cellular redox balance (Noctor et al. 2007). To mitigate these effects, plants utilize integrated defense systems comprising enzymatic antioxidants (Mittler et al. 2004), the AOX pathway to prevent ETC over-reduction (Millar et al. 2011), uncoupling proteins for proton gradient dissipation (Vercesi et al. 2006), and selective autophagy for damaged organelle removal (Yoshimoto et al. 2004).

Recent studies highlight the dual role of mitochondrial ROS as both damaging agents and important signaling molecules in stress responses (Rhoads and Subbaiah 2007). Understanding these processes is crucial for developing stress-tolerant crops through manipulation of mitochondrial antioxidant systems and energy metabolism (Van Aken et al. 2009).

### 10.4.3 Reactive Oxygen Species Production in Peroxisomes Under Stress Conditions: Energy Imbalance and Oxidative Stress

Peroxisomes are major sites of ROS production in plant cells, particularly under stress conditions that disrupt cellular energy homeostasis. These single membrane-bound organelles generate significant amounts of hydrogen peroxide ($H_2O_2$) as an inevitable byproduct of their metabolic processes, with production rates increasing dramatically under various abiotic and biotic stresses (del Río and López-Huertas 2016). The primary enzymatic sources of ROS in peroxisomes include glycolate oxidase (GOX) in the photorespiratory pathway, fatty acid β-oxidation, and flavin oxidase reactions (Corpas et al. 2020). Under normal physiological conditions, the $H_2O_2$ produced is efficiently scavenged by catalase and the ascorbate-glutathione cycle, maintaining redox balance (Mhamdi et al. 2012). However, under stress conditions such as drought, salinity, extreme temperatures, heavy metal toxicity, or pathogen attack, the balance between ROS production and scavenging is disrupted, leading to oxidative stress (Sandalio and Romero-Puertas 2015).

The photorespiratory pathway becomes particularly significant for peroxisomal ROS production under stress conditions that limit $CO_2$ availability, such as drought-induced stomatal closure or high temperature stress (Voss et al. 2013). Under these conditions, the oxygenase activity of Rubisco increases, leading to elevated production of glycolate that must be metabolized in peroxisomes through the GOX reaction, which directly generates $H_2O_2$ (Foyer et al. 2009). Similarly, during fatty acid β-oxidation, which is enhanced during stress-induced senescence or lipid mobilization, acyl-CoA oxidase produces $H_2O_2$ as a byproduct (Graham 2008). Additional ROS sources in peroxisomes include xanthine oxidase, urate oxidase, and polyamine oxidases, all of which contribute to increased ROS production under stress conditions (Corpas and Barroso 2014).

The consequences of peroxisomal ROS overproduction under stress are multifaceted. Excessive $H_2O_2$ can lead to oxidative damage of peroxisomal proteins, including catalase itself, creating a vicious cycle of increasing ROS accumulation (Nyathi and Baker 2006). Peroxisomal ROS also diffuse to other cellular compartments, contributing to overall oxidative stress and potentially triggering programmed cell death (PCD) when antioxidant defenses are overwhelmed (Distéfano et al. 2017). However, peroxisomal ROS also serve important signaling functions, participating in retrograde signaling to the nucleus and s with other organelles to coordinate stress responses (Sewelam et al. 2016). The dual role of peroxisomal ROS as both damaging agents and signaling molecules makes their regulation particularly crucial for plant stress adaptation (Noctor et al. 2018).

Plants have evolved multiple mechanisms to regulate peroxisomal ROS homeostasis under stress conditions. These include the induction of antioxidant enzymes such as catalase, superoxide dismutase (SOD), and components of the ascorbate-glutathione cycle (Lismont et al. 2019). Peroxisomal membrane proteins, including PEX genes, help maintain organelle integrity and prevent ROS leakage (Kao et al. 2018). Additionally, peroxisome proliferation and turnover are dynamically regulated under stress to manage ROS production (Rodríguez-Serrano et al. 2016). Recent studies have highlighted the importance of peroxisome-mitochondria interactions in coordinating cellular responses to energy imbalance under stress (Castro et al.2021), suggesting that these organelles function as an integrated redox network rather than isolated compartments.

### 10.4.4 Reactive Oxygen Species (ROS) Production in the Apoplast Under Stress Conditions: RBOH-Mediated Oxidative Burst in Defense

The apoplast, comprising the cell wall and intercellular spaces, serves as a critical site for ROS generation during plant stress responses, particularly through the activity of plasma membrane-localized NADPH oxidases (respiratory burst oxidase homologs, RBOHs). Under both abiotic and biotic stress conditions, RBOH enzymes catalyze the rapid production of superoxide ($O_2^-$) by transferring electrons from cytoplasmic NADPH to extracellular oxygen, initiating the so-called "oxidative burst" that plays pivotal roles in stress signaling and defense (Kadota et al. 2015). This RBOH-dependent ROS production is tightly regulated by calcium ($Ca^{2+}$) signaling, phosphorylation cascades, and interactions with other signaling molecules such as nitric oxide (NO) and phytohormones (Suzuki et al. 2011). During pathogen attack, pattern recognition receptors (PRRs) perceive microbial-associated molecular patterns (MAMPs), triggering a signaling cascade that activates RBOHs within minutes, leading to $O_2^-$ production that subsequently dismutates to hydrogen peroxide ($H_2O_2$) either spontaneously or via superoxide dismutase (SOD) activity in the apoplast (Torres et al. 2006). The resulting $H_2O_2$ serves multiple functions: it directly inhibits pathogen growth, cross-links cell wall proteins (e.g., extensins) to reinforce structural integrity, and acts as a secondary messenger to activate downstream defense genes (Boller and Felix 2009). Abiotic stresses such as drought, salinity, and mechanical wounding similarly activate RBOHs, though often through distinct upstream signals involving abscisic acid (ABA) or jasmonic acid (JA) (Mittler et al. 2011). For instance, drought-induced ABA accumulation promotes RBOHF phosphorylation via calcium-dependent protein kinases (CDPKs), thereby enhancing apoplastic ROS production that facilitates stomatal closure (Drerup et al. 2013). The spatial and temporal regulation of this oxidative burst is critical, as uncontrolled ROS accumulation can lead to cellular damage; thus, plants employ apoplastic antioxidants like ascorbate peroxidases (APXs) and peroxiredoxins (PRXs) to fine-tune ROS levels (Passardi et al. 2004). Recent studies highlight the crosstalk between apoplastic ROS and organellar ROS (e.g., from chloroplasts or mitochondria) in amplifying stress signals, suggesting an integrated network for systemic stress responses (Gilroy et al. 2016). Moreover, RBOH-derived ROS participate in long-distance signaling via the xylem and phloem, coordinating whole-plant defenses (Miller et al. 2009). Genetic studies in Arabidopsis and crop plants demonstrate that RBOH knockout mutants exhibit heightened susceptibility to both biotic and abiotic stresses, underscoring the non-redundant roles of these enzymes in stress adaptation (Marino et al. 2012). Emerging evidence also implicates RBOHs in symbiotic interactions, where controlled ROS production facilitates root nodulation and mycorrhizal associations (Marino et al. 2011). The dual role of apoplastic ROS as toxic molecules and signaling agents exemplifies the "ROS paradox" in plant biology, where spatial and temporal confinement determines their functional outcomes (Mittler 2017). Future research directions include elucidating RBOH-isoform-specific functions, post-translational regulation mechanisms, and engineering RBOH activity for improved crop resilience without fitness trade-offs (Chapman and Muday 2019).

### 10.4.5 Reactive Oxygen Species (ROS) Production in the Endoplasmic Reticulum Under Stress Conditions

The endoplasmic reticulum (ER) serves as a significant site for ROS generation during cellular stress responses, primarily through protein folding and oxidative processes inherent to its function as a protein processing organelle. Under normal physiological conditions, the ER maintains redox homeostasis through the coordinated action of oxidoreductases and antioxidant systems, but various abiotic and biotic stressors disrupt this balance, leading to elevated ROS production (Ozgur et al. 2014). The primary sources of ER-derived ROS include the electron transport chain of cytochrome P450 enzymes, which can leak electrons to molecular oxygen during catalytic cycles, and the protein folding machinery, particularly the activity of protein disulfide isomerases (PDIs) and endoplasmic reticulum oxidoreductin 1 (ERO1) (Zeeshan et al. 2016). During stress conditions such as drought, heat, or pathogen attack, the accumulation of misfolded proteins triggers the unfolded protein response (UPR), which increases oxidative protein folding activity and consequently ROS generation (Cao and Kaufman 2014). The ERO1-PDI system generates hydrogen peroxide ($H_2O_2$) as a byproduct of disulfide bond formation during protein folding, with ERO1 directly transferring electrons to molecular oxygen (Sevier and Kaiser 2008). Additionally, NADPH oxidase 4 (NOX4), localized to the ER membrane, contributes to ROS production under stress conditions, particularly in response to calcium signaling (Chen et al. 2015). The resulting oxidative environment within the ER lumen can lead to the oxidation of ER-resident proteins, further exacerbating protein misfolding and creating a feed-forward cycle of ER stress and ROS production (Malhotra and Kaufman 2007). ER-derived ROS also function as signaling molecules, participating in calcium release from ER stores and activating stress-responsive transcription factors such as NF-κB and Nrf2 (Santos et al. 2009). The close physical and functional association between the ER and mitochondria, through mitochondria-associated ER membranes (MAMs), facilitates ROS-mediated communication between these organelles during stress responses (Raturi and Simmen 2013). To counteract ER oxidative stress, plants employ several protective mechanisms, including the upregulation of ER-resident antioxidants such as glutathione peroxidases (GPXs) and peroxiredoxins (PRXs), the activation of the UPR to restore protein folding capacity, and the selective autophagy of damaged ER (ER-phagy) (Liu and Howell 2016). The dual role of ER-derived ROS as both deleterious byproducts and

**Table 10.1** Key ROS produced in chloroplasts under stress

| ROS type | Formation site | Primary mechanism | Consequences |
|---|---|---|---|
| Superoxide ($O_2^{\cdot-}$) | PSI (ferredoxin) | Electron leakage to $O_2$ | Converted to $H_2O_2$ by SOD |
| Singlet Oxygen ($^1O_2$) | PSII (reaction center) | Chlorophyll triplet state reacts with $O_2$ | Oxidizes lipids, proteins, D1 protein damage |
| Hydrogen Peroxide ($H_2O_2$) | Stroma | SOD converts $O_2^{\cdot-}$ to $H_2O_2$ | Signals stress responses or causes oxidative damage |

**Table 10.2** Sources of ROS production in plants: normal vs. stress conditions

| ROS source | Normal conditions (Low/Controlled ROS) | Stress conditions (High ROS) | Key ROS produced |
|---|---|---|---|
| Chloroplasts | Photosynthesis (electron leakage from PSI & PSII) | Over-reduction of ETC due to light stress, drought, cold | $^1O_2$ (singlet oxygen), $O_2^-$ (Superoxide), $H_2O_2$ |
| Mitochondria | Respiration (ETC leakage in Complex I & III) | Enhanced ETC disruption under hypoxia, heat, salinity | $O_2^-$, $H_2O_2$ |
| Peroxisomes | Photorespiration (glycolate oxidase) & β-oxidation | Increased photorespiration under drought, high light | $H_2O_2$ |
| Plasma Membrane (NADPH Oxidases, RBOHs) | Low activity for signaling (growth, defense priming) | Activated by pathogens, drought, wounding | $O_2^-$ (later converted to $H_2O_2$) |
| Cell Wall (Peroxidases, Oxidases) | Lignification, growth regulation | Enhanced under pathogen attack, mechanical stress | OH (Hydroxyl radical), $H_2O_2$ |
| Endoplasmic reticulum | Electron leakage from cytochrome P450 | Stress-induced ER dysfunction (e.g., heat, toxins) | $O_2^-$, $H_2O_2$ |
| Apoplast | Low-level ROS from extracellular enzymes | Stress-induced apoplastic ROS burst (e.g., fungal elicitors) | $H_2O_2$, OḤ |

crucial signaling molecules exemplifies the complex regulation of redox homeostasis during stress adaptation (Hetz et al. 2020). Recent studies have highlighted the importance of ER redox systems in plant responses to climate change-associated stresses, suggesting potential targets for improving crop resilience through manipulation of ER stress pathways (Howell 2021) (Tables 10.1 and 10.2).

## 10.5 Conclusion

Reactive oxygen species (ROS) are inevitable byproducts of aerobic metabolism in plants, playing a dual role as both signaling molecules and agents of oxidative damage. Under normal physiological conditions, ROS are primarily generated in chloroplasts, mitochondria, and peroxisomes as a result of photosynthetic and respiratory electron transport, photorespiration, and enzymatic reactions. However, under stress conditions—including abiotic factors such as drought, salinity, extreme temperatures, and heavy metals, as well as biotic stressors like pathogens—ROS production is significantly amplified due to disrupted cellular homeostasis and enhanced enzymatic activities (e.g., NADPH oxidases, peroxidases, and xanthine oxidases).

While excessive ROS accumulation can lead to oxidative stress and cellular damage, plants have evolved sophisticated antioxidant systems to maintain redox balance. The interplay between ROS production and scavenging mechanisms determines the fate of plant cells, influencing adaptation, defense responses, and programmed cell death when necessary. Understanding the sources and regulation of ROS in both normal and stressed conditions is crucial for developing strategies to enhance plant stress tolerance and improve crop productivity in changing environments. Future research should focus on the spatiotemporal dynamics of ROS signaling and the genetic engineering of antioxidant pathways to bolster plant resilience against escalating environmental challenges.

This chapter highlights the complexity of ROS metabolism in plants, emphasizing its pivotal role in growth, stress responses, and survival. A deeper comprehension of ROS sources and their modulation will pave the way for innovative approaches in plant biotechnology and sustainable agriculture.

## References

Apel K, Hirt H (2004) Reactive oxygen species: metabolism, oxidative stress, and signal transduction. Annu Rev Plant Biol 55:373–399. https://doi.org/10.1146/annurev.arplant.55.031903.141701

Asada K (1999) The water-water cycle in chloroplasts. Annu Rev Plant Biol 50:601–639. https://doi.org/10.1146/annurev.arplant.50.1.601

Asada K (2006) Production and scavenging of reactive oxygen species in chloroplasts and their functions. Plant Physiol 141(2):391–396. https://doi.org/10.1104/pp.106.082040

Atkin OK, Macherel D (2009) The crucial role of plant mitochondria in orchestrating drought tolerance. Ann Bot 103(4):581–597

Baker NR (2008) Chlorophyll fluorescence: a probe of photosynthesis in vivo. Annu Rev Plant Biol 59:89–113. https://doi.org/10.1146/annurev.arplant.59.032607.092759

Bienert GP, Chaumont F (2014) Aquaporin-facilitated transmembrane diffusion of hydrogen peroxide. Biochim Biophys Acta 1840(6):1596–1604. https://doi.org/10.1016/j.bbagen.2013.09.017

Birben E, Sahiner UM, Sackesen C, Erzurum S, Kalayci O (2012) Oxidative stress and antioxidant defense. World Allergy Org J 5(1):9–19. https://doi.org/10.1097/WOX.0b013e3182439613

Blokhina O, Fagerstedt KV (2010) Reactive oxygen species and nitric oxide in plant mitochondria: origin and redundant regulatory systems. Ann Bot 105(6):839–850. https://doi.org/10.1093/aob/mcq089

Boller T, Felix G (2009) A renaissance of elicitors: perception of microbe-associated molecular patterns and danger signals by pattern-recognition receptors. Annu Rev Plant Biol 60:379–406. https://doi.org/10.1146/annurev.arplant.57.032905.105346

Cao SS, Kaufman RJ (2014) Endoplasmic reticulum stress and oxidative stress in cell fate decision and human disease. Antioxid Redox Signal 21(3):396–413. https://doi.org/10.1089/ars.2014.5851

Chapman JM, Muday GK (2019) RBOH-dependent ROS synthesis and ROS scavenging by plant specialized metabolites to modulate plant development and stress responses. Chem Res Toxicol 32(3):370–396. https://doi.org/10.1021/acs.chemrestox.9b00028

Castro B, Citterico M, Kimura S, Stevens DM, Wrzaczek M, Coaker G (2021) Stress-induced reactive oxygen species compartmentalization, perception and signalling. Nat Plants 7(4):403–412

Chen SR, Dunigan DD, Dickman MB (2003) Bcl-2 family members inhibit oxidative stress-induced programmed cell death in Saccharomyces cerevisiae. Free Radic Biol Med 34(11):1315–1325

Chen S, Dickman MB (2004) Bcl-2 family members localize to tobacco chloroplasts and inhibit programmed cell death induced by chloroplast-targeted herbicides. J Exp Bot 55(407):2617–2623

Chen K, Kirber MT, Xiao H, Yang Y, Keaney JF (2015) Regulation of ROS signal transduction by NADPH oxidase 4 localization. J Cell Biol 181(7):1129–1139. https://doi.org/10.1083/jcb.200709049

Choudhury S, Panda P, Sahoo L, Panda SK (2013) Reactive oxygen species signaling in plants under abiotic stress. Plant Signal Behav 8(4):e23681

Corpas FJ, Barroso JB (2014) Peroxisomal plant metabolism—an update on nitric oxide, $Ca^{2+}$ and the NADPH recycling network. J Cell Sci 127(18):4081–4088. https://doi.org/10.1242/jcs.151175

Corpas FJ, González-Gordo S, Cañas A, Palma JM (2020) Nitric oxide and hydrogen sulfide in plants: which comes first? J Exp Bot 71(15):4391–4404. https://doi.org/10.1093/jxb/eraa031

del Río LA, López-Huertas E (2016) ROS generation in peroxisomes and its role in cell signaling. Plant Cell Physiol 57(7):1364–1376. https://doi.org/10.1093/pcp/pcw076

Distéfano AM, López GA, Setzes N, Marchetti F, Cainzos M, Cascallares M, Zabaleta E, Pagnussat GC (2017) Ferroptosis in plants: triggers, proposed mechanisms, and the role of iron in modulating cell death. J Exp Bot 68(10):2665–2678. https://doi.org/10.1093/jxb/erx161

Drerup MM, Schlücking K, Hashimoto K, Manishankar P, Steinhorst L, Kuchitsu K, Kudla J (2013) The calcineurin B-like calcium sensors CBL1 and CBL9 together with their interacting protein kinase CIPK26 regulate the Arabidopsis NADPH oxidase RBOHF. Mol Plant 6(2):559–569. https://doi.org/10.1093/mp/sst009

Fahad S, Yaqoob M, Khedher NB et al (2017) Drought and heat stress in plants: a review. J Plant Growth Regul 36(3):588–609. https://doi.org/10.1007/s00344-016-9663-5

Foyer CH, Noctor G (2003) Redox sensing and signalling associated with reactive oxygen in chloroplasts, peroxisomes and mitochondria. Physiol Plant 119(3):355–364. https://doi.org/10.1034/j.1399-3054.2003.00223.x

Foyer CH, Noctor G (2005) Redox homeostasis and antioxidant signaling: a metabolic interface between stress perception and physiological responses. Plant Cell 17(7):1866–1875. https://doi.org/10.1105/tpc.105.033589

Foyer CH, Bloom AJ, Queval G, Noctor G (2009) Photorespiratory metabolism: genes, mutants, energetics, and redox signaling. Annu Rev Plant Biol 60:455–484. https://doi.org/10.1146/annurev.arplant.043008.091948

Foyer CH, Shigeoka S (2011) Understanding oxidative stress and antioxidant functions to enhance photosynthesis. Plant Physiol 155(1):93–100. https://doi.org/10.1104/pp.110.166181

Gill SS, Tuteja N (2010) Reactive oxygen species and antioxidant machinery in abiotic stress tolerance in crop plants. Plant Physiol Biochem 48(12):909–930. https://doi.org/10.1016/j.plaphy.2010.08.016

Gilroy S, Białasek M, Suzuki N, Górecka M, Devireddy AR, Karpiński S, Mittler R (2016) ROS, calcium, and electric signals: key mediators of rapid systemic signaling in plants. Plant Physiol 171(3):1606–1615. https://doi.org/10.1104/pp.16.00434

Graham IA (2008) Seed storage oil mobilization. Annu Rev Plant Biol 59:115–142. https://doi.org/10.1146/annurev.arplant.59.032607.092938

Halliwell B, Gutteridge JMC (2015) Free radicals in biology and medicine, 5th edn. Oxford University Press. https://doi.org/10.1093/acprof:oso/9780198717478.001.0001

Hetz C, Zhang K, Kaufman RJ (2020) Mechanisms, regulation and functions of the unfolded protein response. Nat Rev Mol Cell Biol 21(8):421–438. https://doi.org/10.1038/s41580-020-0250-z

Heyno E, Mary V, Schopfer P, Krieger-Liszkay A (2008) Oxygen activation at the plasma membrane: relation between superoxide and hydroxyl radical production by isolated membranes. Plant Physiol 147(1):206–216. https://doi.org/10.1104/pp.107.115063

Higuchi T (2006) Look back over the studies of lignin biochemistry. J Wood Sci 52(1):2–8

Howell SH (2021) Endoplasmic reticulum stress responses in plants. Annu Rev Plant Biol 72:477–504. https://doi.org/10.1146/annurev-arplant-081720-090528

Ho LH, Giraud E, Uggalla V, Lister R, Clifton R, Glen A, Thirkettle-Watts D, Van Aken O, Whelan J (2008) Identification of regulatory pathways controlling gene expression of stress-responsive mitochondrial proteins in Arabidopsis. Plant Physiol 147(4):1858–1873

Hu X, Jiang M, Zhang A, Lu J (2006) Abscisic acid-induced apoplastic $H_2O_2$ accumulation up-regulates the activities of chloroplastic and cytosolic antioxidant enzymes in maize leaves. Planta 223(1):57–68

Imlay JA (2013) The molecular mechanisms and physiological consequences of oxidative stress: lessons from a model bacterium. Nat Rev Microbiol 11(7):443–454. https://doi.org/10.1038/nrmicro3032

Kadota Y, Shirasu K, Zipfel C (2015) Regulation of the NADPH oxidase RBOHD during plant immunity. Plant Cell Physiol 56(8):1472–1480. https://doi.org/10.1093/pcp/pcv063

Karuppanapandian T, Moon JC, Kim C, Manoharan K, Kim W (2011) Reactive oxygen species in plants: their generation, signal transduction, and scavenging mechanisms. Aust J Crop Sci 5(6):709–725

Kao YT, Gonzalez KL, Bartel B (2018) Peroxisome function, biogenesis, and dynamics in plants. Plant Physiol 176(1):162–177. https://doi.org/10.1104/pp.17.01050

Krieger-Liszkay A (2005) Singlet oxygen production in photosynthesis. J Exp Bot 56(411):337–346. https://doi.org/10.1093/jxb/erh237

Kwak JM, Mori IC, Pei ZM, Leonhardt N, Torres MA, Dangl JL, Bloom RE, Bodde S, Jones JD, Schroeder JI (2003) NADPH oxidase AtrbohD and AtrbohF genes function in ROS-dependent ABA signaling in Arabidopsis. EMBO J 22(11):2623–2633

Lismont C, Revenco I, Fransen M (2019) Peroxisomal hydrogen peroxide metabolism and signaling in health and disease. Int J Mol Sci 20(15):3673. https://doi.org/10.3390/ijms20153673

Liu JX, Howell SH (2016) Managing the protein folding demands in the endoplasmic reticulum of plants. New Phytol 211(2):418–428. https://doi.org/10.1111/nph.13915

Liu Y, Fiskum G, Schubert D (2002) Generation of reactive oxygen species by the mitochondrial electron transport chain. J Neurochem 80(5):780–787

Lee KP, Kim C, Lee DW, Apel K, Tigrina D (2007) Required for regulating the biosynthesis of tetrapyrroles in barley, is an ortholog of the FLU gene of Arabidopsis thaliana. FEBS Lett 581(16):2921–2926

Luis A, Sandalio LM, Corpas FJ, Palma JM, Barroso JB (2006) Reactive oxygen species and reactive nitrogen species in peroxisomes. Production, scavenging, and role in cell signaling. Plant Physiol 141(2):330–335

Lushchak VI (2014) Free radicals, reactive oxygen species, oxidative stress and its classification. Chem Biol Interact 224:164–175. https://doi.org/10.1016/j.cbi.2014.10.016

Malhotra JD, Kaufman RJ (2007) Endoplasmic reticulum stress and oxidative stress: a vicious cycle or a double-edged sword? Antioxid Redox Signal 9(12):2277–2294. https://doi.org/10.1089/ars.2007.1782

Marino D, Dunand C, Puppo A, Pauly N (2011) A burst of plant NADPH oxidases. Trends Plant Sci 17(1):9–15

Marino D, Dunand C, Puppo A, Pauly N (2012) A burst of plant NADPH oxidases. Trends Plant Sci 17(1):9–15. https://doi.org/10.1016/j.tplants.2011.10.001

Mhamdi A, Noctor G, Baker A (2012) Plant catalases: peroxisomal redox guardians. Arch Biochem Biophys 525(2):181–194. https://doi.org/10.1016/j.abb.2012.04.015

Miller G, Suzuki N, Ciftci-Yilmaz S, Mittler R (2010) Reactive oxygen species homeostasis and signalling during drought and salinity stresses. Trends Plant Sci 15(5):300–309. https://doi.org/10.1016/j.tplants.2010.03.007

Mittler R (2002) Oxidative stress, antioxidants and stress tolerance. Trends Plant Sci 7(9):405–410

Mittova V, Tal M, Volokita M, Guy M (2003) Up-regulation of the leaf mitochondrial and peroxisomal antioxidative systems in response to salt-induced oxidative stress in the wild salt-tolerant tomato species Lycopersicon pennellii. Plant Cell Environ 26(4):845–856

Miller G, Schlauch K, Tam R, Cortes D, Torres MA, Shulaev V, Dangl JL, Mittler R (2009) The plant NADPH oxidase RBOHD mediates rapid systemic signaling in response to diverse stimuli. Sci Signal 2(84):ra45

Millar AH, Whelan J, Soole KL, Day DA (2011) Organization and regulation of mitochondrial respiration in plants. Physiol Plant 142(2):157–164. https://doi.org/10.1111/j.1399-3054.2011.01460.x

Mittler R (2017) ROS are good. Trends Plant Sci 22(1):11–19

Mittler R, Vanderauwera S, Gollery M, Van Breusegem F (2004) Reactive oxygen gene network of plants. Trends Plant Sci 9(10):490–498. https://doi.org/10.1016/j.tplants.2004.08.009

Mittler R, Vanderauwera S, Suzuki N, Miller G, Tognetti VB, Vandepoele K, Van Breusegem F (2011) ROS signaling: the new wave? Trends Plant Sci 16(6):300–309. https://doi.org/10.1016/j.tplants.2011.03.007

Mittler R, Zandalinas SI, Fichman Y, Van Breusegem F (2022) Reactive oxygen species signalling in plant stress responses. Nat Rev Mol Cell Biol 23(10):663–679. https://doi.org/10.1038/s41580-022-00499-2

Miyake C (2010) Alternative electron flows (water-water cycle and cyclic electron flow around PSI) in photosynthesis. Plant Cell Physiol 51(12):1951–1963. https://doi.org/10.1093/pcp/pcq173

Møller IM (2001) Plant mitochondria and reactive oxygen species. Plant Physiol 127(3):1045–1050. https://doi.org/10.1104/pp.010365

Møller IM, Kristensen BK (2004) Protein oxidation in plant mitochondria as a stress indicator. Physiol Plant 120(3):312–318. https://doi.org/10.1111/j.0031-9317.2004.0235.x

Møller IM, Jensen PE, Hansson A (2007) Oxidative modifications to cellular components in plants. Annu Rev Plant Biol 58:459–481

Müller P, Li XP, Niyogi KK (2001) Non-photochemical quenching. A response to excess light energy. Plant Physiol 125(4):1558–1566. https://doi.org/10.1104/pp.125.4.1558

Murphy MP (2009) How mitochondria produce reactive oxygen species. Biochem J 417(1):1–13. https://doi.org/10.1042/BJ20081386

Murata N, Takahashi S, Nishiyama Y, Allakhverdiev SI (2007) Photoinhibition of photosystem II under environmental stress. Biochim Biophys Acta 1767(6):414–421. https://doi.org/10.1016/j.bbabio.2006.11.019

Nishiyama Y, Allakhverdiev SI, Murata N (2006) Photosynthetic electron transport and the proton motive force generation. Annu Rev Plant Biol 57:73–100. https://doi.org/10.1146/annurev.arplant.57.032905.105316

Noctor G, De Paepe R, Foyer CH (2007) Mitochondrial redox biology and homeostasis in plants. Annu Rev Plant Biol 58:459–481. https://doi.org/10.1146/annurev.arplant.58.032806.103847

Noctor G, Reichheld JP, Foyer CH (2018) ROS-related redox regulation and signaling in plants. Semin Cell Dev Biol 80:3–12. https://doi.org/10.1016/j.semcdb.2017.07.013

Palma JM, Corpas FJ, del Río LA (2009) Proteome of plant peroxisomes: new perspectives on the role of these organelles in cell biology. Proteomics 9(9):2301–2312

Pastore D, Trono D, Laus MN, Di Fonzo N, Flagella Z (2007) Possible plant mitochondria involvement in cell adaptation to drought stress. A case study: durum wheat mitochondria. J Exp Bot 58(2):195–210

Nyathi Y, Baker A (2006) Plant peroxisomes as a source of signalling molecules. Biochim Biophys Acta 1763(12):1478–1495. https://doi.org/10.1016/j.bbamcr.2006.08.031

Navrot N, Rouhier N, Gelhaye E, Jacquot JP (2007) Reactive oxygen species generation and antioxidant systems in plant mitochondria. Physiol Plant 129(1):185–195

Noctor G, Veljovic-Jovanovic S, Driscoll S, Novitskaya L, Foyer CH (2002) Drought and oxidative load in the leaves of $C_3$ plants: a predominant role for photorespiration? Ann Bot 89(7):841–850

Ozgur R, Turkan I, Uzilday B, Sekmen AH (2014) Endoplasmic reticulum stress triggers ROS signalling, changes the redox state, and regulates the antioxidant defence of Arabidopsis thaliana. J Exp Bot 65(5):1377–1390. https://doi.org/10.1093/jxb/eru034

Passardi F, Cosio C, Penel C, Dunand C (2004) Peroxidases have more functions than a Swiss army knife. Plant Cell Rep 24(5):255–265. https://doi.org/10.1007/s00299-005-0972-6

Pfannschmidt T (2003) Chloroplast redox signals: how photosynthesis controls its own genes. Trends Plant Sci 8(1):33–41. https://doi.org/10.1016/S1360-1385(02)00005-5

Raturi A, Simmen T (2013) Where the endoplasmic reticulum and the mitochondrion tie the knot: the mitochondria-associated membrane (MAM). Biochim Biophys Acta 1833(1):213–224. https://doi.org/10.1016/j.bbamcr.2012.04.013

Rhoads DM, Subbaiah CC (2007) Mitochondrial retrograde regulation in plants. Plant Physiol 144(2):1346–1361. https://doi.org/10.1104/pp.107.100602

Rasmusson AG, Geisler DA, Møller IM (2008) The multiplicity of dehydrogenases in the electron transport chain of plant mitochondria. Mitochondrion 8(1):47–60

Rhoads DM, Umbach AL, Subbaiah CC, Siedow JN (2006) Mitochondrial reactive oxygen species. Contribution to oxidative stress and interorganellar signaling. Plant Physiol 141(2):357–366

Rodríguez-Serrano M, Romero-Puertas MC, Sanz-Fernández M, Huertas E, Sandalio LM (2016) Peroxisomes extend peroxules in a fast response to stress via a reactive oxygen species-mediated induction of the peroxin PEX11a. Plant Physiol 171(3):1665–1674. https://doi.org/10.1104/pp.16.00648

Sandalio LM, Romero-Puertas MC (2015) Peroxisomes sense and respond to environmental cues by regulating ROS and RNS signalling networks. Ann Bot 116(4):475–485. https://doi.org/10.1093/aob/mcv074

Santos CX, Tanaka LY, Wosniak J, Laurindo FR (2009) Mechanisms and implications of reactive oxygen species generation during the unfolded protein response: roles of endoplasmic reticulum oxidoreductases, mitochondrial electron transport, and NADPH oxidase. Antioxid Redox Signal 11(10):2409–2427. https://doi.org/10.1089/ars.2009.2625

Sharma P, Jha AB, Dubey RS, Pessarakli M (2012) Reactive oxygen species, oxidative damage, and antioxidative defense mechanism in plants under stressful conditions J Bot. 217037

Schieber M, Chandel NS (2014) ROS function in redox signaling and oxidative stress. Curr Biol 24(10):R453–R462. https://doi.org/10.1016/j.cub.2014.03.034

Sevier CS, Kaiser CA (2008) Ero1 and redox homeostasis in the endoplasmic reticulum. Biochim Biophys Acta 1783(4):549–556. https://doi.org/10.1016/j.bbamcr.2007.12.011

Sewelam N, Kazan K, Schenk PM (2016) Global plant stress signaling: reactive oxygen species at the cross-road. Front Plant Sci 7:187. https://doi.org/10.3389/fpls.2016.00187

Sies H, Berndt C, Jones DP (2017) Oxidative stress. Annu Rev Biochem 86:715–748. https://doi.org/10.1146/annurev-biochem-061516-045037

Sies H, Jones DP (2020) Reactive oxygen species (ROS) as pleiotropic physiological signalling agents. Nat Rev Mol Cell Biol 21(7):363–383. https://doi.org/10.1038/s41580-020-0230-3

Suzuki N, Koussevitzky S, Mittler R, Miller G (2012) ROS and redox signaling in the response of plants to abiotic stress. Plant Cell Environ 35(2):259–270. https://doi.org/10.1111/j.1365-3040.2011.02336.x

Suzuki N, Miller G, Morales J, Shulaev V, Torres MA, Mittler R (2011) Respiratory burst oxidases: the engines of ROS signaling. Curr Opin Plant Biol 14(6):691–699. https://doi.org/10.1016/j.pbi.2011.07.014

Torres MA, Jones JD, Dangl JL (2006) Reactive oxygen species signaling in response to pathogens. Plant Physiol 141(2):373–378. https://doi.org/10.1104/pp.106.079467

Triantaphylidès C, Havaux M (2009) Singlet oxygen in plants: production, detoxification and signaling. Trends Plant Sci 14(4):219–228. https://doi.org/10.1016/j.tplants.2009.01.008

Triantaphylidès C, Krischke M, Hoeberichts FA, Ksas B, Gresser G, Havaux M, Van Breusegem F, Mueller MJ (2008) Singlet oxygen is the major reactive oxygen species involved in photooxidative damage to plants. Plant Physiol 148(2):960–968

Tseng TS, Whippo C, Hangarter RP, Briggs WR (2007) The role of a 14-3-3 protein in stomatal opening mediated by PHOT2 in Arabidopsis. Plant Cell 19(5):1515–1526

Turrens JF (2003) Mitochondrial formation of reactive oxygen species. J Physiol 552(Pt 2):335–344

Van Aken O, Giraud E, Clifton R, Whelan J (2009) Alternative oxidase: a target and regulator of stress responses. Physiol Plant 137(4):426–439. https://doi.org/10.1111/j.1399-3054.2009.01240.x

Van Breusegem F, Dat JF (2006) Reactive oxygen species in plant cell death. Plant Cell Environ 29(6):1091–1105. https://doi.org/10.1111/j.1365-3040.2006.01500.x

Vercesi AE, Borecký J, Maia IG, Arruda P, Cuccovia IM, Chaimovich H (2006) Plant uncoupling mitochondrial proteins. Physiol Plant 129(1):422–432. https://doi.org/10.1111/j.1399-3054.2006.00856.x

Voss I, Sunil B, Scheibe R, Raghavendra AS (2013) Emerging concept for the role of photorespiration as an important part of abiotic stress response. Plant Biol 15(4):713–722. https://doi.org/10.1111/j.1438-8677.2012.00710.x

Wang Y, Zhang R, Jiao S, Wang X, Lu Y, Du Y, Zhang W, Kang Y, Liu Y, Qin S (2025) Physiological and transcriptomic analyses reveal the regulatory mechanism of tuber growth in potato (Solanum tuberosum L.) under drought, alkaline salt, and combined stresses. Sci Hortic 337:113307

Waszczak C, Carmody M, Kangasjärvi J (2018) Reactive oxygen species in plant signaling. Annu Rev Plant Biol 69:209–236. https://doi.org/10.1146/annurev-arplant-042817-040322

Yoshida K, Noguchi K (2009) Differential gene expression profiles of the mitochondrial respiratory components in illuminated Arabidopsis leaves. Plant Cell Physiol 50(4):740–750. https://doi.org/10.1093/pcp/pcp035

Yoshimoto K, Hanaoka H, Sato S, Kato T, Tabata S, Noda T, Ohsumi Y (2004) Processing of ATG8s, ubiquitin-like proteins, and their deconjugation by ATG4s are essential for plant autophagy. Plant Cell 16(11):2967–2983. https://doi.org/10.1105/tpc.104.025395

Zeeshan HM, Lee GH, Kim HR, Chae HJ (2016) Endoplasmic reticulum stress and associated ROS. Int J Mol Sci 17(3):327. https://doi.org/10.3390/ijms17030327

# Osmolytes in ROS Scavenging and Maintaining Homeostasis Under Abiotic Stress

11

Gholamreza Abdi, Ananya Baidya, Md Sabir Ahmed Mondol, Ubaida Akbar, Aritra Mukherjee, Tadela Susmitha, and Anilava Chatterjee

**Abstract**

Weather fluctuation triggers a suite of abiotic stress factors to induce the production of reactive oxygen species (ROS). The phenomenon has worldwide relevance to food availability and food security, as well as plant development, growth, and survival. Under weather fluctuation, osmolytes are produced in response by plants to act as compatible solutes. Besides inducing homeostasis, osmolytes increase the water potential of the plant, adjust intracellular osmotic pressure to keep cell turgidity in check, and adjust redox metabolism to trap reactive oxygen species (ROS) excess to regulate oxidative as well as osmotic stress

G. Abdi (✉)
Department of Biotechnology, Persian Gulf Research Institute, Persian Gulf University, Bushehr, Iran
e-mail: abdi@pgu.ac.ir

A. Baidya
Department of Plant Physiology, Bidhan Chandra Krishi Viswavidyalaya, Mohanpur, Nadia, West Bengal, India

M. S. A. Mondol · A. Chatterjee
Department of Agricultural Biochemistry, Bidhan Chandra Krishi Viswavidyalaya, Mohanpur, Nadia, West Bengal, India

U. Akbar
Department of Food Technology and Nutrition, Lovely Professional University, Phagwara, Punjab, India

A. Mukherjee
Department of Seed Science and Technology, Bidhan Chandra Krishi Viswavidyalaya, Mohanpur, Nadia, West Bengal, India

T. Susmitha
Department of Agricultural Biochemistry, School of Agricultural Sciences, Mallareddy University, Hyderabad, Telangana, India



N. Iqbal et al. (eds.), *Plant Signaling Molecules in Regulation of ROS-Scavenging System*, https://doi.org/10.1007/978-981-95-3377-0_11

imposed on cellular processes. Carbohydrates, proline, glycine-betaine, and polyamines form a leading group of osmolytes. The main role of these osmolytes is to stabilize intracellular as well as external osmotic gradients. Furthermore, they reverse oxidative damage caused by other reactive oxygen species (ROS) on plant cellular machinery, i.e., hydroxyl ions, superoxide ions, and hydrogen peroxide. Plant developmental regulators such as abscisic acid, brassinosteroids, cytokinins, ethylene, jasmonates, and salicylic acid regulate osmolyte production. It is therefore of essential interest to analyze the manner by which plants subjected to abiotic stress use phytohormones to regulate the production of osmolytes. This chapter explores the molecular processes and possible roles of osmolytes, such as their connection to plants' resilience to abiotic stress, the dynamics of their behavior in reaction to changes in their environment, and the resolution of signal transduction pathways. Furthermore, transgenic plants expressing genes for certain osmolytes demonstrated enhanced resilience to various forms of abiotic stresses.

**Keywords**

Osmotic stress · Osmotic regulation · ROS Scavenging · Cellular homeostasis · Abiotic stress · Hormonal regulation

## 11.1 Introduction

Abiotic stress factors are always present in plants, and they include things such as salinity, drought, oxidative stress, metal toxicity, and severe temperatures. The polyphasic and far-reaching impacts of these abiotic stresses on plant growth, development, productivity, and survival have substantial implications for global food security. Crop yields are already negatively impacted by abiotic stressors, but as climate variability intensifies and environmental quality continues to decline, these stresses will become more frequent and more severe (Fig. 11.1). Mittler et al. (2004) state that oxidative stress is essential for controlling metabolic and physiological processes in plants and for keeping cell homeostasis. Important metabolic and physiological activities that generate reactive oxygen species (ROS) include photosynthesis, respiration, and photorespiration. When present in normal physiological conditions and at physiological concentrations, reactive oxygen species (ROS) such as superoxide anion ($O_2^{\cdot -}$), hydroxyl radical ($\cdot OH$), hydrogen peroxide ($H_2O_2$), and singlet oxygen ($^1O_2$) significantly influence signal transduction and a multitude of physiological and developmental processes. When plants are subjected to abiotic stimuli, they undergo oxidative stress because their ROS synthesis exceeds the capacity of their antioxidant defence mechanisms. Excess reactive oxygen species (ROS) can damage proteins and nucleic acids, which are crucial components of cells. High amounts of reactive oxygen species (ROS) can disrupt metabolic balance, harm membrane integrity, and induce enzyme malfunction (Gill and Tuteja

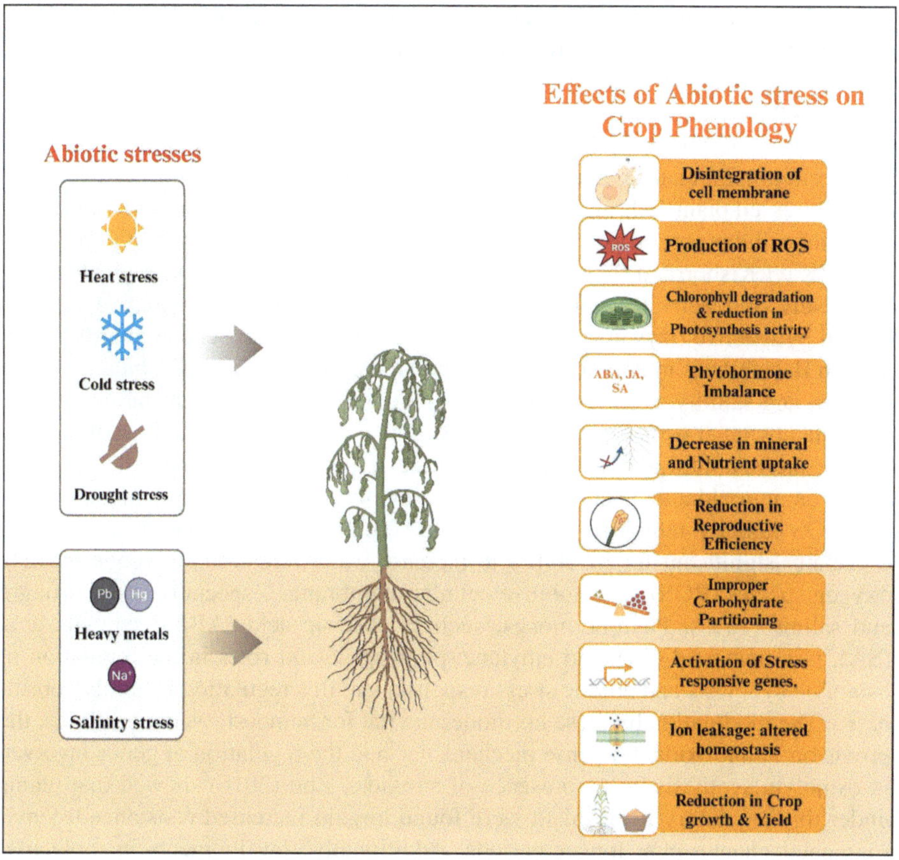


**Fig. 11.1** Different abiotic stress and their effects on crop phenology

2010). Complex antioxidant defence systems have developed in plants to defend them from ROS's cytotoxicity and to restore cellular homeostasis. These safeguards consist of antioxidant molecules that are enzymatic and those that are not. Superoxide dismutase, glutathione reductase, peroxidase, and catalase are enzymatic antioxidants. Collectively, these components protect cells from oxidative stress by neutralizing reactive oxygen species (ROS). Plants must create osmolytes—compatible solutes that aid in stabilizing macromolecular structure, fine-tuning osmotic balance, and scavenging reactive oxygen species (ROS)—if they are to survive abiotic stress (Yancey 2005).

As a defence mechanism against environmental stresses, plant cells synthesize osmolytes, which are very small organic compounds. One distinctive feature of the compounds is their inertness to the body's inherent metabolic mechanisms. Instead, they help plants adapt to different environmental stresses by acting as protective chemicals. The most significant groups of osmolytes that contribute to enhanced stress tolerance include sugar alcohols, proline, glycine betaine, trehalose, and polyamines. In maintaining homeostasis of redox balance and shielding cellular components from oxidative stress, these chemicals perform separate but complementary biochemical and physiological functions (Szabados and Savouré 2010). Proline is the most researched osmolyte because of its many roles in osmotic regulation, membrane integrity, reactive oxygen species (ROS) scavenging, and redox equilibrium maintenance, among many others. Proline has been shown to accumulate in stressed plant cells and acts as an osmoregulatory compound, thus sustaining essential cellular processes like turgor pressure, enzymatic activity, and protein stability (Kavi Kishor and Sreenivasulu 2014). It scavenges hydroxyl radicals (·OH) effectively and activates antioxidant enzymes, reversing the oxidative damage caused by reactive oxygen species (ROS). Likewise, glycine betaine guards photosystem II, stabilizes membrane structure, and increases cellular redox balance under drought and salinity stress. Acclimatization of stress and ROS scavenging are also promoted by other osmolytes, such as polyamines like putrescine and spermine, and sugars like mannitol and sorbitol. Polyamines play multifunctional roles, such as acting as antioxidants, gene expression regulators, antioxidant enzyme activators, and ROS toxicity inhibitors (Alcázar et al. 2010). The disaccharide trehalose is especially significant for its ability to increase water retention, scavenge reactive oxygen species (ROS), and stabilize cellular membranes, especially under drought and salinity stress. Phytohormones, such as abscisic acid (ABA), salicylic acid (SA), jasmonic acid (JA), and ethylene, play significant roles in the regulation of osmolyte synthesis and abiotic stress response, and this regulation is highly coordinated. The modulation by these hormones is vital for homeostasis maintenance, the provision of antioxidant defense mechanisms, and the regulation of genes involved in osmolyte synthesis in the presence of stressors. Zhu (2016) showed that plants under this regulatory mechanism were found to gain increased resistance to environmental change over longer periods. Additionally, developments in molecular biology and technological advancements in genetic transformation have made the process of developing transgenic plants easier. These developments increase the expression of genes involved in osmolyte synthesis significantly, thus increasing the resilience of various crop species to abiotic stressors. This chapter presents an overview of the whole scenario regarding the function of osmolytes in ROS scavenging and homeostasis during abiotic stress. The next section will discuss the biosynthetic pathways in the context of these compounds, outline the molecular mechanisms that regulate their function, present the interactions with antioxidant defense regimes, and investigate the regulation of their activity by hormone regulation. The chapter will also present recent developments in biotechnology and genetic engineering that have been utilized to improve crop resistance to stress by increasing the osmolyte content. Elucidation of molecular and physiological pathways that regulate the induction of osmolytes by stress and ROS scavenging is important for the creation of crop varieties with high stress resistance. This creation is an important step toward the improvement of sustainable agricultural production in the context of climatic and environmental stress.

## 11.2 Abiotic Stress, ROS Accumulation, and Osmoregulation

Among the most essential environmental factors threatening global food security, abiotic stress adversely affects plant growth, productivity, and viability. Owing to variable environmental conditions, plants produce soluble organic osmolytes as a survival strategy. These metabolites, which are electrically neutral and nontoxic, play a vital role in stress adaptation as they prevent cytosolic as well as membrane-bound protein denaturation. In attempts to maintain cellular homeostasis, osmolytes control cell turgor through osmotic adjustment, change redox equilibrium, and promote plant resistance by lowering the concentrations of reactive oxygen species (ROS), re-establishing redox balance, and protecting cellular constituents from osmotic and oxidative stress. Soil salinity, moisture, and temperature are conditions leading to osmotic stress, one of the most essential risks to heterogeneous organisms, from bacteria, fungi, and other lower plants to higher plants. In most cases, the adverse effect of this stress on plant growth, cellular metabolism, and overall viability is expressed as reduced yields. Plants use complex physiological and molecular processes to regulate the expression of genes associated with stress in response to osmotic stress. The regulatory processes involved are directed toward the restoration of normal development, initiating detoxification of toxic metabolites, and normalization of cell metabolism. Physiological, pharmacological, and molecular research provides information on the signal transduction pathways that govern the complex adaptive response system. The accumulation of inorganic and organic osmolytes enables plants to gain the osmotic adjustment necessary. Ion absorption and transport systems regulate the balance of ions in cell membranes and, thus, regulate inorganic osmolytes such as potassium ($K^+$), sodium ($Na^+$), and chloride ($Cl^-$). Organic osmolytes function to increase tolerance to stress by functioning as dual scavengers of reactive oxygen species (ROS) and as osmoprotectants of membrane proteins. The ion transporter family, which contains transmembrane proteins, is responsible for active transport mechanisms that enable ions to pass through biological membranes against the concentration gradient. One of the potential strategies in the development of resistance to osmotic stress is the utilization of plant genetic engineering, specifically through the introduction of some genes of ion transporters into transgenic lines. This strategy is directed toward the preservation of ionic homeostasis and exacerbating acute physiological interruptions from external stimuli. In this chapter, we present a detailed review of the current knowledge on ion transporters and the possibility of the use of genetic engineering to develop crop varieties that show resistance to various stress factors. Previous studies have determined that osmoprotectants include Late Embryogenesis Abundant (LEA) proteins as well as ion transporters. Desiccation and water stress resistance in seeds is mainly due to LEA proteins, which form a family of high-molecular-weight proteins formed during the late embryogenesis stage (Galau et al. 1987). These proteins are molecular chaperones that provide cellular protection against the damaging effects of freezing and dehydration by preventing enzyme aggregation, which would result in deleterious effects (Goyal et al. 2005). Additionally, phytohormones play a central role in coordinating the overall stress response, which includes the strengthening of the plant's defense mechanism when it is exposed to unfavorable conditions. For the facilitation of physiological processes, osmolytes also have phytohormonal functions, which are vital for the activation of regulatory processes. Under unfavorable environmental conditions, osmolytes can regulate phytohormonal activity, allowing the plant to grow and develop without any restriction. Amongst the osmolytes used by the cytosol to establish osmotic balance internally and externally are sugars, glycine betaine, proline, and polyamines. Aside from their osmoregulatory action, these substances protect against oxidative stress by effectively scavenging reactive oxygen species (ROS) like ·OH, $O_2^+$, and $H_2O_2$. Abiotic stress tolerance in plants is maintained through a complex regulatory network involving phytohormones like salicylic acid (SA), abscisic acid (ABA), brassinosteroids, cytokinins, ethylene, and jasmonates (Chakraborty et al. 2024). Reactive oxygen species (ROS) are integral to the mechanism of plant stress response; however, they are involved in oxidative damage and cell harm (Pramanik et al. 2023a). The membrane-associated NADPH oxidases facilitate the production and function of reactive oxygen species (ROS) and act as a signaling response to stress. Excessive production of reactive oxygen species (ROS) causes oxidative damage in an organism, although it is an indispensable component of the mechanisms of oxidative signaling. Under controlled levels, reactive oxygen species (ROS) are signaling compounds that induce protective responses to abiotic and biotic stress. Recent studies have greatly enhanced what is known to date on the mechanisms of plant response to stress, especially the interactions among phytohormones, and functions of reactive oxygen species receptors (ROS) and other regulatory elements. Another important manner in which reactive oxygen species (ROS) are improving our understanding of plant response to environmental stress agents is through their functions as organelle-to-cell signaling mediators. A considerable amount of advancements have been accomplished in modulating reactive oxygen species (ROS) for the purpose of enhancing the tolerance of the plant to stress, from varying the ROS generation, detoxification, and transporting mechanisms. Osmotic stress interferes with cellular homeostasis and induces enormous structural and functional changes in the cells of plants. Accumulation and synthesis of osmoprotectants constitute the mechanism of adaptation as shown by stressed-tolerant plant species. The most critical building blocks for osmoprotectants include sugars and sugar alcohols, amino acids like proline, glycine betaine, polyamines, and a host of amines such as quaternary amines (Vinocur and Altman 2005). Although these crop plants are of immense economic significance, relatively fewer have a low intrinsic capacity to produce these specific osmolytes at physiologically meaningful concentrations. The effective utilization and overexpression of osmoregulatory genes by transgenic means have improved drought tolerance, high temperature tolerance, and high salinity tolerance.

Osmoregulation hypothesis has been an effective way of transferring abiotic stress tolerance. Researchers have effectively enhanced stress tolerance in crops through genetic manipulation of critical pathways to produce osmolytes. The level of control on osmolyte content dependent on abiotic conditions is in line with advances in climate-tolerant crops. Results of current research on the regulatory mechanisms of osmolyte activities, interaction with reactive oxygen species (ROS), and phytohormone control will be of paramount significance to enhance agricultural sustainability in the context of climate change and environmental unpredictability.

## 11.3 Role of Osmolytes in ROS Scavenging

The cellular organelles chloroplasts, mitochondria, and peroxisomes are responsible for storing reactive oxygen species (ROS), which are produced as a result of aerobic respiration in plants (Dietz et al. 2016a, b; Huang et al. 2016; Sandalio and Romero-Puertas 2015). ROS accumulation causes a cascade of cellular harmful events, including DNA destruction, lipid peroxidation, protein oxidation, and enzyme inactivation. As an intricate antioxidant defence system, osmolytes are essential for plants in preserving redox balance and eliminating ROS.

### 11.3.1 Types of Osmolytes

Major groups of osmolytes are categorized into ammonium compounds (polyamines, betaines), sugar and sugar alcohols, and amino acids (proline).

#### 11.3.1.1 Ammonium Compound

- **Polyamines**
  The control of pH in cellular structures is greatly influenced by polyamines, which are nitrogen-incorporated aliphatic low-molecular-weight compounds (Dubey et al. 2021). Important for stress tolerance are the polycationic chemicals spermidine, putrescine, and spermine. The compounds' detoxification of reactive oxygen species (ROS) is an important mechanism by which they regulate ionic equilibrium, stabilize macromolecular structures, and reduce oxidative stress.
- **Betaines**
  Betaines, which are quaternary ammonium compounds, play an essential role in osmosis-mediated intercellular equilibrium maintenance and membrane stabilization during stressful conditions. Betaines directly deplete ROS and activate antioxidant enzymes such as superoxide dismutase (SOD) and catalase (CAT). Transcriptome comparison studies across different crops exposed to different abiotic stress conditions have shown that genes involved in polyamine metabolism are regulated differently (Ebeed 2022). Tomatoes, wheat, rice, and Arabidopsis are among these crops. Polyamines and their regulatory role in modifying abiotic stress-activated transcription factors are the primary foci of this study.
- **Proline**
  In reaction to various abiotic stresses, proline accumulates, making it a common osmolyte. Stabilizes proteins and cellular membranes, scavenges reactive oxygen species (ROS), and maintains redox equilibrium. The osmotic potential of

halophytes is controlled by the accumulation of several osmolytes, most notably proline. In the aftermath of a flood, plants usually see less sunlight, less oxygen, and less carbon dioxide. According to Chanu and Sarangthem (2015), rice plants have been shown to have higher quantities of proline, which helps them modify their osmotic potential and reduce negative effects.

#### 11.3.1.2 Glycine-Betaine

Quaternary ammonium compound glycine-betaine enhances cellular stability under stress by preserving protein tertiary and quaternary structures. Glycine betaine is known to build up in sunflowers and barley, among other crop species, when drought stress is present, according to a previous study. At the intercellular level, these serve as an osmoticum due to the substantial requirement for this osmolyte's manufacture, which can be sustained by fundamental plant metabolism. They identified this as an effective strategy to mitigate drought stress. The application of glycine betaine to sugar beet plants can effectively alleviate the detrimental impacts of lead and cadmium toxicity by enhancing metabolic indices and augmenting the activities of both enzymatic and non-enzymatic antioxidants (Badawy et al. 2024).

#### 11.3.1.3 Sugars

Soluble sugars like trehalose, sucrose, and raffinose act as osmoprotectants and ROS scavengers. Among them, trehalose is particularly notable for its role as a signalling molecule that regulates stress-responsive gene expression and strengthens cellular defence responses. Raffinose is a promising candidate involved in osmoprotection of the plants exposed to abiotic stresses, especially heat stress (Serrano et al. 2019). Studies on wheat seedlings that are drought-tolerant show higher synthesis of sucrose at the metabolic level and also increased expression levels of genes related to sucrose synthesis as compared to the drought-sensitive ones. This is required to save energy during stressful conditions and the regulation of cell metabolism.

### 11.3.2 Mechanisms of ROS Scavenging

Dietz et al. (2016a, b), Huang et al. (2016), and Sandalio and Romero-Puertas (2015) indicate that reactive oxygen species (ROS) are unavoidable by-products of aerobic plant metabolism and are sequestered in different cellular organelles, such as chloroplasts, mitochondria, and peroxisomes. Excessive reactive oxygen species (ROS) can cause DNA, protein, and lipid damage, resulting in programmed cell death (PCD). In addition, reactive oxygen species (ROS) are a vital class of signaling molecules and play an essential role in controlling plant development and stress responses; this is notable. Regulation of both the biosynthesis and catabolism of reactive oxygen species (ROS) is needed to ensure the well-being of plants and facilitate their response to stress because ROS have a dual function. Excess reactive oxygen species (ROS) under changing stress conditions cause a cascade of harmful effects on nucleic acids, proteins, and lipids by inducing oxidative reactions and enzyme inhibition. However, reactive oxygen species (ROS) are also involved in physiological processes, such as metabolic control, defense signaling, and perception of stress. This dual aspect highlights the necessity of a perfectly regulated mechanism for maintaining ROS homeostasis within plant cells. While the removal of excess reactive oxygen species (ROS) would cause oxidative stress and lead to cell death, the removal of ROS would destroy the vital signal transduction pathways. To preclude the damage caused by ROS and conserve the signal pathways, plants have evolved sophisticated antioxidant defense mechanisms. Owing to ROS detoxification in plants, majorly, reactive oxygen species (ROS) are detoxified through the use of both enzymatic and non-enzymatic antioxidant systems, which act together to effectively regulate ROS levels. Enzymic antioxidant mechanisms are pivotal in detoxifying reactive oxygen species (ROS) and include a number of proteins, i.c., glutathione peroxidase (GPX), ascorbate peroxidase (APX), monodehydroascorbate reductase (MDHAR), dehydroascorbate reductase (DHAR), and glutathione reductase (GR). The most important defense mechanism against the generation of superoxide anions ($O_2$·⁚·), hydrogen peroxide ($H_2O_2$), and hydroxyl radicals (·OH) is based on the synergistic action of these enzymes, which act together to prevent oxidative damage. Plants employ a range of non-enzymic antioxidants, i.e., ascorbate (AsA), glutathione (GSH), carotenoids, tocopherols, and phenolic compounds. These molecules are pivotal in protecting cellular structures by efficiently neutralizing toxic reactive oxygen species (ROS). It is increasingly assumed that the enhancement of antioxidant enzymes and metabolites in plants may boost their oxidative stress tolerance and facilitate ROS detoxification. As a result of the relationship between antioxidant activity and stress resistance, it is possible to assume that the capacity of plants to acclimate to stress-environment conditions may be largely enhanced by amplifying their natural defense machinery. The evolution of next-generation genetic transformation technology to overexpress key antioxidant genes may lead to the production of transgenic crops with improved environmental stress resistance and variability in response (Flora 2009). This observation underscores the need to adopt antioxidant-based technologies in crop improvement programs to produce more stress-tolerant varieties, thus ensuring enhanced agricultural productivity and food security.

#### 11.3.2.1 Enzymatic Antioxidant Defence Mechanism

Catalase (CAT), glutathione peroxidase (GPX), superoxide dismutase (SOD), and an enzyme known as APP are the five enzymes that constitute an enzymatic antioxidant defence system in plants. We still know very little about how these enzymes control ROS levels and gene expression, even though their function in ROS detoxification has been known and researched for quite some time. The quick dismutation of osmolytes into hydroxyl radicals is catalysed by superoxide dismutase (SOD), which in turn is broken down into water and oxygen by catalases like CAT and others (Mittler 2017). Plants not only contain enzymatic defences, but also non-enzymatic antioxidant systems that are mostly composed of low-molecular-weight antioxidant metabolites such as flavonoids, glutathione (GSH), and ascorbic acid (AsA). The effective scavenging of ·OH and ´$O_2$ is primarily facilitated by these metabolites. When the production of reactive oxygen species (ROS) exceeds the effectiveness of these protective mechanisms to eliminate them, a condition known as oxidative stress develops, leading to oxidative damage to cellular components and structures and the activation of programmed cell death (PCD). Nevertheless, at low concentrations, reactive oxygen species (ROS) have important signalling roles, regulating fundamental biological processes like stem cell maintenance, cell cycle progression, differentiation, organogenesis, and the plant's response to biotic and abiotic stresses (Zeng et al. 2017). The maintenance of cellular redox homeostasis depends on keeping ROS levels under control. However, ROS can become toxic rather than cytostatic if their concentrations become too high or too low. As a defensive mechanism against oxidative stress and ROS-induced cell damage, plants have developed a highly structured and integrated antioxidant system. Enzymatic and non-enzymatic antioxidants, such as CAT, POD, SOD, GR, GST, GPX, MDHAR, and DHAR, are all part of the system. The following are examples of antioxidants that do not need enzymes: glutathione (GSH), ascorbate (AsA), α-tocopherol, carotenoids, and flavonoids. In order to increase plant stress tolerance, the identified components work together to keep the redox balance in check and avoid ROS formation. The functioning of the antioxidant system is fundamentally linked to stress tolerance among crop species, which explains its necessity to investigate while developing plans for improving adaptation among plants subjected to stressful environmental conditions. Furthermore, osmolytes contribute to regulating ROS detoxification by promoting critical antioxidant enzymes like SOD, POD, and GR activity, thus supporting cellular redox homeostasis. Their function in stress response mechanisms further signifies their role in plant adaptation, thus as prospective targets for metabolic engineering and genetic enhancement aimed at improving tolerance to stress.

#### 11.3.2.2 Non-Enzymatic Antioxidant Defence Mechanism

ROS overload induces extensive cellular injury, such as lipid peroxidation, glycation, oxidation, protein nitration, enzyme inactivation, DNA mutation, and disruption of subcellular organelle membranes. Osmolytes are crucial in the detoxification of ROS through their function in preserving nonenzymatic antioxidants, including ascorbate and glutathione, which are crucial for preventing oxidative stress and ensuring cellular integrity.

##### Ascorbic Acid (Vitamin C)

One kind of ascorbate is reduced ascorbic acid, while another is an oxidized derivative called dehydroascorbic acid (DHA), and the third is monodehydro ascorbic acid (MDHA), which is present in many eukaryotic plants (Fig. 11.2). According to Cruz-Rus et al. (2012), plants regulate their resistance to oxidative stress by keeping the ratio of oxidized to reduced forms under check. Ascorbate is essential for several processes, including acting as an antioxidant, a cofactor for enzymes, and a building block for tartrate and oxalate. Photosynthesis, photoprotection, cell wall production, and plant growth are just a few of the many metabolic and physiological processes in which it is involved. The abiotic stress tolerance is, in essence, vital and is

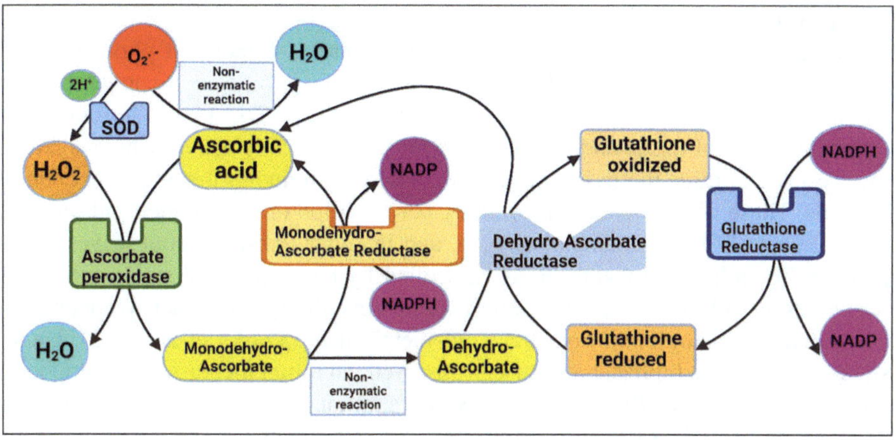


**Fig. 11.2** The Asada-Halliwell pathway for the regeneration of vitamin C and hydrogen peroxide scavenging, which involves the participation of different antioxidant enzymes

linked to the modulation of key signaling molecules such as hydroxyproline, gibberellins, anthocyanins, and ethylene (Mellidou et al. 2012). Ascorbate provides the detoxification of reactive oxygen species (ROS) by way of the glutathione-ascorbate cycle, apart from its inherent scavenging activity. The cycle is an important redox reaction that regulates redox signaling, enzymatic control, and gene expression. In the cycle, monodehydroascorbate (MDHA) is generated from the reactive oxygen species (ROS) generated in the oxidative breakdown of ascorbate. Reduction of MDHA generates dehydroascorbate (DHA), also known as ascorbate at times. The ascorbate–glutathione cycle depends on two important enzymes, monodehydroascorbate reductase (MDHAR) and dehydroascorbate reductase (DHAR), involved in reducing DHA back to ascorbate. The activity of the cycle is pH-dependent, DHA glutathione-dependent reduction being optimal at pH 8.0, a condition akin to the chloroplast stroma of light-exposed plants (Gallie 2013). Ascorbate concentration is highly variable due to numerous environmental and physiological factors such as light intensity, circadian regulation, tissue specificity, and intracellular compartmentalization. Out of the many organelles found in the cell, chloroplasts contain the highest concentration of ascorbate, above 20 mM, while other organelles, such as mitochondria and vacuoles, have relatively low levels (Zechmann 2011).

Ascorbate determination in whole tissues is hindered by organellar-specific variability elicited by abiotic stress stimuli. Aside from its presence in the plasma and nuclear membranes of some plant types, ascorbate also exists in other compartments of cells, such as the chloroplasts, mitochondria, peroxisomes, vacuoles, and apoplastic space.

Ascorbate homeostasis in plant cells is controlled by recycling, transport, and biosynthesis. These activities ensure a steady supply of ascorbate, which plays a crucial role in plant defence and growth by promoting redox balance, stress tolerance, and metabolic stability.

### Tocopherols

Plants, algae, and cyanobacteria are the only creatures capable of photosynthetic lipid biosynthesis, which involves the synthesis of antioxidant tocopherols. The tocopherol content of plant tissue varies considerably. α-Tocopherol is found predominantly in leaf tissue and is termed the most biologically active, which is due to the presence of three methyl groups. γ-Tocopherol is found as the major tocopherol in seeds. These are involved in preventing peroxidation of membrane lipids and the generation of ROS, e.g., singlet oxygen, and in sugar transport from leaves to the phloem. Tocopherol biosynthesis is controlled by environmental stress and stress hormones like JA and ABA. Biosynthesis occurs in chloroplasts, where α-tocopherol is located in plastoglobuli and thylakoid membranes (Szarka et al. 2012). Under stress conditions, a biphasic response is observed: an initial increase in α-tocopherol levels protects against ROS, followed by degradation during severe stress, with this degradation being less pronounced in stress-tolerant plants.

### Glutathione

Glycine and γ-glutamylcysteine are the building blocks of the tripeptide glutathione (GSH), an essential molecule. Glutathione is an essential functional metabolite for aerobic life. The importance of glutathione in maintaining redox equilibrium is demonstrated by its high intracellular content, particularly in chloroplasts (14 mM) (Szarka et al. 2012). You can find glutathione in two different forms: reduced (GSH) and oxidized (GSSG). If you want to fight ROS-mediated oxidative stress, you need a high GSH/GSSG ratio. The ascorbate-glutathione cycle includes glutathione (GSH), which has a direct role in ROS detoxification (Fig. 11.2). Mittova et al. (2003) noted that GR converts oxidized GSSG to reduced glutathione (GSH) by using NADPH as an electron acceptor. Two ATP-dependent activities that result in the synthesis of glutathione are carried out by the enzymes γ-glutamylcysteine ligase (GSH1) and glutathione synthetase (GSH2). According to Shu et al. (2011), the two enzymes mentioned before play a crucial role in the production of glutathione and γ-glutamylcysteine. Sulphate transport, signal transduction, phytochelatin production, and xenobiotic detoxification are among the many functions that GSH plays. It is also an antioxidant.

### Carotenoids

Carotenoids, a group of isoprenoid-derived pigments, are synthesized in plastids and serve multiple functions in photosynthesis, photomorphogenesis, and stress responses. Primarily found in chloroplasts and chromoplasts, carotenoids form pigment-protein complexes that are crucial for photosynthetic processes. As antioxidants, they neutralize singlet oxygen and reduce oxidative damage caused by photoexcitation (Sandmann 2015). The composition and accumulation of carotenoids vary across different tissues and species, with green tissues typically exhibiting conserved profiles, while non-green tissues display greater variability. The regulation of carotenoid levels is governed by transcriptional control and metabolic turnover, facilitated by carotenoid cleavage dioxygenases. In addition to their antioxidant roles, carotenoids influence various aspects of plant development, including plastid biogenesis, flowering, and fruit maturation, and contribute to pigmentation in fruits, flowers, and seeds.

#### Chelating Metal Ions

Heavy metal contamination, including elements such as Cd, Cu, Zn, Ni, Co, Cr, Pb, and As, is commonly observed in agricultural fields worldwide. Major sources of this pollution include the prolonged use of phosphatic fertilizers, the discharge of industrial wastewater and sludge, emissions from human activities, and improper irrigation practices. Heavy metals can be lessened by using osmolytes, which chelate metal ions and decrease the Fenton reaction, which produces hydroxyl radicals (Qadir et al. 2021). Plants release reactive oxygen species (ROS) in response to heavy metal exposure. Indirectly, by encouraging ROS overproduction, or directly, via the Haber–Weiss reaction, these metals contribute to ROS creation. The resulting oxidative stress is a key consequence of heavy metal toxicity in plants.

## 11.4 Osmolytes in Maintaining Cellular Homeostasis

Conserving cellular osmotic balance and ion homeostasis is a fundamental adaptation strategy that plants employ to endure abiotic stress. Controlling intracellular water transport, activating antioxidant defence mechanisms (enzymatic and non-enzymatic), synthesizing osmolytes in the cytoplasm, creating heat shock proteins, and securing harmful ions into vacuoles are all part of this process. Reducing stress-induced cellular damage, the conserved responses under study keep metabolic processes going and cells intact even when exposed to harsh environmental circumstances. Given the well-established role of osmolytes in stress tolerance enhancement, this chapter reviews and assesses current knowledge on their functional importance in the adaptability and survival of plants in response to different abiotic stress stimuli.

As a consequence of osmotic stress, plants produce an excess of osmolytes, such as soluble sugars, proline, and betaine, which are essential for preserving osmotic balance (Zhao et al. 2022). Tolerance to abiotic stress is the result of a complicated molecular, biochemical, and physiological cascade. Utilizing multi-omics approaches that integrate genomes, transcriptomics, proteomics, metabolomics, and epigenomics is crucial for delving profoundly into these intricate systems.

A cascade of complex biochemical and physiological responses, including ion imbalances, ROS production, elevated osmolyte levels, and hormone pathway reorganization, are set off by salt stress. Disruption of K+ homeostasis and normal cellular metabolism occurs when cells absorb an excess of Na+. Detailed research on the involvement of osmolytes in abiotic stress responses has been made possible via bioengineering of genes related to osmolyte production and metabolism. Increased resistance to abiotic stimuli has been directly connected with the accumulation of non-toxic, compatible osmolytes, ROS scavengers, and other stress-specific transgenic products. Cloning and genetic engineering have made it possible to turn plants with the ability to produce osmolytes into excellent agricultural crop plants. This is all because to developments in cellular and molecular biology. Specifically, studies have demonstrated that transgenic tobacco's tolerance to drought stress is enhanced by overexpressing genes associated with proline biosynthesis. Tobacco plants have much greater resistance to cold, drought, and salt when researchers find an overexpression of MfMIPS1 (Medicago falcata myo-inositol phosphate synthase) (Tan et al. 2013). Fan Wei Juan et al. (2012) found that transgenic sweet potatoes expressing the BADH gene had an increased tolerance to several abiotic stress conditions. Once again, transgenic Brassica plants that express the codA choline oxidase gene have increased their resistance to most abiotic stresses by producing an excess of glycine betaine. The otsA and otsB genes, which originate from Escherichia coli, confer increased drought resistance in transgenic tobacco plants, according to many studies. Drought stress sets off a complicated chain reaction on a physiological and biochemical level. First, less transpiration and, by extension, less absorption of carbon dioxide ($CO_2$), result from a decrease in stomatal conductance. In addition, the buildup of reactive oxygen species (ROS) causes oxidative stress, which in turn damages DNA, proteins, and cell membranes (Pramanik et al. 2023b). To restore osmotic balance, rice plants respond to these unfavorable circumstances by increasing the manufacture of osmolytes such as proline, betaine, and soluble sugars. In addition, when plants are stressed by water, their levels of abscisic acid (ABA) rise, which regulates the closing of stomata and encourages the growth of longer roots. To further lessen oxidative damage and counteract the elevated levels of reactive oxygen species (ROS), the body produces more antioxidants, both enzymes and non-enzymic, such as glutathione and ascorbic acid.

## 11.5 Crosstalk Between Osmolytes and Stress Signaling Pathways

The capacity of a plant to develop and live can be drastically impacted by the ever-present environmental stresses. The accumulation of osmolytes is an integral part of the intricate adaptation processes that these organisms developed in response to environmental threats. Osmolytes are organic, low-molecular-weight solutes that sense stresses, kick off signalling cascades that include adaptive responses, and aid in protein stabilization and osmotic correction. In order to detect osmotic stresses like dehydration and salt, the regulation of cellular osmotic balance is critical. As a crucial regulator of oxidative stress, osmolytes can scavenge reactive oxygen species (ROS) generated during stress, protecting cell components from oxidative damage. Plants are able to improve resource utilization and metabolic regulation under extreme adaptation, according to new findings, in part because osmolytes regulate genes that are triggered in reaction to stress.

The osmolytes that plants make are compatible solutes that are essential for signal transduction. Disruption of cellular homeostasis, inhibition of photosynthesis, and oxidative damage caused by the overproduction of ROS are all ways in which abiotic stimuli, such as drought, salt, extreme temperatures, and heavy metal toxicity, affect plant growth and productivity. By stabilizing membrane structure and proteins, for example, and by scavenging reactive oxygen species (ROS) and adjusting osmotic pressure, osmolytes mitigate these harmful consequences. The production and buildup of osmolytes are controlled by a network of stress response pathways, including calcium signaling, ROS-mediated signaling, ABA signaling, mitogen-activated protein kinase (MAPK) signaling, and others. In order for plants to survive and adapt to harsh climatic circumstances, they have developed a coordinated response that increases their tolerance to stress, makes acclimatization easier, and improves detoxifying processes.

## 11.5.1 Hormonal Interactions

There is a complicated interaction between phytohormone signaling and osmotic regulation, as phytohormones are also important in controlling stress-induced osmolyte buildup. Most phytohormones, including ethylene, ABA, SA, and brassinosteroids, cause plants to produce more osmolytes, which boost their ability to withstand stress. These include proline, glycine betaine, soluble sugar, and polyamines.

### 11.5.1.1 Ethylene

Among the many physiological processes that plants rely on for regulation are seed germination, development, growth, withering of flowers and leaves, ripening of fruits, and tolerance to stress. One such hormone is ethylene, a phytohormonal gas. According to Waheed et al. (2024), ethylene is a molecule that modulates abiotic stress responses by way of stress-responsive gene expression. This means that it may increase plants' adaptability and resistance to things like salt, heat, drought, and ozone treatment. Furthermore, ethylene is involved in stress-related osmotic adjustment via regulating osmoprotectant accumulation. The evidence suggests that ethylene production is related to the buildup of osmoprotectants, which in turn protect the plant from environmental stressors. Evidence from experiments points to ethylene's function in the abiotic stress response, namely in the buildup of osmoprotectants.

### 11.5.1.2 Salicylic Acid (SA)

All plant species contain SA, a phenolic hormone that has recently been recognized as an important regulator of stress reactivity. Abiotic stress leads plants to produce reactive oxygen species (ROS), thus they've evolved a variety of adaptation mechanisms to regulate ionic and osmotic imbalances. The processes at hand involve osmoregulatory mechanisms, which involve the accumulation of osmolytes such as proline, soluble sugars, amines, and glycine betaine. To make plants more resistant to abiotic stress, SA affects important metabolic processes as photosynthesis, metabolite accumulation, redox homeostasis, and gene expression regulation. Thorough genomic and proteomic investigations are necessary to fully comprehend the salicylic acid-responsive networks and the functions they play in stress-induced protein and gene regulation (Kaur et al. 2022).

#### 11.5.1.3 Abscisic Acid (ABA)

As a plant responds to salinity, drought, and temperature extremes, ABA is essential for regulating osmolyte buildup. The mechanisms that lead to the accumulation of soluble sugars, proline, and glycine betaine, as well as the control of stress perception, are regulated by the well-known signalling molecule ABA. To illustrate the point, ABA controls the expression of a key enzyme in the proline biosynthetic pathway—the P5CS gene—and activates transcription, therefore starting the biosynthesis of proline. Glycine betaine is an important part of cellular defence and osmotic balance, and the ABA-directed BADH (betaine aldehyde dehydrogenase) gene controls its production by lowering water potential and preventing cellular dehydration. Stress response genes like these are activated when ABA binds to ABA-responsive elements (ABREs) in gene promoters. Calcium ions and reactive oxygen species (ROS) are secondary messengers in the ABA signal transduction pathway that help build up osmolytes. In addition, ABA initiates the antioxidant activities that mitigate oxidative stress, thus establishing a defence mechanism against adverse environmental conditions. By controlling the expression of certain genes, such as P5CS and BADH, ABA generates many environmental signals that influence the synthesis of osmolytes. Osmolytes are crucial for plants to survive and respond to abiotic challenges. Some of the fundamental physiological processes that ABA controls in plants include osmotic adjustment, seed germination, stomata closure, and leaf senescence (Chen et al. 2020). Some of the mechanisms via which plants regulate water potential and turgor in leaves in reaction to various stressors include stomatal closure, osmotic solute buildup, and antioxidant defence system strengthening (Pitaloka et al. 2022).

#### 11.5.1.4 Brassinosteroids (BRs)

Seed germination, cell elongation, photomorphogenesis, and xylem differentiation are just a few of the many plant development processes that brassinosteroids (BRs) help to regulate. According to research by Shu et al. (2016), BRs significantly improve exogenous treatment photochemical efficiency (PSII), quantum yield, and photochemical quenching coefficient (photochemical quenching coefficient), which in turn reduces photoinhibition. Reduced light intensity, low temperature, and calcium nitrate stress caused by exogenously administered BRs all enhance plant photosynthetic efficiency.

### 11.5.2 Stress-Signaling Pathways

A network of signalling channels is responsible for controlling synthesis and osmolyte activity. Protecting cells from oxidative damage, keeping ions in equilibrium, and assuring osmotic balance are the primary functions of the routes. A coordinated response, including kinase pathways, ion-based signalling, and phytohormone activity, regulates environmental stress.

#### 11.5.2.1 MAPK Pathway

Even under isotonic settings, cells still have a chronic sensitivity to water entry and swelling, which is regulated by their structural and physical features. They strictly regulate the transmembrane ion distribution to prevent excessive oedema (Hoffmann et al. 2009). However, when the intracellular and extracellular milieus are out of osmosis balance, the cellular volume varies due to water migration in or out of the cell, which impacts cellular structure and function. The term for this phenomenon is osmotic stress. For volume regulation, cells employ both short-term mechanisms, such as changes in intracellular ion concentration and organic osmolytes, and longer-term mechanisms, like the transcriptional activation of osmoregulatory genes, as cells inflate or shrink. To adapt to their surroundings, the majority of creatures rely on the mitogen-activated protein kinase (MAPK) cascade to keep their growth and development under control (Widmann et al. 1999; Ichimura et al. 2002).

#### 11.5.2.2 Calcium-Dependent Signaling

In both intracellular and extracellular signaling, calcium plays a key role as a secondary messenger, regulating, to put it simply, cellular survival and death. With the use of intracellular compartment exchangers, pumps, and buffers, the calcium signalling system precisely modulates cellular processes via a web of interconnected pathways. Many cellular activities, particularly those involving reactive oxygen species (ROS), are tightly linked to calcium signalling. While reactive oxygen species (ROS) have long been thought of as harmful waste products of aerobic respiration, new research has revealed that, at sub-toxic concentrations, ROS play a crucial role as signalling molecules in fundamental cellular processes, including development and death. Görlach et al. (2015) and others have shown that proper control of cellular signal networks depends on the complex interplay of calcium and reactive oxygen species (ROS) signalling pathways. The Golgi apparatus, the nucleus, and mitochondria are just a few of the organelles that are linked by calcium transport systems. The regulation of calcium homeostasis in mitochondria is regulated by electrogenic uptake. This process is facilitated by the proton gradient that is maintained along the respiratory chain, which creates a huge voltage gradient across the inner mitochondrial membrane. Energy generation, intracellular calcium signalling, and apoptotic processes are all regulated by mitochondrial calcium, making it an essential regulator in cellular physiology and disease. In addition, reactive oxygen species (ROS) like hydroxyl radicals, superoxide anion, and hydrogen peroxide are involved in calcium signaling. These radicals play important roles in the oxidation of proteins, lipids, and polynucleotides, and are involved in many physiological processes when present in moderate amounts (Ravi et al. 2023).

#### 11.5.2.3 Gene Regulation

Olmolytes are essential for improving plants' ability to withstand abiotic stress because they control the expression of stress-responsive transcription factors such as Heat Shock Factors (HSFs) and Dehydration-Responsive Element Binding (DREB) proteins. Osmolytes, which include proline, glycine betaine, trehalose, and polyamines, stabilize proteins, regulate osmotic equilibrium, and prevent ROS oxidative damage. In stress-inducible gene promoters, DREB transcription factors interact with Dehydration-Responsive Elements (DREs), which are essential regulators of drought and salt stress responses. They belong to the family of proteins called AP2/ERF. Late Embryogenesis Abundant (LEA), dehydrins, aquaporins, and antioxidant enzymes are the genes that code for the proteins that help the body deal with osmosis and make efficient use of water. Osteomytes enhance DREB expression by stabilizing protein structure, enhancing DNA-binding affinity, changing signal transduction pathways (especially those not dependent on ABA), and regulating epigenetic changes that make chromatin more accessible. To a similar extent, heat shock proteins (HSPs) like HSP70 and HSP90 control the aggregation of proteins and keep proteostasis in check; HSFs are essential for thermotolerance because they regulate these processes. In addition to stabilizing HSPs, osmolytes modulate MAPK signaling pathways, alter epigenetic modifications (such as histone acetylation and SUMOylation), and prevent reactive oxygen species (ROS) from building up. The crosstalk between the DREB and HSF pathways enhances stress tolerance since both transcription factors employ ROS, ABA-independent regulation, and MAPK cascades as signalling mechanisms. Additionally, several target genes overlap to create protective proteins. Notably, a feedback mechanism is set up when HSFA1 responds to heat and drought stress by increasing DREB expression, and DREB2A upregulates certain HSF genes. The plant's resistance to different stresses is enhanced as a result.

Improving plants' ability to withstand abiotic stress requires careful regulation of osmolytes over stress-inducible transcription factors, including Heat Shock Factors (HSFs) and Dehydration-Responsive Element Binding (DREB) activities. Proteins are stabilized, the osmotic equilibrium is maintained, and oxidative damage caused by ROS is shielded by osmolytes such as polyamines, glycine betaine, trehalose, and osmoprotectants. Major regulators of drought and salt stress responses, DREB transcription factors are members of the AP2/ERF family. They bind to stress-inducible promoters of genes through Dehydration-Responsive Elements (DREs). Osmoprotection and efficient water usage are supported by the antioxidant enzymes, Late Embryogenesis Abundant (LEA) proteins, dehydrins, and aquaporins encoded by these genes. To activate DREB, osmolytes stabilize protein structure, increase DNA-binding affinity, modify signal transduction pathways (especially those that are not dependent on ABA), and change epigenetic marks to make more chromatin accessible. So, HSFs play a pivotal role in thermotolerance by regulating the production of heat shock proteins (HSPs) such as HSP70 and HSP90, which prevent proteins from clumping together and maintain proteostasis. In order to enhance the activity of HSF, osmolytes stabilize HSPs, regulate MAPK signalling pathways, prevent ROS buildup, and modulate gene expression through epigenetic markers such as histone acetylation and SUMOylation. Additionally, stress tolerance is improved via DREB-HSF pathway crosstalk due to the fact that both transcription factors share comparable signalling pathways, target genes that encode protective proteins, ROS, ABA-independent regulation, and MAPK cascades. An interesting feedback loop that strengthens plants' resistance to many stresses is established when HSFA1 increases DREB expression in response to heat and drought stress, and DREB2A activates certain HSF genes.

Incorporating osmolytes into transcriptional networks opens up exciting possibilities for improving crops through genetic engineering, molecular breeding, and the use of exogenous compounds. Rice, wheat, and maize, three essential staple crops, have been genetically modified to be more resistant to heat, drought, and salt by increasing osmolyte production and overexpressing the DREB and HSF genes. Increasing stress tolerance through enhanced transcription factor activity and suppression of stress-responsive protein aggregation has also been demonstrated with the exogenous administration of osmolytes such proline, glycine betaine, and trehalose. Exploring epigenetic changes, protein interactions, and metabolic engineering advancements as potential molecular pathways by which osmolyte regulates DREB and HSF should be the primary goal of future study. Even more exciting is the prospect of designing climate-resilient crops that can maintain increased production under harsh climatic circumstances, which might be achieved by a mix of synthetic biology techniques with CRISPR-based genome editing within the framework of osmolyte-induced stress control. One important tactic for developing climate-resilient agriculture and ensuring food security throughout the world is the way osmolytes interact with stress-responsive transcription factors.

Restoring ionic and osmotic equilibrium under stress is another important technique for increasing plant resilience to abiotic stress. Consequently, in the field of salt tolerance genetic engineering, a lot of attention has gone into finding ways to make plant roots expel more Na + or to trap it in vacuoles. By increasing the expression of transport-related proteins, such as vacuolar antiporters, some transgenic plants have been developed to enhance salt tolerance. Take transgenic melon and tomato plants as an example. These plants showed enhanced salt tolerance when exposed to saline environments, as they were able to withstand greater levels of $K^+$ than control plants.

Furthermore, Arabidopsis was able to improve its salt tolerance through increased Na + sequestration in vacuoles by overexpressing and cloning the genes for vacuolar chloride channel (AtCLCd) for cation detoxification and AtNHX1, an ortholog of NhxI found in yeast. Rajappa et al. (2024) found that transgenic tomato and Arabidopsis plants overexpressing AtNHX1 had significantly improved salt stress tolerance due to an increase in tonoplast transporters.

Similar to the $Na^+/H^+$ antiporters seen in bacteria and fungi that live in the plasma membrane, the Salt Overly Sensitive 1 (SOS1) locus in Arabidopsis thaliana was cloned and overexpressed using the caMV 35S promoters. A critical metric for the normal functioning of sodium/hydrogen antiporters, the proton motive force, is increased by overexpressing SOS1, which improves sodium tolerance (Zhang et al. 2023a, b).

## 11.6 Engineering Stress-Tolerant Crops with Osmolyte Pathways

According to the research, transgenic plants that have the genes for osmolyte biosynthesis overexpressed are better able to withstand various abiotic challenges (Table 11.1). Among the many adaptation systems that plants have evolved to deal with environmental challenges is the osmoprotection system. The system employs a diverse range of osmoprotectants, which are molecules with low molecular weight,

**Table 11.1** Development of transgenic plants resistant to abiotic stresses

| Gene | Transgenic host | Improved Traits | References |
|---|---|---|---|
| *BADH* | *Hordeum vulgare* | Improved ability to withstand cold, heat, and salt; greater amount of glycine betaine; more biomass in the roots and shoots; enhanced tillering | Kishitani et al. (2000) |
| *CBF1* | *Solanum lycopersicum* | Improvements in drought resistance, dwarfism, proline buildup, and catalase activity | Hsieh et al. (2002) |
| *P5Cs* | *S. tuberosum* | Improved salt tolerance; higher proline levels; minimal impact on tuber yield and weight | Hmida-Sayari et al. (2005a) |
| *CMO* | *Spinacia oleracea* | Enhanced heat and salt tolerance at the seedling stage; increased glycine betaine production | Shirasawa et al. (2006) |
| *OsTPS1* | *Oryza sativa* | Improved drought tolerance; elevated trehalose and proline levels | Li et al. (2011) |
| *ONAC022* | *O. sativa* | Enhanced drought tolerance; improved growth, transpiration rate, and proline content; reduced water loss | Hong et al. (2016) |
| *GhABF2* | *Gossypium hirsutum* | Increased drought and salinity tolerance; improved fiber yield; higher proline and catalase activity | Liang et al. (2016) |
| *OsNCED5*, *OsNCED3* | *O. sativa* | Greater drought tolerance; accelerated leaf senescence; increased ABA levels | Huang et al. (2016, 2019) |
| *P5CS* | *Saccharum officinarum* | Enhanced drought resistance; elevated ABA and proline content | Li et al. (2018) |
| *StDREB2* | *S. tuberosum* | Increased drought tolerance; higher biomass, chlorophyll, and proline levels; improved relative water content | El-Esawi and Alayafi (2019) |
| *OsMYB6* | *O. sativa* | Improved salt and drought resistance; increased proline accumulation and antioxidant enzyme activity (CAT, SOD) | Tang et al. (2019) |
| *AsHSP70* | *Agave sisalana* | Enhanced ability to withstand high temperatures; elevated quantities of chlorophyll, proline, soluble sugars, and boll production | Batcho et al. (2021) |

to mitigate the detrimental impacts of stress. The class of substances known as osmoprotectants includes sugars, polyamines, quaternary ammonium compounds, and aspartame. These chemicals are essential for cellular integrity preservation, stress-induced ROS scavenging, metabolic signalling, and enzyme function stabilization; they are non-toxic. Because of the growing impact of climate change, there is an immediate need to develop crops with enhanced resistance to abiotic stress. To circumvent the tedious process of traditional breeding, plants can be genetically modified to incorporate genes that produce osmoprotectants. This tactic is among the more attractive choices. Table 11.1 displays the outcomes of transgenic studies that have improved plant stress tolerance by manipulating osmolytes and associated traits through the use of several genes. One example is sorghum, which, after transgenically expressing the BADH gene, improved in heat, salt, and cold tolerance, glycine betaine accumulation, and biomass characteristics such as root and shoot dry weight and tiller number (Kishitani et al. 2000). In a similar vein, Hmida-Sayari et al. 2005b found that expressing the P5CS gene in potatoes greatly increased their salt tolerance and proline accumulation without drastically reducing tuber output or weight. According to Li et al. (2011), the OsTPS1 gene increased proline and trehalose concentration, which in turn increased rice's drought resistance. In addition, Liang et al. (2016) discovered that the cotton gene GhABF2 increased fibre output, proline content, and catalase activity, as well as drought and salt tolerance. Taken together, this research proves that agricultural plants can be genetically engineered to have better stress tolerance using osmolytes.

Evidence suggests that transgenic plants enhanced with overexpressed genes for osmolyte synthesis are better able to endure abiotic stresses. Among the many ways in which plants have adapted to cope with environmental stress, osmoprotection stands out. The class of substances known as osmoprotectants includes sugars, polyamines, quaternary ammonium compounds, and aspartame. Among their many roles, they aid in cellular homeostasis, control enzyme activity, transmit metabolic signals, and neutralize stress-induced reactive oxygen species (ROS). Crop engineering to boost tolerance to abiotic stress is becoming more urgent as the effects of climate variability continue to worsen. One practical alternative to the time-consuming traditional breeding methods is the genetic alteration of crops by expressing the genes responsible for the manufacture of osmoprotectants. Through the control of osmolytes and associated properties, transgenic research has used these genes to improve crops' abiotic stress tolerance (Table 11.1). Sorghum, for example, exhibited improved heat, salt, and cold stress tolerance, higher glycine betaine levels, and improved biomass parameters, including root and shoot dry matter and tiller number when BADH genes were expressed (Kishitani et al. 2000). To improve salt tolerance and proline content in potatoes without lowering tuber output and weight at specified levels, researchers targeted the Arabidopsis thaliana P5CS gene (Hmida-Sayari et al. 2005b). According to Li et al. (2011), the OsTPS1 gene is responsible for increasing proline and trehalose levels, which in turn boost rice's resistance to drought stress. But cotton's GhABF2 gene improved fibre output, resistance to salt and drought, proline concentration, and catalase activity (Liang et al. 2016). All things considered, the results show that genetic engineering techniques can improve many agricultural plants' osmolyte-mediated stress tolerance.

## 11.7 Future Perspectives and Challenges

Future studies on abiotic stress scavenging by ROS and homeostasis through osmolytes must be metabolic optimization- and genetic engineering-led to maximize their biosynthesis, leading to stress-resilient crops. Access to efficient genome-editing technology like CRISPR-Cas9 enables site-specific editing of osmolyte biosynthesis genes to strengthen plant defence against abiotic stresses. Integration of multi-omics approaches—genomics, transcriptomics, proteomics, and metabolomics—will uncover novel pathways and regulatory circuits that govern osmolyte-mediated stress responses. Research on plant-microbe interactions, particularly with endophytic and rhizosphere-associated microbes, can naturally enhance osmolyte biosynthesis and enable microbial-based alleviation of stresses. The production of biofertilizers that include osmolyte-producing beneficial microbes is an eco-friendly means to enhance plant resilience. Exogenous application of osmolytes via seed priming, foliar spraying, and soil application offers immediate, non-genetic solutions to recover from stress damage. Future research must also look into the synergistic interaction of osmolytes with other stress adaptation mechanisms like antioxidant defence mechanisms, phytohormonal signaling, and secondary metabolite pathways to achieve more efficient stress-mitigation strategies. Machine learning and AI-based predictive models can fine-tune osmolyte applications by adjusting critical parameters like concentration, application timing, and environmental conditions to enhance their protective role.

Marker-assisted selection (MAS) and quantitative trait loci (QTL) mapping for the design of climate-resilient crops are anticipated to facilitate the incorporation of osmolyte-related traits into commercial crop cultivars, thereby enhancing their adaptability and yield potential. To confirm laboratory findings and evaluate the practical viability of osmolyte-based stress tolerance approaches, large-scale field trials will be necessary. Interdisciplinary collaborative work by molecular biologists, agronomists, biotechnologists, and computational scientists will be necessary to allow the transition from theoretical research to practical applications in agriculture. Future research needs to try to elucidate the intricate interactions between osmolytes and phytohormones in a systems biology framework, hence better understanding plant adaptation to stress.

## 11.8 Conclusion

Osmolytes are bioactive chemicals that regulate cellular homeostasis and ROS scavenging, allowing organisms, particularly plants, to survive abiotic stress. Environmental factors, including dryness, salt, temperature fluctuations, and oxidative damage, can lead to abiotic stress. Reactive oxygen species (ROS) damage cells and critical biomolecules when their levels reach dangerously high levels. Glycine betaine, proline, trehalose, polyols, and sugar alcohols are osmoprotectants, which are substances that maintain cellular structure and enhance antioxidant defence mechanisms. The activity of important antioxidant enzymes such as SOD, CAT, and POD is controlled by these substances to prevent oxidative stress and programmed cell death. Osmolytes are crucial in the regulation of stress hormones, including ABA, JA, and SA, which influence gene expression and activate protective mechanisms. Additionally, osmolytes aid in stress-induced protein misfolding and aggregation avoidance by maintaining membrane and protein structural integrity through molecular chaperone function. As osmolytes play an essential role in ROS detoxification, they also control metabolic pathways that produce secondary metabolites that increase stress tolerance. In order to help plants survive environmental stresses, osmolytes play an important role in controlling stress reactions on molecular, biochemical, and physiological levels. Despite significant advancements in our knowledge of how osmolytes control stress responses, a pressing need has arisen to elucidate the complex regulatory networks that govern their production and accumulation. Cooperation in research is necessary to comprehend the unique interplay of osmolytes, antioxidant systems, transcription factors, and metabolic processes under various stress situations. Another way to learn more about how osmolytes help plants adapt to stress is to look at their distribution in different parts of the plant throughout time. Biotechnology platforms such as metabolic engineering and synthetic biology hold great potential for improving crop stress resistance. By utilizing osmolytes in stress-resistant crop production, we can ensure sustainable agriculture in the face of the threat of climate change.

**Authors Contribution** The development of this book chapter was a collaborative effort with the following individual contributions:

**Gholamreza Abdi**: Provided overall supervision, conceptualized the chapter framework, and led critical revisions of the manuscript, literature synthesis, and writing—original draft.
**Ananya Baidya**: Conducted primary literature review, contributed to content development, and assisted in drafting key sections and writing—original draft.
**Md Sabir Ahmed Mondol**: Performed data collection and analysis, contributed to methodology development, and participated in manuscript writing and writing—original draft.
**Ubaida Akbar**: Designed technical components, prepared visualizations (figures/tables), and contributed to specialized content creation, and writing—original draft.
**Aritra Mukherjee**: Coordinated chapter organization, synthesized interdisciplinary perspectives, and led the editing process.
**Tadela Susmitha**: Assisted in data interpretation, validated analytical approaches, and contributed to manuscript refinement, and writing—original draft.
**Anilava Chatterjee**: Provided expert guidance on theoretical aspects, reviewed content for accuracy, and contributed to final revisions.

All authors participated in multiple rounds of discussion, provided critical feedback, and approved the final version of the chapter.

## References

Alcázar R, Altabella T, Marco F, Bortolotti C, Reymond M, Koncz C et al (2010) Polyamines: molecules with regulatory functions in plant abiotic stress tolerance. Planta 231:1237–1249

Badawy AA, Alamri AA, Hussein HAA, Salem NF, Mashlawi AM, Kenawy SK, El-Shabasy A (2024) Glycine betaine mitigates heavy metal toxicity in beta vulgaris (L.): an antioxidant-driven approach. Agronomy 14(4):797

Batcho AA, Sarwar MB, Rashid B, Hassan S, Husnain T (2021) Heat shock protein gene identified from Agave sisalana (As HSP70) confers heat stress tolerance in transgenic cotton (Gossypium hirsutum). Theor Exp Plant Physiol 33(2):141–156

Chakraborty D, Ciceu A, Ballian D, Benito Garzón M, Bolte A, Bozic G et al (2024) Assisted tree migration can preserve the European forest carbon sink under climate change. Nat Clim Chang 14(8):845–852

Chanu WS, Sarangthem K (2015) Changes in proline accumulation, amino acid, sugar and chlorophyll content in leaf and culm of Phourel-amubi, a rice cultivar of Manipur in response to flash flood. Indian J Plant Physiol 20:10–13

Chen K, Li GJ, Bressan RA, Song CP, Zhu JK, Zhao Y (2020) Abscisic acid dynamics, signaling, and functions in plants. J Integr Plant Biol 62(1):25–54

Cruz-Rus E, Amaya I, Valpuesta V (2012) The challenge of increasing vitamin C content in plant foods. Biotechnol J 7(9):1110–1121

Dietz KJ, Mittler R, Noctor G (2016a) Recent progress in understanding the role of reactive oxygen species in plant cell signaling. Plant Physiol 171(3):1535–1539

Dietz S, Bowen A, Dixon C, Gradwell P (2016b) 'Climate value at risk'of global financial assets. Nat Clim Chang 6(7):676–679

Dubey V, Gupta R, Pathania R (2021) Targeting superoxide dismutase confers enhanced reactive oxygen species-mediated eradication of polymyxin B-induced Acinetobacter baumannii persisters. Antimicrob Agents Chemother 65(5):e02180

Ebeed HT (2022) Genome-wide analysis of polyamine biosynthesis genes in wheat reveals gene expression specificity and involvement of STRE and MYB-elements in regulating polyamines under drought. BMC Genomics 23(1):734

El-Esawi MA, Alayafi AA (2019) Overexpression of StDREB2 transcription factor enhances drought stress tolerance in cotton (Gossypium barbadense L.). Genes 10(2):142

Fan Wei Juan FW, Zhang Min ZM, Zhang Hong Xia ZH, Zhang Peng ZP (2012) Improved tolerance to various abiotic stresses in transgenic sweet potato (Ipomoea batatas) expressing spinach betaine aldehyde dehydrogenase. PLoS One. https://doi.org/10.1371/journal.pone.0037344

Flora SJ (2009) Structural, chemical and biological aspects of antioxidants for strategies against metal and metalloid exposure. Oxidative Med Cell Longev 2(4):191–206

Galau GA, Bijaisoradat N, Hughes DW (1987) Accumulation kinetics of cotton late embryogenesis-abundant mRNAs and storage protein mRNAs: coordinate regulation during embryogenesis and the role of abscisic acid. Dev Biol 123(1):198–212

Gallie DR (2013) L-Ascorbic acid: a multifunctional molecule supporting plant growth and development. Scientifica 2013(1):795964

Gill SS, Tuteja N (2010) Reactive oxygen species and antioxidant machinery in abiotic stress tolerance in crop plants. Plant Physiol Biochem 48(12):909–930

Görlach A, Bertram K, Hudecova S, Krizanova O (2015) Calcium and ROS: a mutual interplay. Redox Biol 6:260–271

Goyal S, Dhull SK, Kapoor KK (2005) Chemical and biological changes during composting of different organic wastes and assessment of compost maturity. Bioresour Technol 96(14):1584–1591

Hmida-Sayari A, Costa A, Leone A, Jaoua S, Gargouri-Bouzid R (2005a) Identification of salt stress-induced transcripts in potato leaves by cDNA-AFLP. Mol Biotechnol 30:31–39

Hmida-Sayari A, Gargouri-Bouzid R, Bidani A, Jaoua L, Savouré A, Jaoua S (2005b) Overexpression of Δ1-pyrroline-5-carboxylate synthetase increases proline production and confers salt tolerance in transgenic potato plants. Plant Sci 169(4):746–752

Hoffmann EK, Lambert IH, Pedersen SF (2009) Physiology of cell volume regulation in vertebrates. Physiol Rev 89(1):193–277

Hong Y, Zhang H, Huang L, Li D, Song F (2016) Overexpression of a stress-responsive NAC transcription factor gene ONAC022 improves drought and salt tolerance in rice. Front Plant Sci 7:4

Hsieh TH, Lee JT, Charng YY, Chan MT (2002) Tomato plants ectopically expressing Arabidopsis CBF1 show enhanced resistance to water deficit stress. Plant Physiol 130(2):618–626

Huang S, Van Aken O, Schwarzländer M, Belt K, Millar AH (2016) The roles of mitochondrial reactive oxygen species in cellular signaling and stress response in plants. Plant Physiol 171(3):1551–1559

Huang J, Hu T, Yasir M, Gao Y, Chen C, Zhu R et al (2019) Root growth dynamics and yield responses of rice (Oryza sativa L.) under drought—flood abrupt alternating conditions. Environ Exp Bot 157:11–25

Ichimura K, Shinozaki K, Tena G, Sheen J, Henry Y, Champion A et al (2002) Mitogen-activated protein kinase cascades in plants: a new nomenclature. Trends Plant Sci 7(7):301–308

Kaur G, Tak Y, Asthir B (2022) Salicylic acid: a key signal molecule ameliorating plant stresses. Cereal Res Commun 50(4):617–626

Kavi Kishor PB, Sreenivasulu NESE (2014) Is proline accumulation per se correlated with stress tolerance or is proline homeostasis a more critical issue? Plant Cell Environ 37(2):300–311

Kishitani S, Takanami T, Suzuki M, Oikawa M, Yokoi S, Ishitani M et al (2000) Compatibility of glycinebetaine in rice plants: evaluation using transgenic rice plants with a gene for peroxisomal betaine aldehyde dehydrogenase from barley. Plant Cell Environ 23(1):107–114

Li Y, Li C, Xia J, Jin Y (2011) Domestication of transposable elements into microRNA genes in plants. PLoS One 6(5):e19212

Li J, Phan TT, Li YR, Xing YX & Yang LT (2018) Isolation, transformation and overexpression of sugarcane SoP5CS gene for drought tolerance improvement. Sugar Tech 20(4):464–473

Liang C, Meng Z, Meng Z, Malik W, Yan R, Lwin KM & Zhang R (2016) GhABF2, a bZIP transcription factor, confers drought and salinity tolerance in cotton (Gossypium hirsutum L.). Sci Rep 6(1):35040

Mellidou I, Keulemans J, Kanellis AK, Davey MW (2012) Regulation of fruit ascorbic acid concentrations during ripening in high and low vitamin C tomato cultivars. BMC Plant Biol 12:1–19

Mittler R (2017) ROS are good. Trends Plant Sci 22(1):11–19

Mittler R, Vanderauwera S, Gollery M, Van Breusegem F (2004) Reactive oxygen gene network of plants. Trends Plant Sci 9(10):490–498

Mittova V, Tal M, Volokita M, Guy M (2003) Up-regulation of the leaf mitochondrial and peroxisomal antioxidative systems in response to salt-induced oxidative stress in the wild salt-tolerant tomato species Lycopersicon pennellii. Plant Cell Environ 26(6):845–856

Pitaloka MK, Caine RS, Hepworth C, Harrison EL, Sloan J, Chutteang C et al (2022) Induced genetic variations in stomatal density and size of rice strongly affects water use efficiency and responses to drought stresses. Front Plant Sci 13:801706

Pramanik B, Sar P, Bharti R, Gupta RK, Purkaystha S, Sinha S, Chattaraj S, Mitra D (2023a) Multifactorial role of nanoparticles in alleviating environmental stresses for sustainable crop production and protection. Plant Physiol Biochem 201:107831. https://doi.org/10.1016/j.plaphy.2023.107831

Pramanik B, Sar P, Biswas S, Mondal PC (2023b) Arbuscular mycorrhiza and its role in rice production under drought stress. In: Panneerselvam P, Mohapatra PKD, Nayak AK, Mitra D, Velmourougane K, Santos-Villalobos SDL (eds) Arbuscular Mycorrhizal fungi: for nutrient, abiotic and biotic stresses management in rice. CRC Press/Taylor and Francis Group, pp 67–75. https://doi.org/10.1201/9781003354086-9

Qadir M, Hussain A, Hamayun M, Shah M, Iqbal A, Irshad M et al (2021) Phytohormones producing Acinetobacter bouvetii P1 mitigates chromate stress in sunflower by provoking host antioxidant response. Antioxidants 10(12):1868

Rajappa S, Krishnamurthy P, Huang H, Yu D, Friml J, Xu J, Kumar PP (2024) The translocation of a chloride channel from the Golgi to the plasma membrane helps plants adapt to salt stress. Nat Commun 15(1):3978

Ravi B, Foyer CH, Pandey GK (2023) The integration of reactive oxygen species (ROS) and calcium signalling in abiotic stress responses. Plant Cell Environ 46(7):1985–2006

Sandalio LM, Romero-Puertas MC (2015) Peroxisomes sense and respond to environmental cues by regulating ROS and RNS signalling networks. Ann Bot 116(4):475–485

Sandmann G (2015) Carotenoids of biotechnological importance. In: Biotechnology of isoprenoids, pp 449–467

Serrano N, Ling Y, Bahieldin A, Mahfouz MM (2019) Thermopriming reprograms metabolic homeostasis to confer heat tolerance. Sci Rep 9(1):181

Shirasawa K, Takabe T, Takabe T, Kishitani S (2006) Accumulation of glycinebetaine in rice plants that overexpress choline monooxygenase from spinach and evaluation of their tolerance to abiotic stress. Ann Bot 98(3):565–571

Shu DF, Wang LY, Duan M, Deng YS, Meng QW (2011) Antisense-mediated depletion of tomato chloroplast glutathione reductase enhances susceptibility to chilling stress. Plant Physiol Biochem 49(10):1228–1237

Shu S, Tang Y, Yuan Y, Sun J, Zhong M, Guo S (2016) The role of 24-epibrassinolide in the regulation of photosynthetic characteristics and nitrogen metabolism of tomato seedlings under a combined low temperature and weak light stress. Plant Physiol Biochem 107:344–353

Szabados L, Savouré A (2010) Proline: a multifunctional amino acid. Trends Plant Sci 15(2):89–97

Szarka A, Tomasskovics B, Bánhegyi G (2012) The ascorbate-glutathione-α-tocopherol triad in abiotic stress response. Int J Mol Sci 13(4):4458–4483

Tan J, Wang C, Xiang BIN, Han R, Guo Z (2013) Hydrogen peroxide and nitric oxide mediated cold-and dehydration-induced myo-inositol phosphate synthase that confers multiple resistances to abiotic stresses. Plant Cell Environ 36(2):288–299

Tang J, Wang SQ, Hu KD, Huang ZQ, Li YH, Han Z et al (2019) Antioxidative capacity is highly associated with the storage property of tuberous roots in different sweetpotato cultivars. Sci Rep 9(1):11141

Vinocur B, Altman A (2005) Recent advances in engineering plant tolerance to abiotic stress: achievements and limitations. Curr Opin Biotechnol 16(2):123–132

Waheed A, Zhou L, Wang M, Hailiang X, Tong Z, Wang C, Aili A (2024) Integrative mechanisms of plant salt tolerance: biological pathways, phytohormonal regulation, and technological innovations. Plant Stress 14:100652

Widmann C, Gibson S, Jarpe MB, Johnson GL (1999) Mitogen-activated protein kinase: conservation of a three-kinase module from yeast to human. Physiol Rev 79(1):143–180

Yancey PH (2005) Organic osmolytes as compatible, metabolic and counteracting cytoprotectants in high osmolarity and other stresses. J Exp Biol 208(15):2819–2830

Zechmann B (2011) Subcellular distribution of ascorbate in plants. Plant Signal Behav 6(3):360–363

Zeng J, Dong Z, Wu H, Tian Z, Zhao Z (2017) Redox regulation of plant stem cell fate. EMBO J 36(19):2844–2855

Zhang XY, Tang LH, Nie JW, Zhang CR, Han X, Li QY et al (2023a) Structure and activation mechanism of the rice Salt Overly Sensitive 1 (SOS1) Na+/H+ antiporter. Nat Plants 9(11):1924–1936

Zhang Y, Zhou J, Ni X, Wang Q, Jia Y, Xu X et al (2023b) Structural basis for the activity regulation of Salt Overly Sensitive 1 in Arabidopsis salt tolerance. Nat Plants 9(11):1915–1923

Zhao X, Ma X, Chen B, Shang Y, Song M (2022) Challenges toward carbon neutrality in China: strategies and countermeasures. Resour Conserv Recycl 176:105959

Zhu JK (2016) Abiotic stress signaling and responses in plants. Cell 167(2):313–324

# Signaling Cascades, Transcription Factors, and Gene Expression in ROS-Mediated Abiotic Stress Tolerance

12

Alok Ranjan, Gopal Kalwan, Shruti Kaushik, Diksha Taneja, Priya Yadav, Ashima Nehra, Radha Jangra, Narendra Tuteja, Ritu Gill, and Sarvajeet Singh Gill

**Abstract**

Reactive oxygen species (ROS) are redox biochemical molecules of oxygen, which affect and control several cellular and biophysiochemical processes during environmental stresses in plants. Although excessive ROS generation due to stress conditions can damage biomolecules and cellular organelles, leading to oxidative stress in plants. ROS-producing enzymes such as NADPH oxidases are precisely regulated by several posttranslational modifications. ROS are also perceived by specific redox-sensitive sensors that activate downstream signal transduction pathways. ROS play a key role in sensing environmental stresses, integrating various environmental signals, and triggering stress-responsive networks, thus stimulating plant defense mechanism and enhancing plant resilience.

A. Ranjan
Department of Biotechnology, Patna Women's College, Patna University, Patna, Bihar, India

G. Kalwan
ICAR-National Institute for Plant Biotechnology, New Delhi, India

PG School, ICAR-Indian Agricultural Research Institute, New Delhi, India

S. Kaushik
ICAR – Central Soil Salinity Research Institute, Karnal, India

D. Taneja · A. Nehra · R. Jangra · R. Gill · S. S. Gill (✉)
Centre for Biotechnology, Maharshi Dayanand University, Rohtak, Haryana, India
e-mail: ssgill14@mdurohtak.ac.in

P. Yadav
Department of Botany, Zakir Husain Delhi College, University of Delhi, New Delhi, India

N. Tuteja
International Centre for genetic Engineering and Biotechnology (ICGEB), Aruna Asaf Ali Marg, New Delhi, India



N. Iqbal et al. (eds.), *Plant Signaling Molecules in Regulation of ROS-Scavenging System*, https://doi.org/10.1007/978-981-95-3377-0_12

These pathways transmit signals to the nucleus via redox-mediated mechanisms and Mitogen-Activated Protein Kinase (MAPK) cascade. Transcription factors act as intracellular signal integrators and mediators, playing a pivotal role in orchestrating cellular responses to oxidative stress. However, the precise mechanisms underlying ROS-mediated stress alleviation in plants remain incompletely understood. In this study, we present our current knowledge on ROS-associated signaling pathways, gene expression, and regulatory mechanisms of transcription factors within plant cells under various stress conditions.

**Keywords**

Abiotic stress · Reactive oxygen species · Oxidative stress · ROS Signaling · Transcription factors

## 12.1 Introduction

Reactive oxygen species [ROS, e.g., superoxide radicals ($O_2^{\bullet-}$), hydroxyl radicals ($OH^{\bullet}$), singlet oxygen ($^1O_2$), hydrogen peroxide ($H_2O_2$), etc.) are oxygen-derived-redox molecules, produced at threshold concentrations within plant cells under normal physiological conditions (Wang et al. 2024; Panda et al. 2024). However, harsh abiotic conditions like heat, cold, drought, waterlogging, salinity, and heavy metals toxicity lead to an overproduction of ROS (Apel and Hirt 2004; Mishra et al. 2023). Consequently, higher plants have evolved various effective mechanisms to deal with the continuous production of ROS, including the development of specialized antioxidative systems for ROS detoxification, as well as mechanisms for ROS-mediated signal transduction (Dumanović et al. 2021). Depending on their intracellular concentrations, ROS exhibit a dual role under stress conditions, functioning both as harmful agents and as a key signaling molecules in stress responses (Gill and Tuteja 2010). In the absence of efficient regulatory mechanisms, elevated intracellular ROS levels can lead to extensive oxidative damage, including lipid peroxidation of cellular membranes, oxidation of proteins and carbohydrates, and nucleic acid damage (Wang et al. 2024). Persistent ROS accumulation may lead to oxidative stress, ultimately resulting in cellular dysfunction and plant death (Singh et al. 2019). Conversely, at low-to-moderate concentrations, ROS function as integral signaling molecules in association with other factors that activate antioxidative defense pathways, upregulate the expression of several transcription factors (TFs) and genes, thereby enhancing cellular tolerance to oxidative stress (Mittler et al. 2022). ROS are critical regulators in the perception and transduction of stress signals, the integration of diverse stress-responsive signaling pathways, ultimately leading to the activation of plant defense responses and acclimatization mechanisms in response to environmental stresses (Ranjan et al. 2021). Under stress conditions, the accumulation of ROS is perceived by redox-sensitive sensors, which initiate intracellular signaling cascades predominantly involving redox modifications and the activation of the Mitogen-Activated Protein Kinase (MAPK) pathway (Jalmi and Sinha 2015;

Bhardwaj et al. 2025). These intracellular signaling cascades are regulated by various proteins and enzymes, which ultimately leads to the activation of stress-responsive TFs. TFs are regulatory proteins that bind to specific *cis*-regulatory elements (CREs) present within the promoter regions of the target genes, leading to the modulation of gene expression patterns (Mittler et al. 2022). The resulting expression of stress-related genes contributes to the synthesis of new proteins and posttranslational modifications of existing proteins, ultimately enhancing cellular adaptation to oxidative stress (Huang et al. 2019). ROS are important signaling molecules, acting both as key mediators and ultimate signaling producers for stress responses in plants (Castro et al. 2021). Precise regulation of ROS production facilitates intercellular communication by amplifying stress signals through the activity of NADPH (nicotinamide adenine dinucleotide phosphate) oxidases, also known as respiratory burst oxidase homologs (RBOHs) (Kimura et al. 2020). This ROS-mediated signaling modulates protein conformation and activates defense-related gene expression, thus contributing to an effective stress response (Liebthal et al. 2020). Numerous transcriptomic and transgenic studies have investigated the roles of various TFs and stress-responsive genes, antioxidant enzymes, and MAPK in plants exposed to abiotic stress conditions (Gadjev et al. 2006; Xu et al. 2019; Mishra et al. 2024). Several TFs families in plants, such as ERF (Ethylene Response Factors), AP2 (APETALA2), WRKY, NAC (No Apical Meristem), and MYB (myeloblastosis), play crucial roles in regulating ROS metabolism under stress conditions (Mizoi et al. 2012; Li et al. 2015a; Yao et al. 2017; Liu et al. 2022). They function as central hubs in intracellular signaling networks, integrating diverse environmental cues, and translating them into coordinated gene expression responses that drive appropriate cellular adaptations (Liu et al. 2022). In this chapter, we have explored the role of ROS molecules in regulating defense-related signaling cascades and expression of various stress-responsive TFs, and how these processes contribute in enhancing the stress tolerance in plants.

## 12.2 ROS Signaling Under Abiotic Stresses

Extensive evidence indicates that abiotic stresses trigger the generation of oxygen-free radicals in plant cells. The accumulation of these free radicles under stress conditions disrupts the cellular homeostasis by altering the redox state of various crucial biomolecules, including proteins, enzymes, and receptors, which further activates, modulates, or integrates several stress-responsive signaling pathways (Choudhury et al. 2017). In contrast to traditional signaling molecules like phytohormones, peptides, or nitric oxide (NO), which usually function through specific receptors, whereas variations in ROS level can directly modify the structure and activity of various proteins, thereby impacting a wide range of signaling networks. This ability of ROS to engage in multiple pathways is mainly mediated by oxidative posttranslational modifications, allowing them to act as versatile and dynamic regulators of various stress responses (Mohanta et al. 2018).

## 12.3 ROS-induced Posttranslational Modifications and Signaling

ROS-induced oxidative posttranslational modifications lead to conformational changes in proteins, particularly in kinases, phosphatases, and transcription factors (Martí-Guillén et al. 2022). Such modifications can alter the activity, specificity, and subcellular localization of the affected proteins, thereby modulating stress-responsive signal transduction pathways through activation or repression mechanisms. For example, thiol groups in cysteine and methionine residues of numerous proteins are particularly susceptible to nucleophilic attack by ROS (Alvarez et al. 2011; Waszczak et al. 2015). However, their reactivity is significantly influenced by the surrounding protein microenvironment, including factors such as nearby positively charged residues and hydrogen bonding interactions (Kärkönen and Kuchitsu 2015). Among the most common reactions associated with ROS signaling are those involving sulfenic acids (–SOH), which readily react with proximal protein thiols either intramolecularly or intermolecularly to form disulfide bonds with glutathione (GSH) (Gill et al. 2013; Corpas and Barroso 2015). On the other hand, the reversibility of ROS-induced oxidative post-translational modifications is a critical characteristic of ROS signaling, allowing for dynamic regulation and fine-tuning of cellular responses to various environmental challenges (Corpas et al. 2008; Martí-Guillén et al. 2022). The ability to reversibly modulate oxidative post-translational modifications in a controlled manner enhances the flexibility of ROS signaling under stress, particularly in the context of integrating various stress signals and facilitating recovery from stress (Noctor and Foyer 2016). For instance, glutathionylation is typically reversed by glutaredoxins (GRXs), which restores the original thiol groups, whereas protein disulfide bonds are primarily reduced by thioredoxins (TRXs) (Friso and Van Wijk 2015; Vašková et al. 2023).

ROS-induced oxidative posttranslational modifications regulates a diverse range of metabolic processes in cells and modulate the function and activity of various kinases, phosphatases, TFs, receptors, and ion channels under stress conditions (Martí-Guillén et al. 2022). Prominent examples such as inhibition of protein tyrosine phosphatases (PTPs) and class 2 serine/threonine protein phosphatases (PP2A), both of which play key roles in hormone signaling (Chan et al. 2016). However, these modifications can also activate MAPK cascades like MKK1/MKK and MPK4/MPK6 pathway, along with serine/threonine kinases that are essential for complete activation of MPK3 and MPK6, which are key players in stress-responsive pathways (Matsushita et al. 2020).

## 12.4 ROS-mediated Signaling Pathways in Plants

In plant cells, the detection of ROS and the activation of downstream signaling pathways occur within specific subcellular compartments, influenced by the intricate cellular architecture and microenvironment (Tripathy and Oelmüller 2012; Castro et al. 2021). ROS signaling is generally classified into three main types:

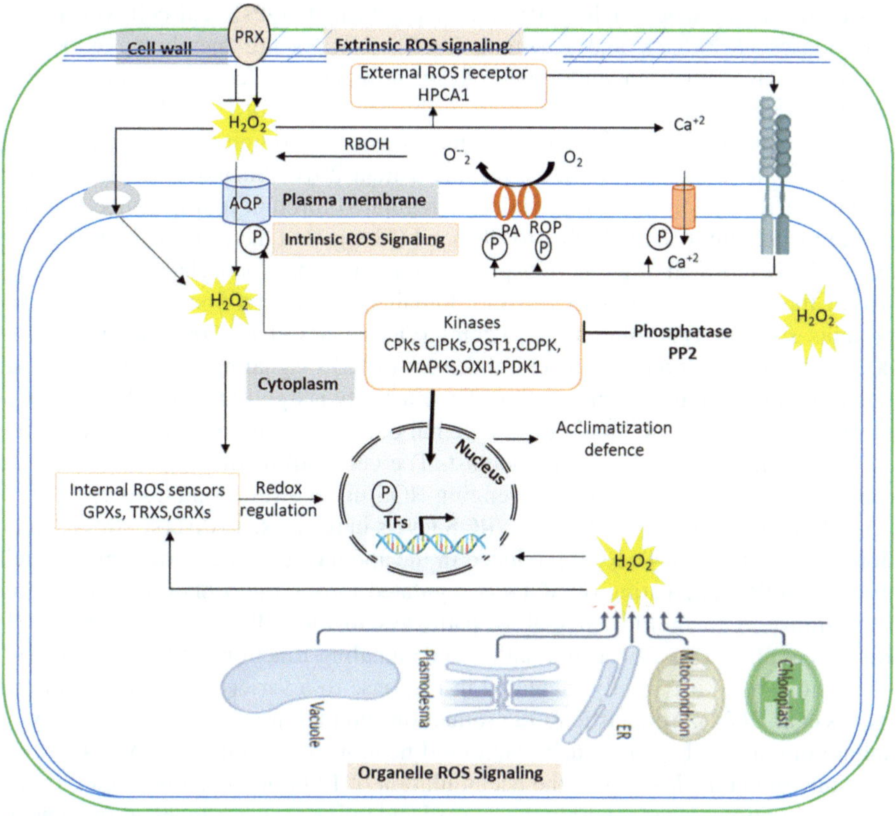


**Fig. 12.1** The extrinsic, intrinsic, and organelle-localized ROS signaling pathways. These signaling pathways are integrated through RBOHs, aquaporins (AQPs), $Ca^{2+}$ channels, receptors, and different enzymes, which ultimately trigger transcriptional responses to stress. ER endoplasmic reticulum, HPCA1 HYDROGEN, OST1 OPEN STOMATA 1, P phosphate, PRX peroxidase, PrxR peroxiredoxin, PDK1 3-PHOSPHOINOSITIDE-DEPENDENT PROTEIN KINASE, TF transcription factor, TRX thioredoxin

extrinsic (occurring in the apoplast and cell wall), intrinsic (within the cytosol and nucleus), and organelle-based (involving chloroplasts, mitochondria, peroxisomes, and other organelles) as shown in Fig. 12.1. Emerging evidence suggests that these signaling mechanisms may operate independently or engage in dynamic cross-talk, especially during stress responses (Ali et al. 2023).

## 12.5 Extrinsic (Apoplast and Cell Wall) ROS Signaling

The apoplast and cell wall serve as important sites for the generation and detoxification of ROS, facilitated by a range of enzymatic systems and non-enzymatic antioxidants (Fallah et al. 2024). Key components involved in ROS dynamics within this extracellular space include RBOHs, aquaporins (AQPs), and cell wall-bound peroxidases (Kärkönen and Kuchitsu 2015). RBOHs are membrane-associated enzymes, which play an important role by transferring electrons from cytosolic NADPH to molecular oxygen, leading to the production of $O_2^{\bullet-}$. These are subsequently converted into $H_2O_2$, either spontaneously or through the catalytic activity of superoxide dismutase (SOD) (Gill and Tuteja 2010; Suzuki et al. 2011). The activity of RBOHs in ROS production is regulated through multiple mechanisms, including calcium ($Ca^{2+}$) binding to EF-hand (Elongation factor-hand) motifs in the cytosolic N-terminal region, phosphorylation or dephosphorylation of residues within the N- or C-terminal cytosolic domains, and interactions with regulatory proteins such as GTP-binding molecules (Ogasawara et al. 2008; Kimura et al. 2020). RBOHs are considered central drivers of ROS signaling due to their crucial function in initiating ROS production. Their activity is tightly regulated in response to various environmental and developmental cues, enabling the generation of specific ROS signatures within the apoplast. The cell wall-mediated peroxidases are capable of both producing and scavenging ROS under stress conditions and play a central role in modulating apoplastic ROS levels in response to diverse stress conditions (Rivas et al. 2024). ROS produced in the apoplast are transported into the cell through AQPs. The gating of AQPs is regulated through phosphorylation, linking ROS transport to multiple signaling pathways in the cell (Apel and Hirt 2004; Kapilan et al. 2018). The accumulation and signaling functions of ROS in the apoplast and cytosol are tightly regulated in response to a variety of stress conditions (Rivas et al. 2024). Apoplastic ROS production and their translocation into the cytosol are modulated by posttranslational modifications of RBOHs and AQPs at their cytosolic domains. Moreover, the accumulation of ROS in the apoplast can trigger phosphorylation cascades within the cytosol and influence $Ca^{2+}$ fluxes across plasma membrane through channels (Fig. 12.1). Consequently, the apoplast–cytosol interface is increasingly recognized as a critical nexus for ROS-mediated signaling under stress conditions (Ogasawara et al. 2008).

## 12.6 Intrinsic (Cytosol and Nucleus) ROS Signaling

The cytosol contains a few ROS-producing enzymes along with several ROS-scavenging mechanisms. The cytosol also contains various signaling networks, including calcium-dependent protein kinases, MAPK cascades, PP2A (protein phosphatase 2A), PP2C (protein phosphatase 2C), PTPs (Phosphatases), and ROP/RAC (Rho-like GTPases) GTPases and redox sensing networks like PrxRs (peroxiredoxins), GRXs (glutathione peroxidases), and TRXs (Waszczak et al. 2018). These signaling networks help integrate ROS signals with other molecules, like $Ca^{2+}$ and phytohormones. The cytosol helps interpret and integrate ROS signals from different cell compartments and sends this information to the nucleus. The plant cell nucleus contains various ROS and redox-regulating proteins, including GRXs, TRXs, and PrxRs. These proteins regulate oxidative posttranslational modifications of transcription factors and help attenuate ROS signals within the nucleus (Waszczak et al. 2018; Liu et al. 2022).

## 12.7 Organelle (Chloroplasts, Mitochondria, and Peroxisomes) ROS Signaling

Plant cell organelles, including chloroplasts, mitochondria, and peroxisomes, possess a range of ROS-scavenging and ROS-producing systems that regulate ROS signaling within the organelles and enable cross-talk between organelles and the nucleus (Dietz et al. 2016). Responses to stressors that predominantly activate extrinsic or intrinsic ROS signaling pathways may occur spatially and/or temporally distinct from those mediated by organelle-to-organelle or organelle-to-nucleus ROS signaling networks. This segregation may serve as a mechanism by which ROS convey specific signals to the nucleus, thereby informing the plant of the type of stress it is experiencing (Dietz et al. 2016).

## 12.8 ROS Signaling and Transcriptional Control During Abiotic Stresses

Redox-sensitive TFs are key regulators in coordinating plant responses to abiotic stress. Various families of TFs, including AP2/ERF (Abiri et al. 2017), WRKY (Ma and Hu 2024), NAC (Fang et al. 2015), and MYB (Li et al. 2015a), are involved in regulating ROS metabolism. Importantly, many of these TFs possess the ability to regulate both enzymatic and non-enzymatic antioxidant pathways (Fang et al. 2015; Abiri et al. 2017; Ma and Hu 2024). The AP2/ERF, WRKY, and NAC TF families are key regulators of the enzymatic antioxidant mechanism, which helps alleviate oxidative stress triggered by abiotic stresses. Moreover, other TFs, such as bHLH (Qian et al. 2021), MYB (Li et al. 2015a), bZIP (Ranjan et al. 2012; Yang et al. 2019), and Dof (Yang et al. 2017), contribute to the regulation of non-enzymatic antioxidant mechanism such as biosynthesis of secondary metabolites (tocopherols, flavonoids, proline, and carotenoids) in plant cell. This dual regulation allows plants to efficiently mitigate oxidative stress by balancing the activity of antioxidant enzymes and non-enzymatic scavengers, thereby enhancing their resilience to environmental stressors. The coordination of these pathways is critical for optimizing redox homeostasis, particularly under conditions of abiotic stress (Gill and Tuteja 2010; Hasanuzzaman et al. 2021). In addition to ROS scavenging, TFs are also involved in ROS production. For example, bHLH123 in tobacco and ERF74 in *Arabidopsis thaliana* have been shown to promote ROS generation by activating the expression of *Nt*RbohE and RbohD, respectively (Liu et al. 2021). Overexpression of TFs in different plant such as *At*WRKY30 (*A. thaliana)* (Scarpeci et al. 2013), *Bd*WRKY36 (*Brachypodium distachyon*) (Sun et al. 2015), *Sb*WRKY30 (*Sorghum bicolor*) (Yang et al. 2020), *Sp*WRKY1 (*Solanum pimpinellifolium*) (Li et al. 2015a), *Zm*WRKY40 (*Zea mays*) (Wang et al. 2018b), and *Zm*WRKY106 (*Z. mays*) (Wang et al. 2018a) have been shown to enhance the activities of several key antioxidant enzymes, including peroxidase (PRX), catalase (CAT), and SOD under drought stress. These TFs also upregulate the expression of several stress-responsive genes, such as RD29A, HSP90 (heat shock protein 90), DREB2A, DREB2B, Cu/ZnSOD,

LEA5, and NtP5CS. Simultaneously, overexpression of these TFs results in a substantial decrease in ROS accumulation, indicating their critical role in enhancing drought tolerance by strengthening antioxidant defense mechanisms and regulating stress-responsive genes. The WRKY TFs enhance plant drought tolerance primarily by increasing the capacity for ROS scavenging. In tomato, the TFs JUNGBRUNNEN1 and SlNAC2 have been shown to promote the expression of SlDREB1, SlDREB2, SlDELLA, and genes involved in GSH biosynthesis, thereby reducing ROS accumulation under drought stress (Thirumalaikumar et al. 2018). In *Brassica napus* L. *Bna*NAC55 modulates the expression of a diverse set of genes involved in both the generation and detoxification of ROS, indicating that NAC TFs contribute to drought resistance through modulation of ROS homeostasis (Fan et al. 2018). Moreover, studies have revealed that *Nt*ERF172 (*Nicotiana tabacum*) and JERF3 increase the expression of *Nt*CAT and SOD, improving drought stress tolerance in tobacco by maintaining CAT-mediated $H_2O_2$ homeostasis (Zhao et al. 2020b). Together, these findings highlight the pivotal roles of various TFs, including WRKY, NAC, and ERF, in coordinating antioxidant defense to mitigate drought-induced oxidative stress.

Salt stress significantly impairs plant growth and development, primarily through the disruption of cellular homeostasis and the accumulation of ROS. Overexpression of several WRKY TFs, such as *Ah*WRKY75 (Zhu et al. 2021), *Hb*WRKY82 (Kang et al. 2021), *Mf*WRKY70 (Xiang et al. 2021), *Ft*WRKY46 (Lv et al. 2020), and *Pc*WRKY33 (Bao et al. 2018), have been shown to enhance salt tolerance in peanut and *A. thaliana*. This enhanced tolerance is largely attributed to the activation of antioxidant enzymes, including SOD, PRX, and CAT. In another study on *Hevea brasiliensis*, overexpression of *Hb*WRKY82 leads to the upregulation of ROS-associated genes (such as RbohD, CSD1, CSD2, and FSD3) along with hormone signaling genes (EIN3, ABF3, ABF4), indicating a potential interplay between ROS and abscisic acid (ABA) signaling pathways in regulating antioxidant enzyme activity (Kang et al. 2021). Similarly, NAC TFs such as *Ca*NAC46 (Ma et al. 2021b), *Gm*NAC06 (Li et al. 2015b), *Gm*NAC065 (Melo et al. 2021), *Mb*NAC25 (Han et al. 2020), and *Rt*NAC100 (Ma et al. 2021a) play critical roles in mitigating ROS damage under salt stress. In *A. thaliana*, soybean, and *Reaumuria trigyna*, overexpression of these TFs enhances the expression of genes encoding antioxidant enzymes, increases antioxidant activity, and promotes proline accumulation, thus contributing to improved salt tolerance. Furthermore, several ERF TFs, such as *Lc*ERF056 (*Lotus corniculatus*) (Wang et al. 2021b), *Gh*ERF13.12 (*Gossypium hirsutum*) (Lu et al. 2021b), *Zm*EREB20 (*Z. mays*) (Fu et al. 2021), and SlERF84 (*Solanum lycopersicum*) (Li et al. 2018), have been implicated in mitigating oxidative stress induced by salt stress. In *Lotus corniculatus*, *Lc*ERF056 has been shown to interact with ROS-associated genes through *cis*-regulatory elements, including GCC box and DRE motifs. Meanwhile, *Gh*ERF13.12 in upland cotton and ZmEREB20 in *A. thaliana* contribute to enhanced proline biosynthesis and upregulation of ROS-scavenging gene expression. Transgenic tomato plants overexpressing SlERF84 exhibit improved capacity for ROS detoxification, highlighting the critical involvement of ERF TFs in regulating antioxidant responses during salt stress. MYB TFs such as MYB49, SlMYB102, and TaMYB86B have also been reported to improve ROS scavenging efficiency in tomato and tobacco under salt stress (Song et al. 2020; Zhang et al. 2020a). Likewise, another study revealed the role of bHLH TFs in ROS regulation; for example, *Nt*bHLH123 directly activates RbohE expression, thus acting as a molecular switch in *Nicotiana tabacum*, while *Ah*bHLH112 in peanut activates PRX gene expression by directly binding to its promoter region (Li et al. 2015b; Zhao et al. 2018). In *A. thaliana*, *Bv*bHLH93 promotes expression of SOD and PRX while repressing RbohD and RbohF, further supporting the dual regulatory role of bHLH TFs in ROS generation and detoxification (Wang et al. 2021a, b, c, d). In addition to this, in poplar, *Pe*STZ1 contributes to salt stress tolerance by modulating the expression of *Pe*ZAT12 and *Pe*APX2, key genes associated with reactive ROS metabolism (He et al. 2020). Collectively, these findings demonstrate that a wide array of TFs contribute to salt stress responses by modulating ROS production and scavenging (Table 12.1). Despite functional studies highlighting their role in regulating ROS levels, the exact molecular mechanisms by which these TFs control ROS metabolism under abiotic stresses remain largely unknown. Under temperature stress conditions, ROS are generated and accumulate in the cytosol and nuclei of plant cells. The elevated ROS levels trigger the redox-dependent activation and subsequent translocation of Heat Stress Transcription Factor 1 (HSF1), a key regulator of the heat stress response (HSFA1) (Giesguth et al. 2015) and multiprotein-bridging factor1c (MBF1C) (Suzuki et al. 2008) from the cytosol to the nucleus and BZIP28 translocated from the endoplasmic reticulum to the nucleus (Srivastava et al. 2013). Combined action of HSFA1 and MBF1C activates the genes for HSPs (heat shock proteins) and other TFs, like dehydration-responsive binding element (DRBE) factors. Both are crucial for enabling plants to develop tolerance to elevated temperatures. The bZIP28 functions in coordination with bZIP60 to initiate the transcription of unfolded protein response (UPR) genes during heat stress. Additionally, various TFs have been recognized as central regulators of antioxidant defense mechanisms under elevated temperature condition. For instance, *Zm*WRKY106, ERF74, and MYB44 activate antioxidant enzymes and contribute to ROS scavenging in maize, *A. thaliana*, and *Xanthoceras sorbifolium*, respectively (Yao et al. 2017; Wang et al. 2018a). Similarly, overexpression of *Ec*DREB2A in tobacco led to increased antioxidant enzyme activity and reduced ROS accumulation under heat stress conditions. Heat shock transcription factors (HSFs) play a pivotal role in the heat stress response by regulating the expression of heat shock proteins (HSPs), which function as molecular chaperones to protect cellular proteins under elevated temperatures. In lilies, *Ll*HsfA4 has been demonstrated to enhance heat stress tolerance by upregulating the expression of APX2, a key gene involved in ROS scavenging (Wang et al. 2022). Conversely, *Zm*Hsf08 appears to function as a negative regulator of abiotic stress responses in maize (Wang et al. 2021a, b, c, d). Additionally, other TFs such as BZR1, OsSPL7, and *Zm*WRKY106 contribute to maintaining ROS homeostasis under high temperatures (Wang et al. 2018a; Hoang et al. 2019). These findings highlight the potential role of TFs to mitigate oxidative damage during heat stress by activating antioxidant defense pathways. The *Ko*WRKY40 overexpressing *A. thaliana* plants displayed increased

**Table 12.1** ROS-mediated transcription factors involved in abiotic stress tolerance

| TFs | ROS involvement | Plant species | Effect on the plant | References |
|---|---|---|---|---|
| CaNAC46 | ROS promotes expression of SOD and PRX | *Capsicum annuum /A. thaliana* | Drought and salt tolerance | Ma et al. (2021a) |
| CaNAC064 | ROS/MDA content reduced | *Capsicum annuum* | Cold tolerant | Hou et al. (2020) |
| GmNAC06 | Decrease ROS levels | Soybean | Drought tolerance | Li et al. (2015b) |
| GmNAC065 | Lower $H_2O_2$ content | Soybean | Drought and heat tolerant | Melo et al. (2021) |
| GmNAC085S | Enhanced antioxidant enzyme activity | Soybean/*A. thaliana* | Drought tolerance | Melo et al. (2021) |
| MbNAC25 | Increase activities of antioxidant enzymes | *Malus baccata /A. thaliana* | Drought, salinity, and cold stress tolerant | Han et al. (2020) |
| OsNAC015 | Decreased accumulation of ROS | Rice | Drought tolerance | Ao et al. (2024) |
| RtNAC100 | Promote ROS accumulation | *Reaumuria trigyna* | Susceptible to salt stress | Ma et al. (2021a) |
| ERF6 | Decreased ROS content | *A. thaliana* | Drought stress tolerance | Dubois et al. (2013) |
| ERF96 | Decrease ROS content | *A. thaliana* | Drought and salt tolerance | Wang et al. (2015) |
| ERF74 | Mediate RbohD-ROS signal pathway | *A. thaliana* | Drought and heat tolerance | Yao et al. (2017) |
| LcERF056 | Modulating ROS-related genes | *Lotus corniculatus* | Salt stress tolerance | Wang et al. (2021b) |
| GhERF13.12 | Enhanced ROS scavenging | *Gossypium hirsutum* | Salt stress tolerance | Lu et al. (2021a) |
| ZmEREB20 | Enhanced ROS scavenging | Maize | Salt tolerance | Fu et al. (2021) |
| SlERF84 | Decrease ROS accumulation | *A. thaliana* | Drought and salt stress tolerance | Li et al. (2018) |
| bHLH123 | Mediate RbohE signaling pathway | Tobacco | Salt stress tolerance | Liu et al. (2021) |
| BvbHLH93 | Enhance CAT and PRX activity | Sugar beet/*A. thaliana* | Salt stress tolerance | Wang et al. (2021a, b, c, d) |
| MdbHLH130 | Lower ROS Content | Apple/Tobacco | Drought stress tolerance | Zhao et al. (2020a) |
| MfbHLH38 | Lower ROS Content | *M. flabellifolia/ A. thaliana* | Drought and salt stress tolerance | Qiu et al. (2020) |
| OsWIH2 | Decrease ROS accumulation | Rice | Drought tolerance | Gu et al. (2021) |

(continued)

**Table 12.1** (continued)

| TFs | ROS involvement | Plant species | Effect on the plant | References |
|---|---|---|---|---|
| CaSBP12 | Decreased accumulation of ROS | Capsicum annuum | Salt stress tolerance | Zhang et al. (2020b) |
| CsHB5 | Increase ROS content | Citrus calli | Sensitive to drought and salt stress | Zhang et al. (2021) |
| MdZAT10 | Accumulate ROS | Apple | Sensitivity to drought stress | Yang et al. (2021) |
| MdHB7-like | Reduce ROS content | Apple | Salt stress tolerance | Yang et al. (2023) |
| ZAT18 | Lower content of ROS | *A. thaliana* | Drought tolerance | Yin et al. (2017) |
| OsMADS57 | Increase activities of antioxidative enzymes | Rice/Arachis hypogaea | Drought tolerance | Wu et al. (2020) |
| WRKY75 | Improve ROS scavenging system | *A. thaliana/Arachis hypogaea* | Salt tolerance in transgenic peanut lines | Devaiah et al. (2007) |
| EjWRKY17 | Decrease ROS accumulation | *Eriobotrya japonica* | Drought tolerance in transgenic Arabidopsis | Wang et al. (2021a) |
| FtWRKY46 | ROS clearance | *Fagopyrum esculentum* (Buckwheat) | Enhanced salt stress tolerance | Lv et al. (2020) |
| HbWRKY82 | Regulate RbohD gene expression | *Hevea brasiliensis* | Salt and drought stress tolerance | Kang et al. (2021) |
| KoWRKY40 | Lower $H_2O_2$ content | *A. thaliana* | Cold stress tolerance | Fei et al. (2022) |
| MfWRKY70 | Maintain ROS homeostasis | *Myrothamnus flabellifolia* | Drought and salinity tolerance | Xiang et al. (2021) |
| PcWRKY33 | Increasing level of cellular ROS | *Polygonum cuspidatum* | Negatively regulates the tolerance | Bao et al. (2018) |
| WRKY33 | ROS clearance | *A. thaliana* | Increase hypoxia tolerance | Dubois et al. (2013) |
| ZmWRKY40 | Increase PRX and CAT activity | *A. thaliana* | Drought tolerance | Wang et al. (2018b) |
| ZmWRKY79 | Boost ROS scavenging | *A. thaliana* | Drought stress tolerance | Gulzar et al. (2021) |
| ZmWRKY106 | Reduces ROS content | Maize | Drought and heat tolerant | Wang et al. (2018a) |

proline accumulation and higher activities of SOD, PRX, and CAT, along with reduced $H_2O_2$ levels compared to wild-type plants under cold stress (Fei et al. 2022). Similarly, overexpression of *Ca*NAC064 and *Mb*NAC25 in pepper and *A. thaliana* significantly enhanced antioxidant enzyme activities and improved ROS-scavenging capacity (Han et al. 2020; Hou et al. 2020). In poplar, *Pe*STZ1 contributes to improved freezing tolerance by directly activating *Pe*APX2, thereby promoting efficient ROS detoxification (He et al. 2020). Moreover, TFs such as MYB, ERF, and bHLH have been shown to regulate the expression of genes involved in the biosynthesis of ascorbic acid, flavonoids, phenolics, and anthocyanins, all of which contribute to alleviating oxidative stress under low-temperature conditions. Importantly, the expression of these cold-responsive TFs is often modulated by hormonal signals, including ethylene, underscoring the complex cross-talk between hormonal pathways and ROS signaling in the regulation of cold stress responses. However, while the functional roles of many of these TFS have been described, the underlying molecular mechanisms by which they regulate ROS signaling and antioxidant responses remain largely unresolved. Further studies are required to develop a comprehensive understanding of these complex regulatory networks and stress tolerance mechanisms in plants.

## 12.9 Conclusion

Redox-sensitive transcription factors play a crucial role in the intricate regulatory network for tolerance against various abiotic stresses, mainly by modulating ROS signaling. Key transcription factors families, such as AP2/ERF, WRKY, NAC, MYB, bHLH, bZIP, and Dof exhibit diverse functions in response to abiotic stresses by coordinating enzymatic and non-enzymatic antioxidant defense system of ROS detoxification and ROS generation by the activation of various defense-related signaling pathways, including MAPKs playing a central role in plant cell signaling. The complexity of ROS responses under diverse environmental conditions may, in part, arise from the differential regulation of ROS production via NADPH oxidases (respiratory burst oxidase homologs, RBOHs), which function across various tissues and developmental stages. Plants have evolved various sophisticated mechanisms to ensure temporal and spatial coordination between ROS and other signaling pathways activated in distinct tissues and at different time points. The potential of TFs such as WRKY, NAC, and ERF to improve stress tolerance by modulating antioxidant enzyme activity and activating stress-responsive genes that includes the increased production of enzymes like SOD, POD, CAT, molecular chaperones like HSPs, as well as synthesis of secondary metabolites like flavonoids and proline that helps in ROS scavenging, especially under drought, salt, heat, and cold stress conditions. A deeper level of intricacy, including hormone signaling cross-talk, as shown with the ABA and ethylene pathways, is also revealed by the dynamic control of ROS by transcription factors. The development of climate-resilient crops completely depends on our understanding of TFs, in terms of their activation mechanism, interaction with *cis*-regulatory elements, and their coordination with complex stress signals. Future research should be focused on dissection of the intricate molecular network using an integrated approach of genomics, transcriptomics, proteomics, and gene editing technologies to elucidate how transcriptional networks modulate ROS homeostasis and signaling under stress conditions.

**Acknowledgments** Work on plant abiotic stress tolerance in SSG laboratory was partially supported by the University Grants Commission (UGC), Science and Engineering Research Board (SERB), Council of Scientific & Industrial Research (CSIR), Govt. of India. SSG and RG also acknowledge partial support from DBT-BUILDER grant (No. BT/INF/22/SP43043/2021). We sincerely apologize to our contemporaries whose work could not be discussed in this article due to space restrictions.

**Conflicts of Interest** Authors declare no Conflicts of interest/Competing interests.

**Author Contribution** NT, RG, SSG, AN; conceptualized the idea; AR, GK, SK, DT, PY, RJ; wrote the first draft of the chapter; GK, SK, DT, PY; drawn the figure; AR, NT, RG, SSG, AN finally edited and finalized the manuscript.

## References

Abiri R, Shaharuddin NA, Maziah M, Yusof ZNB, Atabaki N, Sahebi M, Valdiani A, Kalhori N, Azizi P, Hanafi MM (2017) Role of ethylene and the APETALA 2/ethylene response factor superfamily in rice under various abiotic and biotic stress conditions. Environ Exp Bot. https://doi.org/10.1016/j.envexpbot.2016.10.015

Ali S, Tyagi A, Bae H (2023) ROS interplay between plant growth and stress biology: challenges and future perspectives. Plant Physiol Biochem. https://doi.org/10.1016/j.plaphy.2023.108032

Alvarez C, Lozano-Juste J, Romero LC, García I, Gotor C, León J (2011) Inhibition of Arabidopsis O-acetylserine(thiol)lyase A1 by tyrosine nitration. J Biol Chem. https://doi.org/10.1074/jbc.M110.147678

Ao C-W, Xiang G, Wu Y-F, Wen Y, Zhu Z-L, Sheng F, Du X (2024) OsNAC15 regulates drought and salt tolerance in rice. Physiol Mol Biol Plants 30:1–11

Apel K, Hirt H (2004) Reactive oxygen species: metabolism, oxidative stress, and signal transduction. Annu Rev Plant Biol. https://doi.org/10.1146/annurev.arplant.55.031903.141701

Bao W, Wang X, Chen M, Chai T, Wang H (2018) A WRKY transcription factor, PcWRKY33, from Polygonum cuspidatum reduces salt tolerance in transgenic Arabidopsis thaliana. Plant Cell Rep. https://doi.org/10.1007/s00299-018-2289-2

Bhardwaj D, Sharma S, Sharma A, Gill R, Gill SS, Verma R, Kaul T, Tuteja N (2025) Decoding the signaling triad: molecular interactions of G-proteins, MAP kinases, and helicases in environmental stress responses. Plant Sci:112514. https://doi.org/10.1016/j.plantsci.2025.112514

Castro B, Citterico M, Kimura S, Stevens DM, Wrzaczek M, Coaker G (2021) Stress-induced reactive oxygen species compartmentalization, perception and signalling. Nat Plants 7(4):403–412. https://doi.org/10.1038/s41477-021-00887-0

Chan KX, Mabbitt PD, Phua SY, Mueller JW, Nisar N, Gigolashvili T, Stroeher E, Grassl J, Arlt W, Estavillo GM, Jackson CJ, Pogson BJ (2016) Sensing and signaling of oxidative stress in chloroplasts by inactivation of the SAL1 phosphoadenosine phosphatase. Proc Natl Acad Sci USA. https://doi.org/10.1073/pnas.1604936113

Choudhury FK, Rivero RM, Blumwald E, Mittler R (2017) Reactive oxygen species, abiotic stress and stress combination. Plant J. https://doi.org/10.1111/tpj.13299

Corpas FJ, Barroso JB (2015) Reactive sulfur species (RSS): possible new players in the oxidative metabolism of plant peroxisomes. Plant Sci, Front. https://doi.org/10.3389/fpls.2015.00116

Corpas FJ, del Río LA, Barroso JB (2008) Post-translational modifications mediated by reactive nitrogen species. Plant Signal Behav. https://doi.org/10.4161/psb.3.5.5277

Devaiah BN, Karthikeyan AS, Raghothama KG (2007) WRKY75 transcription factor is a modulator of phosphate acquisition and root development in Arabidopsis. Plant Physiol. https://doi.org/10.1104/pp.106.093971

Dietz KJ, Mittler R, Noctor G (2016) Recent progress in understanding the role of reactive oxygen species in plant cell signaling. Plant Physiol. https://doi.org/10.1104/pp.16.00938

Dubois M, Skirycz A, Claeys H, Maleux K, Dhondt S, De Bodt S, Vanden Bossche R, De Milde L, Yoshizumi T, Matsui M, Inzé D (2013) ETHYLENE RESPONSE FACTOR6 acts as a central regulator of leaf growth under water-limiting conditions in Arabidopsis. Plant Physiol. https://doi.org/10.1104/pp.113.216341

Dumanović J, Nepovimova E, Natić M, Kuča K, Jaćević V (2021) The significance of reactive oxygen species and antioxidant defense system in plants: a concise overview. Front Plant Sci. https://doi.org/10.3389/fpls.2020.552969

Fallah S, Yusefi-Tanha E, Peralta-Videa JR (2024) Interaction of nanoparticles and reactive oxygen species and their impact on macromolecules and plant production. Plant Nano Biol:100105. https://doi.org/10.1016/j.plana.2024.100105

Fan ZQ, Tan XL, Chen JW, Liu ZL, Kuang JF, Lu WJ, Shan W, Chen JY (2018) BrNAC055, a novel transcriptional activator, regulates leaf senescence in Chinese flowering cabbage by modulating reactive oxygen species production and chlorophyll degradation. J Agric Food Chem. https://doi.org/10.1021/acs.jafc.8b02309

Fang Y, Liao K, Du H, Xu Y, Song H, Li X, Xiong L (2015) A stress-responsive NAC transcription factor SNAC3 confers heat and drought tolerance through modulation of reactive oxygen species in rice. J Exp Bot. https://doi.org/10.1093/jxb/erv386

Fei J, Wang YS, Cheng H, Su YB, Zhong YJ, Zheng L (2022) The Kandelia obovata transcription factor KoWRKY40 enhances cold tolerance in transgenic Arabidopsis. BMC Plant Biol. https://doi.org/10.1186/s12870-022-03661-2

Friso G, Van Wijk KJ (2015) Posttranslational protein modifications in plant metabolism. Plant Physiol. https://doi.org/10.1104/pp.15.01378

Fu J, Zhu C, Wang C, Liu L, Shen Q, Xu D, Wang Q (2021) Maize transcription factor ZmEREB20 enhanced salt tolerance in transgenic Arabidopsis. Plant Physiol Biochem. https://doi.org/10.1016/j.plaphy.2020.12.027

Gadjev I, Vanderauwera S, Gechev TS, Laloi C, Minkov IN, Shulaev V, Apel K, Inzé D, Mittler R, Van Breusegem F (2006) Transcriptomic footprints disclose specificity of reactive oxygen species signaling in Arabidopsis. Plant Physiol. https://doi.org/10.1104/pp.106.078717

Giesguth M, Sahm A, Simon S, Dietz KJ (2015) Redox-dependent translocation of the heat shock transcription factor AtHSFA8 from the cytosol to the nucleus in Arabidopsis thaliana. FEBS Lett. https://doi.org/10.1016/j.febslet.2015.01.039

Gill SS, Tuteja N (2010) Reactive oxygen species and antioxidant machinery in abiotic stress tolerance in crop plants. Plant Physiol Biochem. https://doi.org/10.1016/j.plaphy.2010.08.016

Gill SS, Anjum NA, Hasanuzzaman M, Gill R, Trivedi DK, Ahmad I, Pereira E, Tuteja N (2013) Glutathione and glutathione reductase: a boon in disguise for plant abiotic stress defense operations. Plant Physiol Biochem 70:204–212. https://doi.org/10.1016/j.plaphy.2013.05.032

Gu X, Gao S, Li J, Song P, Zhang Q, Guo J, Wang X, Han X, Wang X, Zhu Y, Zhu Z (2021) The bHLH transcription factor regulated gene OsWIH2 is a positive regulator of drought tolerance in rice. Plant Physiol Biochem. https://doi.org/10.1016/j.plaphy.2021.11.031

Gulzar F, Fu J, Zhu C, Yan J, Li X, Meraj TA, Shen Q, Hassan B, Wang Q (2021) Maize WRKY transcription factor ZmWRKY79 positively regulates drought tolerance through elevating ABA biosynthesis. Int J Mol Sci. https://doi.org/10.3390/ijms221810080

Han D, Du M, Zhou Z, Wang S, Li T, Han J, Xu T, Yang G (2020) Overexpression of a malus baccata NAC transcription factor gene MBNAC25 increases cold and salinity tolerance in arabidopsis. Int J Mol Sci. https://doi.org/10.3390/ijms21041198

Hasanuzzaman M, Raihan MRH, Masud AAC, Rahman K, Nowroz F, Rahman M, Nahar K, Fujita M (2021) Regulation of reactive oxygen species and antioxidant defense in plants under salinity. Int J Mol Sci 22(17):9326. https://doi.org/10.3390/ijms22179326

He F, Niu MX, Feng CH, Li HG, Su Y, Su WL, Pang H, Yang Y, Yu X, Wang HL, Wang J, Liu C, Yin W, Xia X (2020) PeSTZ1 confers salt stress tolerance by scavenging the accumulation of ROS through regulating the expression of PeZAT12 and PeAPX2 in Populus. Tree Physiol. https://doi.org/10.1093/treephys/tpaa050

Hoang TV, Vo KTX, Rahman MM, Choi SH, Jeon JS (2019) Heat stress transcription factor OsSPL7 plays a critical role in reactive oxygen species balance and stress responses in rice. Plant Sci. https://doi.org/10.1016/j.plantsci.2019.110273

Hou X, Zhang H, Liu S, Wang X, Zhang Y, Meng Y, Luo D, Chen R (2020) The NAC transcription factor CaNAC064 is a regulator of cold stress tolerance in peppers. Plant Sci. https://doi.org/10.1016/j.plantsci.2019.110346

Huang J, Willems P, Wei B, Tian C, Ferreira RB, Bodra N, Martínez Gache SA, Wahni K, Liu K, Vertommen D, Gevaert K, Carroll KS, Van Montagu M, Yang J, Van Breusegem F, Messens J (2019) Mining for protein S-sulfenylation in Arabidopsis uncovers redox-sensitive sites. Proc Natl Acad Sci USA. https://doi.org/10.1073/pnas.1906768116

Jalmi SK, Sinha AK (2015) ROS mediated MAPK signaling in abiotic and biotic stress- striking similarities and differences. Front Plant Sci. https://doi.org/10.3389/fpls.2015.00769

Kang G, Yan D, Chen X, Yang L, Zeng R (2021) HbWRKY82, a novel IIc WRKY transcription factor from Hevea brasiliensis associated with abiotic stress tolerance and leaf senescence in Arabidopsis. Physiol Plant. https://doi.org/10.1111/ppl.13238

Kapilan R, Vaziri M, Zwiazek JJ (2018) Regulation of aquaporins in plants under stress. Biol Res 51:1–11. https://doi.org/10.1186/s40659-018-0152-0

Kärkönen A, Kuchitsu K (2015) Reactive oxygen species in cell wall metabolism and development in plants. Phytochemistry. https://doi.org/10.1016/j.phytochem.2014.09.016

Kimura S, Hunter K, Vaahtera L, Tran HC, Citterico M, Vaattovaara A, Rokka A, Stolze SC, Harzen A, Meißner L, Wilkens MMT, Hamann T, Toyota M, Nakagami H, Wrzaczek M (2020) CRK2 and C-terminal phosphorylation of NADPH oxidase RBOHD regulate reactive oxygen species production in arabidopsis. Plant Cell. https://doi.org/10.1105/tpc.19.00525

Li J, Luan Y, Liu Z (2015a) Overexpression of SpWRKY1 promotes resistance to Phytophthora nicotianae and tolerance to salt and drought stress in transgenic tobacco. Physiol Plant. https://doi.org/10.1111/ppl.12315

Li Z, Tian Y, Xu J, Fu X, Gao J, Wang B, Han H, Wang L, Peng R, Yao Q (2018) A tomato ERF transcription factor, SlERF84, confers enhanced tolerance to drought and salt stress but negatively regulates immunity against Pseudomonas syringae pv. tomato DC3000. Plant Physiol Biochem. https://doi.org/10.1016/j.plaphy.2018.08.022

Li M, Chen R, Jiang Q, Sun X, Zhang H, Hu Z (2015b) GmNAC06, a NAC domain transcription factor enhances salt stress tolerance in soybean. Plant Mol Biol. https://doi.org/10.1007/s11103-020-01091-y

Liebthal M, Schuetze J, Dreyer A, Mock HP, Dietz KJ (2020) Redox conformation-specific protein–protein interactions of the 2-cysteine peroxiredoxin in arabidopsis. Antioxidants. https://doi.org/10.3390/antiox9060515

Liu D, Li YY, Zhou ZC, Xiang X, Liu X, Wang J, Hu ZR, Xiang SP, Li W, Xiao QZ, Wang Y, Hu RS, Zhao Q (2021) Tobacco transcription factor bHLH123 improves salt tolerance by activating NADPH oxidase NtRbohE expression. Plant Physiol. https://doi.org/10.1093/PLPHYS/KIAB176

Liu P, Wu X, Gong B, Lü G, Li J, Gao H (2022) Review of the mechanisms by which transcription factors and exogenous substances regulate ROS metabolism under abiotic stress. Antioxidants. https://doi.org/10.3390/antiox11112106

Lu L, Qanmber G, Li J, Pu M, Chen G, Li S, Liu L, Qin W, Ma S, Wang Y, Chen Q, Liu Z (2021a) Identification and characterization of the ERF subfamily B3 group revealed GhERF13.12 improves salt tolerance in upland cotton. Front Plant Sci. https://doi.org/10.3389/fpls.2021.705883

Lu L, Qanmber G, Li J, Pu M, Chen G, Li S, Liu L, Qin W, Ma S, Wang Y et al (2021b) Identification and characterization of the ERF subfamily B3 group revealed GhERF13. 12 improves salt tolerance in upland cotton. Front Plant Sci 12:705883
Lv B, Wu Q, Wang A, Li Q, Dong Q, Yang J, Zhao H, Wang X, Chen H, Li C (2020) A WRKY transcription factor, FtWRKY46, from Tartary buckwheat improves salt tolerance in transgenic Arabidopsis thaliana. Plant Physiol Biochem. https://doi.org/10.1016/j.plaphy.2019.12.004
Ma Z, Hu L (2024) WRKY transcription factor responses and tolerance to abiotic stresses in plants. Int J Mol Sci 25. https://doi.org/10.3390/ijms25136845
Ma B, Liu X, Guo S, Xie X, Zhang J, Wang J, Zheng L, Wang Y (2021a) RtNAC100 involved in the regulation of ROS, Na+ accumulation and induced salt-related PCD through MeJA signal pathways in recretohalophyte Reaumuria trigyna. Plant Sci. https://doi.org/10.1016/j.plantsci.2021.110976
Ma J, Wang L, Dai J, Wang Y, Lin D (2021b) The NAC-type transcription factor CaNAC46 regulates the salt and drought tolerance of transgenic Arabidopsis thaliana. BMC Plant Biol. https://doi.org/10.1186/s12870-020-02764-y
Martí-Guillén JM, Pardo-Hernández M, Martínez-Lorente SE, Almagro L, Rivero RM (2022) Redox post-translational modifications and their interplay in plant abiotic stress tolerance. Front Plant Sci. https://doi.org/10.3389/fpls.2022.1027730
Matsushita M, Nakamura T, Moriizumi H, Miki H, Takekawa M (2020) Stress-responsive MTK1 SAPKKK serves as a redox sensor that mediates delayed and sustained activation of SAPKs by oxidative stress. Sci Adv. https://doi.org/10.1126/sciadv.aay9778
Melo BP, Lourenço-Tessutti IT, Fraga OT, Pinheiro LB, de Jesus Lins CB, Morgante CV, Engler JA, Reis PAB, Grossi-de-Sá MF, Fontes EPB (2021) Contrasting roles of GmNAC065 and GmNAC085 in natural senescence, plant development, multiple stresses and cell death responses. Sci Rep. https://doi.org/10.1038/s41598-021-90767-6
Mishra N, Jiang C, Chen L, Paul A, Chatterjee A, Shen G (2023) Achieving abiotic stress tolerance in plants through antioxidative defense mechanisms. Front Plant Sci 14:1110622. https://doi.org/10.3389/fpls.2023.1110622
Mishra S, Ganapathi TR, Pandey GK, Foyer CH, Srivastava AK (2024) Meta-analysis of antioxidant mutants reveals common alarm signals for shaping abiotic stress-induced transcriptome in plants. Antioxidants Redox Signal. https://doi.org/10.1089/ars.2023.0361
Mittler R, Zandalinas SI, Fichman Y, Van Breusegem F (2022) Reactive oxygen species signalling in plant stress responses. Nat Rev Mol Cell Biol. https://doi.org/10.1038/s41580-022-00499-2
Mizoi J, Shinozaki K, Yamaguchi-Shinozaki K (2012) AP2/ERF family transcription factors in plant abiotic stress responses. Biochim Biophys Acta - Gene Regul Mech. https://doi.org/10.1016/j.bbagrm.2011.08.004
Mohanta TK, Bashir T, Hashem A, Abd Allah EF, Khan AL, Al-Harrasi AS (2018) Early events in plant abiotic stress signaling: interplay between calcium, reactive oxygen species and phytohormones. J Plant Growth Regul. https://doi.org/10.1007/s00344-018-9833-8
Noctor G, Foyer CH (2016) Intracellular redox compartmentation and ROS-related communication in regulation and signaling. Plant Physiol. https://doi.org/10.1104/pp.16.00346
Ogasawara Y, Kaya H, Hiraoka G, Yumoto F, Kimura S, Kadota Y, Hishinuma H, Senzaki E, Yamagoe S, Nagata K, Nara M, Suzuki K, Tanokura M, Kuchitsu K (2008) Synergistic activation of the arabidopsis NADPH oxidase AtrbohD by Ca 2+ and phosphorylation. J Biol Chem. https://doi.org/10.1074/jbc.M708106200
Panda SK, Gupta D, Patel M, Vyver CVD, Koyama H (2024) Functionality of reactive oxygen species (ROS) in plants: toxicity and control in Poaceae crops exposed to abiotic stress. Plants 13(15):2071. https://doi.org/10.3390/plants13152071
Qian Y, Zhang T, Yu Y, Gou L, Yang J, Xu J, Pi E (2021) Regulatory mechanisms of bHLH transcription factors in plant adaptive responses to various abiotic stresses. Front Plant Sci. https://doi.org/10.3389/fpls.2021.677611
Qiu JR, Huang Z, Xiang XY, Xu WX, Wang JT, Chen J, Song L, Xiao Y, Li X, Ma J, Cai SZ, Sun LX, Jiang CZ (2020) MfbHLH38, a Myrothamnus flabellifolia bHLH transcription factor,

confers tolerance to drought and salinity stresses in Arabidopsis. BMC Plant Biol. https://doi.org/10.1186/s12870-020-02732-6

Ranjan A, Pandey N, Lakhwani D, Dubey NK, Pathre UV, Sawant SV (2012) Comparative transcriptomic analysis of roots of contrasting Gossypium herbaceum genotypes revealing adaptation to drought. BMC Genomics. https://doi.org/10.1186/1471-2164-13-680

Ranjan A, Sinha R, Sharma TR, Pattanayak A, Singh AK (2021) Alleviating aluminum toxicity in plants: implications of reactive oxygen species signaling and crosstalk with other signaling pathways. Physiol Plant. https://doi.org/10.1111/ppl.13382

Rivas FJM, Fernie AR, Aarabi F (2024) Roles and regulation of the RBOHD enzyme in initiating ROS-mediated systemic signaling during biotic and abiotic stress. Plant Stress. https://doi.org/10.1016/j.stress.2023.100327

Scarpeci TE, Zanor MI, Mueller-Roeber B, Valle EM (2013) Overexpression of AtWRKY30 enhances abiotic stress tolerance during early growth stages in Arabidopsis thaliana. Plant Mol Biol. https://doi.org/10.1007/s11103-013-0090-8

Singh A, Kumar A, Yadav S, Singh IK (2019) Reactive oxygen species-mediated signaling during abiotic stress. Plant Gene. https://doi.org/10.1016/j.plgene.2019.100173

Song Y, Yang W, Fan H, Zhang X, Sui N (2020) TaMYB86B encodes a R2R3-type MYB transcription factor and enhances salt tolerance in wheat. Plant Sci. https://doi.org/10.1016/j.plantsci.2020.110624

Srivastava R, Deng Y, Shah S, Rao AG, Howell SH (2013) Binding protein is a master regulator of the endoplasmic reticulum stress sensor/transducer bZIP28 in Arabidopsis. Plant Cell. https://doi.org/10.1105/tpc.113.110684

Sun J, Hu W, Zhou R, Wang L, Wang X, Wang Q, Feng Z, Li Y, Qiu D, He G, Yang G (2015) The Brachypodium distachyon BdWRKY36 gene confers tolerance to drought stress in transgenic tobacco plants. Plant Cell Rep. https://doi.org/10.1007/s00299-014-1684-6

Suzuki N, Bajad S, Shuman J, Shulaev V, Mittler R (2008) The transcriptional co-activator MBF1c is a key regulator of thermotolerance in Arabidopsis thaliana. J Biol Chem. https://doi.org/10.1074/jbc.M709187200

Suzuki N, Miller G, Morales J, Shulaev V, Torres MA, Mittler R (2011) Respiratory burst oxidases: the engines of ROS signaling. Curr Opin Plant Biol. https://doi.org/10.1016/j.pbi.2011.07.014

Thirumalaikumar VP, Devkar V, Mehterov N, Ali S, Ozgur R, Turkan I, Mueller-Roeber B, Balazadeh S (2018) NAC transcription factor JUNGBRUNNEN1 enhances drought tolerance in tomato. Plant Biotechnol J. https://doi.org/10.1111/pbi.12776

Tripathy BC, Oelmüller R (2012) Reactive oxygen species generation and signaling in plants. Plant Signal Behav 7(12):1621–1633. https://doi.org/10.4161/psb.22455

Vašková J, Kočan L, Vaško L, Perjési P (2023) Glutathione-related enzymes and proteins: a review. Molecules 28(3):1447. https://doi.org/10.3390/molecules28031447

Wang X, Liu S, Tian H, Wang S, Chen JG (2015) The small ethylene response factor ERF96 is involved in the regulation of the abscisic acid response in arabidopsis. Front Plant Sci. https://doi.org/10.3389/fpls.2015.01064

Wang CT, Ru JN, Liu YW, Li M, Zhao D, Yang JF, Fu JD, Xu ZS (2018a) Maize wrky transcription factor zmwrky106 confers drought and heat tolerance in transgenic plants. Int J Mol Sci. https://doi.org/10.3390/ijms19103046

Wang CT, Ru JN, Liu YW, Yang JF, Li M, Xu ZS, Fu JD (2018b) The maize WRKY transcription factor zmWRKY40 confers drought resistance in transgenic arabidopsis. Int J Mol Sci. https://doi.org/10.3390/ijms19092580

Wang J, Chen L, Long Y, Si W, Cheng B, Jiang H (2021a) A novel heat shock transcription factor (Zmhsf08) negatively regulates salt and drought stress responses in maize. Int J Mol Sci. https://doi.org/10.3390/ijms222111922

Wang Y, Wang S, Tian Y, Wang Q, Chen S, Li H, Ma C, Li H (2021b) Functional characterization of a sugar beet bvbhlh93 transcription factor in salt stress tolerance. Int J Mol Sci. https://doi.org/10.3390/ijms22073669

Wang D, Chen Q, Chen W, Liu X, Xia Y, Guo Q, Jing D, Liang G (2021c) A WRKY transcription factor, EjWRKY17, from eriobotrya japonica enhances drought tolerance in transgenic arabidopsis. Int J Mol Sci. https://doi.org/10.3390/ijms22115593

Wang D, Sun Z, Hu X, Xiong J, Hu L, Xu Y, Tang Y, Wu Y (2021d) The key regulator LcERF056 enhances salt tolerance by modulating reactive oxygen species-related genes in Lotus corniculatus. BMC Plant Biol. https://doi.org/10.1186/s12870-021-03336-4

Wang C, Zhou Y, Yang X, Zhang B, Xu F, Wang Y, Song C, Yi M, Ma N, Zhou X, He J (2022) The heat stress transcription factor LlHsfA4 enhanced basic thermotolerance through regulating ROS metabolism in Lilies (Lilium Longiflorum). Int J Mol Sci. https://doi.org/10.3390/ijms23010572

Wang P, Liu WC, Han C, Wang S, Bai MY, Song CP (2024) Reactive oxygen species: multidimensional regulators of plant adaptation to abiotic stress and development. J Integr Plant Biol. https://doi.org/10.1111/jipb.13601

Waszczak C, Akter S, Jacques S, Huang J, Messens J, Van Breusegem F (2015) Oxidative post-translational modifications of cysteine residues in plant signal transduction. J Exp Bot. https://doi.org/10.1093/jxb/erv084

Waszczak C, Carmody M, Kangasjärvi J (2018) Reactive oxygen species in plant signaling. Annu Rev Plant Biol 69:209–236

Wu J, Yu C, Hunag L, Wu M, Liu B, Liu Y, Song G, Liu D, Gan Y (2020) Overexpression of MADS-box transcription factor OsMADS25 enhances salt stress tolerance in Rice and Arabidopsis. Plant Growth Regul. https://doi.org/10.1007/s10725-019-00539-6

Xiang XY, Chen J, Xu WX, Qiu JR, Song L, Wang JT, Tang R, Chen D, Jiang CZ, Huang Z (2021) Dehydration-induced wrky transcriptional factor mfwrky70 of myrothamnus flabellifolia enhanced drought and salinity tolerance in arabidopsis. Biomolecules. https://doi.org/10.3390/biom11020327

Xu Y, Magwanga RO, Cai X, Zhou Z, Wang X, Wang Y, Zhang Z, Jin D, Guo X, Wei Y, Li Z, Wang K, Liu F (2019) Deep transcriptome analysis reveals reactive oxygen species (ROS) network evolution, response to abiotic stress, and regulation of fiber development in cotton. Int J Mol Sci. https://doi.org/10.3390/ijms20081863

Yang G, Yu L, Wang Y, Wang C, Gao C (2017) The translation initiation factor 1A (TheiF1A) from tamarix hispida is regulated by a dof transcription factor and increased abiotic stress tolerance. Front Plant Sci. https://doi.org/10.3389/fpls.2017.00513

Yang S, Xu K, Chen S, Li T, Xia H, Chen L, Liu H, Luo L (2019) A stress-responsive bZIP transcription factor OsbZIP62 improves drought and oxidative tolerance in rice. BMC Plant Biol. https://doi.org/10.1186/s12870-019-1872-1

Yang Z, Chi X, Guo F, Jin X, Luo H, Hawar A, Chen Y, Feng K, Wang B, Qi J, Yang Y, Sun B (2020) SbWRKY30 enhances the drought tolerance of plants and regulates a drought stress-responsive gene, SbRD19, in sorghum. J Plant Physiol. https://doi.org/10.1016/j.jplph.2020.153142

Yang K, Li CY, An JP, Wang DR, Wang X, Wang CK, You CX (2021) The C2H2-type zinc finger transcription factor MdZAT10 negatively regulates drought tolerance in apple. Plant Physiol Biochem. https://doi.org/10.1016/j.plaphy.2021.08.014

Yang J, Qiu L, Mei Q, Sun Y, Li N, Gong X, Ma F, Mao K (2023) MdHB7-like positively modulates apple salt tolerance by promoting autophagic activity and Na+ efflux. Plant J. https://doi.org/10.1111/tpj.16395

Yao Y, He RJ, Xie QL, Zhao X, Deng X, He J, Song L, He J, Marchant A, Chen X-Y et al (2017) ETHYLENE RESPONSE FACTOR 74 (ERF74) plays an essential role in controlling a respiratory burst oxidase homolog D (RbohD)-dependent mechanism in response to different stresses in Arabidopsis. New Phytol 213:1667–1681

Yin M, Wang Y, Zhang L, Li J, Quan W, Yang L, Wang Q, Chan Z (2017) The Arabidopsis Cys2/His2 zinc finger transcription factor ZAT18 is a positive regulator of plant tolerance to drought stress. J Exp Bot. https://doi.org/10.1093/jxb/erx157

Zhang X, Chen L, Shi Q, Ren Z (2020a) SlMYB102, an R2R3-type MYB gene, confers salt tolerance in transgenic tomato. Plant Sci. https://doi.org/10.1016/j.plantsci.2019.110356

Zhang HX, Zhu WC, Feng XH, Jin JH, Wei AM, Gong ZH (2020b) Transcription factor CaSBP12 negatively regulates salt stress tolerance in pepper (Capsicum annuum L.). Int J Mol Sci. https://doi.org/10.3390/ijms21020444

Zhang Y, Zhang Y, Sun Q, Lu S, Chai L, Ye J, Deng X (2021) Citrus transcription factor CsHB5 regulates abscisic acid biosynthetic genes and promotes senescence. Plant J. https://doi.org/10.1111/tpj.15431

Zhao Q, Xiang X, Liu D, Yang A, Wang Y (2018) Tobacco transcription factor NtbHLH123 confers tolerance to cold stress by regulating the NtCBF pathway and reactive oxygen species homeostasis. Front Plant Sci. https://doi.org/10.3389/fpls.2018.00381

Zhao Q, Fan Z, Qiu L, Che Q, Wang T, Li Y, Wang Y (2020a) MdbHLH130, an apple bHLH transcription factor, confers water stress resistance by regulating stomatal closure and ROS homeostasis in transgenic tobacco. Front Plant Sci. https://doi.org/10.3389/fpls.2020.543696

Zhao Q, Hu RS, Liu D, Liu X, Wang J, Xiang XH, Li YY (2020b) The AP2 transcription factor NtERF172 confers drought resistance by modifying NtCAT. Plant Biotechnol J. https://doi.org/10.1111/pbi.13419

Zhu H, Jiang Y, Guo Y, Huang J, Zhou M, Tang Y, Sui J, Wang J, Qiao L (2021) A novel salt inducible WRKY transcription factor gene, AhWRKY75, confers salt tolerance in transgenic peanut. Plant Physiol Biochem. https://doi.org/10.1016/j.plaphy.2021.01.014

# Mitigating ROS Levels Under Abiotic Stress: The Synergistic Role of PGPR and Plant Growth Regulators

13

Nitika Kapoor, Amandeep Singh, Manu Sharma, Tamanna Bhardwaj, Nahida Rehman Mir, Kamini Devi, Jaspreet Kour, Pooja Sharma, Aman Kumar, and Renu Bhardwaj

**Abstract**

Abiotic stresses such as salinity, drought, light intensity, heavy metals, and pesticides significantly affect global crop production and food security. These stresses can lead to the production of reactive oxygen species (ROS), including superoxide anion, hydrogen peroxide, and hydroxyl radicals, which damage cellular components like lipids, proteins, and DNA. However, ROS also play a role in triggering adaptive responses in plants. Abiotic stresses impact the rhizosphere, affecting root growth, nutrient uptake, and plant–microbe interactions, ultimately influencing plant health and productivity. Plants have developed antioxidant defense systems, including enzymes like superoxide dismutase (SOD), catalase (CAT), peroxidase (POD), and ascorbate peroxidase (APX), which help to protect against oxidative stress by maintaining cellular balance. Plant growth-promoting rhizobacteria (PGPRs) and plant growth regulators (PGRs) further support plants by producing stress-related proteins, regulating osmotic balance, and scavenging ROS. The presence of PGPRs in the rhizosphere enhances plant

N. Kapoor · A. Singh · M. Sharma · T. Bhardwaj · N. R. Mir · K. Devi · J. Kour
Department of Botanical and Environmental Sciences, Guru Nanak Dev University, Amritsar, Punjab, India

P. Sharma
D.A.V College Amritsar, Amritsar, Punjab, India

A. Kumar
College of Biotechnology, CCS HAU Hisar, Hisar, Haryana, India

R. Bhardwaj (✉)
Department of Botanical and Environmental Sciences, Guru Nanak Dev University, Amritsar, Punjab, India

University Centre for Research and Development, Chandigarh University, Mohali, Punjab, India



N. Iqbal et al. (eds.), *Plant Signaling Molecules in Regulation of ROS-Scavenging System*, https://doi.org/10.1007/978-981-95-3377-0_13

growth by improving nutrient availability and by boosting nutrient uptake. The present chapter explores the synergistic effects of PGPRs and PGRs,offering protection against various environmental stresses, thereby further promoting sustainable agriculture.

**Keywords**

Abiotic stress · Antioxidants · Heat stress · Pesticides · PGPR

## 13.1 Introduction

Abiotic stresses, such as drought, salinity, and high light intensity, lead to ROS production in plants, triggering oxidative stress (Yadav et al. 2020; Fadiji et al. 2023). ROS accumulation can cause cellular damage, but plants have evolved complex defense mechanisms to mitigate this damage (Zhang et al. 2023b). These mechanisms include enzymatic antioxidants like superoxide dismutase (SOD), catalase (CAT), and peroxidase (POD), ascorbate peroxide (APOX), glutathione reductase (GR), dehydroascorbate reductase (DHAR), and monodehydroascorbate reductase (MDHAR) as well as non-enzymatic antioxidants such as ascorbate, glutathione, and tocopherols (Munns and Millar 2023). Additionally, plants adapt to stress by adjusting stomatal closure, altering photosynthetic pathways, and producing osmoprotectants like proline and trehalose to stabilize proteins and cellular structures (Li et al. 2024). Under stress, plants also upregulate genes related to antioxidants, heat shock proteins (HSPs), and protective proteins, with transcription factors like *MYB* and *AP2/ERF* etc. (Li et al. 2024). Drought stress can stimulate deeper root growth, improving water and nutrient uptake. Furthermore, the synergistic use of plant growth-promoting rhizobacteria (PGPRs) and plant growth regulators (PGRs) enhances plant resilience by promoting stress tolerance, improving root development, and optimizing nutrient uptake (Gupta and Pandey 2023). PGPRs also benefit plants by improving soil conditions through mechanisms like phosphate solubilization and nitrogen fixation. These bacteria can enhance plant growth by producing hormones such as indole-3-acetic acid (IAA), cytokinins (CKs), and gibberellins (GAs), which stimulate root growth and nutrient uptake (de Andrade et al. 2023; Ranjan et al. 2024; Jan et al. 2024). PGPRs improve root architecture, increasing root surface area and length, which boosts the plant's ability to absorb water and nutrients and increases resilience to both abiotic and biotic stresses (Singh et al. 2024; Nazim and Bano 2024). Moreover, some PGPRs suppress plant pathogens through competitive interactions, antimicrobial production, and induction of systemic resistance (Madhumitha and Sundaram 2024). Overall, PGPRs and PGRs are essential for enhancing plant growth, yield, and resilience, offering promising solutions for sustainable agriculture (Kumar et al. 2023).

## 13.2 Abiotic Stress and Reactive Oxygen Species (ROS)

Plants naturally produce ROS such as superoxide radicals, hydrogen peroxide, and hydroxyl radicals during metabolism, particularly in chloroplasts and mitochondria (Mehrabi et al. 2024). These molecules can cause oxidative damage if levels are excessive. ROS production increases under abiotic stress (e.g., drought, salinity, temperature extremes), which stimulates antioxidant systems, including enzymes (SOD, CAT, POD, APOX, GR) and non-enzymatic molecules (ascorbic acid, glutathione), to mitigate damage (Karnwal et al. 2024). Stress responses also involve gene regulation by transcription factors. Plants adapt by adjusting metabolic pathways and enhancing ROS scavenging. Excessive ROS damages cellular structures, but plants manage stress by modifying growth and activating protective mechanisms (Fig. 13.1).

### 13.2.1 Drought Stress

Drought is among the most severe challenges to agricultural productivity that disrupts the plant's water balance at the cellular level, which in turn impacts the entire plant system (Nagarajan and Nagarajan 2010). It significantly alters the biochemistry, morphology, and physiology of plants. It also diminishes soil fertility, reduces crop yields, and negatively affects economic outcomes (Kour and Yadav 2022). Also, the exposure to drought stress triggers excessive production of ROS in plants. However, growing evidence suggests that stress-induced ROS may play a part in signaling within plants (Hasan et al. 2020). ROS can help plants make proline and

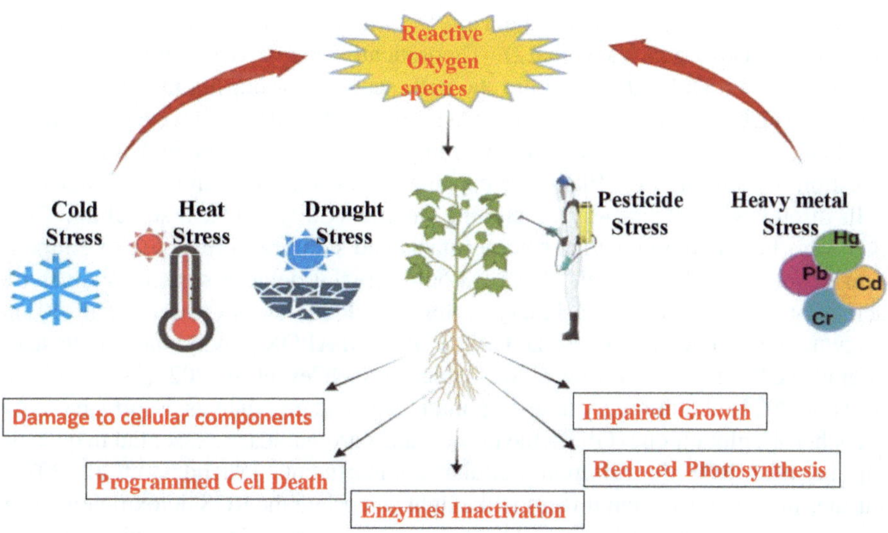


**Fig. 13.1** Effect of abiotic stress on physiological functions of plants

other important molecules when they are stressed, and certain types of ROS, like $H_2O_2$, can trigger signals that protect plant cells and help them grow by influencing gene activity (Hasan et al. 2021). For instance, in a study, Shi et al. (2020) demonstrate that *OsRbohB* supports rice plant growth and stress responses. When this gene is dysfunctional, the plants face greater challenges during drought due to reduced ROS production and increased water loss. Earlier research revealed that the wheat *bZIP* transcription factor gene *TaFD-Like2-1A (TaFDL2-1A)* can enhance drought tolerance in *Arabidopsis* and to investigate the function of *TaFDL2-1A* in wheat, and other researchers conducted research where they concluded that elevating *TaFDL2-1A* levels in wheat enhances the plant's drought resistance by increasing its sensitivity to the hormone ABA, which regulates water loss and stress responses. This also improves the plant's capacity to cope with harmful substances generated during drought (Wang et al. 2022). Similarly, Chun et al. (2021) demonstrated that the *CCoAOMT1* gene contributes to the plant's response to drought stress. Their study revealed that the *CCoAOMT1* mutant has more $H_2O_2$ in its leaves and lower expression levels, showing that *CCoAOMT1* helps plants respond better to drought by managing $H_2O_2$ and ABA signals. Likewise, another study to find out the effect of drought stress Meng et al. (2021) isolated a gene *CaDHN3* from pepper and verified the role of *CaDHN3* in salt and drought stress by overexpressing it (*CaDHN3*) in Arabidopsis and silencing in pepper using virus-induced gene silencing. This experiment indicated that *CaDHN3* interacts with *CaHIRD11* protein and helps plants manage stress by increasing certain gene expressions related to osmotic pressure through the ROS signaling pathway.

### 13.2.2 Thermal Stress

Plants have evolved intricate mechanisms to swiftly detect and respond to temperature stress (Ding and Yang 2022). Theoretically, plants can sense temperature changes at various locations within the cell, such as the plasma membrane or the nucleus, since temperature is a physical stimulus (Zhu 2016). However, the precise mechanisms by which plants perceive temperature signals remain a fundamental question in plant biology. Plants are capable of transmitting heat or cold signals to cells through several pathways, including calcium ($Ca^{2+}$) and ROS signaling. These pathways help maintain cellular homeostasis and stability in response to temperature fluctuations (Ding et al. 2020). Another crucial defense system involves the activation of antioxidant machinery to regulate ROS levels (Ansari et al. 2021). Enzymatic antioxidants like SOD, CAT, POD, and APOX play a vital role in scavenging ROS and maintaining redox balance (Sachdev et al. 2023; Shehzad and Mustafa 2023). Additionally, non-enzymatic antioxidants such as ascorbate (AsA), tocopherols, glutathione (GSH), flavonoids, and carotenoids are essential in quenching harmful ROS in different cellular compartments (Sachdev et al. 2023). Furthermore, environmental factors like light can prime the ROS detoxification process, enhancing the plant's thermotolerance by activating ROS-scavenging enzymes like APX2 (Han et al. 2019). Understanding these mechanisms is crucial for plants to survive and thrive under thermal stress conditions. Studies have been conducted from time to time to understand the effect of thermal stress on plants. Pea, being a cool season crop, is highly affected by heat. To improve the capability of pea in tolerating heat stress, heat shock factor from *Arabidopsis thaliana*, namely, *HsfA1d*, was incorporated. The study concluded that the *HsfA1d* from *Arabidopsis* was shown to increase heat stress resistance in pea plants by boosting antioxidant enzyme activity and reducing hydrogen peroxide levels compared to wild-type plants (Shah et al. 2020; Zhang et al. 2023a, b). Likewise, in another study, in sugar beet plant, high temperature (42 °C) increases MDA and $H_2O_2$ levels in stressed plants as compared to control plants (Aldaby et al. 2025). According to Kumar et al. (2025), heat stress decreases height, number of leaf, leaf area, shoot and root length, chlorophyll, and total soluble sugar content in stressed plant as compared to control sweet potato plant.

### 13.2.3 Heavy Metal Stress

Heavy metals are present in the earth's crust in natural conditions. Metals like Mo, Mn, Co, Fe, Cu, Se, and Zn are considered as essential metals as they are required in small concentrations in plants for their growth and metabolism. These metals are involved in redox reactions and also act as cofactors, which make these metals vital for metabolic activities (Ghori et al. 2019). However, when the concentration of metals exceeds that required by plants, it shows toxic effects. Non-essential metals like Cr, Cd, As, Pb, and Hg do not possess any metabolic role in plants and show toxicity even at low concentrations (Bharti et al. 2022; Edelstein and Ben-Hur 2018). Heavy metals are nondegradable and persist in agricultural soils. These metals are absorbed by the plants, which may or may not be required by the plants. Toxic metals alter the physiological, biochemical, and molecular processes in plants. Accretion of reactive oxygen species is the most significant toxic effects of heavy metals on plants (Noor et al. 2025). Increased production of ROS also enhances the phytotoxicity. As a result, plant growth and metabolism are negatively impacted, which leads to lesser yield of crop (Mansoor et al. 2022). Superoxide anion ($O_2^-$), hydrogen peroxide ($H_2O_2$), singlet oxygen ($^1O_2$), hydroxyl ion ($HO^-$), alkoxy ($RO^-$), peroxyl ($ROO^-$), and organic hydroperoxide (ROOH) are the various forms of ROS that are present in plants (Wakeel et al. 2020). Accumulation of superoxide anion and hydrogen peroxide has been observed in *Brassica juncea* grown under Cr stress (Sharma et al. 2022). Enhanced ROS production led to the generation of oxidative stress in *Morus alba* in the presence of Cd and Pb. Ni-induced oxidative damage was evidenced by higher levels of stress markers in rice seedlings (Rizwan et al. 2024). A significant increase in electrolyte leakage (EL), $H_2O_2$, and MDA content was observed in Cd-stressed *Sorghum bicolor* (Hassan et al. 2020). Also, elevated levels of $H_2O_2$ and MDA were observed in maize plants under arsenic toxicity (Rahman et al. 2023).

### 13.2.4 Salinity Stress

Upon exposure to salinity, there is an increase in concentration of $Na^+$ and $Cl^-$ ions in soil at higher rate, which reduces the essential ions in plants and causes disruption of plant–water relation generating drought-like conditions resulting in osmotic stress, which is further responsible for the generation of ROS in plants. Salinity stress also hampers the electron transport to oxygen reduction pathways in various organelles, which results in over production of ROS in plants (Hasanuzzaman et al. 2021). The effect of salt stress on *Oenanthe javanica* cultivars was observed. Kumar et al. (2021) documented that tolerant cultivar (V11E0022) exhibited less MDA and $H_2O_2$ levels as compared to sensitive cultivar (V11E0135). Siddiqui et al. (2020) documented an increase in MDA, $H_2O_2$, $O_2^{\bullet-}$ levels in salt-stressed *S. lycopersicum* seedlings. The degree of damage due to salinity-induced oxidative stress is observed to vary in different regions of plants. Salinity alters root morphology and trait (Arif et al. 2019). Furthermore, Jabeen et al. (2021) detected a higher concentration of MDA content in the roots than in the leaves in *Glycine max* plants exposed to 100 mM NaCl.

### 13.2.5 Pesticide Stress

Extensive application of pesticides increases the risk of oxidative burst i.e., overproduction of ROS in plants (Zhang et al. 2024). Application of chlorpyrifos to *B. juncea* seedlings resulted in increased levels of MDA, $H_2O_2$, and $O_2^{\bullet-}$ (Bakshi et al. 2021). It is also documented that the application of diazinon, mancozeb, and imidacloprid to *S. lycopersicum* enhanced the MDA content (Hatamleh et al. 2022). However, a decrease in $H_2O_2$ and MDA content was observed in *S. lycopersicum* a day after treatment with thiram. While a rise in $H_2O_2$ and MDA content was seen after 5th and 11th day after treatment (Yüzbaşıoğlu and Dalyan 2019). It was documented that the *Phaseolus leptostachyus* treated with endosulfan showed elevated levels of $H_2O_2$ and MDA (Landa-Faz et al. 2021).

## 13.3 Potential Role of PGPR and PGRs in Growth and Development of Plants

### 13.3.1 Plant Growth-Promoting Rhizobacteria (PGPR)

Biostimulants are a group of different chemicals or microorganisms that increase plant productivity through organic processes (Hasan et al. 2024). PGPR are free-living bacteria that invade plant roots and stimulate plant development. They are classified as biostimulants and further recognized as potential biofertilizers, biopesticides, and phyto-stimulators (Maitra et al. 2021). PGPR promotes plant growth by solubilizing phosphates, fixing nitrogen, or by producing phytohormones. It has been reported that PGPR boosts plant metabolism directly by increasing uptake of water, minerals, improving root architecture, etc. (de Andrade et al. 2023). Additionally, these beneficial microorganisms curb the growth of plant pathogens (Díaz-Rodríguez et al. 2024). Apart from these benefits, PGPR plays a crucial role in improving plant fitness against abiotic stressors (Bhat et al. 2023). Considering all the enlisted benefits, the application of PGPR can be approved as a sustainable option for long-term farming practices. This will certainly reduce the dependency on chemical pesticides and fertilizers, which have negative impacts on human health and the environment (Sinha et al. 2024). Furthermore, the text reveals the PGPR-mediated benefits and their interactions with plant system (Fig. 13.2).

#### 13.3.1.1 PGPR-Mediated Nitrogen Fixation

Nitrogen (N) is an essential and primary element for efficient crop quality and yield. N input drastically regulates grain productivity (Govindasamy et al. 2023). Crops such as rice, cotton, maize, potatoes, and wheat benefit from uplifted application of nitrogen fertilizers (Khakbazan et al. 2023). However, the reservoir of nitrogen is lost due to some processes including leaching and ammonia volatilization. Therefore, nitrogen is the most essential macronutrient in plant physiology including formation of chlorophyll, amino acids, adenosine triphosphate (ATP), and nucleic acids (Wickramasinghe et al. 2021). It has been reported that application of nitrogen-fixing PGPR strains reduces the consumption of chemical fertilizer by 25%.

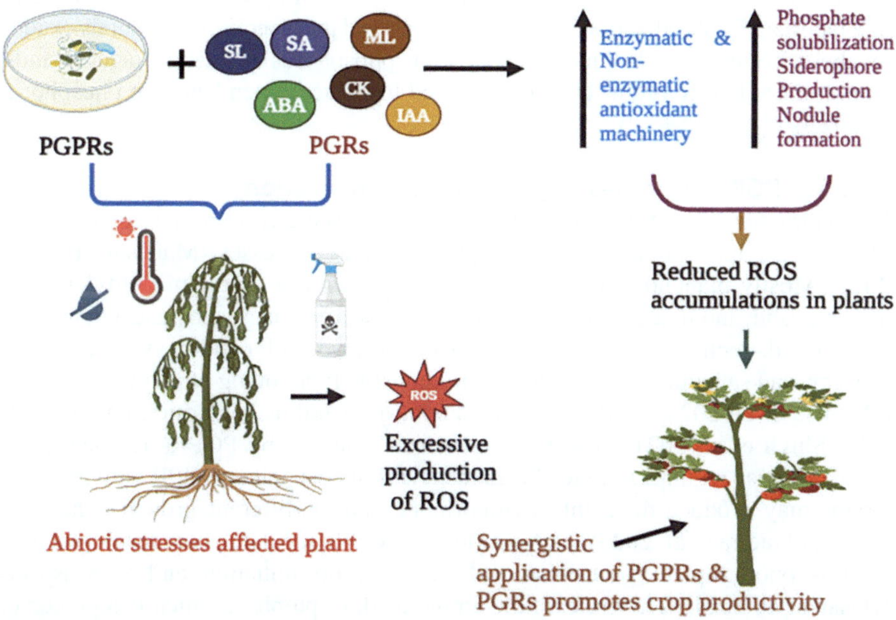


**Fig. 13.2** Represented that plants grown under abiotic stresses condition and synergistic application of PGPRs, and PGRs promotes enzymatic and non-enzymatic antioxidant machinery, nitrogen fixation, mineral phosphate solubilization, and siderophore production, which increase total plant growth and development and further enhance crop productivity

Tapia-García et al. (2020) conducted an experiment and isolated the most versatile nitrogen-fixing endophyte, *Burkholderia,* which enhanced the overall production of maize by densely colonizing host tissues. Likewise, another group experimented with the application of *Azotobacter chroococcum* and *Azospirillum brasilense* strains to local maize variety. They found that maize productivity was enhanced by 1–1.5 fold (Rajeshkumar 2023). When nitrogen-fixing PGPR is sprayed onto crops and crop fields, it promotes growth, fights disease, and maintains a steady nitrogen level in the soil for farming (Hasan et al. 2024).

#### 13.3.1.2 PGPR-Mediated Mineral Phosphate Solubilization

One essential macronutrient needed for plant growth and metabolism is phosphorus (P). However, P is immediately bonded to mineral surfaces or immobilized by metal cations (such Ca, Fe, and Al) when applied to soil, limiting the amount of P available for plant absorption (Ibrahim et al. 2022). Phosphorous participates in vital processes including root and stem development, photosynthesis, plant disease resistance, formation of flower and seed (Fathi and Afra 2023). A study conducted by Wang et al. (2020) and result demonstrated that bacterial genera including *Acinetobacter, Pseudomonas, Massilia, Bacillus, Arthrobacter, Stenotrophomonas, Ochrobactrum,* and *Cupriavidus* solubilize phosphorous. It was found that *Acinetobacter* is a promising candidate in enhancing soil fertility and exhibited a remarkable ability to solubilize P (Wan et al. 2020). Later Liu et al. 2020 reported that P solubilizing bacteria have the capability to form light-density organic acids that facilitate P solubilization. Also, it was observed that specific compounds exuded by these bacteria aid in unlocking existing phosphorus reservoirs in croplands. Hence, these PGPR strains can be employed to reduce dependency on phosphorus fertilizers.

#### 13.3.1.3 PGPR Promotes Phytohormone Production

Phytohormones are defined as organic substances secreted in minute quantities in plant's body that influences specific physiological processes (Maheshwari et al. 2015). Mostly, plant hormones such as auxins, gibberellins, cytokinin, ethylene, and abscisic acid, modulate plant development. However, many substances have also been found, such as brassinosteroids, jasmonic acid, polypeptide systemin, plant steroids, and salicylic acid, which also promote the functioning of plant metabolism (Oyedele et al. 2024). PGPR secretes auxin, gibberellins, cytokinin, and abscisic acid (Singh et al. 2021). The phytohormones released by PGPRs regulate plant health and also regulate secreted hormone levels of host plants. PGPR with the same strains may produce different phytohormones under different growth conditions (Pantoja-Guerra et al. 2023). IAA-producing PGPRs regulate plant development at the tips, phototropism, geotropism, cell division, root initiation, and other aspects (Hasan et al. 2024). They release the amino acid, tryptophan, which is a precursor molecule in bacteria for the production of IAA (Tang et al. 2023). It was found that IAA-producing *Mycobacterium, Azospirillum, Azotobacter, Cellulomonas, Mycoplasma,* and *Rahnella* species were reported in wheat rhizosphere. Furthermore, plant cytokinins inhibit the growth of the main root while promoting cell division, lateral root development, and root hair growth. Certain PGPRs, including *Bacillus cereus*, *Bacillus megaterium*, *Bacillus subtilis*, *Escherichia coli*, *Halomonas desiderata*, and *P. fluorescens* G20–18, produce cytokinins (Ahmad et al. 2023). Notably, *Bradyrhizobium japonicum* E109 also synthesizes several phytohormones, including IAA, GA, ethylene, ABA, and zeatin, as well as small amounts of jasmonic and salicylic acids (Ahmad et al. 2023). These PGPRs positively influence plant growth, suppress disease-causing microorganisms, and help to maintain soil nutrient levels (Gómez-Godínez et al. 2023).

### 13.3.2 Plant Growth Regulators (PGRs)

Plant growth regulators (PGRs) play crucial role in regulating various physiological processes that promote plant growth and development and make plant resilience toward environmental stresses (Suman et al. 2017; Zhumanova et al. 2024). PGRs influence different aspects including root development, flowering and fruit development, senescence and abscission, tissue growth and differentiation, cell division and elongation, nutrient mobilization and utilization, and stress response (Khan et al. 2020). Likewise, application of auxin promotes cell elongation and division, especially in stems and roots by influencing the direction of growth by stimulating cell elongation on one side of the plant, leading to phototropism (growth toward light) and gravitropism (growth in response to gravity). Auxin also enhanced root formation and development, encouraging adventitious roots and a stronger root system (Liu et al. 2024). On the other hand, cytokinin work in conjunction with auxins to balance root and shoot growth, often promoting lateral root development, whereas gibberellins stimulate flowering in some plants and are involved in fruit development, assisting in the elongation of fruit and seed development. Abscisic Acid (ABA) helps plants respond to stress conditions like drought by closing stomata to reduce water loss. It also plays a role in seed dormancy and germination control (Brookbank et al. 2021). A multitude of research studies indicated the effect of PGRs on the growth and development of plants under abiotic stress condition. A research experiment conducted on tomato plants showed that exogenous application of 1-naphthylacetic acid (NAA) significantly increased lateral root development of plant grown under the cadmium (Cd) stress and also reduced $Cd^{2+}$ accumulation in root cell (Liu et al. 2024). Likewise, in another study conducted on rice plants, application of salicylic acids (SA) promotes antioxidant enzyme activities, proline content, gene expression, and increased overall growth of plant grown under drought and salt stresses (Shan et al. 2024). Supplementation of Melatonin (Mel) reduced ROS accumulation and increased antioxidant enzymes activities in Indian mustard plants grown under chromium stress (Kour et al. 2024). Understanding the role of plant growth regulators allows for more precise agricultural and horticultural practices, ultimately enhancing plant health, productivity, and resilience to environmental stressors (Table 13.1).

**Table 13.1** Synergistic effect of PGPR and PGR under abiotic stress in plants

| Sr. No | Type of Abiotic stress | Plant species | PGPR | PGR | Synergistic effects | References |
|---|---|---|---|---|---|---|
| 1. | Drought | *Brassica juncea* | *Pseudomonas fluorescens* | Methyl jasmonate | Improvement in plant growth including root and shoot biomass, and photosynthetic pigments was observed upregulated antioxidant activity confirms reduction in drought-induced oxidative stress | Khan et al. (2024) |
| 2. | Drought | *Brassica Rapa* subsp. *pekinensis* | *Bacillus butanolivorans KJ40* | Salicylic acid | Improvement in plant growth factors like fresh weight was observed. Enhanced antioxidant enzyme activity like CAT and POD and reduced lipid peroxidation and MDA content further ameliorated oxidative stress | Kim and Sang (2024) |
| 3. | Drought | *Glycine max L.* | *Bacillus* sp. *strain IPR-4* | Melatonin | Plant growth parameters like height, biomass, chlorophyll content, and activity of antioxidant enzymes like APX, SOD, and GR were enhanced. In a while, MDA and $H_2O_2$ content decreased | Peter et al. (2024) |

| | | | | | | |
|---|---|---|---|---|---|---|
| 4. | Cd | *Brassica juncea* L. | *Pseudomonas putida* and *Pseudomonas fluorescens* | Melatonin | Enhanced photosynthetic pigment and secondary metabolite content and increased expression of photosynthetic genes (*psb A, psb B, CHS, PAL,* and *PSY*) | Bhardwaj et al. (2024) |
| 5. | | *Brassica oleracea* | *Klebsiella* sp. (SS7) and *Klebsiella pneumoniae* (SS8) | Jasmonic acid | Chlorophyll content, fresh and dry biomass of stem, root length, fresh and dry weight of roots, and activity of antioxidants enzymes including SOD, POX, CAT, and POD was observed to be elevated after the synergistic application | Shahid et al. (2024) |

(continued)

**Table 13.1** (continued)

| Sr. No | Type of Abiotic stress | Plant species | PGPR | PGR | Synergistic effects | References |
|---|---|---|---|---|---|---|
| 6. | Cold | *Lycopersicon esculentum* | *Bacillus cereus* (ATCC 14579) and *Pseudomonas putida* (KX574857) | Salicylic acid | An increase in plant height, root fresh and dry weight, total dissolved solids, chlorophyll and carotenoid content, flavonoid content, and antioxidant enzyme activity, and reduction in electrolyte leakage and regulation of proline production was observed | Nazim and Bano (2024) |
| 7. | Salinity | *Triticum aestivum* | *Bacillus velezensis* UTB96 | GR24 (Strigolactone) | Reduction in electrolyte leakage, $H_2O_2$ MDA content in leaves, salt content in both root and leaves and, oxidative stress was observed by enhancing antioxidants like CAT, APX, and POX activity, chlorophyll content, relative water content, and plant growth attributes | Mehrabi et al. (2024) |

| | | | | | | |
|---|---|---|---|---|---|---|
| 8. | Cd | *Spinacia oleracea* | *Bacillus licheniformis* | Melatonin | An increase in plant height, root length, dry and fresh weight, chlorophyll, and activity of antioxidant enzymes (SOD, CAT, POD) was observed after combined application | Asif et al. (2023) |
| 9. | Drought | *Brassica napus* | *Pseudomonas putida* | Salicylic acid | Improved germination parameters, shoot and root length, leaf area, relative water content, membrane stability index, proline, flavonoids, phenol, and antioxidants activity (SOD, GSH, and APX) | Tanveer et al. (2023) |
| 10. | As and cd | *Origanum vulgare L.* | *Azotobacter chroococcum*, *Azospirillium brasilense*, and *Pseudomonas fluorescens* | Methyl Jasmonate | All the physiological and biochemical attributes were positively affected, and DPPH radical scavenging ability and *C4H* and *RAS* gene expression were found to be increased | Jafari et al. (2023) |

(continued)

**Table 13.1** (continued)

| Sr. No | Type of Abiotic stress | Plant species | PGPR | PGR | Synergistic effects | References |
|---|---|---|---|---|---|---|
| 11. | Pesticide | *Brassica juncea* L. | *Pseudomonas putida* | Melatonin | Increase in the content of organic acids like malic acid, succinic acid, and fumaric acid which provide defense against stress | Jan et al. (2022) |
| 12. | Drought | *Cicer arietinum L.* | *Bacillus subtilis*, *Bacillus thuringiensis* and *bacillus megaterium* | Salicylic acid and putrescine | The amelioration of oxidative stress was achieved by maintaining osmoregulation, and decreasing lipid peroxidation. The plant showed overall improved growth | Khan et al. 2018a, c) |
| 13. | Drought | *Helianthus annus* | *Planomicrobium chinense strain* P1 and *Bacillus cereus strain* P2 | Salicylic acid | Enhanced accumulation of cu, co, and Zn, increased growth in the form of root and shoot length, fresh and dry weight of root and increased content of Chl, carotenoid, sugar and phenolics. Reduced proline and MDA content was also observed | Khan et al. 2018b |

## 13.4 Synergistic Role of PGPRs and PGRs in Regulation of Abiotic Stress-Induced Oxidative Stress in Plants

### 13.4.1 Drought Stress

Though plants face several stresses in their life span due to their sessile nature, drought stress is considered a very important abiotic stress. It has been observed to decrease the growth of plants, yield, and distribution (Dai 2013). For proper growth and development, plants require both nutrition and water throughout their life. However, drought can be responsible for causing nutrient deficiency in soil and thereby decrease the movement and absorption of nutrients to the roots from soil (Silva et al. 2009, 2011). In addition to this, drought is also responsible for causing oxidative stress and also excess production of ROS molecules. Soil bacteria are responsible for the maintenance of mutual interactions with the roots of plants that is responsible for enhanced growth of plants. In addition, it is also beneficial for plant tolerance against various abiotic stresses, including drought. This could be achieved through enhanced levels of cellular metabolites and also influence the plants microbiome (Saharan and Nehra 2011). Soil bacteria are also helpful in the downregulation and upregulation of different genes that are associated with the plant metabolites under stress. This further alters the expression of metabolic genes and provides tolerance to plants under drought stress (Sziderics et al. 2007). The application of PGPRs and plant growth regulators (PGRs) is also combat with abiotic stresses in plants (Jiménez 2005).

In wheat plant, when drought-stressed plants were treated with *Bacillus megaterium,* oxidative stress was declined by improving the activity of antioxidative enzymes, declining the lipid peroxidation and electrolyte leakage (Rashid et al. 2022). In another study by He et al. (2021), the activity of SOD, CAT, and POD was enhanced along with declining the MDA content and the accumulation of $H_2O_2$ in plants of *Lolium perenne* under drought stress with the application of *Bacillus* sp. WM13-24 and *Pseudomonas* sp. M30-35. Moreover in reports by Azizi et al. (2021), application of *Pseudomonas fluorescence* and *Pseudomonas putida* in plants of *Myrtus communis* L. reduced the oxidative stress by improving the activity of both enzymatic and non-enzymatic antioxidants.

According to the reports of Aslam et al. 2021, it has been observed that in Canola plants grown under drought stress, the activity of enzymatic antioxidants (SOD, POD, PPO and CAT) was improved due to the application of plant growth regulators (IAA, gibberellic acid) both individually and in combination with NPK fertilizers, thereby resulting in decrease of oxidative stress. It has been observed that in chickpea and wheat plant, individual application of both PGRs and PGPRs causes enhancement in the growth of plants; however, their combined application was more beneficial in ameliorating the drought stress in these plants (Khan et al. 2017, 2018a). According to the reports of Khan et al. 2018b, when PGPRs and PGRs were applied to drought-stressed plants, chlorophyll content was increased, new proteins were induced along with the amelioration of oxidative stress. According to the observations from the study, the combined application of PGPRs and PGRs was responsible for reducing lipid peroxidation more efficiently and increasing the leaf area of the plants. Moreover, the combined application of these PGPRs and PGRs was also responsible for improving the relative water content in leaves of chickpea under drought stress.

Further, in the study conducted by Khan et al., 2019a, it was observed that combined application of PGPRs (*Bacillus thuringiensis, Bacillus megaterium,* and *Bacillus subtilis*) along with PGRs (salicylic acid) was responsible for reducing the oxidative stress through the improvement in antioxidative defense system, protein content, and reducing lipid peroxidation in chickpea plants. In another study conducted by Batool et al. 2020, it was observed that due to application of *Bacillus subtilis* HAS31 on plants of *Solanum tuberosum* L. grown under drought stress, amelioration of oxidative stress occurred. This was achieved by increasing the activity of enzymatic antioxidants including POD, CAT, and SOD in plants, which were treated with PGPRs as compared to stressed plants.

### 13.4.2 Thermal Stress

Thermal stress is considered as an emerging threat for growth and yield of plants, specifically in the tropical and sub-tropical areas around the world. It has been observed that due to global warming, crops have been exposed to elevated temperature conditions, thereby resulting in the decline in crop yield (Kumari et al. 2022). Different plants show different patterns of growth under temperature stress depending upon the variety of crop, duration of exposure, and extremity of stress. Due to extreme temperature, alteration in the protein content and stability of membrane has been resulted. Moreover, in plants under high temperature, death of the plant cells has been observed, thereby resulting in cellular destructions (Ahuja et al. 2010).

It has been observed that the growth and development of plants are dependent on photoperiod and temperature. According to reports of Yin et al. (1996), it has been observed that rice plants show maximum growth at 27–32 °C. Plants have been observed to minimize the oxidative damage caused in plants under heat stress. This could be achieved by the activation of antioxidative enzymes as CAT, SOD, POD, and APX, which are responsible for scavenging the ROS generated by oxidative stress (Howarth 2005). In the study conducted by Al-Zahrani et al. (2022), it has been observed that certain plant growth regulators such as BRs, methyl jasmonate, ascorbic acid, and antioxidant like α tocopherol have been associated with providing tolerance and protection to plants against oxidative damage generated by heat stress (Mohammed and Tarpley 2011). Ascorbic acid, which is a antioxidative metabolite, has been observed to play an important role in protecting plants against various stresses, including heat stress. It is also involved in reducing the production of ROS by moderating glutathione in ascorbate–glutathione cycle (Guo et al. 2005; Noctor et al. 2002).

In addition to PGRs, microbes are also very beneficial in protecting plants against stresses including heat stress. It has been observed that PGPRs help in both plant growth promotion and reducing stress (Tanaka and Van Houtan 2022). It has been reported that *Bacillus safensis* ameliorate the heat stress in plants by regulating the production of ROS and activities of antioxidative enzymes. Furthermore, these PGPRs reduce the damage to chloroplast and cell membrane structure, thereby improving resilience of plants to heat stress (Sarkar et al. 2018). In another study by Meena et al. (2017), it was reported that due to application of *Septoglomus deserticola* and *Septoglomus constrictum* in tomato plants under heat and drought stress, oxidative stress was reduced.

### 13.4.3 Salinity Stress

Salinity stress is a significant environmental factor that can severely affect plant growth and development. High salt concentrations in soil can lead to a range of physiological and biochemical responses in plants. There are various means by which salinity stress impacts plants including reduced water uptake, reduced turgor pressure, sodium adsorption, potassium deficiency, nutrient uptake inhibition, damage chlorophyll content, growth retardation, and enhanced production of ROS. Further, high salt levels create a water potential gradient, making it difficult for roots to absorb water. This can lead to symptoms of drought even when water is available (Sarma and Deka 2024).

High sodium levels can disrupt enzyme activity and interfere with nutrient uptake. Moreover, high salinity can hinder potassium ($K^+$) absorption, leading to deficiencies that affect protein synthesis and overall plant health. Salinity can interfere with the uptake of other essential nutrients, leading to imbalances that affect plant health and productivity. According to Tolay (2021) in Basil plant, 1.5% NaCl concentration reduced nutrient uptake (K, Na/k ratio, Zn, Cu and Mn contents) in plants. Salinity stress could diminish chlorophyll content and photosynthetic efficiency, reducing energy availability for growth. Overall growth may be stunted, with effects on both root and shoot development. Salinity stress leads to oxidative stress, prompting plants to activate antioxidant defense mechanisms. High salinity influences flowering time, fruit set, and seed quality, impacting crop yields (Hasana and Miyake 2017).

The plant rhizosphere holds a diversity of living biota in which PGPRs hold a significant position. By colonizing the rhizosphere and adopting a wide array of tolerance mechanisms, PGPR plays a crucial role in enhancing crop resilience to abiotic stresses including salinity stress (Sarma and Deka 2024). These mechanisms include the synthesis of hormones, iron-cheating molecules including siderophores, organic acids, and stress-induced metabolites (Subiramani et al. 2020). PGRs are plant signaling molecules that also modulate essential elements of growth, development, and reactions to environmental stress (Kaya et al. 2023). Phytohormones produced by PGPR also modulate root development, promoting lateral root proliferation and enhanced nutrient uptake. Many studies have reported the synergistic role of PGPR and PGR in mitigating salinity stress in plants by regulating ROS induced oxidative stress. The combination of *Pseudomonas fluorescens* and melatonin enhances antioxidant enzyme activity, proline content, and flavonoid production in mustard, leading to improved salt tolerance (Khan et al. 2023). Similarly, melatonin and putrescine interact to increase plant growth attributes like fresh and dry weight of shoot, photosynthetic pigments including chlorophyll (Chl a), Chl a + b, and carotenoids, proline, total soluble sugars, and cell membrane stability index (CMSI) in snap bean (*Phaseolus vulgaris* L.) seedlings under salt stress. Reduction in salt content, lipid peroxidation indicated by MDA content, methylglyoxal, and hydrogen peroxide ($H_2O_2$) were also reported, which reduced salinity-induced oxidative stress and inhibit polyamine catabolism (El-Beltagi et al. 2023). In rice *(Oryza sativa*), the inoculation of salt-tolerant bacteria (*Brevibacterium frigoritolerans* and *Bacillus safensis*) along with melatonin application improves growth in the form of plant height, biomass accumulation, chlorophyll content, antioxidant enzyme activity, and gene expression (*SOD*1, *CAT*a, and *PAL*1) related to salt stress tolerance (Gupta et al. 2023). In maize (*Zea mays*), combined PGPR and salicylic acid (SA) application significantly improved growth parameters (root and shoot dry weight), grain yield, and physiological attributes (relative water content, Chl a and Chl b) under high salinity conditions (Ali et al. 2023). In chickpea (*Cicer arietinum* L.), the combination of PGPR (*Rhizobium* sp. and *Azotobacter* sp.) and SA resulted in improved root growth, nodulation (nodules and nodular mass), grain yield, and increased uptake of nutrients in the form of nitrogen and phosphorus under salt stress (Riaz et al. 2019). Similar results were also obtained in wheat, where co-application of PGPR and brassinosteroid improved plant growth, enhanced photosynthetic pigment content, regulated antioxidant system, and improved osmolyte accumulation, leading to reduced oxidative damage (Mehrabi et al. 2024).

### 13.4.4 Heavy Metal Stress

The impact of heavy metals on microbial phytohormone production varies, with both positive and negative effects observed. Carlos et al. (2016) found that As, Pb, and Cu enhanced IAA production in certain bacterial strains, while Mn, Cd, and Ni generally inhibited it. Notably, *Serratia* K120 showed a positive correlation between IAA production and ACC deaminase activity under heavy metal stress, promoting *Helianthus annuus* growth and suggesting potential for phytoremediation. Under stressful conditions, excessive production of ROS disrupts redox balance, damages DNA, proteins, and membrane fluidity, ultimately leading to cellular death. Numerous species of growth-enhancing rhizobacteria have been shown to withstand oxidative stress through the action of antioxidative enzymes. Seneviratne et al. (2016) reported that Pb and Ni exposure reduced both growth and IAA production in *Bradyrhizobium japonicum*, while Cu stress slightly increased IAA production. Another study indicated that phytohormone-producing bacteria exhibited enhanced heavy metal accumulation and growth in contaminated media (Hrynkiewicz et al. 2015).

The combined effect of PGPR and PGRs is crucial in managing oxidative stress in plants exposed to heavy metal stress. Plants treated with PGPR demonstrate notably enhanced biomass and chlorophyll levels under stress conditions, accompanied by improved water potential and reduced electrolytic leakage. PGPR application alleviates oxidative stress by decreasing the activities of catalase, peroxidase, polyphenol oxidase, and total polyphenol in comparison to untreated plants (Kang et al. 2014). Notably, PGPR treatment influences the concentrations of plant hormones, such as abscisic acid (ABA), gibberellic acid (GA), and indole-3-acetic acid (IAA), which are essential for stress response (Kang et al. 2014; Khan and Bano 2019). The synergistic action of PGPR and PGRs in regulating oxidative stress encompasses various mechanisms. PGPR boosts plant growth, enhances nutrient absorption, and regulates antioxidant enzyme activities (Subiramani et al. 2020). Concurrently, they affect the synthesis and equilibrium of plant hormones, which are vital for stress signaling and response. This integrated action of PGPR and PGRs assists plants in better managing heavy metal stress by minimizing oxidative damage and fostering overall plant vigor and development (Khan and Bano 2019).

Under heavy metal stress and nitrogen-poor conditions, endophytic bacteria can produce nitrogenase enzymes, supplying ample nitrogen to their host plants. A study by Doty et al. (2009) identified endophytic genera (*Acinetobacter, Burkholderia, Rahnella*, and *Sphingomonas*) in *Salix sitchensis* and *Populus trichocarpa* that could synthesize nitrogenase enzymes and fix atmospheric nitrogen (Santos et al. 2014). Additionally, rhizobacteria's production of citric, gluconic, and oxalic acids plays a crucial role in heavy metal mobilization, thereby shielding plant roots from the harmful effects of heavy metals (Rosier et al. 2018; Hassan et al. 2019). In a study by Paredes-Páliz et al. (2018), rhizobacteria capable of forming biofilms were chosen based on their metal tolerance and introduced to *Spartina densiflora*. The plants were cultivated for 4 months before being harvested. Using the frozen plant parts, researchers conducted enzyme assays and analyzed gene expression. The findings revealed enhanced activity of SOD, CAT, and APOX, along with a doubled concentration of thiobarbituric acid reactive substances (TBARs), leading to cellular and membrane damage. Despite this, the oxidative stress index (OSI) showed a significant reduction (>50%) in inoculated plants. Studies have shown that plants inoculated with rhizobacteria exhibit alterations in gene expression related to ethylene biosynthesis when subjected to abiotic stresses (Parmar and Dufresne 2011; Mhlongo et al. 2018; Igiehon and Babalola 2018). While ethylene (ET) is essential for normal plant development and fruit maturation, its production increases significantly under stressful conditions, negatively impacting seed germination and root development (Majeed et al. 2015). Fortunately, rhizobacteria containing ACC deaminase can break down ACC, the ET precursor, thereby reducing excessive ethylene production during stress and protecting plants from its inhibitory effects (Fahr et al. 2013; Singh et al. 2016). Beneficial rhizobacteria also promote proline synthesis in plants experiencing abiotic stress (Ahmed et al. 2019) (Table 13.1).

### 13.4.5 Pesticide Stress

The impact of pesticides on plants varies based on dosage, leading to disruptions in physiological and biochemical processes that ultimately reduce growth and yield (Shahzad et al. 2018; Sharma et al. 2018a, b). However, plants possess mechanisms to combat pesticide stress through detoxification systems. Among the various strategies employed by plants to mitigate the adverse effects of pesticide exposure, PGRs play a crucial role in enabling plants to withstand stressful conditions. PGRs modulate signaling pathways and enhance plant resistance. They constitute a fundamental framework that encompasses metabolic reactions essential for various plant life processes, thereby enhancing crop quality and productivity (Varshney et al. 2015). When plants absorb pesticides, it not only diminishes their vitality but also poses risks to humans and animals consuming them (Ahmed et al. 2018). Non-target plants are adversely affected through the generation of reactive oxygen species (Soares et al. 2018). Nucleic acids, lipids, and proteins are oxidatively damaged by these extremely reactive chemicals, which change the biochemical and physiological functions of plant cells (Sharma et al. 2019).

Various phytohormones/ PGRs are reported to alleviate pesticide stress in plants by regulating ROS levels. Brassinosteroids enhance resistance by activating antioxidant defenses and accelerating pesticide breakdown, significantly mitigating pesticide-induced biomass reduction (Ahammed et al. 2012). Applying 24-epibrassinolide enhances the production of metabolic enzymes essential for pesticide degradation. As a pre-soak treatment, it significantly affects the transcription of antioxidants and defense-related genes in tomato leaves exposed to chlorpyrifos, indicating its beneficial impact on gene expression (Sharma et al. 2017; Wang et al. 2017). Kaya and Doganlar (2016) reported an increase in chlorophyll levels when tomato plants exposed to imazapic were treated with exogenous jasmonic acid. Treating tomato plants with SA increased the activity of glutathione S-transferase (GST) by enhancing the expression of *GST1/2/3* and *P450* genes when exposed to pesticide stress (Yüzbaşıoğlu and Dalyan 2019). The application of SA mitigates the impact of glyphosate on growth and hydrogen peroxide levels. In barley plants, SA stimulates the production of non-protein thiols and boosts the activities of several enzymes, including SOD, APOX, CAT, and glutathione S-transferase (GST) (Spormann et al. 2019).

Research has shown that certain persistent pesticides can alter soil microbial communities, potentially inhibiting the beneficial effects of PGPR. However, some microorganisms have demonstrated the ability to either withstand or break down various pesticides. These include several PGPR species, such as *Bacillus, Alcaligenes, Pseudomonas, Azotobacter, Acetobacter*, *Flavobacteria, Arthrobacter, Enterobacter*, and *Klebsiella*. Many species of *Pseudomonas, Bacillus*, and *Azotobacter* can endure and thrive in harsh environmental conditions by forming cysts. *Azotobacter* species, in particular, can tolerate extreme environments through cyst production (Chennappa et al. 2014). Another study conducted on *Vigna radiata* L. *Pseudomonas* sp. PGR-11 enhanced dry biomass, germination rate grown under metalaxyl, carbendazim, and tebuconazole stress (Al-Enazi et al. 2022). The same study reported that several PGPR *Azotobacter* species exhibited resistance to four commonly used paddy cultivation pesticides: *Pendimethalin, Chloropyrifos, Glyphosat*, and *Phorate*, at concentrations of 1% to 5%. The presence of these PGPR in agricultural fields may decrease the need for chemical fertilizers and pesticides, which are both expensive and potentially harmful. To mitigate the toxic and hazardous effects of costly fertilizers and pesticides or to reduce nitrogen and phosphorus fertilizer application rates for crops, effective PGPR such as *Azotobacter. Acetobacter* can be utilized to promote sustainable agriculture (Rani et al. 2019).

## 13.5 Conclusion and Future Perspectives

Abiotic stresses, such as drought, salinity, thermal stress, heavy metals, and pesticides, lead to excessive production of ROS, causing oxidative damage to proteins and DNA in plants. This negatively impacts plant growth, development, and yields. However, PGPRs and PGRs help to mitigate the detrimental effects of various environmental stresses by enhancing plants' antioxidant systems, both enzymatic and non-enzymatic to combat ROS. Additionally, PGPRs improve soil biodiversity, fertility, and plant growth by supporting functions like nitrogen fixation, mineral phosphate solubilization, phytohormone, and siderophore production. The synergistic effect of PGPR and PGRs holds great promise for advancing agricultural practices by enhancing plant growth, stress tolerance, nutrient uptake, disease resistance, and sustainability. Future research in this area will likely lead to more innovative and eco-friendly solutions that increase productivity while maintaining environmental health. As our understanding of plant–microbe interactions and plant physiology deepens, the potential for synergistic PGPR–PGR applications will continue to expand. Further advances in biotechnology would allow for tailored solutions, enhancing crop quality and efficiency. This synergy offers a promising path for sustainable, high-quality, and resilient agriculture in the future.

## References

Ahammed GJ, Gao CJ, Ogweno JO, Zhou YH, Xia XJ, Mao WH et al (2012) Brassinosteroids induce plant tolerance against phenanthrene by enhancing degradation and detoxification in *Solanum lycopersicum* L. Ecotoxicol Environ Saf 80:28–36

Ahmad M, Khalid A, Zahir ZA (2023) Phytohormones: microbial production and applications. In: Biological approaches to regenerative soil systems, pp 205–220

Ahmed A, Shamsi A, Bano B (2018) Deciphering the toxic effects of iprodione, a fungicide and malathion, an insecticide on thiol protease inhibitor isolated from yellow Indian mustard seeds. Environ Toxicol Pharmacol 61:52–60

Ahmed S, Roy Choudhury A, Chatterjee P, Samaddar S, Kim K, Jeon S, Sa T (2019) The role of plant growth-promoting rhizobacteria to modulate proline biosynthesis in plants for salt stress alleviation. In: Plant growth promoting rhizobacteria for sustainable stress management: volume 1: rhizobacteria in abiotic stress management, pp 1–20

Ahuja I, de Vos RC, Bones AM, Hall RD (2010) Plant molecular stress responses face climate change. Trends Plant Sci 15(12):664–674

Aldaby E, Abdel-motagaly FM, Elzohari MH, Amro AM (2025) Thriving under heat stress: exploring ecophysiological and agronomic dynamics of sugar beet (Beta vulgaris L.) across varied thermal regimes. Assiut Univ J Multidiscip Sci Res 2(1):20–42

Al-Enazi NM, AlTami MS, Alhomaidi E (2022) Unraveling the potential of pesticide-tolerant Pseudomonas sp. augmenting biological and physiological attributes of Vigna radiata (L.) under pesticide stress. RSC Adv 12(28):17765–17783

Ali Q, Ahmad M, Kamran M, Ashraf S, Shabaan M, Babar BH et al (2023) Synergistic effects of Rhizobacteria and salicylic acid on maize salt-stress tolerance. Plants 12(13):2519

Al-Zahrani HS, Alharby HF, Fahad S (2022) Antioxidative defense system, hormones, and metabolite accumulation in different plant parts of two contrasting rice cultivars as influenced by plant growth regulators under heat stress. Front Plant Sci 13:911846

Ansari MI, Jalil SU, Ansari SA, Hasanuzzaman M (2021) GABA shunt: a key-player in mitigation of ROS during stress. Plant Growth Regul 94:131–149

Asif M, Ahmad R, Pervez A, Al Farraj DA, Elshikh MS, Shahzad M et al (2023) Combination of melatonin and plant growth promoting rhizobacteria improved the growth of Spinacia oleracea L. under the arsenic and cadmium stresses. Physiol Mol Plant Pathol 127:102097

Aslam MM, Farhat F, Siddiqui MA, Yasmeen S, Khan MT, Sial MA, Khan IA (2021) Exploration of physiological and biochemical processes of canola with exogenously applied fertilizers and plant growth regulators under drought stress. PLoS One 16(12):e0260960

Arif MR, Islam MT, Robin AHK (2019) Salinity Stress Alters Root Morphology and Root Hair Traits in Brassica napus. Plants (Basel) 8(7):192. https://doi.org/10.3390/plants8070192 PMID: 31252515; PMCID: PMC6681291

Azizi S, Kouchaksaraei MT, Hadian J, Abad ARFN, Sanavi SAMM, Ammer C, Bader MKF (2021) Dual inoculations of arbuscular mycorrhizal fungi and plant growth-promoting rhizobacteria boost drought resistance and essential oil yield of common myrtle. For Ecol Manag 497:119478

Bakshi P, Chouhan R, Sharma P, Mir BA, Gandhi SG, Landi M et al (2021) Amelioration of chlorpyrifos-induced toxicity in Brassica juncea L. by combination of 24-epibrassinolide and plant-growth-promoting rhizobacteria. Biomolecules 11(6):877

Batool T, Ali S, Seleiman MF et al (2020) Plant growth promoting rhizobacteria alleviates drought stress in potato in response to suppressive oxidative stress and antioxidant enzymes activities. Sci Rep 10:16975. https://doi.org/10.1038/s41598-020-73489-z

Bhardwaj T, Kour J, Chouhan R, Devi K, Singh H, Gandhi SG et al (2024) Integrated transcriptomic and physio-molecular studies unveil the melatonin and PGPR induced protection to photosynthetic attributes in Brassica juncea L. under cadmium toxicity. J Hazard Mater 476:134875

Bharti R and Sharma R (2022) Effect of heavy metals: An overview. Proc 51:880–885

Bhat MA, Mishra AK, Jan S, Bhat MA, Kamal MA, Rahman S et al (2023) Plant growth promoting rhizobacteria in plant health: a perspective study of the underground interaction. Plants 12(3):629

Brookbank BP, Patel J, Gazzarrini S, Nambara E (2021) Role of basal ABA in plant growth and development. Genes 12(12):1936

Carlos MHJ, Stefani PVY, Janette AM, Melani MSS, Gabriela PO (2016) Assessing the effects of heavy metals in ACC deaminase and IAA production on plant growth-promoting bacteria. Microbiol Res 188:53–61

Chennappa G, Adkar-Purushothama CR, Naik MK, Suraj U, Sreenivasa MY (2014) Impact of pesticides on PGPR activity of *Azotobacter* sp. isolated from pesticide flooded paddy soils. Greener J Agric Sci 4(4):117–129

Chun HJ, Lim LH, Cheong MS, Baek D, Park MS, Cho HM et al (2021) Arabidopsis CCoAOMT1 plays a role in drought stress response via ROS-and ABA-dependent manners. Plants 10(5):831

Dai A (2013) Increasing drought under global warming in observations and models. Nat Clim Chang 3(1):52–58

de Andrade LA, Santos CHB, Frezarin ET, Sales LR, Rigobelo EC (2023) Plant growth-promoting rhizobacteria for sustainable agricultural production. Microorganisms 11(4):1088

Díaz-Rodríguez AM, Chávez-Luzanía RA, Montoya-Martínez AC, de los Santos Villalobos, S. (2024) Plant growth-promoting bacteria as a sustainable agricultural strategy. In: New insights, trends, and challenges in the development and applications of microbial inoculants in agriculture. Academic Press, pp 37–50

Ding Y, Yang S (2022) Surviving and thriving: how plants perceive and respond to temperature stress. Dev Cell 57(8):947–958

Ding Y, Shi Y, Yang S (2020) Molecular regulation of plant responses to environmental temperatures. Mol Plant 13(4):544–564

Doty SL, Oakley B, Xin G, Kang JW, Singleton G, Khan Z et al (2009) Diazotrophic endophytes of native black cottonwood and willow. Symbiosis 47:23–33

Edelstein M, Ben-Hur M (2018) Heavy metals and metalloids: Sources, risks and strategies to reduce their accumulation in horticultural crops. Sci Hortic 234:431–44

El-Beltagi HS, El-Yazied AA, El-Gawad HGA, Kandeel M, Shalaby TA, Mansour AT et al (2023) Synergistic impact of melatonin and putrescine interaction in mitigating salinity stress in snap bean seedlings: reduction of oxidative damage and inhibition of polyamine catabolism. Horticulturac 9(2):285

Fadiji AE, Yadav AN, Santoyo G, Babalola OO (2023) Understanding the plant-microbe interactions in environments exposed to abiotic stresses: an overview. Microbiol Res 271:127368

Fahr M, Laplaze L, Bendaou N, Hocher V, Mzibri ME, Bogusz D, Smouni A (2013) Effect of lead on root growth. Front Plant Sci 4:175

Fathi A, Afra JM (2023) Plant growth and development in relation to phosphorus: a review. Bull Univ Agric Sci Vet Med Cluj-Napoca Agric 80(1):1–7

Gómez-Godínez LJ, Aguirre-Noyola JL, Martínez-Romero E, Arteaga-Garibay RI, Ireta-Moreno J, Ruvalcaba-Gómez JM (2023) A look at plant-growth-promoting bacteria. Plants 12(8):1668

Ghori NH, Ghori T, Hayat M, Imadi S, Gul A, Altay V et al. (2019) Heavy metal stress and responses in plants Int J Environ Sci Technol (Tehran) 16:1807–1828

Govindasamy P, Muthusamy SK, Bagavathiannan M, Mowrer J, Jagannadham PTK, Maity A et al (2023) Nitrogen use efficiency—a key to enhance crop productivity under a changing climate. Front Plant Sci 14:1121073

Guo Z, Tan H, Zhu Z, Lu S, Zhou B (2005) Effect of intermediates on ascorbic acid and oxalate biosynthesis of rice and in relation to its stress resistance. Plant Physiol Biochem 43(10–11):955–962

Gupta S, Pandey S (2023) Plant growth promoting rhizobacteria to mitigate biotic and abiotic stress in plants. In: Sustainable agriculture reviews 60: microbial processes in agriculture, pp 47–68

Gupta A, Tiwari RK, Shukla R, Singh AN, Sahu PK (2023) Salinity alleviator bacteria in rice (Oryza sativa L.), their colonization efficacy, and synergism with melatonin. Front Plant Sci 13:1060287

Han SH, Park YJ, Park CM (2019) Light primes the thermally induced detoxification of reactive oxygen species during development of thermotolerance in Arabidopsis. Plant Cell Physiol 60(1):230–241

Hasana R, Miyake H (2017) Salinity stress alters nutrient uptake and causes the damage of root and leaf anatomy in maize. KnE Life Sciences 219–225. https://doi.org/10.18502/kls.v3i4.708

Hasan MM, Alharby HF, Uddin MN, Ali MA, Anwar Y, Fang XW et al (2020) Magnetized water confers drought stress tolerance in Moringa biotype via modulation of growth, gas exchange, lipid peroxidation and antioxidant activity. Pol J Environ Stud 1:29

Hasan MM, Alabdallah NM, Alharbi BM, Waseem M, Yao G, Liu XD et al (2021) GABA: a key player in drought stress resistance in plants. Int J Mol Sci 22(18):10136

Hasanuzzaman M, Raihan MRH, Masud AAC, Rahman K, Nowroz F, Rahman M et al (2021) Regulation of Reactive Oxygen Species and Antioxidant Defense in Plants under Salinity. Int J Mol Sci 22(17):9326. https://doi.org/10.3390/ijms22179326 PMID: 34502233; PMCID: PMC8430727

Hasan A, Tabassum B, Hashim M, Khan N (2024) Role of plant growth promoting rhizobacteria (PGPR) as a plant growth enhancer for sustainable agriculture: a review. Bacteria 3(2):59–75

Hassan MK, McInroy JA, Kloepper JW (2019) The interactions of rhizodeposits with plant growth-promoting rhizobacteria in the rhizosphere: a review. Agriculture 9(7):142

Hatamleh AA, Danish M, Al-Dosary MA, El-Zaidy M, Ali S (2022) Physiological and oxidative stress responses of Solanum lycopersicum (L.)(tomato) when exposed to different chemical pesticides. RSC advances 12(12):7237–52

He W, Yan K, Zhang Y, Bian L, Mei H, Han G (2021) Contrasting photosynthesis, photoinhibition and oxidative damage in honeysuckle (Lonicera japonica Thunb.) under iso-osmotic salt and drought stresses. Environ Exp Bot 182:104313

Howarth CJ (2005) Genetic improvements of tolerance to high temperature. In: Abiotic stresses: plant resistance through breeding and molecular approaches. Howarth Press Inc, New York, pp 277–300

Hrynkiewicz K, Złoch M, Kowalkowski T, Baum C, Niedojadło K, Buszewski B (2015) Strain-specific bioaccumulation and intracellular distribution of $Cd^{2+}$ in bacteria isolated from the rhizosphere, ectomycorrhizae, and fruitbodies of ectomycorrhizal fungi. Environ Sci Pollut Res 22:3055–3067

Ibrahim M, Iqbal M, Tang YT, Khan S, Guan DX, Li G (2022) Phosphorus mobilization in plant–soil environments and inspired strategies for managing phosphorus: a review. Agronomy 12(10):2539

Igiehon NO, Babalola OO (2018) Below-ground-above-ground plant-microbial interactions: focusing on soybean, rhizobacteria and mycorrhizal fungi. Open Microbiol J 12:261

Jabeen Z, Fayyaz HA, Irshad F, Hussain N, Hassan MN, Li J et al (2021) Sodium nitroprusside application improves morphological and physiological attributes of soybean (Glycine max L.) under salinity stress 16(4):e0248207

Jafari SH, Arani AM, Esfahani ST (2023) The combined effects of rhizobacteria and methyl jasmonate on rosmarinic acid production and gene expression profile in Origanum vulgare L. under salinity conditions. J Plant Growth Regul 42(3):1472–1487

Jan S, Singh B, Bhardwaj R, Kapoor D, Kour J, Singh R et al (2022) Application of melatonin and PGPR alleviates thiamethoxam induced toxicity by regulating the TCA cycle in Brassica juncea L. Saudi J Biol Sci 29(3):1348–1354

Jan S, Bhardwaj R, Sharma NR, Singh R (2024) Unraveling the role of plant growth regulators and plant growth promoting rhizobacteria in phytoremediation. J Plant Growth Regul 43:1–17

Jawad Hassan M, Ali Raza M, Ur Rehman S, Ansar M, Gitari H, Khan I et al (2020) Effect of Cadmium Toxicity on Growth, Oxidative Damage, Antioxidant Defense System and Cadmium Accumulation in Two Sorghum Cultivars. Plants (Basel) 9(11):1575

Jiménez VM (2005) Involvement of plant hormones and plant growth regulators on in vitro somatic embryogenesis. Plant Growth Regul 47:91–110

Kang SM, Khan AL, Waqas M, You YH, Kim JH, Kim JG et al (2014) Plant growth-promoting rhizobacteria reduce adverse effects of salinity and osmotic stress by regulating phytohormones and antioxidants in *Cucumis sativus*. J Plant Interact 9(1):673–682

Karnwal A, Shrivastava S, Al-Tawaha ARMS, Kumar G, Kumar A, Kumar A (2024) PGPR-mediated breakthroughs in plant stress tolerance for sustainable farming. J Plant Growth Regul 43(9):2955–2971

Kaya A, Doganlar ZB (2016) Exogenous jasmonic acid induces stress tolerance in tobacco (*Nicotiana tabacum*) exposed to imazapic. Ecotoxicol Environ Saf 124:470–479

Kaya C, Ugurlar F, Ashraf M, Ahmad P (2023) Salicylic acid interacts with other plant growth regulators and signal molecules in response to stressful environments in plants. Plant Physiol Biochem 196:431–443

Khakbazan M, Nyiraneza J, Jiang Y, Huang J (2023) Increasing potato profitability through rotation systems and winter cover, nitrogen fertilizer, and supplemental irrigation in Atlantic Canada. Agrosyst Geosci Environ 6(2):e20382

Khan N, Bano A, Babar MA (2017) The root growth of wheat plants, the water conservation and fertility status of sandy soils influenced by plant growth promoting rhizobacteria. Symbiosis 72:195–205

Khan N, Bano A, Zandi P (2018a) Effects of exogenously applied plant growth regulators in combination with PGPR on the physiology and root growth of chickpea (Cicer arietinum) and their role in drought tolerance. J Plant Interact 13(1):239–247

Khan N, Zandi P, Ali S, Mehmood A, Adnan Shahid M, Yang J (2018b) Impact of salicylic acid and PGPR on the drought tolerance and phytoremediation potential of Helianthus annus. Front Microbiol 9:2507

Khan N, Bano A, Rahman MA, Rathinasabapathi B, Babar MA (2018c) UPLC-HRMS-based untargeted metabolic profiling reveals changes in chickpea (*Cicer arietinum*) metabolome following long-term drought stress. Plant Cell Environ 42:115

Khan N, Bano A (2019) Rhizobacteria and abiotic stress management. In Plant Growth Promoting Rhizobacteria for Sustainable Stress Management. Springer: Singapore 65–80

Khan N, Asadullah, Bano A (2019a) Rhizobacteria and abiotic stress management. In: Plant Growth Promoting Rhizobacteria for Sustainable Stress Management: Volume 1: Rhizobacteria in Abiotic Stress Management, pp 65–80

Khan N, Bano A, Rahman MA, Guo J, Kang Z, Babar MA (2019b) Comparative physiological and metabolic analysis reveals a complex mechanism involved in drought tolerance in chickpea (Cicer arietinum L.) induced by PGPR and PGRs. Sci Rep 9(1):2097

Khan N, Bano AM, Babar A (2020) Impacts of plant growth promoters and plant growth regulators on rainfed agriculture. PLoS One 15(4):e0231426

Khan V, Umar S, Iqbal N (2023) Synergistic action of Pseudomonas fluorescens with melatonin attenuates salt toxicity in mustard by regulating antioxidant system and flavonoid profile. Physiol Plant 175(6):e14092

Khan V, Mubashshir M, Umar S, Iqbal N (2024) Methyl Jasmonate and Pseudomonas fluorescens synergistically boost antioxidative defense, secondary metabolites, and osmolyte production to enhance drought resilience in mustard. J Plant Growth Regul:1–19

Kim ST, Sang MK (2024) Synergistic effects of salicylic acid and bacillus butanolivorans KJ40 for enhancing Napa Cabbage (Brassica Napa subsp. pekinensis) resilience to water-deficit stress. Horticulturae 10(6):618

Kour D, Yadav AN (2022) Bacterial mitigation of drought stress in plants: current perspectives and future challenges. Curr Microbiol 79(9):248

Kour J, Bhardwaj T, Chouhan R, Singh AD, Gandhi SG, Bhardwaj R et al (2024) Phytomelatonin maintained chromium toxicity induced oxidative burst in Brassica juncea L. through improving antioxidant system and gene expression. Environ Pollut 124256

Kumar K, Anil A, Geetha P, Kumar T, Pandey B (2023) Effect of plant growth regulators and plant growth promoting rhizobacteria on physio-chemical properties of mung bean under drought stress. Biological Forum 15(4):753–761

Kumari A, Lakshmi GA, Krishna GK, Patni B, Prakash S, Bhattacharyya M et al (2022) Climate change and its impact on crops: a comprehensive investigation for sustainable agriculture. Agron 12(12):3008

Kumar S, Li G, Yang J, Huang X, Ji Q, Liu Z et al (2021) Effect of salt stress on growth, physiological parameters, and ionic concentration of water dropwort (Oenanthe javanica) cultivars. Front Plant Sci 12:660409

Kumar S, Yu R, Liu Y, Liu Y, Khan MN, Liu Y et al (2025) Exogenous melatonin enhances heat stress tolerance in sweetpotato by modulating antioxidant defense system, osmotic homeostasis and stomatal traits. Hortic Plant J 11(1):431–445

Landa-Faz A, González-Orenga S, Boscaiu M, Rodríguez-Vázquez R, Vicente O (2021) Effect of the pesticide endosulfan and two different biostimulants on the stress responses of Phaseolus leptostachyus plants grown in a saline soil. Agronomy 11(6):1208

Li X, Zhang W, Niu D, Liu X (2024) Effects of abiotic stress on chlorophyll metabolism. Plant Sci 342:112030

Liu J, Qi W, Li Q, Wang SG, Song C, Yuan XZ (2020) Exogenous phosphorus-solubilizing bacteria changed the rhizosphere microbial community indirectly. 3 Biotech 10:1–11

Liu H, Wu Y, Cai J, Chen Y, Zhou C, Qiao C et al (2024) Effect of auxin on cadmium toxicity-induced growth inhibition in Solanum lycopersicum. Toxics 12(5):374

Mansoor S, Ali Wani O, Lone JK, Manhas S, Kour N, Alam P et al (2022) Reactive oxygen species in plants: from source to sink. AntioxidS 11(2):225
Madhumitha M, Sundaram V (2024) Plant growth regulators in abiotic stress resilience of plants: a review. J Exp Agric Int 46(8):364–372
Maheshwari DK, Dheeman S, Agarwal M (2015) Phytohormone-producing PGPR for sustainable agriculture. In: Bacterial metabolites in sustainable agroecosystem, pp 159–182
Maitra S, Brestic M, Bhadra P, Shankar T, Praharaj S, Palai JB et al (2021) Bioinoculants—natural biological resources for sustainable plant production. Microorganisms 10(1):51
Majeed A, Abbasi MK, Hameed S, Imran A, Rahim N (2015) Isolation and characterization of plant growth-promoting rhizobacteria from wheat rhizosphere and their effect on plant growth promotion. Front Microbiol 6:198
Meena M, Swapnil P, Zehra A, Aamir M, Dubey MK, Goutam J, Upadhyay RS (2017) Beneficial microbes for disease suppression and plant growth promotion. In: Plant-microbe interactions in agro-ecological perspectives: volume 2: microbial interactions and agro-ecological impacts, pp 395–432
Meng YC, Zhang HF, Pan XX, Chen N, Hu HF, Haq SU et al (2021) CaDHN3, a PepperC(apsicum annuum L.) Dehydrin Gene Enhances the Tolerance against Salt and Drought Stresses by Reducing ROS Accumulation. Int J Mol Sci 22(6):3205
Mehrabi SS, Sabokdast M, Bihamta MR, Dedičová B (2024) The coupling effects of PGPR inoculation and foliar spraying of strigolactone in mitigating the negative effect of salt stress in wheat plants: insights from phytochemical, growth, and yield attributes. Agriculture 14(5):732
Mhlongo MI, Piater LA, Madala NE, Labuschagne N, Dubery IA (2018) The chemistry of plant–microbe interactions in the rhizosphere and the potential for metabolomics to reveal signaling related to defense priming and induced systemic resistance. Front Plant Sci 9:112
Mohammed AR, Tarpley L (2011) High night temperature and plant growth regulator effects on spikelet sterility, grain characteristics and yield of rice (Oryza sativa L.) plants. Can J Plant Sci 91(2):283–291
Munns R, Millar AH (2023) Seven plant capacities to adapt to abiotic stress. J Exp Bot 74(15):4308–4323
Nagarajan S, Nagarajan S (2010) Abiotic tolerance and crop improvement. In: Abiotic stress adaptation in plants: physiological, molecular and genomic foundation, pp 1–11
Nazim K, Bano A (2024) Role of salicylic acid and plant growth promoting rhizobacteria (PGPR) to enhance cold tolerance in tomato (Lycopersicon esculentum Mill.). Pak J Bot 56(3):1193–1207
Noor MA, Hassan MU, Khan TA, Zhou B, Huang G (2025) Biochar and Melatonin Partnership Mitigates Arsenic Toxicity in Rice by Modulating Antioxidant Defense, Phytochelatin Synthesis, and Down-Regulating the Transporters Involved in Arsenic Uptake. Plants 14(15):2453
Noctor G, Gomez L, Vanacker H, Foyer CH (2002) Interactions between biosynthesis, compartmentation and transport in the control of glutathione homeostasis and signalling. J Exp Bot 53(372):1283–1304
Oyedele AO, Ezaka E, Taiwo LB (2024) Microbial biosynthesis of the classical phytohormones by plant growth-promoting microorganisms in plants. In: Microbial biostimulants for plant growth and abiotic stress amelioration. Elsevier, pp 345–366
Pantoja-Guerra M, Valero-Valero N, Ramírez CA (2023) Total auxin level in the soil–plant system as a modulating factor for the effectiveness of PGPR inocula: a review. Chem Biol Technol Agric 10(1):6
Paredes-Páliz K, Rodríguez-Vázquez R, Duarte B, Caviedes MA, Mateos-Naranjo E, Redondo-Gómez S, Caçador MI, Rodríguez-Llorente ID, Pajuelo E (2018) Investigating the mechanisms underlying phytoprotection by plant growth-promoting rhizobacteria in *Spartina densiflora* under metal stress. Plant Biol 20:497–506
Parmar N, Dufresne J (2011) Beneficial interactions of plant growth promoting rhizosphere micro-organisms. In: Bioaugmentation, biostimulation and biocontrol, pp 27–42

Peter O, Imran M, Shaffique S, Kang SM, Rolly NK, Felistus C et al (2024) Combined application of melatonin and Bacillus sp. strain IPR-4 ameliorates drought stress tolerance via hormonal, antioxidant, and physiomolecular signaling in soybean. Front Plant Sci 15:1274964

Rahman MM, Das AK, Sultana S, Ghosh PK, Islam MR, Keya SS et al (2023) Biochar potentially enhances maize tolerance to arsenic toxicity by improving physiological and biochemical responses to excessive arsenate. Biochar 5(1):71

Rajeshkumar PM (2023) The role of plant growth promoting rhizobacteria in sustainable agriculture. Vidhyayana ISSN 2454-8596, 9(si1)

Rani R, Kumar V, Gupta P, Chandra A (2019) Application of plant growth promoting rhizobacteria in remediation of pesticides contaminated stressed soil. In: New and future developments in microbial biotechnology and bioengineering. Elsevier, pp 341–353

Ranjan A, Rajput VD, Prazdnova EV, Gurnani M, Sharma S, Bhardwaj P et al (2024) Augmenting abiotic stress tolerance and root architecture: the function of phytohormone-producing PGPR and their interaction with nanoparticles. S Afr J Bot 167:612–629

Rashid U, Yasmin H, Hassan MN, Naz R, Nosheen A, Sajjad M et al (2022) Drought-tolerant Bacillus megaterium isolated from semi-arid conditions induces systemic tolerance of wheat under drought conditions. Plant Cell Rep:1–21

Rizwan M, Usman K, Alsafran M (2024) Ecological impacts and potential hazards of nickel on soil microbes, plants, and human health. Chemosphere 357:142028

Riaz A, Rafique M, Aftab M, Qureshi MA, Javed H, Mujeeb F, Akhtar S (2019) Mitigation of salinity in chickpea by plant growth promoting rhizobacteria and salicylic acid. Eurasian J Soil Sci 8(3):221–228

Rosier A, Medeiros FH, Bais HP (2018) Defining plant growth promoting rhizobacteria molecular and biochemical networks in beneficial plant-microbe interactions. Plant Soil 428:35–55

Sachdev S, Ansari SA, Ansari MI (2023) Antioxidant defensive mechanisms to regulate cellular redox homeostatic balance. In: Reactive oxygen species in plants: the right balance. Singapore, Springer Nature Singapore, pp 143–172

Saharan BS, Nehra V (2011) Plant growth promoting rhizobacteria: a critical review. Life Sci Med Res 21(1):30

Santos F, Penaflor MFG, Paré PW, Sanches PA, Kamiya AC, Tonelli M et al (2014) A novel interaction between plant-beneficial rhizobacteria and roots: colonization induces corn resistance against the root herbivore *Diabrotica speciosa*. PLoS One 9(11):e113280

Sarkar J, Chakraborty B, Chakraborty U (2018) Plant growth promoting rhizobacteria protect wheat plants against temperature stress through antioxidant signalling and reducing chloroplast and membrane injury. J Plant Growth Regul 37(4):1396–1412

Sarma AK, Deka K (2024) Plant growth promoting rhizobacteria: an option for reducing abiotic stress in plant. Commun Soil Sci Plant Anal 55:1–18

Seneviratne M, Gunaratne S, Bandara T, Weerasundara L, Rajakaruna N, Seneviratne G, Vithanage M (2016) Plant growth promotion by *Bradyrhizobium japonicum* under heavy metal stress. S Afr J Bot 105:19–24

Shah Z, Iqbal A, Khan FU, Khan HU, Durrani F, Ahmad MZ (2020) Genetic manipulation of pea (Pisum sativum L.) with Arabidopsis's heat shock factor HsfA1d improves ROS scavenging system to confront thermal stress. Genet Resour Crop Evol 67:2119–2127

Shahid S, Dar A, Hussain A, Khalid I, Latif M, Ahmad HT et al (2024) Enhancing cauliflower growth under cadmium stress: synergistic effects of Cd-tolerant Klebsiella strains and jasmonic acid foliar application. Front Microbiol 15:1444374

Shahzad B, Tanveer M, Che Z, Rehman A, Cheema SA, Sharma A et al (2018) Role of 24-epibrassinolide (EBL) in mediating heavy metal and pesticide induced oxidative stress in plants: a review. Ecotoxicol Environ Saf 147:935–944

Shan L, Xu Y, Wu D, Hu J, Yu T, Dang C et al (2024) Effects of salicylic acid on growth, physiology, and gene expression in rice seedlings under salt and drought stress. Plant Stress 11:100413

Sharma A, Kumar V, Bhardwaj R, Thukral AK (2017) Seed pre-soaking with 24-epibrassinolide reduces the imidacloprid pesticide residues in green pods of *Brassica juncea* L. Toxicol Environ Chem 99(1):95–103

Sharma P, Chouhan R, Bakshi P, Gandhi SG, Kaur R, Sharma A, Bhardwaj R (2022) Amelioration of chromium-induced oxidative stress by combined treatment of selected plant-growth-promoting rhizobacteria and earthworms via modulating the expression of genes related to reactive oxygen species metabolism in Brassica juncea. Front Microbiol 13:802512

Sharma A, Kumar V, Kumar R, Shahzad B, Thukral AK, Bhardwaj R (2018a) Brassinosteroid-mediated pesticide detoxification in plants: A mini-review. Cogent Food Agric 4(1):1436212

Sharma A, Kumar V, Yuan H, Kanwar MK, Bhardwaj R, Thukral AK, Zheng B (2018b) Jasmonic acid seed treatment stimulates insecticide detoxification in *Brassica juncea* L. Front Plant Sci 9:1609

Sharma A, Yuan H, Kumar V, Ramakrishnan M, Kohli SK, Kaur R et al (2019) Castasterone attenuates insecticide induced phytotoxicity in mustard. Ecotoxicol Environ Saf 179:50–61

Shehzad J, Mustafa G (2023) Mechanism of reactive oxygen species regulation in plants. In: Reactive oxygen species: prospects in plant metabolism. Singapore, Springer Nature Singapore, pp 17–41

Shi Y, Chang YL, Wu HT, Shalmani A, Liu WT, Li WQ et al (2020) OsRbohB-mediated ROS production plays a crucial role in drought stress tolerance of rice. Plant Cell Rep 39:1767–1784

Siddiqui MH, Alamri S, Alsubaie QD, Ali HM (2020) Melatonin and gibberellic acid promote growth and chlorophyll biosynthesis by regulating antioxidant and methylglyoxal detoxification system in tomato seedlings under salinity. J Plant Growth Regul 39(4):1488–502

Silva EC, Nogueira RJ, Vale FH, Melo NFD, Araújo FPD (2009) Water relations and organic solutes production in four umbu tree (Spondias tuberosa) genotypes under intermittent drought. Braz J Plant Physiol 21:43–53

Silva EC, Nogueira RJMC, Silva MA, Albuquerque MB (2011) Drought stress and plant nutrition. Plant Stress 5(1):32–41

Singh S, Parihar P, Singh R, Singh VP, Prasad SM (2016) Heavy metal tolerance in plants: role of transcriptomics, proteomics, metabolomics, and ionomics. Front Plant Sci 6:1143

Singh S, Kumar V, Datta S, Ramamurthy PC, Singh J (2021) Molecular mechanism and signaling pathways interplay between plant hormones during plant-microbe crosstalk. In: Microbial management of plant stresses. Woodhead Publishing, pp 93–105

Singh N, Maurya V, Sharma I, Kumar Y, Kumar R, Sharma A (2024) Combinations of PGPR with other soil micro-and macro-organisms for improved stress tolerance in plants. Nova Science Publishers, USA

Sinha AK, Puri S, Sharma MD (2024) PGPR from its discovery to its commercial product development for sustainable agriculture: a review. J Agric Earth Environ Sci 3(3):8

Soares C, Spormann S, Fidalgo F (2018) Salicylic acid improves the performance of the enzymatic antioxidant system of barley exposed to glyphosate. Free Radic Biol Med 120:S157

Spormann S, Soares C, Fidalgo F (2019) Salicylic acid alleviates glyphosate-induced oxidative stress in *Hordeum vulgare* L. J Environ Manag 241:226–234

Subiramani S, Ramalingam S, Muthu T, Nile SH, Venkidasamy B (2020) Development of abiotic stress tolerance in crops by plant growth-promoting rhizobacteria (PGPR). In: Phytomicrobiome in stress regulation, pp 125–145

Suman M, Sangma PD, Meghawal DR, Sahu OP (2017) Effect of plant growth regulators on fruit crops. J Pharmacogn Phytochem 6(2):331–337

Sziderics AH, Rasche F, Trognitz F, Sessitsch A, Wilhelm E (2007) Bacterial endophytes contribute to abiotic stress adaptation in pepper plants (Capsicum annuum L.). Can J Microbiol 53(11):1195–1202

Tanaka KR, Van Houtan KS (2022) The recent normalization of historical marine heat extremes. PloS Climate 1(2):e0000007

Tang J, Li Y, Zhang L, Mu J, Jiang Y, Fu H et al (2023) Biosynthetic pathways and functions of indole-3-acetic acid in microorganisms. Microorganisms 11(8):2077

Tanveer S, Akhtar N, Ilyas N, Sayyed RZ, Fitriatin BN, Perveen K, Bukhari NA (2023) Interactive effects of Pseudomonas putida and salicylic acid for mitigating drought tolerance in canola (Brassica napus L.). Heliyon 9(3):e14193. https://doi.org/10.1016/j.heliyon.2023.e14193 PMID: 36950648; PMCID: PMC10025117

Tapia-García EY, Hernández-Trejo V, Guevara-Luna J, Rojas-Rojas FU, Arroyo-Herrera I, Meza-Radilla G et al (2020) Plant growth-promoting bacteria isolated from wild legume nodules and nodules of Phaseolus vulgaris L. trap plants in central and southern Mexico. Microbiol Res 239:126522

Tolay I (2021) The impact of different Zinc (Zn) levels on growth and nutrient uptake of Basil (Ocimum basilicum L.) grown under salinity stress. PLoS One 16(2):e0246493

Varshney S, Khan MIR, Masood A, Per TS, Rasheed F, Khan NA (2015) Contribution of plant growth regulators in mitigation of herbicidal stress. J Plant Biochem Physiol 3(2):4

Wakeel A, Xu M, Gan Y (2020) Chromium-induced reactive oxygen species accumulation by altering the enzymatic antioxidant system and associated cytotoxic, genotoxic, ultrastructural, and photosynthetic changes in plants. Int J Mol Sci 21(3):728

Wan W, Qin Y, Wu H, Zuo W, He H, Tan J et al (2020) Isolation and characterization of phosphorus solubilizing bacteria with multiple phosphorus sources utilizing capability and their potential for lead immobilization in soil. Front Microbiol 11:752

Wang Z, Jiang Y, Peng X, Xu S, Zhang H, Gao J, Xi Z (2017) Exogenous 24-epibrassinolide regulates antioxidant and pesticide detoxification systems in grapevine after chlorothalonil treatment. Plant Growth Regul 81:455–466

Wang L, Lin H, Dong Y, Li B, He Y (2020) Effects of endophytes inoculation on rhizosphere and endosphere microecology of Indian mustard (Brassica juncea) grown in vanadium-contaminated soil and its enhancement on phytoremediation. Chemosphere 240:124891

Wang B, Li L, Liu M, Peng D, Wei A, Hou B et al (2022) TaFDL2-1A confers drought stress tolerance by promoting ABA biosynthesis, ABA responses, and ROS scavenging in transgenic wheat. Plant J 112(3):722–737

Wickramasinghe WRKDWKV, Girija D, Gopal KS, Kesevan S (2021) Multi-phasic nitrogen fixing plant growth promoting rhizobacteria as biofertilizer for rice cultivation. Res J Agric Sci 12(2):399–404

Yadav S, Modi P, Dave A, Vijapura A, Patel D, Patel M (2020) Effect of abiotic stress on crops. Sustainable Crop Prod 3(17):5–16

Yin X, Kropff MJ, Goudriaan JAN (1996) Differential effects of day and night temperature on development to flowering in rice. Ann Bot 77(3):203–213

Yüzbaşıoğlu E, Dalyan E (2019) Salicylic acid alleviates thiram toxicity by modulating antioxidant enzyme capacity and pesticide detoxification systems in the tomato (*Solanum lycopersicum* Mill.). Plant Physiol Biochem 135:322–330

Zhang G, Yan P, Leng D, Shang L, Zhang C, Wu Z, Wang Z (2023a) Salicylic acid treatment alleviates the heat stress response by reducing the intracellular ROS level and increasing the cytosolic trehalose content in pleurotus ostreatus. Microbiol Spectr 11(1):e03113–e03122

Zhang Y, Xu J, Li R, Ge Y, Li Y, Li R (2023b) Plants' response to abiotic stress: mechanisms and strategies. Int J Mol Sci 24(13):10915

Zhang B, Lv F, Yang J (2024) Pesticides toxicity, removal and detoxification in plants: A review. Agron 14(6):1260

Zhu JK (2016) Abiotic stress signaling and responses in plants. Cell 167(2):313–324

Zhumanova N, Akimbayeva N, Myrzakhmetova N, Dzhiembaev B, Ku A, Diyarova B et al (2024) A comprehensive review of new generation plant growth regulators. ES Food Agrofor 17:1190

# Salicylic Acid and Potassium Alleviate Salt Stress in Mustard Through Modulation of Antioxidant Enzymes, Biochemical Parameters, and Yield Attributes

**14**

Ragini Singh, Varisha Khan, Anshu Jha, Parul Jain, Shahid Umar, and Noushina Iqbal

**Abstract**

Salt stress is a key abiotic factor that inhibits mustard plant growth and grain yield, owing to oxidative damage, osmotic imbalance, and disturbances in metabolic activities. Salicylic acid (SA) and potassium (K) have been identified as crucial components in regulating salt stress and improving plant tolerance. SA and potassium K function as reactive oxygen species (ROS) scavengers, contributing to cellular water balance by reducing oxidative damage and facilitating osmotic adjustment. Additionally, they improve nitrogen assimilation, which leads to an increase in proline and protein content. This reinforces the plant's ability to withstand salt stress. The present study revealed the significant potential of SA and K, individually or in combination, to alleviate the adverse effects of salt stress on mustard plants. This treatment significantly improved nitrate reductase activity by 27.72%. Additionally, the combination enhanced the activity of the antioxidant enzymes activities, including SOD, CAT, and APX contents increased by 93.94%, 37.77%, and 151.51% respectively. The accumulation of total phenolic compounds (TPC) and total flavonoid contents (TFC) was significantly increased, both essential for preventing oxidative damage. These findings indicate the synergistic effect of SA and K in strengthening different physiological and biochemical processes required for salt tolerance. The increase in these pathways shows that SA and K mitigate adverse stress effects and prepare plants for enhanced metabolic efficiency under stressful conditions. For instance, the stability of cells is enhanced by an increase in proline and protein content, while the increased activity of antioxidant enzymes ameliorates the damage to cells

R. Singh · V. Khan · A. Jha · P. Jain · S. Umar · N. Iqbal (✉)
Department of Botany, School of Chemical and Life Sciences, Jamia Hamdard, New Delhi, India



N. Iqbal et al. (eds.), *Plant Signaling Molecules in Regulation of ROS-Scavenging System*, https://doi.org/10.1007/978-981-95-3377-0_14

caused by reactive oxygen species (ROS). The significant involvement of secondary metabolites in stress adaptation is further emphasized by the increased accumulation of TPC and TFC.

**Keywords**

Abiotic stress · Oxidative stress · Antioxidant · Phenol · Flavonoids · ROS · Secondary metabolites

## 14.1 Introduction

Salinity stress is a significant abiotic stress that results in a metabolic imbalance because of osmotic and oxidative stress. The cytosol accumulates $Na^+$ ions, which disrupts the noncovalent interaction between amino acids, causes ionic and osmotic toxicity, and lowers potassium and calcium concentrations due to excessive salt concentrations (Ali et al. 2011; Johnson et al. 2002). Crop productivity is restricted by salinity, which impacts approximately 10% of the world's agricultural land (Dixon et al. 2002). Significant attention has been directed toward the mitigation of salinity stress in plants. Phytohormones and nutrients are crucial in the modulation of stress adaptation, which is one of the numerous adaptation strategies that plants employ. Salicylic acid and potassium are getting significant attention due to their ability to modulate growth in both stress and non-stress environments. Potassium is a macronutrient that is necessary for the optimal development and productivity of plants. In addition to its role in the regulation of stomata, the activation of numerous enzymes, and the formation of proteins, potassium is also essential for the management of a variety of abiotic stresses (Ul-Allah et al. 2020; Reddy et al. 2004). It also scavenges ROS, which reduces the deleterious impact of salt stress on plant development and yield (Cakmak 2005).

Potassium is also vital in the maintenance of the $Na^+/K^+$ ratio, which aids in osmotic adjustment and the removal of excess ROS production during abiotic stresses (Sangakkara 2001). In addition to reducing the adverse effects of stress, optimal potassium fertilization and proper nutrition also improve crop growth and productivity (Egilla et al. 2001; Shah et al. 2018). It helps plants make dry matter by making it easier for photosynthesized material to get to the roots and sink organs and by improving photosynthesis by controlling stomata (Marschner 2011; Römheld and Kirkby 2010). Plants need to have potassium because it helps keep cell membranes stable and intact during salt stress (Ul-Allah et al. 2020).

Salicylic acid is recognized as one of the most critical phytohormones that regulate the growth and development of plants under stress. A phenolic compound found in plants works as an internal signaling molecule to turn on a stress tolerance mechanism (Ma et al. 2017; Zhou et al. 2021). Its stress tolerance method produces osmolytes and antioxidants that preserve osmotic balance and scavenge ROS (Gautam and Singh 2009; Hayat et al. 2010; Swanson and Gilroy 2010). Nevertheless, the optimal concentration of SA is necessary for a maximal response (Miura and Tada 2014). The growth and yield of crops are significantly influenced by appropriate mineral nutrition, in addition to a variety of agricultural practices, as reported by Ahanger et al. (2017). When the K content is low under salt stress, the rise in $Na^+$ ions and toxicity is greater (Bacha et al. 2015). There is a link between potassium nutrition and nitrogen metabolism, which maintains proline concentration and antioxidants such as GSH and ascorbate for redox potential. In India, mustard is a significant oilseed crop, contributing 24.2% to the country's total edible oil reservoir (DRMR 2013). The mustard seeds contain an oil content of 35–48% and a protein content of 37–42% in the cake (Verma et al. 2018). The impact of SA and K on the alleviation of salt stress in mustard is not adequately supported by scientific research. Despite the extensive research on salt stress tolerance in mustard, the integrated treatments of SA and K in response to salt stress remain to be clarified. This research has focused on a variety of physiological and biochemical characteristics. We hypothesized that the combination of a foliar spray of SA and a soil application of K could improve mustard's tolerance to salt stress by improving photosynthetic pigments and enhancing N-assimilation, proline content, and antioxidative enzymes. Reducing salt stress by reducing MDA and $H_2O_2$ concentration can improve mustard plant development and, as a result, yield.

## 14.2 Material and Methods

### 14.2.1 Experimental Plants and Their Growth Condition

The present study was performed on mustard (*Brassica juncea L.*) cultivated in an herbal garden at Jamia Hamdard, Delhi, India. Healthy mustard seeds were sterilized using 0.01% $HgCl_2$ and subsequently rinsed many times with distilled water. Seeds were sown in earthen pots filled with 8 kg of soil. The plants were treated with a saline solution (100 mM NaCl) of 600 mL in each container after the emergence of leaves. We applied 20 ml of potassium and salicylic acid (0.5 mM) individually and in combination at 20 DAS. The study included eight treatments, each with three replicates: T1 (control), T2 (100 mM salt), T3 (potassium), T4 (0.5 mM salicylic acid), T5 (potassium +0.5 mM salicylic acid), T6 (100 mM salt + potassium), T7 (100 mM salt + salicylic acid), and T8 (100 mM salt + potassium +0.5 mM salicylic acid). We collected data for various attributes at 30 DAS.

### 14.2.2 Assessments of Morphological Parameters

The plants were randomly selected at 30 DAS to measure shoot and root length, plant height, fresh and dried weight of root and shoot, number of leaves, and leaf area. The plants were uprooted from the soil, and their root and shoot lengths were precisely measured using a metric scale. We recorded the weights of fresh shoots and roots using an electric analytical balance (Vibra DJ 1505/Denver Instrument, APX 200). The samples needed to be kept in an oven (Scientific Systems 1.01) at 65 °C to dry out within 72 hours before their constant weight and dry weights were calculated using an electronic balance (Denver Instrument, APX 200). Using a leaf area meter (LA211 Systronic, New Delhi, India), the leaf area was evaluated.

### 14.2.3 Biochemical Parameters

#### 14.2.3.1 Measurement of Photosynthetic Parameters and Soluble Protein

The protocol for the isolation of total chlorophyll and carotenoids was established by Hiscox and Israelstam in 1979 and subsequently modified by Richardson et al. (2002). Total chlorophyll content and carotenoids were determined by utilizing the plant extract in DMSO to record absorbance at 645 and 663 with a UV-visible spectrophotometer (Model 119, Systronics, India). There was only 3 mL of DMSO in the blank tube, and the sample was not measured. The carotenoids and chlorophyll content were quantified in mg/g fresh weight (mg/g FW). We estimated the total soluble protein using Bradford's (1976) protocol. Additional information is accessible in the published paper of Singh et al. (2024a).

#### 14.2.3.2 Estimation of Proline Content and Total Soluble Sugar Content

The proline was estimated using the method outlined by Bates et al. (1973). We measured the absorbance at 520 nm and expressed it as mg g FW, using L-proline (Sigma Aldrich, India) as the standard. Using Dey's method (1990), the total soluble sugar (TSS) was estimated. The absorbance was measured at 485 nm and given as mg $g^{-3}$ FW. D-glucose (Sigma-Aldrich) was used as a standard. The additional information is available in the published paper of Singh et al. (2024b).

#### 14.2.3.3 Estimation of Nitrate Reductase Activity

NR activity in fresh leaves was estimated using Jaworski (1971). A tube containing 0.5 g of fresh leaf material was mixed with 2.5 ml each of sodium phosphate buffer and KNO3. The solution absorbance was measured at 540 nm. Details are available in a published paper by Singh et al. (2024a).

#### 14.2.3.4 Estimation of Total Phenol Content and Flavonoid Extraction

Total phenolic content (TPC) was determined using the Folin–Ciocalteu (F–C) reagent as described by Ainsworth and Gillespie (2007). The extract was analyzed for absorbance at 765 nm, and the phenolic content was quantified as mg GAE $g^{-1}$ FW. The total phenolic content is quantified as gallic acid equivalents (GAE). The total flavonoid content (TFC) was assessed using the aluminum trichloride complex formation assay, as described by Singh et al. (2024a). The absorbance was measured at 510 nm and reported as mg rutin $g^{-1}$ FW. Further information can be found in a published article by Singh et al. (2024a).

#### 14.2.3.5 Estimation of Antioxidant Enzymes and MDA Content

The fresh leaves (0.5 g) were ground with an extraction buffer containing 1 mM ethylenediaminetetraacetic acid (EDTA) and 2% PVPP (polyvinylpolypyrrolidone). The collected supernatant, which had the enzymes that were extracted, was used to study superoxide dismutase (SOD), catalase (CAT), and ascorbate peroxidase (APX). Beauchamp and Fridovich (1971) established a methodology for measuring the activity of superoxide dismutase (SOD). The enzyme activity was measured by its absorbance at 560 nm. This was given in units (U) per mg of protein per minute. A method established by Beers and Sizer (1952) was employed to estimate catalase (CAT) activity. The CAT activity was quantified in units (U) per milligram (mg) of protein per minute (min). The ascorbate peroxidase (APX) activity was assessed using the method outlined by Nakano and Asada 1981. We were able to quantify the APX activity in units per mg of protein per minute by observing the decrease in absorbance at 290 nm. Further information is accessible in a paper published by Khan et al. (2023).

#### 14.2.3.6 Histochemical Staining for ROS Production

The histochemical staining of superoxide ions in mustard plant leaves was performed following the method established by Kumar et al. (2014), with minor modifications. To dissolve 0.2 g of NBT, we used 100 mL of 50 mM sodium phosphate buffer at pH 7.5. We submerged the leaf samples in the freshly prepared NBT (nitro blue tetrazolium) solution and incubated them for 5 hours. Once superoxide markings developed on the leaf, we transferred it to ethanol to remove chlorophyll and took photographs. After that, we boiled the samples in 100% ethanol for 20 minutes. We took photographs of the staining results.

We dissolved DAB (3,3-diaminobenzidine) (50 mg) in 45 ml of distilled water with a pH of 3.8 in an amber-colored vial. We changed the volume to 50 ml to make a 1 mg/ml solution. The leaf was immersed in the solution for 6 hours; when a spot of $H_2O_2$ emerged, it was moved to ethanol to eliminate chlorophyll, and photographed.

#### 14.2.3.7 Analysis of Cell Viability and Cellular Damage in Leaves Using Confocal Laser Scanning Microscopy

The fresh leaves were subjected to dual staining for 15 minutes using a solution containing 30 μM propidium iodide (PI) and 10 μM acridine orange (AO). The leaves were cut into small sections. The leaves were rinsed with a phosphate buffer solution (0.1 M, pH 7.0) and subsequently affixed to glass slides. Following this, the stained roots were examined using a confocal laser scanning microscope (Leica Microsystems TCS-SP5 model). The microscope facilitated the visualization and differentiation of living tissues (represented by AO staining) and dead tissues (represented by PI staining).

## 14.3 Results

### 14.3.1 Analysis of Salicylic Acid and Potassium on Growth Traits of the Plants Under Salt Stress

Salt negatively affected the morphological parameters evaluated. Application of 100 mM salt solution resulted in a reduction of 33.70% in root length, 37.01% in shoot length, 38.73% in plant height, 47.44% in dry weight, and 39.26% in leaf area relative to the control plant (Table 14.1).

Under unstressed conditions, however, the supplementation of salicylic acid proved beneficial in enhancing growth parameters (Fig. 14.1). Notably, plants treated with the combination of both salicylic acids and potassium (k) exhibited the highest trend in promoting the above-mentioned growth parameters by 93.91%, 83%, 79.2%, 44.7%, and 77.6% compared to the control plant. SA treatment with potassium improved plant growth most significantly under stress, increasing root length by 56.6%, shoot length by 88.17%, height by 77.55%, dry weight by 54.97%, and leaf area by 80.66%.

### 14.3.2 Effect of Salicylic Acid and Potassium on Total Chlorophyll and Carotenoid Contents of the Plants Under Salt Stress

Salt stress significantly reduces total chlorophyll by 47.45% and total carotenoid content by 41.83% as compared to control groups (Fig. 14.2). Under normal conditions, the treatment of potassium and salicylic acid individually or in a combination effectively reverses the inhibitory effect of NaCl on total chlorophyll and carotenoid content. Under stressed conditions, plants treated with SA and potassium enhanced the total chlorophyll content by 80.64% and the total carotenoid content by 55.38% relative to the plant treated with a 100-mM salt solution.

### 14.3.3 Effect of Salicylic Acid and Potassium on the Total Protein Content of the Plants Under Salt Stress

Under salt stress, there was a 22.36% decrease in total protein content compared to the control plants. Salicylic acid and potassium independently also increased soluble protein content by 17.40% and 8.86%, respectively, compared to the control plants (Table 14.2). The plants that were treated with both salicylic acid and potassium had 28.9% more total protein than the plants that were only treated with salicylic acid. Under stressed conditions, applying salicylic acid along with potassium showed a maximum increase in total protein content by 27.57% compared to the stressed plants.

**Table 14.1** Effect on root length, shoot length, height of plant, root fresh and dry biomass, shoot fresh and dry biomass, number of leaves, and leaf area in mustard cultivar Pusa Jagannath exposed to control or salt stress with or without salicylic acid and potassium either singly or in combination at 30 days

| Treatment (30 DAS) | Root length (cm) | Shoot length (cm) | Height of plant (cm) | Root FW (g/plant) | Root DW (g/plant) | Shoot FW (g/plant) | Shoot DW(g/plant) | No. of Leaves | Leaf Area ($cm^2$) |
|---|---|---|---|---|---|---|---|---|---|
| T1 (Control) | $7.968 \pm 0.407^{c}$ | $16.653 \pm 0.962^{c}$ | $25.621 \pm 1.350^{c}$ | $2.161 \pm 0.096^{c}$ | $0.078 \pm 0.003^{c}$ | $14.557 \pm 0.713^{c}$ | $2.161 \pm 0.106^{c}$ | $4.367 \pm 0.217^{c}$ | $13.689 \pm 0.614^{c}$ |
| T2 (100 mM NaCl) | $5.283 \pm 0.181^{f}$ | $10.416 \pm 0.518^{f}$ | $15.699 \pm 0.713^{f}$ | $1.145 \pm 0.042^{f}$ | $0.041 \pm 0.002^{f}$ | $8.734 \pm 0.395^{f}$ | $1.124 \pm 0.040^{f}$ | $2.314 \pm 0.105^{f}$ | $8.315 \pm 0.406^{f}$ |
| T3 (K) | $11.563 \pm 0.515^{b}$ | $24.797 \pm 1.176^{b}$ | $38.360 \pm 1.529^{b}$ | $2.907 \pm 0.101^{b}$ | $0.099 \pm 0.004^{b}$ | $16.213 \pm 0.800^{b}$ | $2.503 \pm 0.112^{b}$ | $5.966 \pm 0.243^{b}$ | $19.199 \pm 0.828^{b}$ |
| T4 (0.5 mM SA) | $14.055 \pm 0.658^{b}$ | $26.710 \pm 1.124^{b}$ | $41.766 \pm 1.990^{b}$ | $3.048 \pm 0.137^{b}$ | $0.101 \pm 0.005^{b}$ | $17.469 \pm 0.861^{b}$ | $2.703 \pm 0.125^{b}$ | $6.157 \pm 0.302^{b}$ | $21.122 \pm 0.953^{b}$ |
| T5 (K+ 0.5 mM SA) | $15.451 \pm 0.707^{a}$ | $30.462 \pm 1.274^{a}$ | $45.913 \pm 2.011^{a}$ | $3.350 \pm 0.155^{a}$ | $0.121 \pm 0.005^{a}$ | $18.925 \pm 0.880^{a}$ | $3.122 \pm 0.151^{a}$ | $6.769 \pm 0.316^{a}$ | $24.318 \pm 1.215^{a}$ |
| T6 (NaCl + K) | $6.080 \pm 0.253^{e}$ | $11.988 \pm 0.591^{e}$ | $18.069 \pm 0.955^{de}$ | $1.318 \pm 0.052^{e}$ | $0.047 \pm 0.002^{e}$ | $10.918 \pm 0.513^{e}$ | $1.255 \pm 0.061^{e}$ | $2.464 \pm 0.112^{e}$ | $9.570 \pm 0.371^{e}$ |
| T7 (NaCl + O.5 mM SA) | $6.579 \pm 0.299^{d}$ | $12.970 \pm 0.618^{d}$ | $19.550 \pm 0.910^{e}$ | $1.426 \pm 0.069^{d}$ | $0.051 \pm 0.001^{d}$ | $12.010 \pm 0.593^{d}$ | $1.342 \pm 0.065^{d}$ | $2.882 \pm 0.113^{d}$ | $10.355 \pm 0.443^{d}$ |
| T8 (NaCl + K + 0.5 mm SA) | $8.273 \pm 0.353^{c}$ | $19.6 \pm 0.603^{c}$ | $25.874 \pm 1.196^{c}$ | $1.994 \pm 0.080^{c}$ | $0.071 \pm 0.002^{c}$ | $13.829 \pm 0.686^{c}$ | $2.312 \pm 0.107^{c}$ | $4.124 \pm 0.201^{c}$ | $15.022 \pm 0.649^{c}$ |

The data are presented as the mean of three replicates and represented as mean ± SE ($n = 4$)
Mean ± SE followed by dissimilar lowercase letter/s depicts significant differences among the treatments at $p \leq 0.05$ by DMRT

**Fig. 14.1** Effect of salicylic acid (0.1 mM) and potassium (40 mg $kg^{-1}$ soil) either singly or in combination with each other on the growth of the mustard cultivar. The data are presented as the mean of 4 replicates ($n = 4$) and represented as Mean ± SE. Values of bars with the same letter are not significantly different, $p < 0.05$ Duncan's multiple range

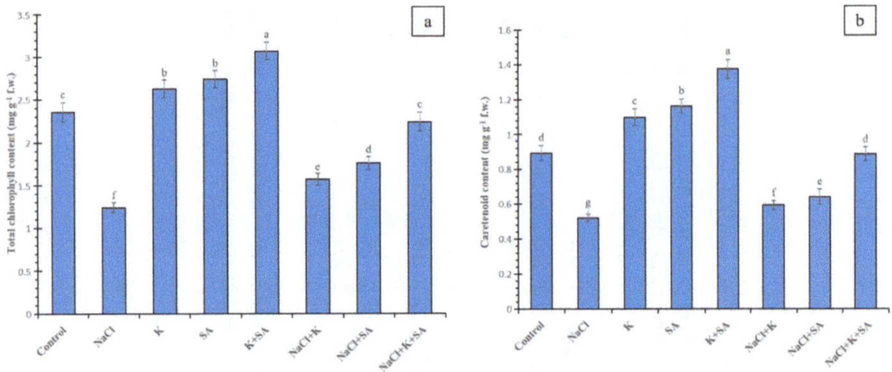


**Fig. 14.2** Effect of salicylic acid (0.1 mM) and potassium (40 mg $kg^{-1}$ soil) either singly or in combination with each other on the (**a**) Total chlorophyll content and (**b**) carotenoid content in the mustard cultivar. The data are presented as the mean of 4 replicates ($n = 4$) and represented as Mean ± SE. Values of bars with the same letter are not significantly different, $P < 0.05$ Duncan's multiple range

### 14.3.4 Impact of Salicylic Acid and Potassium on Proline, Sugar Content, and NR Activity Under Salt Stress

Under salt stress, there was a 62.02% rise in proline content and a 50.83% increase in total soluble sugar content relative to the control groups. Salicylic acid and potassium independently increased proline accumulation by 67.83% and 75.90%,

**Table 14.2** Effect on superoxide dismutase (SOD), catalase (CAT) ascorbate peroxidase (APX), malondialdehyde (MDA), and protein content in mustard cultivar Pusa Jagannath exposed to control or salt-stressed with or without salicylic acid and potassium either singly or in combination

| Treatment (30 DAYS) | SOD (U $mg^{-1}$ Protein $min^{-1}$) | CAT (U $mg^{-1}$ Protein $min^{-1}$) | APX (U $mg^{-1}$ Protein $min^{-1}$) | MDA (n mol $g^{-1}$ F.W.) | Protein Content (mg $g^{-1}$ F.W.) |
|---|---|---|---|---|---|
| T1 (Control) | $8.64 \pm 0.625^{e}$ | $176 \pm 8.112^{e}$ | $4.594 \pm 0.231^{e}$ | $1.254 \pm 0.053^{g}$ | $9.031 \pm 0.462^{c}$ |
| T2 (100mM NaCl) | $14.180 \pm 0.641^{d}$ | $215 \pm 8.820^{d}$ | $8.565 \pm 0.212^{d}$ | $2.209 \pm 0.150^{a}$ | $7.011 \pm 0.351^{f}$ |
| T3 (K) | $11.400 \pm 0.449^{f}$ | $155 \pm 7.900^{f}$ | $3.353 \pm 0.118^{g}$ | $1.583 \pm 0.079^{e}$ | $9.832 \pm 0.487^{b}$ |
| T4 (0.5 mM SA) | $12.410 \pm 0.612^{f}$ | $192 \pm 7.320^{f}$ | $5.351 \pm 0.201^{f}$ | $1.613 \pm 0.079^{e}$ | $10.605 \pm 0.545^{b}$ |
| T5 (K+ 0.5 mM SA) | $6.821 \pm 0.318^{fg}$ | $198 \pm 6.460^{g}$ | $5.443 \pm 0.138^{h}$ | $1.376 \pm 0.066^{f}$ | $11.644 \pm 0.582^{a}$ |
| T6 (NaCl + K) | $15.900 \pm 0.846^{c}$ | $190 \pm 9.420^{c}$ | $5.805 \pm 0.264^{c}$ | $2.811 \pm 0.141^{b}$ | $7.168 \pm 0.398^{e}$ |
| T7 (NaCl + O.5 mM SA) | $17.620 \pm 0.812^{b}$ | $192 \pm 9.120^{b}$ | $6.206 \pm 0.292^{b}$ | $2.709 \pm 0.125^{c}$ | $8.065 \pm 0.408^{d}$ |
| T8 (NaCl + K + 0.5 mM SA) | $22.110 \pm 1.081^{a}$ | $212 \pm 9.200^{a}$ | $8.357 \pm 0.416^{a}$ | $1.897 \pm 0.092^{d}$ | $8.944 \pm 0.383^{c}$ |

The data are presented as the mean of three replicates and represented as Mean ± SE ($n = 4$)
Mean ± SE followed by dissimilar lowercase letter/s depicts significant differences among the treatments at $p \leq 0.05$ by DMRT

respectively, relative to the control groups (Fig. 14.3). Plants exposed to salt when treated with potassium and salicylic acid, individually or in combination, showed an increase in sugar content of 24.74%, 28.63%, and 31.03%, respectively. Salt application was found to decrease NR activity by 25.35% relative to control plants. Under unstressed conditions, the individual application of SA and potassium notably increased the NR activity by 3.48% and 8.55%. SA, along with potassium, increased the NR activity by 12.44% compared to the control plant. SA and potassium application led to a 27.72% increase in NR activity compared to salt-stressed plants.

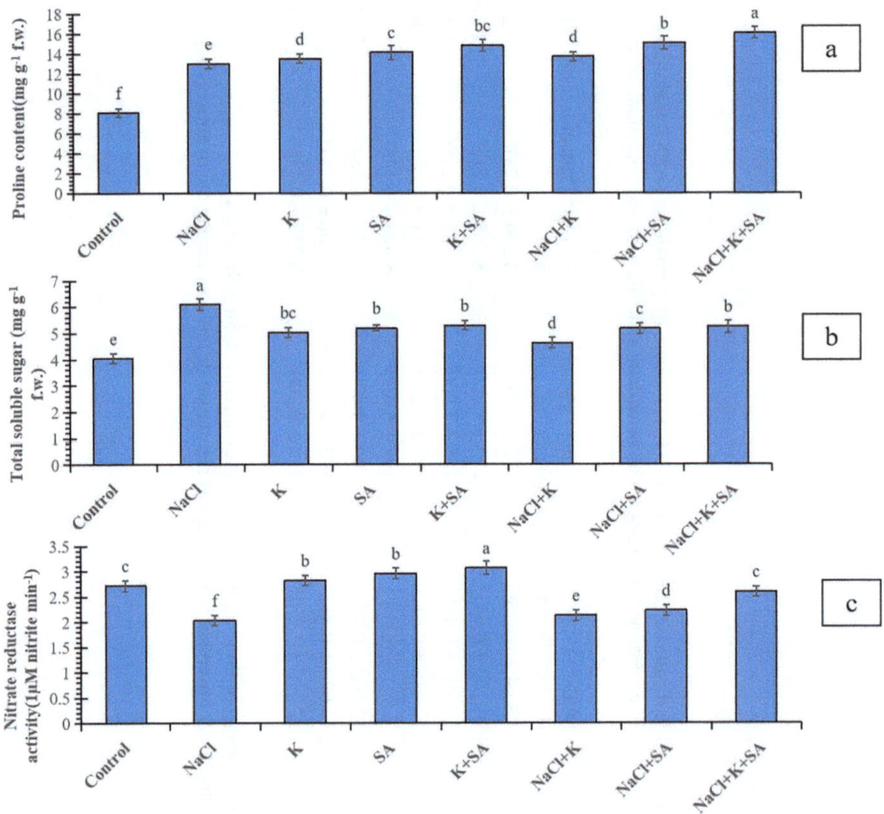


**Fig. 14.3** Effect of salicylic acid (0.1 mM) and potassium (40 mg $kg^{-1}$ soil) either singly or in combination with each other on the (**a**) Proline content, (**b**) Total solubel sugar and (**c**) nitrate reductase activity in the mustard cultivar. The data are presented as the mean of 4 replicates ($n = 4$) and represented as Mean ± SE. Values of bars with the same letter are not significantly different, $P < 0.05$ Duncan's multiple range

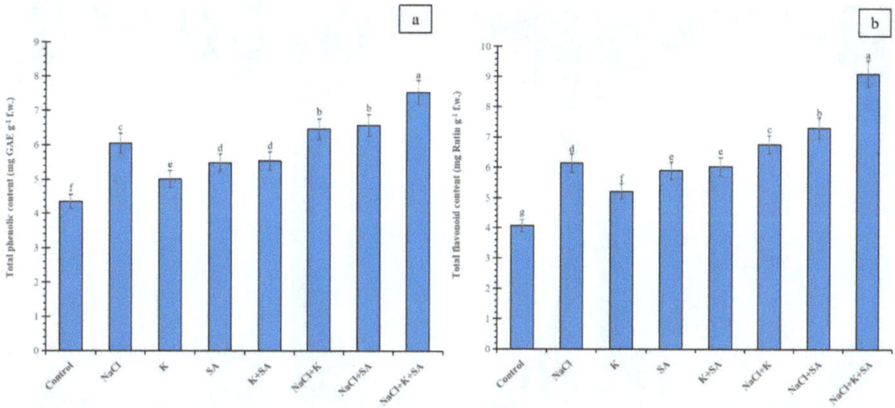


**Fig. 14.4** Effect of salicylic acid (0.1 mM) and potassium (40 mg $kg^{-1}$ soil) either singly or in combination with each other on the (**a**) Total phenolic content and (**b**) Total flavonoid content in the mustard cultivar. The data are presented as the mean of 4 replicates ($n = 4$) and represented as Mean ± SE. Values of bars with the same letter are not significantly different, $P < 0.05$ Duncan's multiple range

### 14.3.5 Effect of Salicylic Acid and Potassium on Total Phenol and Total Flavonoid Content Under Salt Stress

Salt stress led to an increase in total flavonoid content by 51.26% compared to the control plants (Fig. 14.4). However, the treatments with potassium, salicylic acid, and their combination further enhanced the total phenol content in the stressed plants by 6.95%, 8.77%, and 24.66%, respectively. A similar trend has been found in the total flavonoid content.

### 14.3.6 Salicylic Acid and Potassium Improved Antioxidative Enzyme Activity and MDA Content

Plants subjected to salt stress exhibited increased levels of antioxidant enzyme activities, including SOD, CAT, and APX by 64.12%, 22.15%, and 86.43% compared to the control plants. Applying salicylic acid and potassium separately under unstressed conditions increased their activity relative to control plants (Table 14.2). When salicylic acid was applied alone, it resulted in a notable increase in SOD content by 43.63%, CAT content by 9.09%, and APX content by 16.47% compared to the control plants. Moreover, treating the salt-stressed plants with a combination of salicylic acid and potassium enhancer or with potassium enhancer alone led to even more significant improvements in antioxidant content. Specifically, the SOD content increased by 39.47%, 65.96%, and 93.94%, respectively, while the CAT content increased by 13.54%, 25.80%, and 37.77%, respectively. Furthermore, the APX content exhibited a substantial increase of 75.75%, 115.15%, and 151.51%, respectively, when treated with potassium, salicylic acid, and a combination of both.

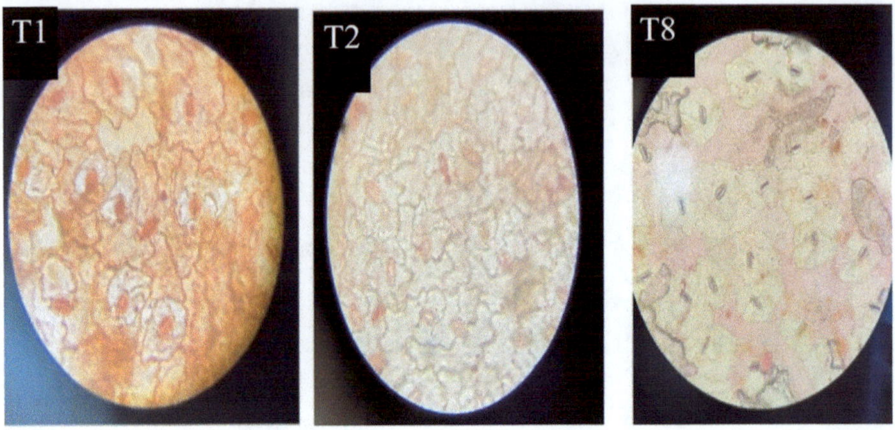


**Fig. 14.5** Stomatal behavior of mustard cultivar under (**a**) control (**b**) 100 mM NaCl + K (40 mg $kg^{-1}$) soil + SA (0.5 mM) at 30 days of sowing. The stomatal aperture was observed under compound microscopy. NaCl sodium chloride, SA salicylic acid, K potassium

Under salt stress, there was a 76.15% increase in MDA content compared to the control plant. Under the stressed conditions, the application of SA with potassium decreases MDA content by 14.12% as compared to the stressed plants (Table 14.2).

### 14.3.7 Effect of Salicylic Acid and Potassium on Stomatal Responses Under Salt Stress

Salt stress severely damaged the stomatal aperture (Fig. 14.5). The light illustrations of salt-stressed plants display morphological abnormalities and significant damage to stomatal cells, resulting in alterations to the stomatal shape and aperture. Applying SA and potassium under unstressed conditions led to reduced stomatal aperture. Additionally, we found that SA under stressed conditions along with potassium, enhances stomata opening.

### 14.3.8 Effects of Potassium and Salicylic Acid on the Histochemical Staining of Superoxide Anion ($O_2$. -) and Hydrogen Peroxide

The NBT and DAB staining methods were used to assess the levels of oxidative stress indicators in the leaf (Fig. 14.6). The color intensity of NBT and DAB staining correlates with the levels of $O_2$ and $H_2O_2$. When compared to the control, the leaves that were exposed to salt-stressed leaves got more dark blue spots that corresponded to $O_2$ because NBT was turned into formazan. However, plants treated with SA and K showed reduced dark blue spots. Plants treated with SA and K represented the greatest reduction in $O_2$ and hence minimal blue spots. On the other

**Fig. 14.6** Visualization of $H_2O_2$ by (**a**) DAB and, (**b**) superoxide ion by NBT in mustard cultivar under control, 100 mM NaCl, K (40 mg $kg^{-1}$) soil, SA (0.5 mM) and 100 mM NaCl + K (40 mg $kg^{-1}$) soil + SA (0.5 mM). NaCl sodium chloride, SA salicylic acid, K potassium

hand, DAB stain was used to measure the intensity of brown spots representing $H_2O_2$ accumulation, as in the presence of peroxidases, DAB reacts with $H_2O_2$ to produce a brown polymerization product. Compared to the control, leaves under salt stress showed significant brown spots. Brown spot reduction was seen in leaves treated with only SA and K, but the maximum reduction was observed in leaves treated with both SA and K together.

### 14.3.9 Cellular Damage to Mustard Leaves Under Salt Stress as Determined via CLSM and Overcoming the Damage Using Salicylic Acid and Potassium

Under salt stress, mustard plant leaves experience oxidative stress, resulting in tissue death. The breakdown of cellular membranes was indicated by an increase in red fluorescence from PI dye in the leaf cells. PI dye can penetrate the membranes of dead cells, causing them to accumulate a red color. Therefore, PI dye serves as a reliable indicator of stressed conditions. However, when mustard leaves were treated with potassium or salicylic acid individually, they exhibited reduced cell membrane damage compared to untreated leaves (Fig. 14.7). Notably, leaves treated with a combination of salicylic acid and potassium showed minimal red fluorescence (indicating less membrane damage) both under stressed and unstressed conditions. The leaves of control plants, on the other hand, had strong green fluorescence from

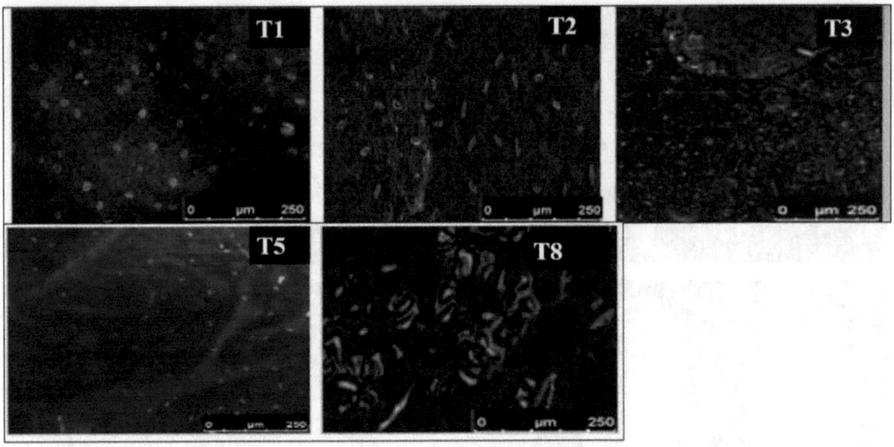


**Fig. 14.7** Live and dead cell imaging of leaves in mustard cultivar under control, 100 mM NaCl, K (40 mg $kg^{-1}$) soil, SA (0.5 mM) and 100 mM NaCl + K (40 mg $kg^{-1}$) soil + SA (0.5 mM). Cell viability observed under confocal microscopy. NaCl sodium chloride, SA salicylic acid, K potassium

acridine orange (AO) and much less PI fluorescence, which suggests that there was less cell damage.

### 14.3.10 Yield Parameters

Under salt stress, there was a decrease observed in 9.43% for pod per plant, 38.23% for pod length, 32.80% for number of seeds per pod, 39.14% for number of seeds per plant, and 27.22% for 1000 seeds weight (Table 14.3). However, we observed notable improvements in the yield parameters when we treated the salt-stressed plants with potassium, salicylic acid, or a combination of both. The application of potassium resulted in a surge of 1.73% in pods per plant, 14% in pod length, 8.77% in the number of seeds per pod, 10.65% in the number of seeds per plant, and 10.15% in 1000 seeds weight. Similarly, the use of salicylic acid led to increases of 5.83% in pods per plant, 37.14% in pod length, 18.25% in the number of seeds per pod, 47.78% in the number of seeds per plant, and 23.27% in 1000 seeds weight. When salicylic acid and potassium were applied together, the results were even better. The number of pods per plant increased by 8.09%, the length of each pod increased by 53.71%, the number of seeds per pod increased by 30.98%, the number of seeds per plant increased by 41.58%, and the weight of 1000 seeds increased by 31.73%.

**Table 14.3** Effect on pod/plant, pod length, pod fresh weight, pod dry weight, number of seeds per pot, fresh weight of seeds per pod, dry weight of seeds per pot, number of seeds per plant, and 1000 seed weight in mustard cultivar Pusa Jagannath exposed to control or salt-stressed with or without salicylic acid and potassium either singly or in combination

| Treatments | Pod/plant | Pod length (cm) | Pod FW (g) | Pod DW (g) | No. of Seeds/ pod | No. of seeds /pod (FW) | seeds/pod (DW) | No. of seeds /plant | 1000 seed weight (g) |
|---|---|---|---|---|---|---|---|---|---|
| T1 (Control) | 63.667 ± 2.551[c] | 5.667 ± 0.265[d] | 0.355 ± 0.016[c] | 0.162 ± 0.008[c] | 22.667 ± 1.112[c] | 0.275 ± 0.015[d] | 0.100 ± 0.004[c] | 1443.139 ± 71.157[c] | 11.370 ± 0.469[c] |
| T2 (100mM Nacl) | 57.663 ± 2.772[f] | 3.500 ± 0.149[g] | 0.253 ± 0.012[f] | 0.120 ± 0.006[f] | 15.231 ± 0.743[f] | 0.180 ± 0.008[g] | 0.050 ± 0.013[f] | 878.265 ± 41.913[f] | 8.274 ± 0.314[f] |
| T3 (K) | 64.665 ± 3.137[b] | 5.800 ± 0.265[c] | 0.395 ± 0.018[b] | 0.169 ± 0.008[b] | 23.350 ± 1.056[b] | 0.280 ± 0.012[c] | 0.108 ± 0.004[b] | 1509.927 ± 72.496[b] | 13.700 ± 0.625[b] |
| T4 (0.5 mM SA) | 68.000 ± 3.141[b] | 6.333 ± 0.302[b] | 0.401 ± 0.012[b] | 0.184 ± 0.008[b] | 27.657 ± 1.266[b] | 0.368 ± 0.016[b] | 0.122 ± 0.004[b] | 1880.676 ± 84.034[b] | 14.333 ± 0.665[b] |
| T5 (K+ 0.5 mM SA) | 72.333 ± 3.410[a] | 7.100 ± 0.326[a] | 0.533 ± 0.017[a] | 0.197 ± 0.009[a] | 30.502 ± 1.475[a] | 0.490 ± 0.024[a] | 0.153 ± 0.007[a] | 2206.301 ± 107.315[a] | 15.900 ± 0.715[a] |
| T6 (NaCl + K) | 58.661 ± 2.548[e] | 3.990 ± 0.175[f] | 0.282 ± 0.016[e] | 0.129 ± 0.006[e] | 16.567 ± 0.813[e] | 0.204 ± 0.011[f] | 0.056 ± 0.002[e] | 971.836 ± 46.592[e] | 9.114 ± 0.426[e] |
| T7 (NaCl + O.5 mM SA) | 59.557 ± 3.978[d] | 4.267 ± 0.208[e] | 0.299 ± 0.012[d] | 0.136 ± 0.006[d] | 17.210 ± 0.845[d] | 0.220 ± 0.011[e] | 0.067 ± 0.003[d] | 1024.975 ± 49.249[d] | 9.805 ± 0.430[d] |
| T8 (NaCl + K + 0.5 mm SA) | 62.331 ± 3.165[c] | 5.380 ± 0.254[d] | 0.340 ± 0.017[c] | 0.159 ± 0.007[c] | 19.950 ± 0.918[c] | 0.256 ± 0.012[d] | 0.094 ± 0.003[c] | 1243.503 ± 60.175[c] | 10.900 ± 0.505[c] |

The data are presented as the mean of four replicates and represented as Mean ± SE ($n = 4$)
Mean ± SE followed by dissimilar lowercase letter/s depicts significant differences among the treatments at $p \leq 0.05$ by DMRT

## 14.4 Discussion

Salinity stress is a substantial hazard to agricultural productivity on a global scale, resulting in damaging effects on the morphological and biochemical functions of plants. It inhibits seed germination, plant growth, development, and, ultimately, agricultural yield (Zhang and Dai 2019). Salicylic acid (SA) is a phenolic compound and endogenous plant hormone regulator that is essential for a variety of biochemical and physiological processes. It has been found to improve plant resilience to both biotic and abiotic stress (Yalpani et al. 1994; Durner and Klessig 1996; Szalai et al. 2000). Salicylic acid acts as a signaling molecule, initiating plant development and defensive mechanisms, including systemic acquired resistance and the hypersensitive response to pathogen assaults (Klessig and Malamy 1994). Previous research showed that salinity stress decreases the total dry matter of plants, which leads to a decrease in crop yield. However, potassium has been shown to reduce the deleterious effects of salt stress on plant development and biomass production. Under saline conditions, the growth and biomass yield of cereals such as barley and legumes are enhanced by potassium application (Kabir et al. 2004; Mahmood 2011; and Dawood et al. 2014). This study investigates the mechanisms by which SA and K mediate the improvement of biochemical attributes, yield, and growth in mustard plants under salt stress by decreasing ROS levels and increasing the activity of critical antioxidant enzymes.

Salt stress has a detrimental effect on mustard growth parameters, including root length, stem length, plant height, and leaf area (Table 14.1). These effects are predominantly associated with impaired cell expansion and division and reduced meristematic activity. Salicylic acid enhances nutrient assimilation, stimulates cell division, regulates stress-responsive genes, and activates defense mechanisms, thereby improving stress tolerance and growth (El-Esawi et al. 2017). Potassium, which is crucial for osmotic regulation, enzyme activation, and protein synthesis, improves stress resilience and growth by improving water-use efficiency, stomatal function, and nutrient transport (Hafsi et al. 2021; Shabala et al. 2016). The findings suggest that the combined application of salicylic acid and potassium yields the most efficacious results in both normal and salt-stressed conditions, as both compounds independently stimulate growth.

Our data indicate that salt stress significantly reduced the total chlorophyll and carotenoid content (Fig. 14.2). Salt stress significantly affects the concentration of pigments, including total chlorophyll and carotenoids, as well as protein content (Aghaleh et al. 2009; Ramzan et al. 2023). Nevertheless, the treatment of potassium, salicylic acid (Ilyas et al. 2024), and their combination significantly mitigated the adverse effects of salt stress, leading to an enhancement of the assessed parameters.

Total soluble sugar accumulation in plants because of salt stress can be attributed to the osmotic adjustment mechanism. When plants are under salt stress, they tend to store compatible solutes, like carbohydrates, to keep their cellular osmotic potential and water balance (Ashraf and Harris 2004). According to Hayat et al. (2010), salicylic acid has been shown to facilitate the metabolism and mobilization of carbohydrates, thereby increasing the accumulation of sugar in plants. Furthermore, potassium plays a regulatory role in the transport and metabolism of sugar (Cakmak 2005). Sugar content is synergistically increased by the combined application of salicylic acid and potassium, which suggests that they have complementary effects on the regulation of carbohydrates in plants (Fig. 14.3). Proline, a critical osmolyte, is also essential for the improvement of plant tolerance to salinity. It functions as an antioxidant and osmoprotectant, thereby assisting plants in the preservation of cellular integrity and their defense against oxidative stress (Szabados and Savouré 2010). In the current investigation, proline content increased in response to salt stress, and additional increases were observed with the application of potassium, salicylic acid, and a combination of both.

Nitrogen is a vital nutrient for plant growth and development, particularly in the early stages. This study found that salt stress reduces nitrate reductase (NR) activity, indicating disruptions in nitrogen assimilation. However, using potassium and salicylic acid (SA), and both together, greatly increased NR activity under salt stress (Fig. 14.3). This is in line with previous research on nitrogen metabolism and salt stress (Munns and Tester 2008). Potassium enhances NR activity by activating enzymes involved in nitrogen assimilation, while salicylic acid modulates enzyme activities and stress-related signaling pathways (Khan et al. 2015; Kaur et al. 2022). These findings underscore the potential of potassium and salicylic acid, individually and in combination, to mitigate salt stress effects on nitrogen metabolism and enhance plant resilience and stress tolerance.

The current study reveals that the total phenol and flavonoid content of the plants significantly increased in comparison to the control plant under salt stress (Fig. 14.4). This result aligns with other studies revealing the important role of phenolic compounds as defense mechanisms against environmental challenges (Dixon et al. 2002). As antioxidants, phenolic substances scavenge ROS and protect against oxidative damage to plant tissues (Fini et al. 2011). Consequently, the noticeable increase in total phenol content under salt stress might be linked to the defensive response of the plant against oxidative stress. Potassium is an essential component of plant stress responses and has been reported to facilitate the synthesis of phenolic compounds (Wang et al. 2013). Additionally, the total phenol content was similarly affected by the application of salicylic acid (SA), suggesting that it is involved in the increase of phenolic compound production. SA, a signaling molecule, is involved in a variety of physiological processes, including the regulation of secondary metabolites, including phenols (Klessig et al. 2000). Salicylic acid's complementary action with potassium in controlling gene expression and metabolic pathways connected to phenolic compound manufacturing may help to explain the synergistic impact (Singh et al. 2019). Similar patterns also applied to the overall flavonoid content. Salt stress led to an increase in the total flavonoid content in comparison to the control group, suggesting that the flavonoid biosynthesis pathways were activated under stress conditions. Treatments with potassium, salicylic acid, or both increased the total flavonoid concentration in salt-stressed plants. Potassium and salicylic acid have been reported to stimulate flavonoid accumulation in plants (Marschner and Cakmak 1989; Hayat et al. 2010).

In our investigation, salt stress caused oxidative stress, as evidenced by the elevated levels of $H_2O_2$, which impeded the correct functioning of the plant (Table 14.2). In agreement with our findings, Ahmad et al. (2016) documented the synergistic effects of salicylic acid and potassium in enhancing antioxidant enzymes and mitigating oxidative stress in plants. The use of SA and K reduces ROS levels and lipid peroxidation by improving the antioxidant enzyme system (Table 14.2). This discovery is in accordance with prior research that has shown the involvement of salt stress in the induction of oxidative stress and lipid peroxidation in plant tissues (Hossain et al. 2015; Rady et al. 2023). Potassium is recognized for its crucial function in plant stress responses, including the control of reactive oxygen species scavenging systems and antioxidant enzyme activities (Sharma et al. 2012). Salicylic acid has been shown to mitigate oxidative stress by increasing antioxidant enzyme activities and scavenging reactive oxygen species (Shang et al. 2022). This synergistic effect implies that salicylic acid and potassium collaborate to improve the antioxidant defense system and mitigate lipid peroxidation in plants during salt-stress conditions. The combined treatment may entail the activation of common signaling pathways and the modulation of gene expression associated with antioxidant metabolism (Shang et al. 2022). The treatment with salicylic acid alone led to a significant elevation in the levels of superoxide dismutase (SOD), catalase (CAT), and ascorbate peroxidase (APX) in comparison to the control plants. This result is in accordance with prior research that showed the role of salicylic acid in improving the antioxidant defense system of plants in response to many stresses (Shang et al. 2022). The antioxidant defense system is more effectively enhanced by the combined treatment of salicylic acid and potassium than by individual treatments, as evidenced by the increase in SOD, CAT, and APX content. Superoxide dismutase is a crucial enzyme that scavenges superoxide radicals, whereas catalase and ascorbate peroxidase facilitate the decomposition of hydrogen peroxide, a detrimental reactive oxygen species. The elevation in enzyme levels signifies an increase in the antioxidant defense mechanism, aiding plants in mitigating the adverse impacts of oxidative stress caused by salinity (Gill and Tuteja 2010).

In our investigation, salt exposure resulted in extensive damage to stomatal cells, including morphological distortions, altered size, and impaired opening in response to salt stress. The administration of salicylic acid or potassium to salt-stressed plants led to significant enhancements in stomatal functioning and plant morphology (Jangra et al. 2022). Both salicylic acid and potassium treatments reduced morphological deformities and increased stomatal openness as compared to untreated salt-stressed plants. The findings confirm the potential of salicylic acid and potassium as effective mitigators of salt-induced stomatal dysfunction and morphological distortions (Johnson et al. 2002).

Salt stress led to a substantial reduction in the number of pods, the length of pods, the number of seeds per pod, the number of seeds per plant, and the weight of 1000 seeds (Table 14.3). This supports earlier research that showed salt adversely affects crop production (Ashraf and Harris 2004). The application of potassium slightly improved these parameters, highlighting its importance in increasing the yields of crops under stress (Shang et al. 2022). These traits were also improved by salicylic acid, which is in line with its known ability to boost growth and yield under stress (Miura and Tada 2014). The combined treatment of potassium and salicylic acid gave the most significant improvements, indicating a synergistic effect. These findings indicate that salicylic acid and potassium modulate physiological processes, alleviate salt stress, and improve productivity by enhancing nutrient absorption, photosynthesis, osmotic regulation, and enzyme activation (Shang et al. 2022; Wang et al. 2013).

## 14.5 Conclusion

The synergistic application of potassium (K) and salicylic acid (SA) is a promising approach to alleviate the detrimental effects of salt stress in mustard plants. This is achieved by enhancing critical physiological and biochemical processes, such as the synthesis of phenolic and flavonoid compounds, proline accumulation, and antioxidative defense. This combination substantially enhances the resilience of plants in the presence of salinity stress. The results highlight the potential of SA and K to enhance crop productivity in challenging environmental conditions. To completely understand the mechanisms of their synergistic effects, it is important to conduct more thorough research on the enzymatic pathways of nitrogen and sulfur assimilation, as well as the hormonal interplay between abscisic acid (ABA) and ethylene. Such insights will facilitate the development of agricultural practices that are optimized to ensure sustainable crop production in saline-prone regions and enhance salt tolerance.

**Authors Contribution** RS: Writing first draft and making figures, VK: experimentation and formatting, AJ: writing first draft and formatting, PJ: experimentation, SU: supervision, editing, NI: conceptualization, editing, supervision.

## References


Aghaleh M, Niknam V, Ebrahimzadeh H, Razavi K (2009) Salt stress effects on growth, pigments, proteins and lipid peroxidation in *Salicornia persica* and *S. europaea*. Biol Plant 53:243–248

Ahanger MA, Tomar NS, Tittal M, Argal S, Agarwal R (2017) Plant growth under water/salt stress: ROS production; antioxidants and significance of added potassium under such conditions. Physiol Mol Biol Plants 23:731–744

Ahmad P, Abdel Latef AA, Hashem A, Abd_Allah EF, Gucel S, Tran LS (2016) Nitric oxide mitigates salt stress by regulating levels of osmolytes and antioxidant enzymes in chickpea. Front Plant Sci 7:347

Ainsworth EA, Gillespie KM (2007) Estimation of total phenolic content and other oxidation substrates in plant tissues using Folin–Ciocalteu reagent. Nat Protoc 2(4):875–877

Ali S, Zeng F, Cai S, Qiu B, Zhang G (2011) The interaction of salinity and chromium in the influence of barley growth and oxidative stress. Plant Soil Environ 57:153–159

Ashraf MP, Harris PJ (2004) Potential biochemical indicators of salinity tolerance in plants. Plant Sci 166(1):3–16

Bacha H, Ródenas R, López-Gómez E, García-Legaz MF, Nieves-Cordones M, Rivero RM, Martínez V, Botella MÁ, Rubio F (2015) High Ca2+ reverts the repression of high-affinity K+

uptake produced by Na+ in *Solanum lycopersycum* L.(var. microtom) plants. J Plant Physiol 180:72–79

Bates LS, Waldren RP, Teare I (1973) Rapid determination of free proline for water-stress studies. Plant Soil 39:205–207

Beauchamp C, Fridovich I (1971) Superoxide dismutase: improved assays and an assay applicable to acrylamide gels. Anal Biochem 44(1):276–287

Beers RF, Sizer IW (1952) A spectrophotometric method for measuring the breakdown of hydrogen peroxide by catalase. J Biol Chem 195(1):133–140

Bradford MM (1976) A rapid and sensitive method for the quantitation of microgram quantities of protein utilizing the principle of protein-dye binding. Anal Biochem 72(1-2):248–254

Cakmak I (2005) The role of potassium in alleviating detrimental effects of abiotic stresses in plants. J Plant Nutr Soil Sci 168(4):521–530

Dawood MG, Abdelhamid MT, Schmidhalter U (2014) Potassium fertiliser enhances the salt-tolerance of common bean (*Phaseolus vulgaris* L.). J Hortic Sci Biotechnol 89(2):185–192

Dey PM (1990) Oligosaccharides. In: Methods in plant biochemistry, vol 2. Academic Press, pp 189–218

Dixon RA, Achnine L, Kota P, Liu CJ, Reddy MS, Wang L (2002) The phenylpropanoid pathway and plant defence—a genomics perspective. Mol Plant Pathol 3(5):371–390

Durner J, Klessig DF (1996) Salicylic acid is a modulator of tobacco and mammalian catalases. J Biol Chem 271(45):28492–28501

Egilla JN, Davies FT, Drew MC (2001) Effect of potassium on drought resistance of *Hibiscus rosa-sinensis* cv. Leprechaun: plant growth, leaf macro-and micronutrient content and root longevity. Plant Soil 229:213–224

El-Esawi MA, Elansary HO, El-Shanhorey NA, Abdel-Hamid AM, Ali HM, Elshikh MS (2017) Salicylic acid-regulated antioxidant mechanisms and gene expression enhance rosemary performance under saline conditions. Front Physiol 8:716

Fini A, Brunetti C, Di Ferdinando M, Ferrini F, Tattini M (2011) Stress-induced flavonoid biosynthesis and the antioxidant machinery of plants. Plant Signal Behav 6(5):709–711

Gautam S, Singh PK (2009) Salicylic acid-induced salinity tolerance in corn grown under NaCl stress. Acta Physiol Plant 31:1185–1190

Gill SS, Tuteja N (2010) Reactive oxygen species and antioxidant machinery in abiotic stress tolerance in crop plants. Plant Physiol Biochem 48(12):909–930

Hafsi C, Bettaib J, Falleh H, Zorrig W, Ksouri R, Abdelly C, Debez A (2021) Phenolic accumulation and related antioxidant capacity in stems and roots of the Tunisian extremophile *Sulla carnosa* as influenced by potassium application under salinity stress. Arab J Geosci 14(1):34

Hayat Q, Hayat S, Irfan M, Ahmad A (2010) Effect of exogenous salicylic acid under changing environment: a review. Environ Exp Bot 68(1):14–25

Hiscox JD, Israelstam GF (1979) A method for the extraction of chlorophyll from leaf tissue without maceration. Can J Bot 57(12):1332–1334

Hossain MA, Bhattacharjee S, Armin SM, Qian P, Xin W, Li HY, Burritt DJ, Fujita M, Tran LS (2015) Hydrogen peroxide priming modulates abiotic oxidative stress tolerance: insights from ROS detoxification and scavenging. Front Plant Sci 6:420

Ilyas M, Maqsood MF, Shahbaz M, Zulfiqar U, Ahmad K, Naz N, Ali MF, Ahmad M, Ali Q, Yong JW, Ali HM (2024) Alleviating salinity stress in canola (*Brassica napus* L.) through exogenous application of salicylic acid. BMC Plant Biol 24(1):611

Jangra M, Devi S, Satpal KN, Goyal V, Mehrotra S (2022) Amelioration effect of salicylic acid under salt stress in Sorghum bicolor L. Appl Biochem Biotechnol 194(10):4400–4423

Jaworski EG (1971) Nitrate reductase assay in intact plant tissues. Biochem Biophys Res Commun 43(6):1274–1279

Johnson R, Vishwakarma K, Hossen MS, Kumar V, Shackira AM, Puthur JT, Abdi G, Sarraf M, Hasanuzzaman M (2002) Potassium in plants: growth regulation, signaling, and environmental stress tolerance. Plant Physiol Biochem 172:56–69

Kabir ME, Karim MA, Azad MA (2004) Effect of potassium on salinity tolerance of mungbean (*Vigana radiata* L. Wilczek), pp 103–110

Kaur G, Tak Y, Asthir B (2022) Salicylic acid: a key signal molecule ameliorating plant stresses. Cereal Res Commun 50(4):617–626

Khan MI, Fatma M, Per TS, Anjum NA, Khan NA (2015) Salicylic acid-induced abiotic stress tolerance and underlying mechanisms in plants. Front Plant Sci 6:462

Khan V, Umar S, Iqbal N (2023) Synergistic action of *Pseudomonas fluorescens* with melatonin attenuates salt toxicity in mustard by regulating antioxidant system and flavonoid profile. Physiol Plant 175(6):e14092

Klessig DF, Malamy J (1994) The salicylic acid signal in plants. Plant Mol Biol 26:1439–1458

Klessig DF, Durner J, Noad R, Navarre DA, Wendehenne D, Kumar D, Zhou JM, Shah J, Zhang S, Kachroo P, Trifa Y (2000) Nitric oxide and salicylic acid signaling in plant defense. Proc Natl Acad Sci U S A 97(16):8849–8855

Kumar D, Yusuf MA, Singh P, Sardar M, Sarin NB (2014) Histochemical detection of superoxide and H2O2 accumulation in Brassica *juncea* seedlings. Bio-protocol 4(8):e1108

Ma X, Zheng J, Zhang X, Hu Q, Qian R (2017) Salicylic acid alleviates the adverse effects of salt stress on *Dianthus superbus* (Caryophyllaceae) by activating photosynthesis, protecting morphological structure, and enhancing the antioxidant system. Front Plant Sci 8:600

Mahmood K (2011) Salinity tolerance in barley (Hordeum vulgare L.): effects of varying NaCl, K?/Na? and NaHCO3 levels on cultivars differing in tolerance. Pak J Bot 43(3):1651–1654

Marschner H (2011) Marschner's mineral nutrition of higher plants. Academic press

Marschner H, Cakmak I (1989) High light intensity enhances chlorosis and necrosis in leaves of zinc, potassium, and magnesium deficient bean (*Phaseolus vulgaris*) plants. J Plant Physiol 4(3):308–315

Miura K, Tada Y (2014) Regulation of water, salinity, and cold stress responses by salicylic acid. Front Plant Sci 5:4

Munns R, Tester M (2008) Mechanisms of salinity tolerance. Annu Rev Plant Biol 59(1):651–681

Nakano Y, Asada K (1981) Hydrogen peroxide is scavenged by ascorbate-specific peroxidase in spinach chloroplasts. Plant Cell Physiol 22(5):867–880

Rady MM, Mossa AT, Youssof AM, Osman AS, Ahmed SM, Mohamed IA (2023) Exploring the reinforcing effect of nano-potassium on the antioxidant defense system reflecting the increased yield and quality of salt-stressed squash plants. Sci Hortic 308:111609

Ramzan T, Shahbaz M, Maqsood MF, Zulfiqar U, Saman RU, Lili N, Irshad M, Maqsood S, Haider A, Shahzad B, Gaafar AR (2023) Phenylalanine supply alleviates the drought stress in mustard (*Brassica campestris*) by modulating plant growth, photosynthesis, and antioxidant defense system. Plant Physiol Biochem 201:107828

Reddy AR, Chaitanya KV, Vivekanandan M (2004) Drought-induced responses of photosynthesis and antioxidant metabolism in higher plants. J Plant Physiol 161(11):1189–1202

Richardson AD, Duigan SP, Berlyn GP (2002) An evaluation of noninvasive methods to estimate foliar chlorophyll content. New Phytol 153(1):185–194

Römheld V, Kirkby EA (2010) Research on potassium in agriculture: needs and prospects. Plant Soil 335:155–180

Sangakkara UR (2001) Plant stress factors: their impact on productivity of cropping systems. In: Crop science: progress and prospects. Papers presented at the third international crop science congress, Hamburg, Germany. CABI Publishing, Wallingford UK, pp 101–117

Shabala L, Zhang J, Pottosin I, Bose J, Zhu M, Fuglsang AT, Velarde-Buendia A, Massart A, Hill CB, Roessner U, Bacic A (2016) Cell-type-specific H+-ATPase activity in root tissues enables K+ retention and mediates acclimation of barley (*Hordeum vulgare*) to salinity stress. Plant Physiol 172(4):2445–2458

Shah T, Khan H, Noor MA, Ghoneim A, Wang X, Sher A, Nasir M, Basahi MA (2018) Effects of potassium on phenological, physiological and agronomic traits of maize (*Zea mays* L.) under high nitrogen nutrition with optimum and reduced irrigation. Appl Ecol Environ Res 16(5)

Shang JX, Li X, Li C, Zhao L (2022) The role of nitric oxide in plant responses to salt stress. Int J Mol Sci 23(11):6167

Sharma P, Jha AB, Dubey RS, Pessarakli M (2012) Reactive oxygen species, oxidative damage, and antioxidative defense mechanism in plants under stressful conditions. J Bot 2012:217037

Singh K, Singh J, Rawat M (2019) Green synthesis of zinc oxide nanoparticles using *Punica Granatum* leaf extract and its application towards photocatalytic degradation of Coomassie brilliant blue R-250 dye. SN Appl Sci 1:1–8

Singh R, Saffeullah P, Ahmad S, Umar S (2024a) Linseed (Linum usitatissimum L.) accessions depict genetic variations in growth and yield characteristics, biochemical parameters, antioxidant potential, and secoisolariciresinol diglucoside content. Genet Resour Crop Evol 1:1–24

Singh R, Saffeullah P, Umar S, Abass S, Ahmad S, Iqbal N (2024b) Comparing the individual and combined effects of nano zinc and conventional zinc fertilization on growth, yield, phytochemical properties, antioxidant activity, and secoisolariciresinol diglucoside content in linseed. Plant Nano Biol 10:100098

Swanson S, Gilroy S (2010) ROS in plant development. Physiol Plant 138(4):384–392

Szabados L, Savouré A (2010) Proline: a multifunctional amino acid. Trends Plant Sci 15(2):89–97

Szalai G, Tari I, Janda T, Pestenacz A, Páldi E (2000) Effects of cold acclimation and salicylic acid on changes in ACC and MACC contents in maize during chilling. Biol Plant 43:637–640

Ul-Allah S, Ijaz M, Nawaz A, Sattar A, Sher A, Naeem M, Shahzad U, Farooq U, Nawaz F, Mahmood K (2020) Potassium application improves grain yield and alleviates drought susceptibility in diverse maize hybrids. Plants 9(1):75

Verma SB, Singh D, Chauhan RM (2018) Effect of neem (*Azadirachta indica* L.), mustard (*Brassica juncea*) de-oiled seed cake and bio-fertilizer on the growth and yield of wheat (*Triticum aestivum* L.). J Pharmacogn Phytochem 7(5):2416–2427

Wang M, Zheng Q, Shen Q, Guo S (2013) The critical role of potassium in plant stress response. Int J Mol Sci 14(4):7370–7390

Yalpani N, Enyedi AJ, León J, Raskin I (1994) Ultraviolet light and ozone stimulate accumulation of salicylic acid, pathogenesis-related proteins and virus resistance in tobacco. Planta 193(3):372–376

Zhang Q, Dai W (2019) Plant response to salinity stress. In: Stress physiology of woody plants. CRC Press, pp 155–173

Zhou X, Joshi S, Khare T, Patil S, Shang J, Kumar V (2021) Nitric oxide, crosstalk with stress regulators and plant abiotic stress tolerance. Plant Cell Rep 40(8):1395–1414

# Abscisic Acid and ROS Accumulation Under Abiotic Stress in Crops

15

Fabio Scotto di Covella and Antonio Ferrante

**Abstract**

Abscisic acid (ABA) and reactive oxygen species (ROS) are crucial signaling molecules in crop plants, playing pivotal roles in regulating responses to various abiotic stresses such as drought, salinity, and high light. Under stress conditions, ABA induces ROS production through the activation of NADPH oxidases while simultaneously promoting the expression of antioxidant enzymes to maintain ROS homeostasis. This delicate balance is essential for stress perception, signal transduction, and the initiation of appropriate adaptive responses. The interplay between ABA and ROS signaling networks is complex and tightly regulated, with variations in sensitivity and tolerance observed among different crop cultivars. ABA triggers stomatal closure by activating the efflux of $K^+$ and $Cl^-$ ions from guard cells, leading to loss of turgor. RBOHD and RBOHF oxidases, regulated by ABA, are involved in the production of ROS accumulation outside guard cells, modulating stomatal response to drought stress. Under salinity stress, plants activate detoxification enzymes and antioxidant compounds as a defense strategy against ROS accumulation. High light stress also induces ROS production, leading to photoinhibition and changes in the relationship between chlorophyll, carotenoids, and ABA. Understanding the molecular mechanisms underlying ABA and ROS accumulation and their downstream effects is crucial for developing stress-tolerant crop varieties and improving agricultural productivity in the face of changing climatic conditions.

**Keywords**

Biostimulants · Crops · Stomatal closure · Antioxidant enzymes · NADPH oxidases · Vegetables

F. Scotto di Covella · A. Ferrante (✉)
Institute of Crop Science, Sant'Anna School of Advanced Studies of Pisa, Pisa, Italy
e-mail: antonio.ferrante@santannapisa.it



N. Iqbal et al. (eds.), *Plant Signaling Molecules in Regulation of ROS-Scavenging System*, https://doi.org/10.1007/978-981-95-3377-0_15

## 15.1 Introduction

Abscisic acid (ABA) and reactive oxygen species (ROS) play crucial roles in plant responses to various abiotic stresses. In crops, these signaling molecules are rapidly produced and accumulated when plants encounter adverse environmental conditions such as drought, salinity, extreme temperatures, or heavy metal toxicity (Chen et al. 2020). ABA acts as a key phytohormone that regulates stomatal closure, gene expression, and physiological adaptations to minimize water loss and maintain cellular homeostasis. Concurrently, ROS accumulation serves as both a stress signal and a potentially damaging agent, triggering antioxidant defense mechanisms and programmed cell death pathways (Mittler et al. 2022).

ROS includes superoxide anion radical ($O_2^-$), hydrogen peroxide ($H_2O_2$), hydroxyl radical ($^{\cdot}OH$), and singlet oxygen ($^1O_2$). These molecules have important regulation action in plants; the negative effects lie in the overaccumulation in cells and impair detoxification systems (Baxter et al. 2014).

The interplay between ABA and ROS signaling networks is complex and tightly regulated in crop plants. Under abiotic stress conditions, ABA induces the production of ROS through the activation of NADPH oxidases, while also promoting the expression of antioxidant enzymes to maintain ROS homeostasis (Fig. 15.1). This delicate balance is essential for stress perception, signal transduction, and the initiation of appropriate adaptive responses. Crop species have evolved diverse strategies to modulate ABA and ROS levels, with variations in sensitivity and tolerance observed among different cultivars. Understanding the molecular mechanisms underlying ABA and ROS accumulation and their downstream effects is crucial for developing stress-tolerant crop varieties and improving agricultural productivity in the face of changing climatic conditions (Fig. 15.1).

## 15.2 Role of Abscisic Acid (ABA) and Plant Hormone Interactions on ROS Accumulation

ABA is a key phytohormone that acts as a signaling molecule during abiotic stress. It is involved in the regulation of various physiological processes, including stomatal closure, which reduces water loss during drought (Wasilewska et al. 2008; Thilakarathne et al. 2025). The biosynthesis of ABA mainly comes from the indirect $C_{40}$ pathway through the degradation of carotenoids via the violaxanthin with the involvement of the key enzyme 9-cis-epoxycarotenoid dioxygenase (NCED, EC 1.13.11.51) that leads to ABA (Oritani and Kiyota 2003; Franzoni et al. 2016).

Exogenous treatments with ABA at 0.1 or 1 mM demonstrated an accumulation of ROS with an increase in membrane degradation in maize leaves after 12 or 24 h (Jiang and Zhang 2001). In ABA-treated leaf maize, the activities of Zn and Cu superoxide dismutase (SOD, EC 1.15.1.1) increased for reducing the hydrogen peroxide. This ROS can also have signaling function and amplify the crop response (Suzuki et al. 2013). The role of ABA in ROS accumulation has been demonstrated

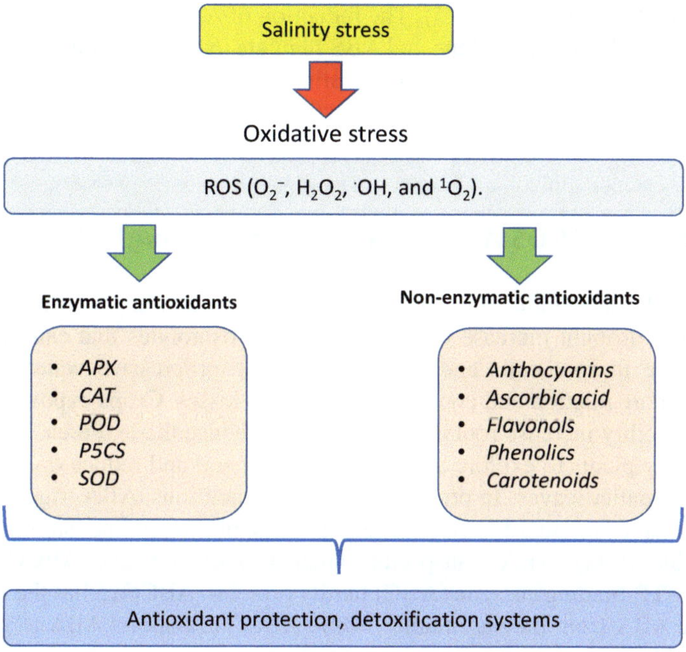


**Fig. 15.1** Enzymatic and non-enzymatic antioxidants in the regulation of oxidative status of plants

with MAPK kinase inhibitors that avoided the ABA-induced hydrogen peroxide ($H_2O_2$) production 2 h after treatment (Zhang et al. 2006).

The ABA is involved in various physiological processes. The most important role of ABA is the regulation of stomata closure (Chen et al. 2020). It is a strategy to avoid excessive water loss through the leaf transpiration. Since the stomata represent the entering point for carbon dioxide ($CO_2$), its closure inhibits photosynthesis and induces various physiological disorders at chloroplast level.

ABA and ROS can have a tight and synergistically action to enhance stress tolerance in plants. In fact, ABA can modulate the production of ROS-scavenging enzymes and prevent oxidative damage. A crosstalk between ABA and ROS signaling pathways helps fine-tune the plant's response to stress, ensuring a coordinated defense mechanism (Qi et al. 2018).

A cross talk between ABA and ethylene has been studied in sunflowers (*Helianthus annuus* L.) seeds germination. It has been found that ABA strongly represses germination, but the exposure of seeds to ethylene greatly recovers the germination ability (Ye et al. 2012). It is well known that ABA is involved in seeds dormancy and inhibits germination. However, gaseous ethylene treatment greatly improves the seeds germination. It seems that they show antagonist response. The negative interaction between ethylene and ABA has been found to be mediated in part by *SWI-INDEPENDENT 3 LIKE1* and *SWI-INDEPENDENT 3 LIKE2* via histone deacetylation, which influences the dormancy of seeds. The concentration of ABA in dormant seeds greatly declined in ethylene-treated seeds

(El-Maarouf-Bouteau et al. 2015). The reduction of ABA seems to be regulated by $H_2O_2$. It seems that both ethylene and ABS regulate the ROS accumulation. At transcriptional level, ethylene and ABA signaling interactions have been mediated by *ETHYLENE-INSENSITIVE3 (EIN3)* and *ABA-INSENSITIVE4* transcription factors (Yu et al. 2019).

## 15.3 ABA and ROS Accumulation Under Drought Stress

The water shortage induces plant strategies for overcoming the drought period. Plants under drought increase the biosynthesis of osmolytes that can improve the water uptake in root cells and whole plant. The progressive water availability induces osmoregulation and counteracts the water losses. Crops exposed to limited water availability increase root development and reduce the leaf area. This strategy is adopted by plants to explore a higher volume of soil and reduce transpiration by producing smaller leaves. In prolonged drought conditions, ABA triggers efflux of $K^+$, $Cl^-$ ions from guard cells, leading to loss of turgor and stomatal closure (MacRobbie 1998). ABA transporters such as ABCG25 and ABCG40, which belong to ATP-binding cassette (ABC) family proteins. ABCG25 has the function to export the ABA from the vasculature, while ABCG40 imports ABA to guard cells (Liu et al. 2022).

Drought is perceived by plants at roots level and activates the accumulation of CLAVATA3/EMBRYO-SURROUNDING REGION RELATED 25 (CLE25) a signal peptide, which is transferred to the outside of guard cells through vascular tissues. CLE25 activates the expression of *NCED3*, via BARELY ANY MERISTEM 1 (BAM1) and *BAM3* receptor kinases, which modulate ABA biosynthesis and stomata at guard cell level (Takahashi et al. 2018).

The Respiratory Burst Oxidase Homolog Protein D (RBOHD) and RBOHF are involved in the production of ROS accumulation outside guard cells. Functional studies using double mutation of RBOHD and RBOHF in Arabidopsis reduced the ROS accumulation and inhibited the stomatal closure, demonstrating the mitigating action against drought (Kwak et al. 2003). Drought stress induces the production of ROS in guard cells based on ABA signaling. It has been demonstrated that ABA activates the ROS accumulation in guard cells mainly by regulating the RBOHD and RBOHF oxidases (Arnaud et al. 2023). These oxidases, RBOHD and RBOHF, have NADPH or NADH binding sites, which are involved in the electron transfer to apoplastic $O_2$ to produce $O_2^-$, which is converted by SOD to $H_2O_2$. Treatment with exogenous $H_2O_2$ modulates the stomatal response through the ABA and calcium channel. These findings connect plant hormones, ROS, and signaling elements.

Therefore, the ROS accumulation in response to drought is a positive function in order to modulate the stomata regulation and avoid excessive water loss by transpiration. Stomata play an important role in the water balance, and the aperture is formed by a pair of guard cells. The stomata represent the direct connection of plants with the environment through the gas exchanges. Unfortunately, the stomata are also the gateway to pathogens. NADPH oxidases localized in the plasmalemma are responsible for the ROS accumulation in leaves. Carotenoids are the main ROS protection and scavengers in the chloroplast. The action of carotenoids is associated with the double bonds in the molecule. Therefore, carotenoids with longer structures and more double bonds have higher efficiency in ROS removal, including singlet oxygen (Widomska et al. 2019).

The effect of ABA under drought stress has been studied using *aba2-1* an Arabidopsis mutant. Osmotic stress induced by polyethylene glycol (PEG) increased ABA concentration in Arabidopsis plants by over 19-fold. On the contrary, mild stress did not induce the endogenous ABA accumulation in the *aba2-1* mutant. The application of exogenous ABA in plants under stress conditions induced the ABA increase and the ROS scavenger enzymes such as SOD, catalase (CAT; EC 1.11.1.6), ascorbate peroxidase (APX; EC 1.11.1.11), and NADPH oxidase (NOX; EC 1.6.3.1). These results demonstrated the role of ABA in the regulation of ROS level under drought stress conditions (Ozfidan et al. 2012).

## 15.4 Salinity, ABA, and ROS

Salinity is an important abiotic stress along the coastal area and is one of the major limitation factors for several crops. High salinity can be a result of salt accumulation in the soils due to the seawater underground infiltration or salinity deposition by marine aerosol (Hossain and Dietz 2016). The effect on crops is the reduction of growth because of lower photosynthesis rate and reduced water uptake. The crop cycle can be prolonged, and under severe salinity with salt concentration above the crop tolerance threshold can lead to lower yield and quality. Plants can have various strategies for counteracting the negative effects of salinity (Munns and Tester 2008). Plants based on their tolerance to salinity can be classified as halophytes or glycophytes. Halophytes can adapt to high levels of salinity in the soil. They can grow fast and withstand 200–500 mM NaCl (tolerant) or grow slowly up to 200 mM NaCl (tolerant with reduced growth). Glycophytes are species sensitive to higher or lower concentrations of salts in the circulating soil solution.

ABA plays an important role in regulating crop response to salinity stress. The role of this plant hormone is involved in osmotic adjustment, improving hydraulic conductance, promoting the accumulation of compatible solutes, and regulation of ROS. ABA function can be associated with stomatal regulation with the involvement of potassium efflux at roots level, and the activation of genes encoding salt overly sensitive (SOS) channels (Fig. 15.2). It has been reported that ABA can affect antioxidants concentration such ascorbic acid and cysteine to keep a redox balance under salt stress conditions (Toscano et al. 2023). The overproduction of ROS under salinity has been considered the major cause of lipid membrane peroxidation, nucleic acids and proteins, carbohydrates oxidation, and DNA damage.

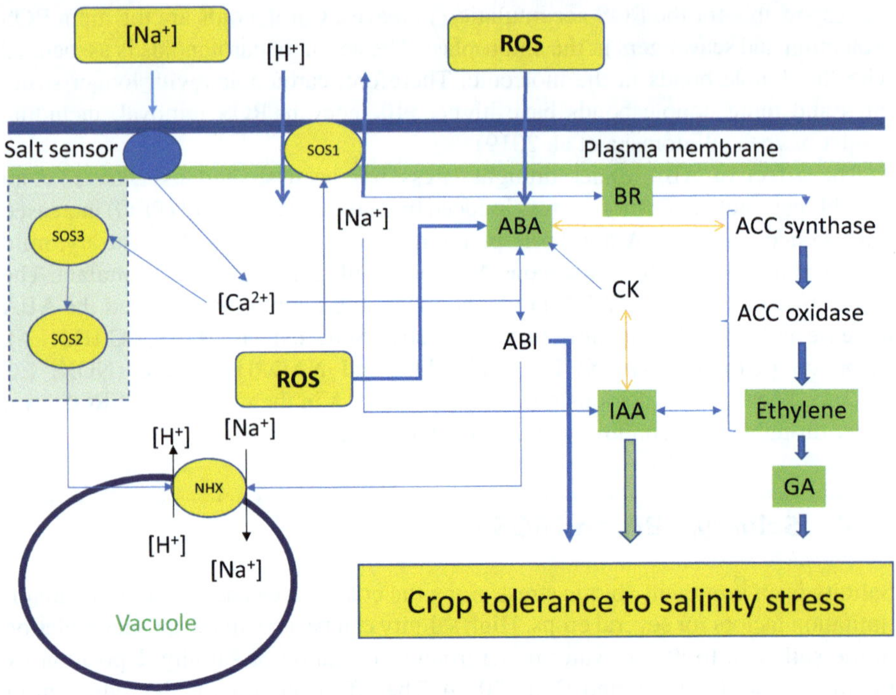


**Fig. 15.2** Effect of salinity stress on ROS accumulation and interaction with ABA and other plant hormones. As defense strategy, plants activate SOSs and accumulation of sodium in vacuole. (Modified from Toscano et al. (2023))

## 15.5 High Light Intensity Stress, ABA, and ROS

The energy absorbed by leaf can be used for the photosynthesis, but if there are stressful conditions, this energy is not used and can cause damage at the chloroplast level.

$$\text{chl} + \text{hv} \rightarrow {}^{1}\text{chl} \rightarrow O_2 \rightarrow {}^{1}O_2 + \text{chl}$$

Chlorophyll absorbing energy can jump from excited to ground state, releasing energy in chloroplasts, transporting energy to oxygen, and produces singlet oxygen.

When plants are exposed to excessive light intensity, a reduction in photosynthetic efficiency—referred to as photoinhibition—is triggered. This occurs when the photon flux density surpasses a critical threshold, saturating the photosystem reaction centers as no further electron acceptors are available. Photosystem II (PSII) is the first to reach saturation, while Photosystem I (PSI) may continue to function via cyclic electron transport. Under these conditions, the primary challenge becomes the dissipation of surplus excitation energy absorbed by the photosynthetic pigments.

If pigments remain in an excited state for prolonged periods, the likelihood of alternative photochemical reactions increases, often resulting in the formation of reactive oxygen species (ROS) (Bruno and Sanità di Toppi 2018). This excess energy, if not properly dissipated, can cause photodamage, compromising plant physiological processes (García-Pérez et al. 2025).

Inefficient dissipation of excitation energy leads to over-reduction of the electron transport chain, which facilitates the linear flow of electrons from water to NADPH. Some of the absorbed energy is safely dissipated as heat via non-photochemical quenching (NPQ) within the PSII light-harvesting complex. Additional dissipation mechanisms include photorespiration and photochemical quenching, although both processes are also associated with ROS production. These reactive molecules are neutralized by the plant's antioxidant system, which includes both hydrophilic and lipophilic components in enzymatic and non-enzymatic forms (Galvez-Valdivieso et al. 2009).

Under high irradiance, ROS significantly inhibit photosynthesis due to their strong oxidative nature, damaging crucial components of the thylakoid membranes. PSII is particularly vulnerable to oxidative stress, as it is both a site of ROS generation and contains the D1 protein, which is highly susceptible to degradation by singlet oxygen.

In these high-light stress scenarios, the delicate balance between light absorption and energy dissipation is disrupted, resulting in the overproduction of ROS (Bruno and Sanità di Toppi 2018). A common consequence is the accumulation of hydrogen peroxide ($H_2O_2$) in chloroplasts. This triggers rapid transcriptional activation of the ASCORBATE PEROXIDASE 2 (APX2) gene (Fryer et al. 2003; Ball et al. 2004).

Interestingly, this gene activation and the associated enhancement in energy dissipation are suppressed under conditions of high ambient humidity. This observation suggests that under low humidity, leaves subjected to intense light undergo water loss, triggering ABA accumulation—a key hormone in the regulation of plant water homeostasis (Fryer et al. 2003; Galvez-Valdivieso et al. 2009).

ABA is central to the plant's adaptive response to abiotic stress. Its levels rise significantly under challenging environmental conditions such as high light exposure (Tuteja 2007). ABA biosynthesis begins from β-carotene and proceeds through zeaxanthin and antheraxanthin, which are converted to all-trans-violaxanthin. The enzyme 9-cis-epoxycarotenoid dioxygenase (NCED) then catalyzes the production of xanthoxin, a precursor of ABA. Upon stress alleviation, ABA is rapidly catabolized into inactive products such as phaseic acid and dihydrophaseic acid (Sah et al. 2016).

Notably, ABA accumulation is transient and closely tied to ROS production, suggesting that high-light-induced ROS—or ROS waves—are required for ABA signaling. This, in turn, mediates transcriptional activation and stomatal responses (Song et al. 2014). There is growing evidence of a tightly regulated feedback mechanism involving stomatal function and ABA levels (Galvez-Valdivieso et al. 2009; Suzuki et al. 2013). Environmental shifts may induce ROS, ABA, and nitric oxide (NO) production, each capable of modulating stomatal aperture depending on the specific stressor. These stomatal changes affect the internal leaf microenvironment, further modulating the generation and perception of signaling molecules and reinforcing the feedback loop.

A rapid ROS wave has been shown to be essential for systemic acquired acclimation (SAA) to excess light stress (Mittler and Blumwald 2015). ROS, thus, play a dual role in plant stress physiology: they are both damaging byproducts of stress and crucial signaling molecules that mediate intra- and intercellular communication. In Arabidopsis, it has been demonstrated that ROS waves are indispensable for long-distance signaling and whole-plant adaptation to high light intensity (Mittler et al. 2011; Suzuki et al. 2013).

Under non-stressed conditions, plants possess a robust ROS-scavenging capacity. However, the propagation of ROS waves across tissues likely involves auto-propagation, where individual cells along the signaling path continuously generate new ROS. Furthermore, systemic signaling induced by localized high light exposure is accompanied by plasma membrane electrical changes in a light wavelength-specific manner (Szechyńska-Hebda et al. 2010). Because membrane potentials can be modulated by ROS and because the propagation rate of electrical signals mirrors that of ROS waves (Miller et al. 2010), it is plausible that ROS wave generation influences the amplitude and dynamics of these electrical signals.

In conclusion, high light stress induces photoinhibition, ROS accumulation, and ABA signaling, which together form an integrated response system for energy dissipation, gene regulation, and systemic acclimation. ROS, far from being merely toxic byproducts, function as vital components of a sophisticated signaling network, underscoring their importance in modulating plant resilience under abiotic stress. Understanding these mechanisms offers potential strategies for improving crop performance in increasingly variable environmental conditions.

## 15.6 Heat Stress, ABA, and ROS

Heat stress negatively influences various physiological and metabolic processes in crops. In various geographical areas, this abiotic stress is the most important constraint for crop yield and produce quality. The heat stress usually induces damage to cell membrane with loss of functionality of integrated proteins and channels. Moreover, the accumulation of ROS induces lipid peroxidation with loss of integrity. High temperatures can cause several physiological alterations in cells with the accumulation of ROS. Crops that suffer from high temperatures and heat stress are macrothermal species that naturally grow in Spring and Summer. In several geographical areas such as the Mediterranean basin, heat stress is the major abiotic stress associated with drought and salinity (Leotta et al. 2023; Diffenbaugh et al. 2007). It has been demonstrated that calcium has a role in heat tolerance in crops. The use of calcium channel blockers increased the heat damage, while exogenous calcium supply reduced the heat stress. ABA interacting with other plant hormones such as ethylene and salicylic acid also has a beneficial effect under high temperature (Larkindale and Knight 2002). In cucumber, heat tolerance was associated with an increase in ABA concentration and $H_2O_2$, along with the upregulation of RBOH and HSP70 gene expression. These conditions decreased the oxidative stress in the leaves of grafted plants. It has been suggested that ABA could be transported from roots to shoot (Li et al. 2014). The ABA concentration measured in leaf and xylem was higher in grafted plants. These results demonstrated the role of ABA in mitigating the heat stress and ROS accumulation (Wang et al. 2023).

In lily, the *LlHsfA4* transcription factor has also been considered a key regulator of heat tolerance (Wang et al. 2022). It seems that ABA-responsive element (ABRE) is connected with *LlHsfA4* for improving the heat tolerance.

## 15.7 Conclusion

In conclusion, ABA and ROS are crucial signaling molecules in crop plants, playing essential roles in regulating responses to various abiotic stresses such as drought, salinity, and high light intensity. ABA induces ROS production through the activation of NADPH oxidases while simultaneously promoting the expression of antioxidant enzymes to maintain ROS homeostasis. This delicate balance is essential for stress perception, signal transduction, and the initiation of appropriate adaptive responses. Under drought stress, ABA triggers stomatal closure by activating the efflux of $K^+$ and $Cl^-$ ions from guard cells, and RBOHD and RBOHF oxidases are involved in the production of ROS accumulation outside guard cells, modulating stomatal response. Salinity stress activates detoxification enzymes and antioxidant compounds as a defense strategy against ROS accumulation. High light stress also induces ROS production, leading to photoinhibition and changes in the relationship between chlorophyll, carotenoids, and ABA.

## References


Arnaud D, Deeks MJ, Smirnoff N (2023) RBOHF activates stomatal immunity by modulating both reactive oxygen species and apoplastic pH dynamics in Arabidopsis. Plant J 116(2):404–415

Ball L et al (2004) Evidence for a direct link between glutathione biosynthesis and stress defense gene expression in Arabidopsis. Plant Cell 16:2448–2462

Baxter A, Mittler R, Suzuki N (2014) ROS as key players in plant stress signalling. J Exp Bot 65(5):1229–1240. https://doi.org/10.1093/jxb/ert375. Epub 2013 Nov 19

Bruno L, Sanità di Toppi L (2018) Interazioni delle piante con i principali fattori di stress abiotico. In: Chapter in Interazioni piante ambiente

Chen K, Li GJ, Bressan RA, Song CP, Zhu JK, Zhao Y (2020) Abscisic acid dynamics, signaling, and functions in plants. J Integr Plant Biol 62(1):25–54

Diffenbaugh NS, Pal JS, Giorgi F, Gao X (2007) Heat stress intensification in the Mediterranean climate change hotspot. Geophys Res Lett 34(11)

El-Maarouf-Bouteau HAYAT, Sajjad Y, Bazin J, Langlade N, Cristescu SM, Balzergue S et al (2015) Reactive oxygen species, abscisic acid and ethylene interact to regulate sunflower seed germination. Plant Cell Environ 38(2):364–374

Franzoni G, Cocetta G, Trivellini A, Angeli S, Ferrante A (2016) Abscisic acid and carotenoids metabolism in tomato during postharvest. Acta Hort 1194:381–388

Fryer MJ, Ball L, Oxborough K, Karpinski S, Mullineaux PM, Baker NR (2003) Control of Ascorbate Peroxidase 2 expression by hydrogen peroxide and leaf water status during excess light stress reveals a functional organisation of Arabidopsis leaves. Plant J 33:691–705

Galvez-Valdivieso G, Fryer MJ, Lawson T, Slattery K, Truman W, Smirnoff N, Asami T, Davies WJ, Jones AM, Baker NR, Mullineaux PM (2009) The high light response in Arabidopsis involves ABA signaling between vascular and bundle sheath cells. Plant Cell 21(7):2143–2162. https://doi.org/10.1105/tpc.108.061507
García-Pérez P, Losada-Barreiro S, Bravo-Díaz C, Gallego PP, Lucini L (2025) Plant oxidative stress and specialized metabolites: a holistic physicochemical and metabolomics perspective. Plant Physiol Biochem 227:110093. https://doi.org/10.1016/j.plaphy.2025.110093
Hossain MS, Dietz KJ (2016) Tuning of redox regulatory mechanisms, reactive oxygen species and redox homeostasis under salinity stress. Front Plant Sci 7:548
Jiang M, Zhang J (2001) Effect of abscisic acid on active oxygen species, antioxidative defence system and oxidative damage in leaves of maize seedlings. Plant Cell Physiol 42(11):1265–1273
Kwak JM, Mori IC, Pei ZM, Leonhardt N, Torres MA, Dangl JL, Bloom RE, Bodde S, Jones JD, Schroeder JI (2003) NADPH oxidase AtrbohD and AtrbohF genes function in ROS-dependent ABA signaling in Arabidopsis. EMBO J 22:2623–2633
Larkindale J, Knight MR (2002) Protection against heat stress-induced oxidative damage in Arabidopsis involves calcium, abscisic acid, ethylene, and salicylic acid. Plant Physiol 128(2):682–695. https://doi.org/10.1104/pp.010320
Leotta L, Toscano S, Ferrante A, Romano D, Francini A (2023) New strategies to increase the abiotic stress tolerance in woody ornamental plants in Mediterranean climate. Plants 12(10):2022. https://doi.org/10.3390/plants12102022
Li HAO, Liu SS, Yi CY, Wang F, Zhou JIE, Xia XJ et al (2014) Hydrogen peroxide mediates abscisic acid-induced HSP 70 accumulation and heat tolerance in grafted cucumber plants. Plant Cell Environ 37(12):2768–2780
Liu H, Song S, Zhang H, Li Y, Niu L, Zhang J, Wang W (2022) Signaling transduction of ABA, ROS, and Ca2+ in plant stomatal closure in response to drought. Int J Mol Sci 23(23):14824
MacRobbie EAC (1998) Signal transduction and ion channels in guard cells. Philos Trans R Soc Lond B Biol Sci 353(1374):1475–1488
Miller G, Suzuki N, Ciftci-Yilmaz S, Mittler R (2010) Reactive oxygen species homeostasis and signalling during drought and salinity stresses. Plant Cell Environ 33(4):453–467. https://doi.org/10.1111/j.1365-3040.2009.02041.x. Epub 2009 Aug 27
Mittler R, Blumwald E (2015) The roles of ROS and ABA in systemic acquired acclimation. Plant Cell 27(1):64–70. https://doi.org/10.1105/tpc.114.133090
Mittler R, Vanderauwera S, Suzuki N, Miller G, Tognetti VB, Vandepoele K, Gollery M, Shulaev V, Van Breusegem F (2011) ROS signaling: the new wave? Trends Plant Sci 16(6):300–309. https://doi.org/10.1016/j.tplants.2011.03.007
Mittler R, Zandalinas SI, Fichman Y, Van Breusegem F (2022) Reactive oxygen species signalling in plant stress responses. Nat Rev Mol Cell Biol 10:663–679. https://doi.org/10.1038/s41580-022-00499-2
Munns R, Tester M (2008) Mechanisms of salinity tolerance. Annu Rev Plant Physiol 59:651–681. https://doi.org/10.1146/annurev.arplant.59.032607.092911
Oritani T, Kiyota H (2003) Biosynthesis and metabolism of abscisic acid and related compounds. Nat Prod Rep 20(4):414–425
Ozfidan C, Turkan I, Sekmen AH, Seckin B (2012) Abscisic acid-regulated responses of aba2-1 under osmotic stress: the abscisic acid-inducible antioxidant defence system and reactive oxygen species production. Plant Biol 14(2):337–346. https://doi.org/10.1111/j.1438-8677.2011.00496.x
Qi J, Song CP, Wang B, Zhou J, Kangasjärvi J, Zhu JK, Gong Z (2018) Reactive oxygen species signaling and stomatal movement in plant responses to drought stress and pathogen attack. J Integr Plant Biol 60(9):805–826
Sah SK, Reddy KR, Li J (2016) Acid and abiotic stress tolerance in crop plants. Front Plant Sci 7:571. https://doi.org/10.3389/fpls.2016.00571
Song Y, Miao Y, Song CP (2014) Behind the scenes: the roles of reactive oxygen species in guard cells. New Phytol 201:1121–1140

Suzuki N, Miller G, Salazar C, Mondal HA, Shulaev E, Cortes DF et al (2013) Temporal-spatial interaction between reactive oxygen species and abscisic acid regulates rapid systemic acclimation in plants. Plant Cell 25(9):3553–3569. https://doi.org/10.1105/tpc.113.114595

Szechyńska-Hebda M, Kruk J, Górecka M, Karpińska B, Karpiński S (2010) Evidence for light wavelength-specific photoelectrophysiological signaling and memory of excess light episodes in Arabidopsis. Plant Cell 22(7):2201–2218. https://doi.org/10.1105/tpc.109.069302

Takahashi F, Suzuki T, Osakabe Y, Betsuyaku S, Kondo Y, Dohmae N, Fukuda H, Yamaguchi-Shinozaki K, Shinozaki K (2018) A small peptide modulates stomatal control via abscisic acid in long-distance signalling. Nature 556:235–238

Thilakarathne AS, Liu F, Zou Z (2025) Plant signaling hormones and transcription factors: key regulators of plant responses to growth, development, and stress. Plants 14(7):1070

Toscano S, Romano D, Ferrante A (2023) Molecular responses of vegetable, ornamental crops, and model plants to salinity stress. Int J Mol Sci 24(4):3190

Tuteja N (2007) Abscisic acid and abiotic stress signaling. Plant Signal Behav 2:135–138. https://doi.org/10.4161/psb.2.3.4156

Wang C, Zhou Y, Yang X, Zhang B, Xu F, Wang Y et al (2022) The heat stress transcription factor LlHsfA4 enhanced basic thermotolerance through regulating ROS metabolism in lilies (*Lilium longiflorum*). Int J Mol Sci 23(1):572

Wang Y, Lei B, Deng H, Liu X, Dong Y, Chen W et al (2023) Exogenous abscisic acid affects the heat tolerance of rice seedlings by influencing the accumulation of ROS. Antioxidants 12(7):1404. https://doi.org/10.3390/antiox12071404

Wasilewska A, Vlad F, Sirichandra C, Redko Y, Jammes F, Valon C et al (2008) An update on abscisic acid signaling in plants and more. Mol Plant 1(2):198–217

Widomska J, Welc R, Gruszecki WI (2019) The effect of carotenoids on the concentration of singlet oxygen in lipid membranes. Biochim Biophys Acta (BBA)-Biomembranes 1861(4):845–851

Ye N, Zhu G, Liu Y, Zhang A, Li Y, Liu R et al (2012) Ascorbic acid and reactive oxygen species are involved in the inhibition of seed germination by abscisic acid in rice seeds. J Exp Bot 63(5):1809–1822. https://doi.org/10.1093/jxb/err336

Yu Y, Wang J, Li S, Kakan X, Zhou Y, Miao Y et al (2019) Ascorbic acid integrates the antagonistic modulation of ethylene and abscisic acid in the accumulation of reactive oxygen species. Plant Physiol 179(4):1861–1875

Zhang A, Jiang M, Zhang J, Tan M, Hu X (2006) Mitogen-activated protein kinase is involved in abscisic acid-induced antioxidant defense and acts downstream of reactive oxygen species production in leaves of maize plants. Plant Physiol 141(2):475–487

# The AsA–GSH Cycle: A Vital Line of Defense Against Heat Stress-Induced Oxidative Damage in Plants

16

Gholamreza Abdi, Anilava Chatterjee, Md Sabir Ahmed Mondol, Ubaida Akbar, Aritra Mukherjee, and Tadela Susmitha

**Abstract**

The Ascorbate–Glutathione (AsA–GSH) cycle plays a critical role in plant defense against oxidative damage from stress conditions of high temperature. Excess generation of reactive oxygen species (ROS) in the form of superoxide anions, hydrogen peroxide, and hydroxyl radicals can cause extensive damage to cellular macromolecules. The cycle above utilizes key antioxidants such as glutathione (GSH) and ascorbate (AsA) and enzymes such as dehydroascorbate reductase (DHAR), glutathione reductase (GR), monodehydroascorbate reductase (MDHAR), and ascorbate peroxidase (APX). The ascorbate molecule plays the role of a direct reactive oxygen species (ROS) scavenger, whereas the enzyme APX, along with ascorbate, catalyzes the reduction of hydrogen peroxide to water. Restoration of the active form of ascorbate (DHA) involves the utilization

G. Abdi (✉)
Department of Biotechnology, Persian Gulf Research Institute, Persian Gulf University, Bushehr, Iran
e-mail: abdi@pgu.ac.ir

A. Chatterjee · M. S. A. Mondol
Department of Agricultural Biochemistry, Bidhan Chandra Krishi Viswavidyalaya, Mohanpur, Nadia, West Bengal, India

U. Akbar
Department of Food Technology and Nutrition, Lovely Professional University, Phagwara, Punjab, India

A. Mukherjee
Department of Seed Science and Technology, Bidhan Chandra Krishi Viswavidyalaya, Mohanpur, Nadia, West Bengal, India

T. Susmitha
Department of Agricultural Biochemistry, School of Agricultural Sciences, Mallareddy University, Hyderabad, Telangana, India



N. Iqbal et al. (eds.), *Plant Signaling Molecules in Regulation of ROS-Scavenging System*, https://doi.org/10.1007/978-981-95-3377-0_16

of DHAR to reduce it to its reduced form from its oxidized form, along with oxidized GSH (GSSG), which is again reduced to its reduced form by the enzyme GR. Ongoing replenishment of active antioxidants is essential to ensure effective defense against reactive oxygen species (ROS), particularly in the case of prolonged exposure to high-temperature stress. Interestingly, however, it has been determined that the activity of these enzymes can be affected by heat stress, which has the potential to shift the dynamics of the AsA-GSH cycle. The current chapter explains the mechanisms through which the AsA-GSH cycle affects the balance of reactive oxygen species (ROS) in the cellular environment of plants. This study offers useful information on the tolerance and acclimatization of plants to high-temperature stress.

**Keywords**

Heat stress · The Ascorbate-Glutathione (AsA-GSH) cycle · ROS

## 16.1 Introduction

The efficiency of antioxidant systems at the metabolic level is closely related to the ability of plants to adapt to challenging environmental conditions. Stress conditions lead to excessive production of reactive oxygen species (ROS), mainly in mitochondria and chloroplasts, which are the main sites of ROS generation (Gill and Tuteja 2010; Lascano et al. 1999). These ROS can cause severe damage to proteins, carbohydrates, lipids (through lipid peroxidation), DNA, and cellular membranes, eventually resulting in cellular disintegration and death (Hussain Shah et al. 2021; Imlay 2003). To combat these harmful effects, plants have developed complex defense mechanisms that are meant to eliminate ROS. These mechanisms include a well-coordinated network of enzymatic and non-enzymatic antioxidants (Hussain Shah et al. 2021). Non-enzymatic antioxidants contain GSH, AsA, tocopherols, carotenoids, and flavonoids, which have the antioxidant activities by diminishing oxidative damage (Gill and Tuteja 2010). Contrarily, enzymatic antioxidant defense comprises of SOD, CAT, GPX, APX, GR, GST, PRX, MDHAR, and DHAR (Gill and Tuteja 2010; Hasanuzzaman et al. 2017b). Among those, the Ascorbate–Glutathione cycle is a major part in the detoxification process against ROS (Li et al. 2010). Ascorbate–Glutathione cycle contains the major constituents viz., AsA, GSH, APX, GR, MDHAR, and DHAR that function in synergy to maintain the level of ROS at nontoxic levels, thereby protecting against oxidative damage (Hasanuzzaman et al. 2019). Ascorbic acid (AsA) is one of the key antioxidant species that acts as a redox compound and plays a significant role in many biological processes in plants (Chaturvedi et al. 2022). These include the protection of α-tocopherol in its reduced form, stimulation of zeaxanthin synthesis as a mechanism for light energy dissipation in thylakoid membranes, and regulation of prosthetic metal ion concentrations for activation of antioxidant enzymes (Hasanuzzaman et al. 2012). Similarly, glutathione (GSH) has a versatile role as an antioxidant. It serves as a substrate for the enzymes GST, which detoxify xenobiotics, and GPX, which neutralizes ROS (Hasanuzzaman et al. 2017c). The glyoxalase system, consisting of glyoxalase I (Gly I) and glyoxalase II (Gly II), also uses GSH to detoxify methylglyoxal, recycling GSH in the process to maintain its antioxidative capacity (Hasanuzzaman et al. 2017a). This evidence has indicated that the GSH/GSSG redox pair has a significant role in stress signaling, maintaining cellular homeostasis, and inducing adaptive responses (Hasanuzzaman et al. 2017a, c). Again, both AsA and GSH have been reported to improve photosynthesis, water-use efficiency, osmoregulation, and overall plant productivity, especially under abiotic stresses. The discovery of the AsA–GSH cycle stimulated extensive interest in understanding the antioxidative defense system. Additional research is required to explain physiological processes leading to the development of oxidative stress and the way components of the AsA-GSH cycle regulate those processes (Hasanuzzaman et al. 2019).

## 16.2 AsA–GSH Pathway: Components

### 16.2.1 Ascorbic Acid (AsA)

Asc, glutathione (GSH) are also the primary components of low-molecular weight nonenzymatic antioxidant in plant cell, which directly participates in ROS scavenging through Asa–GSH cycles (Li et al. 2010). AsA, popularly known as vitamin C, is synthesized either endogenously by living organisms or obtained through diet (Hasanuzzaman et al. 2019). It is known to mainly act as a redox buffer and is involved as a cofactor in most of the biological activities like signal transduction, cell division, growth regulation, and enzyme activity (García-Caparrós et al. 2021). The biologically active form of AsA is ascorbate (Ahmad et al. 2017; El-Beltagi et al. 2022). Structurally, L-ascorbic acid is one of the most studied compounds, simple in structure but closely related to the C6 sugars. It is an aldono-1,4-lactone of a hexonic acid, also known as L-galacturonic or L-gluonic acid. An enediol group is present between C2 and C3 in its chemical structure, through which it can donate one or two electrons; its first oxidized intermediate formed is monodehydroascorbic acid (MDHA), which then converts into DHA. The presence of hydroxyl groups on C5 and C6 imparts alcoholic properties to the molecule, hence forming acetals and ketals upon reaction with aldehydes and ketones, respectively (Tripathi et al. 2009). Its chemical structure presents two asymmetric carbons conferring a positive optical rotation that increases up to +160 degrees in an alkaline pH, such as 2 N NaOH, but remains unaltered in acidic conditions (Tripathi et al. 2009). AsA acts as an efficient antioxidant for detoxification of lipid peroxyl radicals formed during lipid peroxidation and thus regenerates tocopherol from its radical form; it detoxifies ROS such as superoxide and hydroxyl radicals (Munné-Bosch 2005). It further interacts with H2O2 through isoforms of ascorbate peroxidase (APX) by utilizing AsA as a reducing agent in distinct cellular compartments. There is a variation in the concentration and redox state of AsA among different cellular compartments, while its impact is massive on growth and development and also on the resilience of abiotic stress in plants, including heat stress (Veljović-Jovanović et al. 2017). The redox potential of AsA at the tissue and organ level regulates the developmental processes. Seed development might be modulated via alteration in the AsA–DHA redox couple (Miret and Müller 2017; Smirnoff 2018).

Ascorbate biosynthesis is one of the most conserved metabolic pathways and is mainly conducted in the mitochondria. Four pathways are considered for the synthesis of AsA, which are L-glucose pathway, D-glucuronate pathway, D-mannose/L-galactose pathway, and D-galacturonate pathway (Miret and Müller 2017; Smirnoff 2018; Veljović-Jovanović et al. 2017). This is known as the Smirnoff–Wheeler pathway, or the D-mannose/L-galactose pathway, which makes use of L-galactone-1,4-lactone as the key precursor (Yoshimura and Ishikawa 2017). Starting from methyl D-galacturonate, which is the intermediate of pectin degradation when ripening happens, the biosynthesis pathway for D-galacturonate produces AsA (Smirnoff 2018). In D-glucuronate biosynthesis, it would be likely that the transformation process of UDP-glucose or myo-inositol into UDP-D-glucuronate did not involve any inversion of carbon skeleton (Ishikawa et al. 2018; Smirnoff 2018). Interestingly, AsA biosynthesis is rather closely associated with cell wall synthesis (Hasanuzzaman et al. 2019; Ishikawa et al. 2018). The cellular concentration of AsA is determined through a balance of its synthesis and degradation, mirroring the anabolic and catabolic activities. Although AsA is not stable and redox reactions account for only half of its uses, recycling AsA from MDHA and DHA is significant for maintaining redox homeostasis. MDHA is recycled back to AsA through monodehydroascorbate reductase, using NADPH. On the other hand, DHA is recycled back to AsA through dehydroascorbate reductase, using GSH (Asada 2006). These recycling mechanisms maintain a greater pool of AsA, supporting its key roles in redox metabolism and the water–water cycle, optimizing photosynthesis (Miyake 2010). If oxidized forms of AsA are not efficiently recycled, they degrade into oxalic acid or L-tartaric acid (Smirnoff and Wheeler 2024). It, thus, plays a vital role in plant physiology for antioxidant defense, growth regulation, and stress tolerance. Sub-cellular biosynthesis and recycling together with interaction with other cellular components underpin its significant role in preserving cellular homeostasis and in optimizing performance both under normal and stress conditions.

The biosynthesis pathway is shown in Fig. 16.1.

### 16.2.2 Glutathione (GHS)

Glutathione or GSH is one of the most active antioxidants interplaying in the course of regeneration of ascorbic acid within the ascorbic acid–glutathione cycle. Moreover, this chemical constituent takes an active participation in the regulation of intracellular H2O2 content, and thus it is considered relevant in the cellular redox balance (Y. Li et al. 2010). GSH is typically a small tripeptide, comprising γ-L-glutamyl-L-cysteinylglycine (Gill et al. 2013; Hasanuzzaman et al. 2019). Glutathione occurs in two important states: the oxidized form, or GSSG, and its reduced form, GSH. Glutathione reductase is an enzyme responsible for the

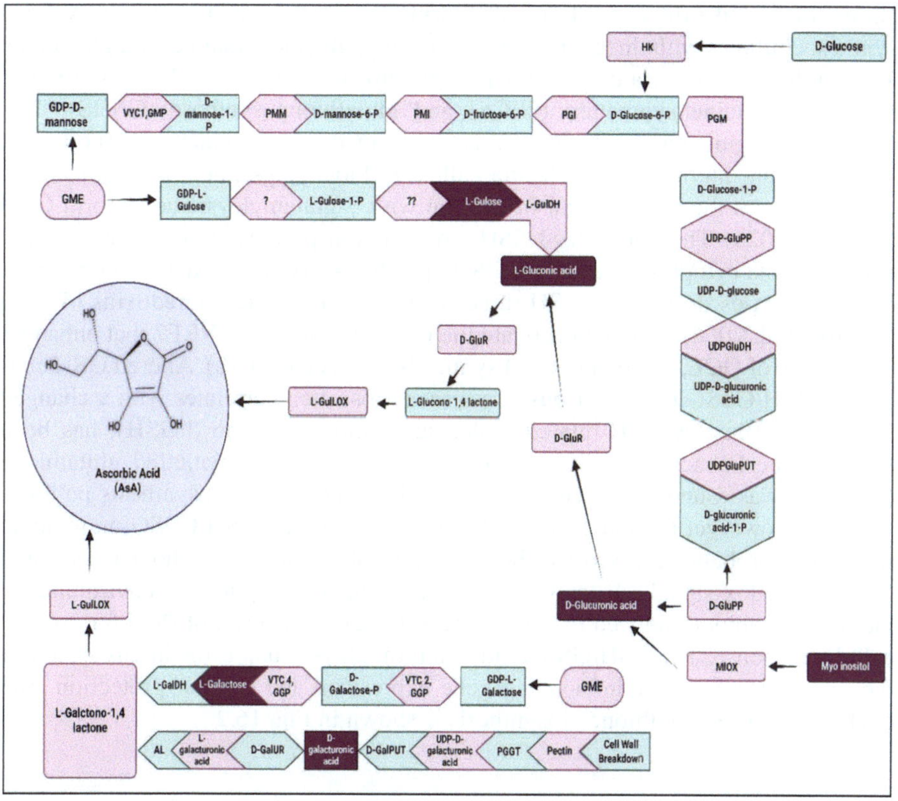


**Fig. 16.1** The metabolites in the violet box denote the names of each metabolic pathway in the biosynthesis of ascorbate, which involves a complicated series of processes, some of which involve undiscovered enzymes (Hasanuzzaman et al. 2019)

reduction of GSSG back into GSH. Among many others, GSH is recognized as one of the most active scavengers of ROS. Under normal physiological conditions, the reduced form of GSH is usually predominant over GSSG, underlining the major importance of this tripeptide in maintaining redox homeostasis (Rai et al. 2023). GSH is regarded as the most essential protector against oxidative damage in cells caused by ROS because of its high cellular concentration and its buffering action on redox potential. It is distributed in three major pools in the cell: the cytoplasm, accounting for 80–85%; the mitochondria, accounting for 10–15% and the endoplasmic reticulum (García-Caparrós et al. 2021). More than 98% of the total pool of glutathione in plants is present in the reduced form, as GSH. Glutathione is one of the most significant biomolecules, besides antioxidant activity; it performs a wide array of biological functions such as detoxification of xenobiotics, regulation of the cell cycle, and apoptosis. Glutathione maintains the cellular redox balance, modulates the immune response, and takes part in fibrogenesis. Glutathione or GSH is also a non-protein reduced sulphur compound. It is synthesized in multiple cellular compartments like the cytosol, mitochondria, peroxisomes, and chloroplasts. GSH synthesis initiates with three amino acids, namely, glycine, cysteine, and glutamate. The synthesis occurs via two ATP-dependent enzymatic reactions by two proteins: γ-glutamylcysteine synthetase (γECS) and glutathione synthase (GSHS). Two amino acids, glutamate and cysteine, are combined by glutamate–cysteine ligase, GCL, into an intermediary molecule called γ-glutamylcysteine. Glutathione synthase catalyzes the combination of glycine with γ-glutamylcysteine to form GSH (Rai et al. 2023). The synthesized GSHs are then transported to other cellular organelles from the cytoplasm. Although GSH-specific peroxidases exist, plant glutathione peroxidases do not use GSH directly but might act via thioredoxins (TRXs) (Navrot et al. 2006). ROS can activate the transcription factor NRF2 that enhances the activity of GCL, enhancing GSH synthesis (Lian et al. 2018). Altered GSH content or GSH/GSSG under various stress scenarios often correlates with a changed biosynthetic flux. A GSH-related molecule, homoglutathione (hGSH), has been found in the Fabaceae family (Klapheck 1988), and hydroxymethyl glutathione (hmGSH) was found in the Gramineae family, although its biosynthesis pathway was not known yet (Klapheck et al. 1992). The complex GSH/GSSG antioxidant machinery probably evolved for the regulation of cellular redox homeostasis and tuning of redox signals. It integrates various metabolic inputs and environmental factors, like abiotic-induced oxidative stress, to adjust the level of ROS for optimal cellular response. The GSH/GSSG ratio, in turn, plays a major role in this system at the fine-tuning level by influencing gene expression and cellular protection (Rai et al. 2023). The glutathione biosynthesis is shown in Fig. 16.2.

### 16.2.3 Ascorbate Peroxidase (APX)

Ascorbate peroxidase (APX, EC 1.11.1.11) belongs to the peroxidase hemoprotein enzymes in plant cells, with different isoforms such as cAPX, mitAPX, peroxisomal and glyoxysomal mAPX, and chAPX. Though they all catalyze the same reaction, there are much differences between their properties. All the isoforms show some specific differences concerning the substrate specificity of molecular weight with the optimum pH for both activities and stabilities generally found between pH 7 and 8 (Anjum et al. 2016; del Río and López-Huertas 2016; Praduman et al. 2014). One peculiar feature of APX isoforms is their dependence on ascorbate as a co-substrate. If the levels of ascorbate become too low, its enzymatic activity is notably reduced, showing just how crucial this molecule is for the proper functioning of APX. The role of APX during ROS detoxification is central to $H_2O_2$ scavenging. This is mediated by the binding of $H_2O_2$ and the generation of an intermediate (I) in which the heme iron (Fe) is oxidized to an oxyferryl species ($Fe_4+ = O$). Below is a representation of the APX catalytic cycle:

(a) APX reacts with $H_2O_2$ to form Intermediate I.

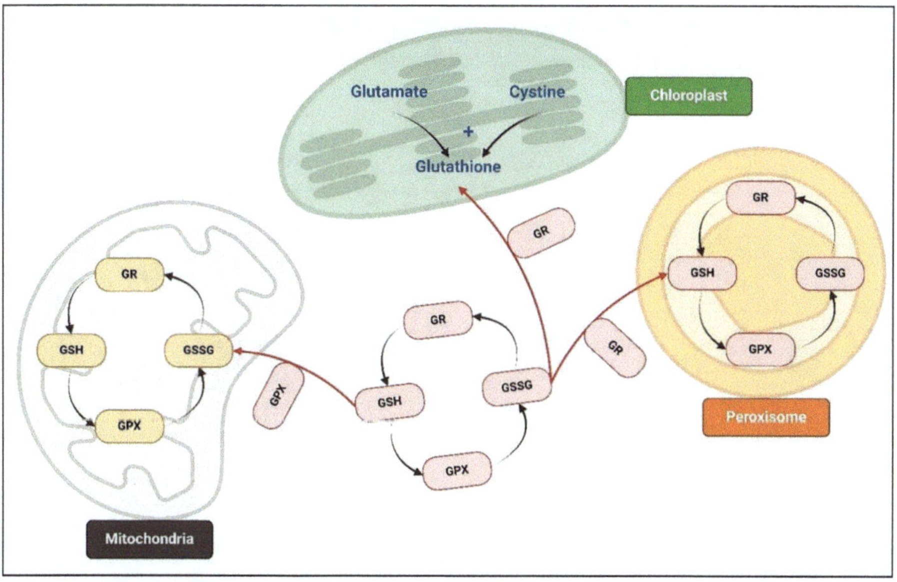


**Fig. 16.2** Diagrammatic representation of the biosynthetic processes of glutathione in chloroplasts, peroxisomes, and mitochondria (Rai et al. 2023)

$$\mathrm{APX} + \mathrm{H_2O_2} \rightarrow \mathrm{Intermediate}\left(\mathrm{I}\right) + \mathrm{H_2O}$$

(b) Intermediate I reacts with a substrate (HS), giving Intermediate II and an oxidized substrate (S).

$$\mathrm{Intermediate}\left(\mathrm{I}\right) + \mathrm{HS} \rightarrow \mathrm{Intermediate}\left(\mathrm{II}\right) + \mathrm{S}$$

(c) Intermediate II reacts with another molecule of substrate (HS), regenerating APX and releasing oxidized substrate and water

$$\mathrm{Intermediate}\left(\mathrm{II}\right) + \mathrm{HS} \rightarrow \mathrm{APX} + \mathrm{S} + \mathrm{H_2O}$$

The APX detoxification role is shown in Fig. 16.3.

It is worth mentioning that all APX isoforms are sensitive to cyanide and azide, which makes their active site highly sensitive to particular inhibitors. Iron is essentially present at the catalytic site, and even in the presence of a sufficient amount of ascorbate, the deficiency of iron strongly diminishes the activities of cAPX (Guo et al. 2017, 2020). APX plays an important role in the AsA–GSH cycle in plant cells, which plays a critical role in the detoxification of H2O2. In the process, the formation of MDHA takes place, which can either be recycled back into AsA or hydrolyzed irreversibly into DHA (Chaudhry and Sidhu 2022; Chen et al. 2025; Ullah et al. 2016).

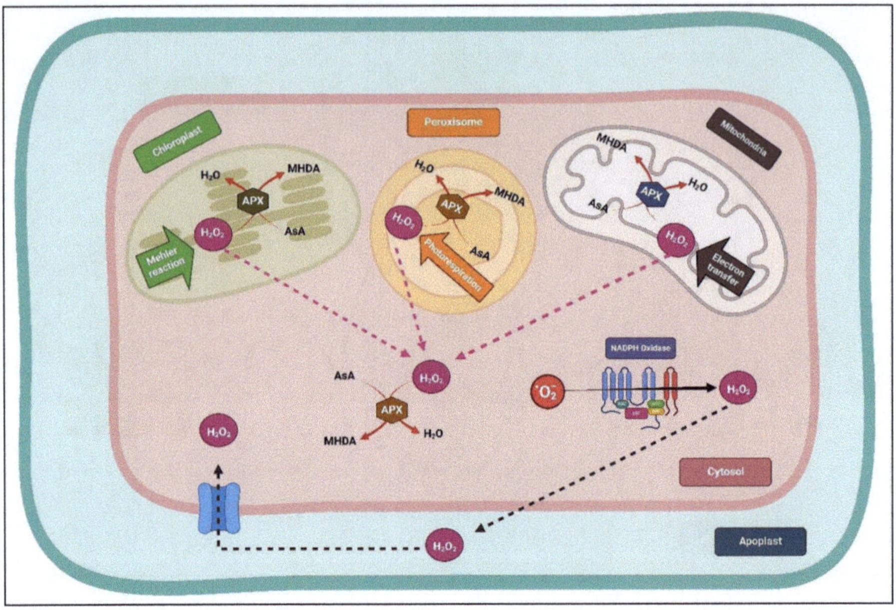


**Fig. 16.3** The role of Ascorbate peroxidase (APX) in excess reactive oxygen species (ROS) elimination in chloroplasts, mitochondria, and peroxisomes (Hasanuzzaman et al. 2019)

This binuclear iron-dependent enzyme has a dual dependency on both AsA and iron for its activity; hence, it is sensitive to the shortage of any one of the mentioned factors. This fact further underlines the complexity in the role that APX plays in the maintenance of cellular redox homeostasis. In this respect, APX has been able to efficiently detoxify H2O2 into water and oxygen, thus saving plant cells from oxidative damage, while keeping the AsA–GSH cycle going. Such features distinguish APX from all other peroxidases and serve to underscore the important role that it plays in the responses of plants to stresses and in the regulation of plant metabolism.

### 16.2.4 Monodehydroascorbate Reductase (MDHAR)

Monodehydroascorbate reductase is a key enzyme to regenerate ascorbate in plants, preventing oxidative stress and contributing to redox balance. The enzyme, known as MDHAR (EC 1.6.5.4), carries out its reductive activity within the ascorbate-glutathione cycle, employing electrons donated either by NADH or NADPH in the reduction of MDHA into AsA. It is a capacity that proves fundamental in keeping AsA levels adequately maintained, above all when depressed by heightened levels of reactive oxygen species. MDHAR occurs in several isoforms that translocate to different organelles of the plant cell, from cytosol to mitochondria, chloroplasts, peroxisomes, and glyoxysomes (Chauhan et al. 2022; Leterrier and Cagnac 2018; Pessarakli 2019). The role of MDHAR as an antioxidant is shown in Fig. 16.4.

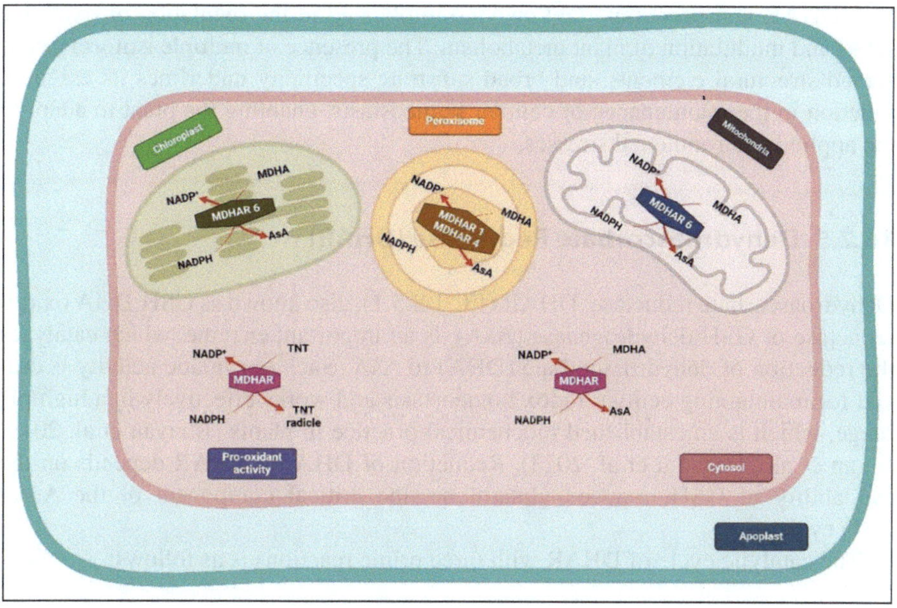


**Fig. 16.4** The antioxidant function of MDHAR in regenerating AsA in chloroplasts, mitochondria, and peroxisomes, and the pro-oxidant role of MDHAR in 2,4,6-trinitrotoluene (TNT) toxicity

Recent studies have enlightened the structural features of MDHAR. In silico approaches allowed (Begara-Morales et al. 2019, 2021) to present a three-dimensional model of peroxisomal MDHAR in peas. For MDHAR structural analysis in *Oryza sativa* (japonica rice), studies were conducted (Park et al. 2016; Sonkar et al. 2023; Tanaka et al. 2021) which revealed that MDHAR is a flavin adenine dinucleotide and pyridine-binding domain-containing protein. The overall architecture of MDHAR showed similarity to Fe-S protein reductases with an α/β fold. Some important conserved residues, Arg320 and Tyr349, are required for catalysis. The FAD-binding domain binds FAD through hydrogen bonding and van der Waals interactions. Conserved residues involved in the binding are Gly13, Gly15, Gly297, Ala122, Ala319, and Thr123. Further, the structural bridge of Lys53, Glu178, and Pro49 underpins the role of residues contributing toward MDHAR enzymatic activity. MDHAR has a high potential to be responsive under abiotic stress, since MDHAR catalyzes the reduction of MDHA that accumulates by the scavenging of ROS (Bhuyan et al. 2019; Imran et al. 2024; Sidhu et al. 2020). The adaptive response was noticed among many plant species, which refers to the importance of MDHAR in the tolerance to stress. The chlMDHAR, however, contributes to photosynthesis, particularly when peroxiredoxin is absent (Chaudhry and Sidhu 2022; Hasanuzzaman et al. 2018). Its activities significantly increase, up to three- to six-fold, along with the ripening of pepper fruits, which suggests a role in metabolic changes related to development (Chen et al. 2024; Mishra et al. 2023).

MDHAR is a very important enzyme implicated in the regulation of oxidative stress and modulation of plant metabolism. The presence of multiple isoforms, conserved structural elements, and broad substrate specificity underlines its essential function in the maintenance of cellular homeostasis, enabling the plant to adapt to the applied environmental stresses.

### 16.2.5 Dehydroascorbate Reductase (DHAR)

Dehydroascorbate reductase, DHAR (EC 1.8.5.1), also known as GSH:DHA oxidoreductase or GSH dehydrogenase (AsA), is an important enzyme, which catalyzes the reduction of dehydroascorbate (DHA) to AsA. Such enzymatic activity is crucial for maintaining cellular redox homeostasis and works effectively in a high pH range, which is an established biochemical practice in plants (Bhuyan et al. 2019; Imran et al. 2024; Lu et al. 2023). Reduction of DHA by DHAR depends on the availability of GSH, reduced glutathione, the critical component of the AsA–GSH cycle.

The catalytic cycle of DHAR with three prime reactions is as follows:

(a) The active site cysteine of DHAR gets oxidized by DHA and consequently forms sulfenic acid and AsA.

$$\mathbf{DHAR-S+DHA \rightarrow DHAR-SOH+AsA}$$

(b) Oxidized DHAR-SOH reacts with GSH to form a mixed disulfide, DHAR-S-SG.

$$\mathbf{DHAR-SOH+GSH \rightarrow DHAR-S-SG}$$

(c) A second molecule of GSH reduces the mixed disulfide to recycle DHAR and produce oxidized glutathione, GSSG.

$$\mathbf{DHAR-S-SG+GSH \rightarrow DHAR-S+GSSG}$$

The net reaction is thus as follows:

$$\mathbf{2GSH+DHA \rightarrow GSSG+AsA}$$

DHAR is a monomeric enzyme belonging to the glutathione S-transferase class. It has an exceptionally conserved three-dimensional structure among several plant species, with only a few minor characteristics, such as short-chain segments in front of the α-helix. The DHAR enzyme contains two special substrates, including glutathione at the G-site and dehydroascorbate at the H-site. The GSH, at the G-site, interacts with Cys20 via the establishment of a disulfide bond. The glutamyl group of GSH is stabilized via hydrogen bonding with a water molecule and the backbones of Ser73 and Asp72. The γ-glutamyl group engages with Phe22 via van der

Waals forces and hydrogen bonding with Lys59, while the core cysteinyl group is reinforced by Val60. In addition, the glycinyl moiety of GSH forms a salt bridge with Lys47 to enhance stability (Hamada et al. 2023; Sigel 2024). On the other hand, the DHAR active site involved in the binding of DHA, or so-called H-site, adopts the structural flexibility favorable for the interactions of substrate and catalysis (Cheng et al. 2024; Foyer and Kunert 2024).

The participation of DHAR in the plant responses to stresses has gained much attention lately, therefore. These stresses inhibit oxidative damage through enzymatic inter-conversions that convert DHA back into ascorbate in order to maintain redox balance. Also, chloroplastic DHAR enhances photosynthetic efficiency significantly under oxidant challenge. Thus, it executes a very important role in plant physiology as well as in the adaptation to various types of stress.

### 16.2.6 Glutathione Reductase (GR)

Glutathione reductase (GR; EC 1.8.1.7), also recognized as GSR or NADP+ oxidoreductase, is an essential flavoprotein oxidoreductase that catalyzes the conversion of oxidized glutathione (GSSG) to its reduced form (GSH). This reaction is crucial for maintaining cellular redox balance and depends on NADPH as an electron donor. Based on its source, GR has been found to occur in different structural forms, mainly monomeric, dimeric, heterodimeric, and heterotetrameric (Gill et al. 2013; Horváth et al. 2023; Manna and Bandyopadhyay 2023). In higher plants, two main genes encode GR: GR1 and GR2. GR1, with its shorter protein sequence, are found mostly in the cytosol or peroxisomes. GR2, with its longer N-terminal sequence, directs the enzyme to the mitochondria or chloroplasts (Cassier-Chauvat et al. 2023; Hossain et al. 2022). Notably, chloroplastic isoforms of GR constitute some 80% of the GR activity in plants (Müller-Schüssele et al. 2021; Raihan et al. 2024). Different plant-based sources, such as spinach and tobacco, have been successfully used to purify and study the different isoforms of GR (Hasanuzzaman et al. 2017c; Roy et al. 2024). Structurally, GR is a flavin-containing enzyme, and its conserved Rossmann folds are important for FAD and NADPH binding (Kesawat et al. 2023; Mishra et al. 2023). The quaternary structure of the enzyme varies between sources, but most studies have identified four key functional domains: the FAD-binding domain, NADPH-binding domain, GSSG-binding domain, and an interface domain (Hanukoglu 2017; Mansouri et al. 2022). Some of these residues are crucial for the function of the enzyme, like Arg287 and Arg293 (Hasanuzzaman et al. 2019). Another key feature is a redox-active disulfide bridge within its active site formed by conserved cysteines. Notably, in higher plants, one of the conserved cysteines is substituted by Ser164 (Sachdev et al. 2023; Sharma et al. 2022). Typically, the catalytic action of GR is expressed in two consecutive steps:

(a) *FAD Reduction:* NADPH reduces the flavin moiety of the enzyme; it generates electrons transferred to the cleavage of the disulfide bridge, producing an anion-thiolate intermediate, which releases a cysteine.

(b) *Reduction of GSSG:* When it binds to the active site, GSSG creates a disulfide bridge with both histidine and cysteine residues, in the process producing two molecules of reduced glutathione, and it re-structures the disulfide bridge at the active site of the enzyme (Ferreira et al. 2023).

The overall form of the reaction catalyzed by GR is

$$\mathbf{GSSG + NADPH + H^{+} \rightarrow 2GSH + NADP^{+}}$$

GR is strongly substrate-specific with considerably greater affinity for NADPH than NADP+ in stoichiometric proportion. The activities of this enzyme are also dependent on pH, NADPH, and GSSG availability. Optimal activity exists between pH 5.5 and 7.0. Inhibitory effect occurs at low or high pH and also under limited supply of NADPH, which could partially be avoided by GSSG as per Foyer et al., (2019) and Tanaka et al., (2021) Glutathione reductase is a crucial enzyme, which is associated with antioxidant defense systems when the plant is exposed to oxidative stress. GR keeps regenerating GSH continuously, thus enabling continuous detoxification of reactive oxygen species and sustaining cellular redox balance. Chloroplastic GR is particularly crucial in photosynthetic tissues due to the fact that it guards photosynthetic machinery from the deleterious effect of ROS. Hence, the significance of this enzyme to plant physiology may be manifested in its isoforms in addition to its structural and functional flexibility and multiple regulation mechanisms. High catalytic efficiency of GR, which is absolutely required for oxidative stress response, makes it one of the pillars of the redox defense system in guarding plant cells against environmental and metabolic stresses.

## 16.3 AsA–GSH Pathway: Enzymatic Reactions

The ascorbate–glutathione (AsA-GSH) cycle is a vital antioxidant defense mechanism in plant cells, mainly responsible for detoxifying hydrogen peroxide (H2O2). It involves the coordinated action of antioxidants such as ascorbate (AsA) and glutathione (GSH) with key enzymes—ascorbate peroxidase (APX), monodehydroascorbate reductase (MDHAR), dehydroascorbate reductase (DHAR), and glutathione reductase (GR). Both AsA and GSH are distributed in all cellular compartments, including the cytosol, nucleus, chloroplast, mitochondria, and peroxisomes, with different isoforms of these enzymes adapted to each organelle (Foyer and Kunert 2024). Present in millimolar concentrations, these antioxidants have very high redox potentials, enabling them to interact with a wide array of cellular components and pathways. Their most important function is to keep the cellular state in reduced condition, a prerequisite for optimum cellular functioning and stress tolerance. In a series of steps, AsA and GSH collectively quench H2O2 while being regenerated. The cycle of reactions underlying this interaction not only illustrates the dynamic nature of these interactions between the antioxidants and their associated enzymes but also reflects the critical importance of such antioxidant systems in protection against oxidative damage in plant cells.

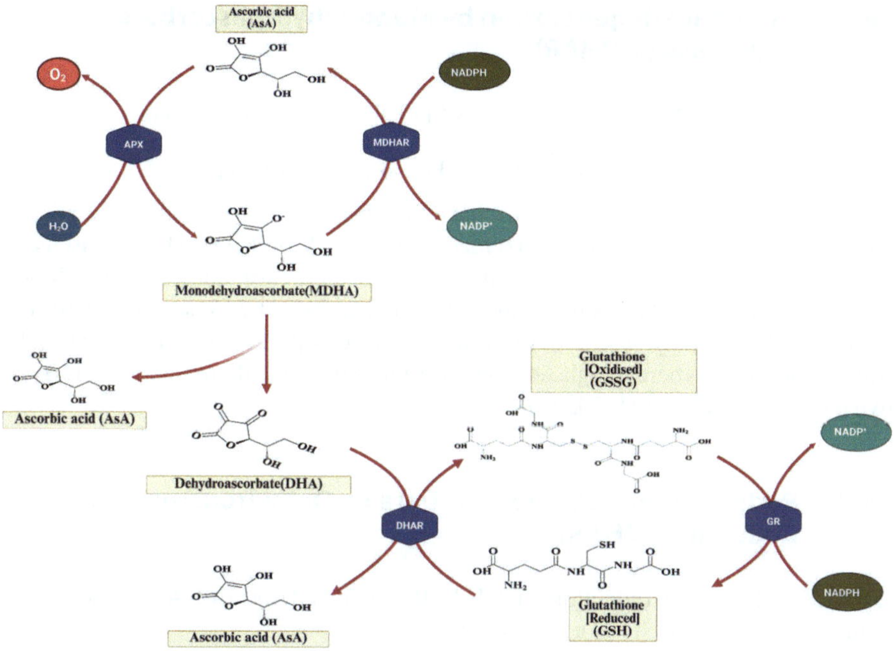


**Fig. 16.5** Ascorbate–Glutathione pathway (Asada–Halliwell pathway)

Figure 16.5 shows the Ascorbate–glutathione pathway.

### 16.3.1 Hydrogen Peroxide Reduction by Ascorbate Peroxidase (APX)

Ascorbate Peroxidase (APX) is the first enzyme of the AsA–GSH cycle. It detoxifies $H_2O_2$. The reaction is summarized below:

$$\mathbf{H_2O_2 + AsA \rightarrow 2H_2O + MDHA}$$

APX reduces $H_2O_2$ to water, oxidizing AsA to MDHA.

A heme prosthetic group is present at the catalytic site of ascorbate peroxidase (APX). It facilitates the transfer of electrons from AsA to $H_2O_2$ for the reduction process. As discussed earlier, there are different isoforms of APX present in cellular compartments like chloroplasts, mitochondria, cytosol, and peroxisomes. These isoforms are major part of the efficient detoxification of $H_2O_2$ generated in these organelles. The activity of APX depends on pH and substrate (AsA) availability. If there is not enough substrate (AsA), APX activity goes down, leading to $H_2O_2$ accumulation and oxidative stress (Li 2023).

### 16.3.2 Ascorbate Regeneration by Monodehydroascorbate Reductase (MDHAR)

MDHAR reduces MDHA to AsA with NAD(P)H as an electron donor

$$\mathbf{MDHA + NAD(P)H + H^{+} \rightarrow AsA + NAD(P)^{+}}$$

MDHA is an unstable intermediate that either disproportionate non-enzymatically into AsA and DHA or reduced enzymatically to AsA by MDHAR. As discussed earlier, MDHAR is localized in multiple cellular compartments, such as chloroplasts, mitochondria, and peroxisomes, which ensures fast regeneration of AsA to maintain APX activity. The activity of this enzyme is particularly important under stress conditions, where a high AsA pool is required for the detoxification of oxidative damage (Madhu et al. 2024b).

### 16.3.3 Reduction of Dehydroascorbate by Dehydroascorbate Reductase (DHAR)

DHAR catalyzes the reduction of DHA, the oxidized form of AsA, back to its reduced form using GSH as a reductant:

$$\mathbf{DHA + 2GSH \rightarrow AsA + GSSG}$$

DHA is highly unstable. If not reduced promptly, it tends to be hydrolyzed irreversibly into 2,3-diketogulonic acid. DHAR ensures the efficient recycling of DHA to AsA. Thus, DHAR is responsible for preventing the loss of the cell's antioxidant ability. The activity of this enzyme is controlled by the cellular redox state (Das et al. 2022).

### 16.3.4 Glutathione Regeneration by Glutathione Reductase (GR)

GR catalyzes the reduction of GSSG to GSH, with NADPH as an electron donor:

$$\mathbf{GSSG + NADPH + H+ \rightarrow 2GSH + NADP +}$$

GR is responsible for the maintenance of high GSH/GSSG ratio, critical for DHAR function, and other plant antioxidant mechanisms. The activity of this enzyme mainly depends upon the availability of NADPH. Here, the NADPH is supplied by the pentose phosphate pathway and the photosynthetic electron transport chain. Cell organelles including chloroplasts, mitochondria, and cytosol, having their own GR isoforms, ensure the redox balance across cellular compartments (Lapenna 2023).

### 16.3.5 Non-Enzymatic Reactions

In addition to enzymatic processes, non-enzymatic reactions are involved in the AsA–GSH cycle that contribute to the restoration of antioxidants and detoxification of ROS. These include

(a) *Disproportionation of MDHA:*

$$\mathbf{2MDHA \rightarrow AsA + DHA}$$

This reaction is a spontaneous process occurring in aqueous environments, and it helps to regenerate AsA from MDHA, although it also gives rise to DHA, which needs to be further recycled by DHAR (Arabia et al. 2024).

(b) *AsA and GSH as Direct Scavengers*

To prevent oxidative damage to biological components, AsA and GSH can both directly react with ROS, including hydroxyl radicals (OH•) and singlet oxygen ($^1O_2$) (Demirci-Çekiç et al. 2022). Table 16.1 shows the localization of different enzymes and respective isoforms involved in AsA–GSH cycle in plants.

## 16.4 Regulation of AsA–GSH Cycle

The regulation of the AsA–GSH cycle is a complex event controlled by numerous factors, such as exogenous treatments, signaling molecules, environmental conditions, and specific stressors. Here, the present review delineates these regulatory aspects, focusing on the recent findings and insight into their mechanisms.

### 16.4.1 Regulation by AsA and GSH

The metabolic profiles and activities of the AsA-GSH cycle are significantly modulated by the availability of its two central metabolites—AsA and GSH. Exogenous application of them has been widely reported to improve the antioxidant enzyme activities, thereby enhancing the resistance of plants toward different environmental stressors. Several studies have demonstrated the efficacy of exogenous treatments with AsA or GSH in boosting plant defense systems. The beneficial roles of external formulations in increasing the activity of antioxidant enzymes, thus preventing oxidative damage, are well established. Exogenous GSH pretreatment significantly restores growth in certain rice cultivars subjected to submergence. The heightened sub-tolerance induced by GSH may be partially ascribed to augmented concentrations of chlorophylls, carotenoids, soluble proteins, and proline. Exogenous GSH also alleviates sub-induced oxidative damage, as indicated by decreased levels of

**Table 16.1** Localization of different enzymes and respective isoforms involved in ascorbate–glutathione (AsA–GSH) cycle in plants

| Enzyme | Location organelle | Isoforms of the enzyme | Associated plant species | Reference |
|---|---|---|---|---|
| Monodehydroascorbate reductase (MDHAR) | Cytosol | MDHAR-2 MDHAR-3 | 1. *Triticum aestivum* 2. *Arabidopsis thaliana* | Strand et al. (1999) Zhou et al. (2021) Lisenbee et al. (2005) |
| | Mitochondria | Mitochondrial MDHAR (MDHAR-5) | 1. *Glycine max* 2. *Oryza sativa* | |
| | Peroxisomes | Peroxisomal MDHAR (MDHAR-1, MDHAR-4) | *Pisum sativum* | |
| | Chloroplast | Chloroplast MDHAR (MDHAR-6) | 1. *Arabidopsis thaliana* 2. Gossypium herbaceum | |
| Glutathione reductase (GR) | Chloroplast | Chloroplast GR (GR-2) | 1. *Spinacia oleracea* 2. *Arabidopsis thaliana* | Delorme-Hinoux et al. (2016), Farooq and Azam (2006), Kataya and Reumann (2010), Li et al. (2022), Marty et al. (2019), Rinne et al. (2001) |
| | Peroxisomes | Peroxisomal GR (GR-1) | *Pisum sativum* | |
| | Cytosol | Cytosolic GR (GR-1) | 1. *Glycine max* 2. *Pisum sativum* | |
| | Mitochondria | Mitochondrial GR (GR-2) | 1. *Oryza sativa* 2. *Triticum aestivum* | |
| Dehydroascorbate reductase (DHAR) | Cytosol | Cytosolic DHAR (DHAR-1, DHAR-2) | 1. *Glycine max* 2. *Arabidopsis thaliana* | Rahantaniaina et al. (2017), Terai et al. (2020) |
| | Mitochondria | Mitochondrial DHAR | *Spinacia oleracea* | |
| | Chloroplast | Chloroplast DHAR (DHAR-2) | 1. *Oryza sativa* 2. *Triticum aestivum* | |

(continued)

**Table 16.1** (continued)

| Enzyme | Location organelle | Isoforms of the enzyme | Associated plant species | Reference |
|---|---|---|---|---|
| Ascorbate peroxidase (APX) | Peroxisomes | Peroxisomal APX (pAPX) (APX-3, APX-4) | 1. *Arabidopsis thaliana* 2. *Pisum sativum* | Asada (1999), Duan et al. (2012), Jardim-Messeder et al. (2022), Kaur et al. (2021), Liu et al. (2018), Maruta et al. (2010), Yamada et al. (2002) |
| | Mitochondria | Mitochondrial APX (mtAPX) | 1. *Nicotiana tabacum* 2. *Glycine max* | |
| | Cytosol | Cytosolic APX (cAPX) (APX-1, APX-2,APX-6) | 1. *Oryza sativa* 2. *Arabidopsis thaliana* | |
| | Chloroplast (stroma, thylakoid) | Stromal APX (sAPX), thylakoid-bound APX (tAPX) | 1. *Spinacia oleracea* 2. *Arabidopsis thaliana* | |

H2O2 and malondialdehyde, corresponding with enhanced activity of antioxidant enzymes (Siddiqui et al. 2021).

Application of exogenous AsA also enhances the antioxidant potential of plants. For instance, Munir et al. 2021 showed that the external application of ascorbic acid amplifies the antioxidant efficacy of *Ocimum sanctum* L. cultivated under saline stress. Similarly, (Hanife et al. 2023) showed exogenous GSH synergistically regulates GSH production, metabolism, and redox homeostasis to improve resistance to oxidative stress induced by lead (Pb) in *Brassica napus* L. All these studies have established AsA and GSH as two key metabolites responsible for keeping up the effective functionality of the AsA–GSH cycle. These two metabolites play a very important role in the maintenance of redox homeostasis and assure plants of an effective ROS detoxification process under stressful conditions, thus enhancing survival chances.

### 16.4.2 Regulation by Salicylic Acid (SA)

Salicylic acid (SA) is a signaling molecule involved in the responses of plants to different abiotic stresses. The regulatory role played by SA on the AsA-GSH cycle has been well documented and involves the modulation of antioxidant enzyme activities. Altaf et al. (2022) showed that Preliminary therapy of salicylic acid elevated the levels of superoxide dismutase (SOD), catalase (CAT), glutathione peroxidase (GPX), glutathione S-transferase (GST), and ascorbate–glutathione

(AsA-GSH) pathway enzymes while also augmenting ascorbate (AsA) and glutathione (GSH) concentrations and reducing hydrogen peroxide ($H_2O_2$) and superoxide anion levels in rice seedlings through the regulation of antioxidant enzyme gene expression. This further proves that SA-mediated regulation of AsA–GSH cycle increases the tolerance to oxidative stress. Besides its role in postharvest stress management, SA has been suggested to be involved in the regulation of photosynthetic efficiency and in stress signaling. It has been shown that SA affects the expression of genes encoding for key enzymes of AsA–GSH cycle and thereby fine-tunes the antioxidant defense system of the plant. All these findings bring forth the multifunctional nature of SA in plant stress physiology.

### 16.4.3 Influence of Nitric Oxide (NO), Cadmium (Cd), Molybdenum (Mo), and Methyl Viologen (MV)

Various elements and compounds, such as nitric oxide (NO), cadmium (Cd), molybdenum (Mo), and methyl viologen (MV), significantly impact the AsA–GSH cycle. These factors exert their effects by modulating the activities of antioxidant enzymes and influencing the redox balance of AsA and GSH. Kaya et al. (2023) demonstrated that exogenous nitric oxide treatment significantly contributes to salt stress tolerance in wheat plants, produced by hydrogen sulfide through the activation of enzymes associated with the ascorbate–glutathione cycle.

In *Camellia sinensis* L. seedlings treated with high concentration (30 mM) cadmium, (Wang et al. 2023) observed that ascorbic acid and glutathione can mitigate cadmium stress in plants by enhancing antioxidant capacity to eliminate reactive oxygen species.

MV treatment in cucumber leaves caused large increases in DHA, GSH, and GSSG levels, though AsA content in chloroplasts and cytosol decreased slightly (Song et al. 2006). Interestingly, mitochondrial GSSG levels and cytosolic GSH were minimally affected by MV treatment. This is consistent with the idea that the AsA–GSH cycle acts in a highly compartmentalized fashion and that some components are tailor-made for particular stressors.

Wu et al. (2014) discussed the resemblances and variances between the Mo-efficient and Mo-inefficient cultivars of wheat in response to Mo application and showed that the activities of antioxidant enzymes such as superoxide dismutase, peroxidase, catalase, ascorbate peroxidase, and the contents of non-enzymatic antioxidants content such as ascorbic acid, reduced glutathione, carotenoid were significantly increased, which further indicates that Mo application can improve abilities of antioxidative defense of wheat. Such results suggest a positive role for Mo in the regulation of AsA–GSH cycle-related antioxidant defenses. All these results clearly indicate that plant antioxidant defense is really complex and related to AsA–GSH cycle regulation.

### 16.4.4 Regulation by Environmental Acclimation

Environmental acclimation by drought or chilling preconditioning can upregulate the AsA–GSH cycle through increasing activities of its associated enzymes. Such acclimation treatments elicit coordinated activation of antioxidant defenses that enhance plant tolerance to oxidative stress. Amoah et al. (2023) showed that drought-acclimated (DA) millets exhibited enhanced drought tolerance, demonstrated by superior growth, photosynthesis, antioxidant defense, polyphenol accumulation, and gene expression associated with antioxidants and polyphenol production. The evidence demonstrated that environmental acclimation-induced systemic responses improve the ability of plants to cope with redox homeostasis. Increased activities in APX, MDAR, DHAR, and GR are also observed on account of various acclimation treatments, enabling the plants to achieve improved capacity to withstand oxidative stresses. The AsA–GSH cycle operates optimally only under favorable environmental management. Suitable light, temperature, and irrigation practices may certainly improve the antioxidative potentials of plants in combating various unfavorable climatic fluctuations, thereby imparting an improvement in resistance in them.

### 16.4.5 Extending the Regulation Network

Besides the above mechanisms, recent research identified other levels of regulation in the AsA–GSH cycle. Some of the mechanisms through which these plants regulate antioxidant responses include hormonal signaling, epigenetic regulation, and the role of transcription factors. For example, ABA and ethylene have been established to regulate APX, GR, and other cycle enzymes' activities (Alwutayd et al. 2024; Jiang et al. 2022). Newer frontiers in the regulation of these antioxidants include histone acetylation and DNA methylation patterns controlling the expression of genes in the AsA–GSH pathway. Recent evidence indicates that ncRNAs are also involved in the functioning of AsA–GSH cycle. The ncRNAs interact with the key components of the antioxidant network, which modifies their enzyme activity and tunes the level of metabolites (Bose Mazumdar and Chattopadhyay 2024; Madhu et al. 2024a). This study proposes new directions of investigation with future options for crop enhancement and stress control.

## 16.5 Interaction of Other Pathways with AsA-GSH Pathways in Modulating ROS Metabolism

The ascorbate–glutathione cycle is a vital mechanism for detoxifying reactive oxygen species, leading to the removal of oxidative stress. Other than its role in the management of oxidative stress, this cycle cross talks with many other processes to modulate ROS metabolism and plant cellular responses. Here in the current review, it has been described how the AsA–GSH cycle cross talks with other processes and plays an aggregate role in stress tolerance of plants.

### 16.5.1 Signaling Role of AsA and GSH and Their Interaction with Other Pathways

AsA acts as a co-factor for many enzymes; it has a very significant role in the xanthophyll cycle, wherein it also functions as a de-epoxidase co-factor of violaxanthin. AsA is used for the production of organic acids like oxalate and tartrate, which are essential for cellular metabolism and defense (Mishra et al. 2024). The oxidative degradation of AsA to dehydroascorbate (DHA) generates an electrochemical gradient across the plasma membrane, with great implications for signaling (Przybyło and Langner 2020). In redox signaling, GSH through glutathionylation can modulate cysteine residues of proteins, which regulate the control of transcription and protein function (Lou 2022). GSH also plays a role in regulating stomatal responses to methyl jasmonate (MeJA) in Arabidopsis by governing harmony between hormonal signaling and environmental response (Foyer and Kunert 2024). GSH activates calcium signaling and specific gene expression in tobacco, indicating that it is an important player in the different pathways of calcium-dependent protein kinase (Lou 2022). It further supports its role in cellular signaling and defense by reducing hydrogen peroxide (H2O2) by Prx, which is GSH dependent (Das et al. 2020).

### 16.5.2 Interplay Between AsA–GSH Cycle and the Nitric Oxide (NO) Metabolic Pathway

It has been well established that NO is one of the primary signaling molecules involved in the AsA–GSH cycle, particularly during stressful conditions (Chauhan et al. 2022; Gan et al. 2024; Hou et al. 2024). Oxidation of GSH by NO results in the production of GSNO, an intermediate that further leads to reactions catalyzed by GSNOR. These reactions will maintain a balance between nitric oxide and nitrosothiols such that redox balance would be given. S-nitrosylation, regulated by nitric oxide, as a post-translational modification, is necessary for controlling the activity of enzymes in the AsA-GSH cycle (Lv et al. 2023; Wei et al. 2022). Among these, ascorbate peroxidase (APX) is particularly responsive to nitric oxide. The nitrosylation of its active-site cysteine (Cys32) might increase its activity under salt stress and improve antioxidant defense mechanism; but oxidation of cysteine would inhibit APX action during intense oxidative stress (Lv et al. 2023). Monodehydroascorbate reductase (MDHAR) was nitrated; a nitrated cofactor binding site was generated, inactivating its activity and hindering recycling of ascorbate; dehydroascorbate reductase (DHAR) appeared also to be inactivated but structural–functional correlation must be drawn (Sougrakpam et al. 2023). Finally, glutathione reductase (GR) is not nitrated in some species like peas but is inactivated under certain conditions.

### 16.5.3 Interplay Between the AsA–GSH Pathway and Glyoxalase System

The glyoxalase system is an important detoxification pathway for methylglyoxal (MG), and its activity is closely related to the AsA–GSH cycle through GSH (Dorion et al. 2021). MG is a byproduct of glycolysis that is detoxified by glyoxalase I (Gly I) and glyoxalase II (Gly II). Gly I catalyzes the conjugation of MG with GSH to form S-D-lactoylglutathione, which is further degraded by Gly II into D-lactate, thereby regenerating GSH (Hossain et al. 2021). This relationship results in effective removal of MG to prevent its toxic effects on cellular environment including antioxidant enzymes. The relationship between the glyoxalase system and GSH is crucial for the functional interaction of these two pathways in maintaining cellular redox balance in stress conditions.

### 16.5.4 Interactions of AsA–GSH Pathway with Xenobiotic Detoxification Pathways

Plants conjugate toxic chemicals with GSH by the use the AsA-GSH cycle during xenobiotic detoxification. Glutathione S-transferases (GSTs) is one of the important enzymes involved in this reaction, which also helps in the sequestration of detoxified products into vacuoles. ATP-driven transporters help in transporting these detoxified products. Through xenobiotics detoxification, this system thus ensures the formation and storage of secondary metabolites (Del Buono et al. 2020; Espinosa-Vellarino et al. 2024). For example, GSTs facilitate the conjugation of herbicides with GSH. This results in vacuolar sequestration of the herbicides, thereby protecting plant tissues from their toxic effects (Gaafar et al. 2022).

### 16.5.5 Interaction of AsA–GSH Cycle with Phytohormone Biosynthesis Pathways

The AsA–GSH cycle interacts closely with the phytohormone production pathways, thus modulating plant development and stress responses. AsA is a cofactor for several dioxygenases participating in the biosynthesis of phytohormones, such as abscisic acid (ABA) and ethylene (Bulley et al. 2021). Abscisic acid is a phytohormone involved in the regulation of development and adaptation to stresses; ascorbic acid is needed for ABA signaling and action. Low AsA triggers PR protein expression (Huang et al. 2019), which may indicate its role in interacting with ABA-dependent defense systems. GSH affects ethylene biosynthesis. Gene expression of ethylene biosynthesis enzymes is positively controlled by GSH for such genes as 1-aminocyclopropane-1-carboxylate (ACC) synthase and ACC oxidase (Min et al. 2020). Further, GSH enhances SAT levels, thus increasing tolerance against nickel toxicity (Datta et al. 2015). In rice, overexpression of SA metabolism genes leads to enhanced levels of SA and GSH and both are implicated in the control of the oxidative stress and hormonal signaling pathways (Liu et al. 2022; Singh et al. 2015).

### 16.5.6 Interaction of AsA–GSH Pathway with Metal Chelation Processes

The AsA–GSH cycle interacts with metal chelation pathways to detoxify the heavy metal and metalloids and brings about cellular homeostasis. GSH is the precursor for the synthesis of phytochelatins (PCs). Phytochelatins, on the other hand, sequester toxic metals to ensure their vacuolar compartmentalization (Ansari et al. 2024; Seregin and Kozhevnikova 2023). Phytochelatin synthase (PCS) is the primary enzyme catalyzing this process by using GSH as its substrate. In addition, glutaredoxins (GRx) and thioredoxins (TRx) are also two very important GSH-dependent enzymes involved in this process. They help to reduce $H_2O_2$ and modulate protein thiols for metal detoxification and redox balance. Metallothioneins (MTs) also cooperate with phytochelatins (PCs) in the metal chelation process (Montes-Rocha et al. 2024). These mechanisms allow for detoxification of heavy metals to protect cellular functions from metal-induced oxidative damage (Montes-Rocha et al. 2024).

## 16.6 Conclusion

The ascorbate–glutathione cycle plays a pivotal role in redox biology and cellular homeostasis in plant systems, mediating the effects of oxidative stress to alleviate injury while simultaneously managing the largest component of hydrogen peroxide signaling for many important physiological processes in plant cells. Therefore, AsA–GSH regulation is believed to play an essential role in balancing detrimental effects by reactive species, while numerous biochemical and signal pathways are induced through ascorbate and glutathione changes for plant growth and stress adaptations. Ascorbate and glutathione are complex and multifunctional antioxidants whose functions appear somewhat unique yet synergistic. Ascorbate efficiently scavenges for superoxide radicals, modulates superoxide-dependent signaling and enhances other antioxidant mechanisms, such as the regeneration of tocopherol in chloroplasts, thereby aiding further in cell wall growth and calcium signaling and in photoprotection through the xanthophyll cycle. Glutathione, on the other hand, maintains redox balance by detoxifying processes, metal chelation, and protein glutathionylation. The two together form a cohesive network that aids the organism in combating both abiotic and biotic stress. The enzymatic components of the AsA-GSH cycle—ascorbate peroxidase, monodehydroascorbate reductase, dehydroascorbate reductase, and glutathione reductase—operate with geographical and regulatory flexibility. Subcellular compartmentalization and metabolite transport systems add complexity, since some processes, such as the reduction of MDHA and DHA next to the photosynthetic electron transport chain, occur in a glutathione-independent manner.

Recent evidence strongly underscores the nuclear roles of the AsA-GSH cycle in redox signaling, stress response, and cell fate determination. The formation of ROS in the nucleus, particularly under stress, underlines an essential function of redox systems; however, their impact on nuclear activities, such as regulation of the cell cycle and transcription, remains largely unexamined. The cycle's incorporation into larger metabolic networks enhances its importance. It affects phytohormone production, post-translational modifications such as S-nitrosylation, and detoxification pathways, underscoring its pivotal role in plant development and defense. In summary, the AsA–GSH cycle plays a crucial role in linking redox control with stress mitigation. Although much has been discovered so far, nuclear functions, compartmentalization, and interconnections with other pathways may continue to improve our understanding and disclose potential biotechnological applications in enhancing plant resistance and productivity.

**Author Contributions** *Gholamreza Abdi*: Conceptualized the chapter, provided supervision, and led the writing and critical revision process, and writing—original draft.

*Anilava Chatterjee*: Developed the theoretical framework, contributed to methodology, and participated in substantive editing.

*Md Sabir Ahmed Mondol*: Conducted research, analyzed data, and co-wrote significant sections, and writing—original draft.

*Ubaida Akbar*: Designed technical components, prepared visualizations, and contributed specialized content.

*Aritra Mukherjee*: Structured the chapter, synthesized interdisciplinary perspectives, and coordinated editing, and writing—original draft.

*Tadela Susmitha*: Validated analyses, assisted in manuscript refinement, and ensured quality control.

All authors engaged in intellectual discussions, reviewed drafts, and approved the final version.

## Bibliography

Ahmad P, Ahanger MA, Alyemeni MN, Wijaya L, Egamberdieva D, Bhardwaj R, Ashraf M (2017) Zinc application mitigates the adverse effects of NaCl stress on mustard [*Brassica juncea* (L.) Czern & Coss] through modulating compatible organic solutes, antioxidant enzymes, and flavonoid content. J Plant Interact 12(1):429–437. https://doi.org/10.1080/17429145.2017.1385867

Altaf MM, Diao X, Wang H, Khan LU, ur Rehman A, Shakoor A, Altaf MA, Farooq TH (2022) Salicylic acid induces vanadium stress tolerance in rice by regulating the AsA-GSH cycle and glyoxalase system. J Soil Sci Plant Nutr 22(2):1983–1999. https://doi.org/10.1007/s42729-022-00788-x

Alwutayd KM, Alghanem SMS, Alwutayd R, Alghamdi SA, Alabdallah NM, Al-Qthanin RN, Sarfraz W, Khalid N, Naeem N, Ali B, Saleem MH, Javed S, Gómez-Oliván LM, Abeed AHA (2024) Mitigating chromium toxicity in rice (Oryza sativa L.) via ABA and 6-BAP: unveiling synergistic benefits on morphophysiological traits and ASA-GSH cycle. Sci Total Environ 908:168208. https://doi.org/10.1016/j.scitotenv.2023.168208

Amoah JN, Adu-Gyamfi MO, Kwarteng AO (2023) Effect of drought acclimation on antioxidant system and polyphenolic content of Foxtail Millet (Setaria italica L.). Physiol Mol Biol Plants 29(10):1577–1589. https://doi.org/10.1007/s12298-023-01366-w

Anjum NA, Sharma P, Gill SS, Hasanuzzaman M, Khan EA, Kachhap K, Mohamed AA, Thangavel P, Devi GD, Vasudhevan P, Sofo A, Khan NA, Misra AN, Lukatkin AS, Singh HP, Pereira E, Tuteja N (2016) Catalase and ascorbate peroxidase—representative H2O2-detoxifying heme enzymes in plants. Environ Sci Pollut Res 23(19):19002–19029. https://doi.org/10.1007/s11356-016-7309-6

Ansari MKA, Iqbal M, Ahmad M, Munir M, Gaffar SA, Chaachouay N (2024) Heavy metal stress and cellular antioxidant systems of plants: a review. Agric Rev. https://doi.org/10.18805/ag.RF-321

Arabia A, Munné-Bosch S, Muñoz P (2024) Ascorbic acid as a master redox regulator of fruit ripening. Postharvest Biol Technol 207:112614. https://doi.org/10.1016/j.postharvbio.2023.112614
Asada K (1999) The water-water cycle in chloroplasts: scavenging of active oxygens and dissipation of excess photons. Annu Rev Plant Physiol Plant Mol Biol 50(1):601–639. https://doi.org/10.1146/annurev.arplant.50.1.601
Asada K (2006) Production and scavenging of reactive oxygen species in chloroplasts and their functions. Plant Physiol 141(2):391–396
Begara-Morales JC, Chaki M, Valderrama R, Mata-Pérez C, Padilla M, Barroso JB (2019) Transcriptional regulation of gene expression related to hydrogen peroxide (H2O2) and nitric oxide (NO). In: Nitric oxide and hydrogen peroxide signalling in higher plants. Springer International Publishing, pp 69–90. https://doi.org/10.1007/978-3-030-11129-8_4
Begara-Morales JC, Mata-Pérez C, Padilla MN, Chaki M, Valderrama R, Aranda-Caño L, Barroso JB (2021) Role of electrophilic nitrated fatty acids during development and response to abiotic stress processes in plants. J Exp Bot 72(3):917–927. https://doi.org/10.1093/jxb/eraa517
Bhuyan MHMB, Hasanuzzaman M, Al Mahmud J, Hossain MS, Bhuiyan TF, Fujita M (2019) Unraveling morphophysiological and biochemical responses of Triticum aestivum L. to extreme pH: coordinated actions of antioxidant defense and glyoxalase systems. Plants 8(1):24. https://doi.org/10.3390/plants8010024
Bose Mazumdar A, Chattopadhyay S (2024) Unfolding the role of glutathione to combat environmental stresses through "omics"-based approaches/studies. J Plant Growth Regul. https://doi.org/10.1007/s00344-024-11476-8
Bulley SM, Cooney JM, Laing W (2021) Elevating ascorbate in Arabidopsis stimulates the production of abscisic acid, phaseic acid, and to a lesser extent auxin (IAA) and jasmonates, resulting in increased expression of DHAR1 and multiple transcription factors associated with abiotic stress tolerance. Int J Mol Sci 22(13):6743. https://doi.org/10.3390/ijms22136743
Cassier-Chauvat C, Marceau F, Farci S, Ouchane S, Chauvat F (2023) The glutathione system: a journey from cyanobacteria to higher eukaryotes. Antioxidants 12(6):1199. https://doi.org/10.3390/antiox12061199
Chaturvedi S, Khan S, Bhunia RK, Kaur K, Tiwari S (2022) Metabolic engineering in food crops to enhance ascorbic acid production: crop biofortification perspectives for human health. Physiol Mol Biol Plants 28(4):871–884. https://doi.org/10.1007/s12298-022-01172-w
Chaudhry S, Sidhu GPS (2022) Climate change regulated abiotic stress mechanisms in plants: a comprehensive review. Plant Cell Rep 41(1):1–31. https://doi.org/10.1007/s00299-021-02759-5
Chauhan P, Mir RA, Khah MA (2022) Ascorbate–Glutathione Cycle. In: Nitric oxide in plants. Wiley, pp 148–178. https://doi.org/10.1002/9781119800156.ch8
Chen L-H, Xu M, Cheng Z, Yang L-T (2024) Effects of nitrogen deficiency on the photosynthesis, chlorophyll a fluorescence, antioxidant system, and sulfur compounds in Oryza sativa. Int J Mol Sci 25(19):10409. https://doi.org/10.3390/ijms251910409
Chen H, Shi Z, Zhong Z, Ji N, Zhou X, Dan Z, Shen X (2025) The effect of glutathione metabolism on the release of wound-induced spores from Pyropia yezoensis. Algal Res 86:103921. https://doi.org/10.1016/j.algal.2025.103921
Cheng Q, Zou X, Wang Y, Yang Z, Qiu X, Wang S, Yang Y, Yang D, Kim HS, Jia X, Li L, Kwak S-S, Wang W (2024) Overexpression of dehydroascorbate reductase gene IbDHAR1 improves the tolerance to abiotic stress in sweet potato. Transgenic Res 33(5):427–443. https://doi.org/10.1007/s11248-024-00408-7
Das S, Reddy RC, Chadchan KS, Patil AJ, Biradar MS, Das KK (2020) Nickel and oxidative stress: cell Signalling mechanisms and protective role of vitamin C. Endocr Metab Immune Disord Drug Targets 20(7):1024–1031. https://doi.org/10.2174/1871530319666191205122249
Das BK, Kumar A, Sreekumar SN, Ponraj K, Gadave K, Kumar S, Murali Achary VM, Ray P, Reddy MK, Arockiasamy A (2022) Comparative kinetic analysis of ascorbate (Vitamin-C) recycling dehydroascorbate reductases from plants and humans. Biochem Biophys Res Commun 591:110–117. https://doi.org/10.1016/j.bbrc.2021.12.103
Datta R, Kumar D, Sultana A, Hazra S, Bhattacharyya D, Chattopadhyay S (2015) Glutathione regulates ACC synthase transcription via WRKY33 and ACC oxidase by modulating mRNA

stability to induce ethylene synthesis during stress. Plant Physiol 169:01543.2015. https://doi.org/10.1104/pp.15.01543
Del Buono D, Terzano R, Panfili I, Bartucca ML (2020) Phytoremediation and detoxification of xenobiotics in plants: herbicide-safeners as a tool to improve plant efficiency in the remediation of polluted environments. A mini-review. Int J Phytoremediation 22(8):789–803. https://doi.org/10.1080/15226514.2019.1710817
del Río LA, López-Huertas E (2016) ROS generation in peroxisomes and its role in cell signalling. Plant Cell Physiol pcw076. https://doi.org/10.1093/pcp/pcw076
Delorme-Hinoux V, Bangash SAK, Meyer AJ, Reichheld J-P (2016) Nuclear thiol redox systems in plants. Plant Sci 243:84–95. https://doi.org/10.1016/j.plantsci.2015.12.002
Demirci-Çekiç S, Özkan G, Avan AN, Uzunboy S, Çapanoğlu E, Apak R (2022) Biomarkers of oxidative stress and antioxidant defense. J Pharm Biomed Anal 209:114477. https://doi.org/10.1016/j.jpba.2021.114477
Dorion S, Ouellet JC, Rivoal J (2021) Glutathione metabolism in plants under stress: beyond reactive oxygen species detoxification. Meta 11(9):641. https://doi.org/10.3390/metabo11090641
Duan M, Ma N-N, Li D, Deng Y-S, Kong F-Y, Lv W, Meng Q-W (2012) Antisense-mediated suppression of tomato thylakoidal ascorbate peroxidase influences anti-oxidant network during chilling stress. Plant Physiol Biochem 58:37–45. https://doi.org/10.1016/j.plaphy.2012.06.007
El-Beltagi HS, Ahmad I, Basit A, Shehata WF, Hassan U, Shah ST, Haleema B, Jalal A, Amin R, Khalid MA, Noor F, Mohamed HI (2022) Ascorbic acid enhances growth and yield of sweet peppers (Capsicum annum) by mitigating salinity stress. Gesunde Pflanzen 74(2):423–433. https://doi.org/10.1007/s10343-021-00619-6
Espinosa-Vellarino FL, Garrido I, Casimiro I, Silva AC, Espinosa F, Ortega A (2024) Enzymes involved in antioxidant and detoxification processes present changes in the expression levels of their coding genes under the stress caused by the presence of antimony in tomato. Plants 13(5):609. https://doi.org/10.3390/plants13050609
Farooq S, Azam F (2006) The use of cell membrane stability (CMS) technique to screen for salt tolerant wheat varieties. J Plant Physiol 163(6):629–637. https://doi.org/10.1016/j.jplph.2005.06.006
Ferreira-Cravo M, Moreira DC & Hermes-Lima M (2023) Glutathione depletion disrupts redox homeostasis in an anoxia-tolerant invertebrate. AntioxidS 12(6):1197
Foyer CH, Kunert K (2024) The ascorbate–glutathione cycle coming of age. J Exp Bot 75(9):2682–2699. https://doi.org/10.1093/jxb/erae023
Foyer CH, Harbinson J (2019) Oxygen metabolism and the regulation of photosynthetic electron transport. In Causes of photooxidative stress and amelioration of defense systems in plants 1–42
Gaafar RM, Osman ME-AH, Abo-Shady AM, Almohisen IAA, Badawy GA, El-Nagar MMF, Ismail GA (2022) Role of antioxidant enzymes and glutathione S-transferase in bromoxynil herbicide stress tolerance in wheat plants. Plants 11(20):2679. https://doi.org/10.3390/plants11202679
Gan X, Qiao S, Liang Y, Li K, Xu H (2024) Ascorbate-glutathione cycle and thioredoxin system are involved in nitric oxide alleviating excess nitrate stress in tomato seedlings. J Hortic Sci Biotechnol 99(1):33–45. https://doi.org/10.1080/14620316.2023.2218381
García-Caparrós P, De Filippis L, Gul A, Hasanuzzaman M, Ozturk M, Altay V, Lao MT (2021) Oxidative stress and antioxidant metabolism under adverse environmental conditions: a review. Bot Rev 87(4):421–466. https://doi.org/10.1007/s12229-020-09231-1
Gill SS, Tuteja N (2010) Reactive oxygen species and antioxidant machinery in abiotic stress tolerance in crop plants. Plant Physiol Biochem 48(12):909–930. https://doi.org/10.1016/j.plaphy.2010.08.016
Gill SS, Anjum NA, Hasanuzzaman M, Gill R, Trivedi DK, Ahmad I, Pereira E, Tuteja N (2013) Glutathione and glutathione reductase: a boon in disguise for plant abiotic stress defense operations. Plant Physiol Biochem 70:204–212. https://doi.org/10.1016/j.plaphy.2013.05.032

Guo K, Tu L, Wang P, Du X, Ye S, Luo M, Zhang X (2017) Ascorbate alleviates Fe deficiency-induced stress in cotton (Gossypium hirsutum) by modulating ABA levels. Front Plant Sci 07. https://doi.org/10.3389/fpls.2016.01997

Guo K, Li Z, Tian H, Du X, Liu Z, Huang H, Wang P, Ye Z, Zhang X, Tu L (2020) Cytosolic ascorbate peroxidases plays a critical role in photosynthesis by modulating reactive oxygen species level in stomatal guard cell. Front Plant Sci 11. https://doi.org/10.3389/fpls.2020.00446

Hamada A, Tanaka Y, Ishikawa T, Maruta T (2023) Chloroplast dehydroascorbate reductase and glutathione cooperatively determine the capacity for ascorbate accumulation under photooxidative stress conditions. Plant J 114(1):68–82. https://doi.org/10.1111/tpj.16117

Hanife S, Namdjoyan S, Kermanian H (2023) Synergistic effects of exogenous glutathione and calcium on ascorbate–glutathione cycle and glutathione-associated enzymes upregulation under lead stress in Brassica napus L. Environ Sci Pollut Res 30(50):108979–108991. https://doi.org/10.1007/s11356-023-30000-1

Hanukoglu I (2017) Conservation of the enzyme–coenzyme interfaces in FAD and NADP binding adrenodoxin reductase—a ubiquitous enzyme. J Mol Evol 85(5–6):205–218. https://doi.org/10.1007/s00239-017-9821-9

Hasanuzzaman M, Hossain MA, da Silva JAT, Fujita M (2012) Plant response and tolerance to abiotic oxidative stress: antioxidant defense is a key factor. In: Crop stress and its management: perspectives and strategies. Springer Netherlands, pp 261–315. https://doi.org/10.1007/978-94-007-2220-0_8

Hasanuzzaman M, Al Mahmud J, Nahar K, Anee TI, Inafuku M, Oku H, Fujita M (2017a) Responses, adaptation, and ROS metabolism in plants exposed to waterlogging stress. In: Reactive oxygen species and antioxidant systems in plants: role and regulation under abiotic stress. Springer Singapore, pp 257–281. https://doi.org/10.1007/978-981-10-5254-5_10

Hasanuzzaman M, Nahar K, Anee TI, Fujita M (2017b) Exogenous silicon attenuates cadmium-induced oxidative stress in Brassica napus L. by modulating AsA-GSH pathway and glyoxalase system. Front Plant Sci 8. https://doi.org/10.3389/fpls.2017.01061

Hasanuzzaman M, Nahar K, Anee TI, Fujita M (2017c) Glutathione in plants: biosynthesis and physiological role in environmental stress tolerance. Physiol Mol Biol Plants 23(2):249–268. https://doi.org/10.1007/s12298-017-0422-2

Hasanuzzaman M, Nahar K, Anee TI, Khan MIR, Fujita M (2018) Silicon-mediated regulation of antioxidant defense and glyoxalase systems confers drought stress tolerance in Brassica napus L. S Afr J Bot 115:50–57. https://doi.org/10.1016/j.sajb.2017.12.006

Hasanuzzaman M, Borhannuddin Bhuyan MHM, Anee TI, Parvin K, Nahar K, Al Mahmud J, Fujita M (2019) Regulation of ascorbate-glutathione pathway in mitigating oxidative damage in plants under abiotic stress. Antioxidants 8(9). MDPI. https://doi.org/10.3390/antiox8090384

Horváth E, Kulman K, Tompa B, Hajnal ÁB, Pelsőczi A, Bela K, Gallé Á, Csiszár J (2023) Glutathione transferases are involved in the genotype-specific salt-stress response of tomato plants. Antioxidants 12(9):1682. https://doi.org/10.3390/antiox12091682

Hossain MA, Hoque TS, Zaid A, Wani SH, Mostofa MG, Henry R (2021) Targeting the ascorbate-glutathione pathway and the glyoxalase pathway for genetic engineering of abiotic stress-tolerance in rice. In: Molecular breeding for rice abiotic stress tolerance and nutritional quality. Wiley, pp 398–427. https://doi.org/10.1002/9781119633174.ch21

Hossain A, Bhutia KL, Pramanick B, Maitra S, Ibrahimova U, Kumari VV, Ahmad Z, Uzair M, Aftab T (2022) Glutathione in higher plants: biosynthesis and physiological mechanisms during heat and drought-induced oxidative stress. In: Antioxidant defense in plants. Springer Nature Singapore, pp 181–214. https://doi.org/10.1007/978-981-16-7981-0_9

Hou X, Dan Y, Qi N, Zhang M, Li C, Li Y, Yao Y, Liao W (2024) Nitric oxide promotes adventitious rooting in cucumber under drought stress through regulating ascorbate glutathione cycle and protein S-nitrosylation. J Plant Growth Regul 43(8):2771–2784. https://doi.org/10.1007/s00344-024-11304-z

Huang D, Huo J, Zhang J, Wang C, Wang B, Fang H, Liao W (2019) Protein S-nitrosylation in programmed cell death in plants. Cell Mol Life Sci 76(10):1877–1887. https://doi.org/10.1007/s00018-019-03045-0

Hussain Shah S, Kumar Barwal S, Yerramilli V (2021) An insight on the ascorbate peroxidase and glutathione reductase activities in plants under salinity stress: mini review C A R A S. https://www.researchgate.net/publication/380319746

Imlay JA (2003) Pathways of oxidative damage. Ann Rev Microbiol 57(1):395–418. https://doi.org/10.1146/annurev.micro.57.030502.090938

Imran M, Widemann E, Shafiq S, Bakhsh A, Chen X, Tang X (2024) Salicylic acid and melatonin synergy enhances boron toxicity tolerance via AsA–GSH cycle and glyoxalase system regulation in fragrant rice. Metabolites 14(10):520. https://doi.org/10.3390/metabo14100520

Ishikawa T, Maruta T, Yoshimura K, Smirnoff N (2018) Biosynthesis and regulation of ascorbic acid in plants. In: Antioxidants and antioxidant enzymes in higher plants. Springer International Publishing, pp 163–179. https://doi.org/10.1007/978-3-319-75088-0_8

Jardim-Messeder D, Zamocky M, Sachetto-Martins G, Margis-Pinheiro M (2022) Chloroplastic ascorbate peroxidases targeted to stroma or thylakoid membrane: the chicken or egg dilemma. FEBS Lett 596(23):2989–3004. https://doi.org/10.1002/1873-3468.14438

Jiang Z, Zhu H, Zhu H, Tao Y, Liu C, Liu J, Yang F, Li M (2022) Exogenous ABA enhances the antioxidant defense system of maize by regulating the AsA-GSH cycle under drought stress. Sustainability 14(5):3071. https://doi.org/10.3390/su14053071

Kataya ARA, Reumann S (2010) Arabidopsis glutathione reductase 1 is dually targeted to peroxisomes and the cytosol. Plant Signal Behav 5(2):171–175. https://doi.org/10.4161/psb.5.2.10527

Kaur S, Prakash P, Bak D-H, Hong SH, Cho C, Chung M-S, Kim J-H, Lee S, Bai H-W, Lee SY, Chung BY, Lee SS (2021) Regulation of dual activity of ascorbate peroxidase 1 from Arabidopsis thaliana by conformational changes and posttranslational modifications. Front Plant Sci 12. https://doi.org/10.3389/fpls.2021.678111

Kaya C, Ugurlar F, Ashraf M, Alam P, Ahmad P (2023) Nitric oxide and hydrogen sulfide work together to improve tolerance to salinity stress in wheat plants by upraising the AsA-GSH cycle. Plant Physiol Biochem 194:651–663. https://doi.org/10.1016/j.plaphy.2022.11.041

Kesawat MS, Satheesh N, Kherawat BS, Kumar A, Kim H-U, Chung S-M, Kumar M (2023) Regulation of reactive oxygen species during salt stress in plants and their crosstalk with other signalling molecules—current perspectives and future directions. Plants 12(4):864. https://doi.org/10.3390/plants12040864

Klapheck S (1988) Homoglutathione: isolation, quantification and occurrence in legumes. Physiol Plant 74(4):727–732. https://doi.org/10.1111/j.1399-3054.1988.tb02044.x

Klapheck S, Chrost B, Starke J, Zimmermann H (1992) γ-Glutamylcysteinylserine — a new homologue of glutathione in plants of the family Poaceae*. Botanica Acta 105(3):174–179. https://doi.org/10.1111/j.1438-8677.1992.tb00284.x

Lapenna D (2023) Glutathione and glutathione-dependent enzymes: from biochemistry to gerontology and successful aging. Ageing Res Rev 92:102066. https://doi.org/10.1016/j.arr.2023.102066

Lascano HR, Gomez LD, Casano LM, Trippi VS (1999) Wheat chloroplastic glutathione reductase activity is regulated by the combined effect of pH, NADPH and GSSG. Plant Cell Physiol 40(7):683–690. https://doi.org/10.1093/oxfordjournals.pcp.a029593

Leterrier M, Cagnac O (2018) Function of the various MDAR isoforms in higher plants. In: Antioxidants and antioxidant enzymes in higher plants. Springer International Publishing, pp 83–94. https://doi.org/10.1007/978-3-319-75088-0_5

Li S (2023) Novel insight into functions of ascorbate peroxidase in higher plants: more than a simple antioxidant enzyme. Redox Biol 64:102789. https://doi.org/10.1016/j.redox.2023.102789

Li Y, Liu Y, Zhang J (2010) Advances in the research on the AsA-GSH cycle in horticultural crops. Front Agric China 4(1):84–90. https://doi.org/10.1007/s11703-009-0089-8

Li Y, Huang F, Tao Y, Zhou Y, Bai A, Yu Z, Xiao D, Zhang C, Liu T, Hou X, Li Y (2022) BcGR1.1, a cytoplasmic localized glutathione reductase, enhanced tolerance to copper stress in Arabidopsis thaliana. Antioxidants 11(2):389. https://doi.org/10.3390/antiox11020389

Lian G, Gnanaprakasam JR, Wang T, Wu R, Chen X, Liu L, Shen Y, Yang M, Yang J, Chen Y, Vasiliou V, Cassel TA, Green DR, Liu Y, Fan TW, Wang R (2018) Glutathione de novo synthesis but not recycling process coordinates with glutamine catabolism to control redox homeostasis and directs murine T cell differentiation. elife 7. https://doi.org/10.7554/eLife.36158

Lisenbee CS, Lingard MJ, Trelease RN (2005) Arabidopsis peroxisomes possess functionally redundant membrane and matrix isoforms of monodehydroascorbate reductase. Plant J 43(6):900–914. https://doi.org/10.1111/j.1365-313X.2005.02503.x

Liu F, Huang N, Wang L, Ling H, Sun T, Ahmad W, Muhammad K, Guo J, Xu L, Gao S, Que Y, Su Y (2018) A novel L-ascorbate peroxidase 6 gene, ScAPX6, plays an important role in the regulation of response to biotic and abiotic stresses in sugarcane. Front Plant Sci 8. https://doi.org/10.3389/fpls.2017.02262

Liu J, Qiu G, Liu C, Li H, Chen X, Fu Q, Lin Y, Guo B (2022) Salicylic acid, a multifaceted hormone, combats abiotic stresses in plants. Life 12(6):886. https://doi.org/10.3390/life12060886

Lou MF (2022) Glutathione and glutaredoxin in redox regulation and cell signalling of the lens. Antioxidants 11(10):1973. https://doi.org/10.3390/antiox11101973

Lu X, Zhang H, Zhang N, Dong C, Ji H, Yu J, Ban Z, Yan R, Zhang T, Chen C, Jiang Y (2023) Effects of ozone treatment on gene profiling involved in ASA-GSH cycle in postharvest cantaloupe. Sci Hortic 312:111843. https://doi.org/10.1016/j.scienta.2023.111843

Lv C, Liang Y, Wang M, Li K, Sun X, Xu H (2023) S-nitrosylation of SlAPX is involved in alleviating oxidative damage in transgenic tobacco under nitrate stress. Agronomy 13(5):1322. https://doi.org/10.3390/agronomy13051322

Madhu, Kaur A, Singh K, Upadhyay SK (2024a) Ascorbate oxidases in bread wheat: gene regulatory network, transcripts profiling, and interaction analyses provide insight into their role in plant development and stress response. Plant Growth Regul 103(1):209–224. https://doi.org/10.1007/s10725-023-01103-z

Madhu, Sharma A, Kaur A, Singh K, Upadhyay SK (2024b) Modulation in gene expression and enzyme activity suggested the roles of monodehydroascorbate reductase in development and stress response in bread wheat. Plant Sci 338:111902. https://doi.org/10.1016/j.plantsci.2023.111902

Manna I, Bandyopadhyay M (2023) The impact of engineered nickel oxide nanoparticles on ascorbate glutathione cycle in Allium cepa L. Physiol Mol Biol Plants 29(5):663–678. https://doi.org/10.1007/s12298-023-01314-8

Mansouri HR, Gracia Carmona O, Jodlbauer J, Schweiger L, Fink MJ, Breslmayr E, Laurent C, Feroz SP, Goncalves LC, Rial DV, Mihovilovic MD, Bommarius AS, Ludwig R, Oostenbrink C, Rudroff F (2022) Mutations increasing cofactor affinity, improve stability and activity of a Baeyer–Villiger monooxygenase. ACS Catal 12(19):11761–11766. https://doi.org/10.1021/acscatal.2c03225

Marty L, Bausewein D, Müller C, Bangash SAK, Moseler A, Schwarzländer M, Müller-Schüssele SJ, Zechmann B, Riondet C, Balk J, Wirtz M, Hell R, Reichheld J, Meyer AJ (2019) Arabidopsis glutathione reductase 2 is indispensable in plastids, while mitochondrial glutathione is safeguarded by additional reduction and transport systems. New Phytol 224(4):1569–1584. https://doi.org/10.1111/nph.16086

Maruta T, Tanouchi A, Tamoi M, Yabuta Y, Yoshimura K, Ishikawa T, Shigeoka S (2010) Arabidopsis chloroplastic ascorbate peroxidase isoenzymes play a dual role in photoprotection and gene regulation under photooxidative stress. Plant Cell Physiol 51(2):190–200. https://doi.org/10.1093/pcp/pcp177

Min D, Li Z, Ai W, Li J, Zhou J, Zhang X, Mu D, Li F, Li X, Guo Y (2020) The co-regulation of ethylene biosynthesis and ascorbate–glutathione cycle by methy jasmonate contributes to aroma formation of tomato fruit during postharvest ripening. J Agric Food Chem 68(39):10822–10832. https://doi.org/10.1021/acs.jafc.0c04519

Miret JA, Müller M (2017) AsA/DHA redox pair influencing plant growth and stress tolerance. In: Ascorbic acid in plant growth, development and stress tolerance. Springer International Publishing, pp 297–319. https://doi.org/10.1007/978-3-319-74057-7_12

Mishra N, Jiang C, Chen L, Paul A, Chatterjee A, Shen G (2023) Achieving abiotic stress tolerance in plants through antioxidative defense mechanisms. Front Plant Sci 14. https://doi.org/10.3389/fpls.2023.1110622

Mishra S, Sharma A, Srivastava AK (2024) Ascorbic acid: a metabolite switch for designing stress-smart crops. Crit Rev Biotechnol 44(7):1350–1366. https://doi.org/10.1080/07388551.2023.2286428

Miyake C (2010) Alternative electron flows (water–water cycle and cyclic electron flow around PSI) in photosynthesis: molecular mechanisms and physiological functions. Plant Cell Physiol 51(12):1951–1963. https://doi.org/10.1093/pcp/pcq173

Montes-Rocha JA, Alonso-Castro AJ, Carranza-Álvarez C (2024) Mechanism of heavy metal-induced stress and tolerance. In: Aquatic contamination. Wiley, pp 61–76. https://doi.org/10.1002/9781119989318.ch4

Müller-Schüssele SJ, Bohle F, Rossi J, Trost P, Meyer AJ, Zaffagnini M (2021) Plasticity in plastid redox networks: evolution of glutathione-dependent redox cascades and glutathionylation sites. BMC Plant Biol 21(1):322. https://doi.org/10.1186/s12870-021-03087-2

Munir N, Khilji SA, Shabir M, Sajid ZA (2021) Exogenous application of ascorbic acid enhances the antimicrobial and antioxidant potential of Ocimum sanctum L. grown under salt stress. J Food Qual 2021:1–8. https://doi.org/10.1155/2021/4977410

Munné-Bosch S (2005) The role of -tocopherol in plant stress tolerance. J Plant Physiol 162(7):743–748. https://doi.org/10.1016/j.jplph.2005.04.022

Navrot N, Collin V, Gualberto J, Gelhaye E, Hirasawa M, Rey P, Knaff DB, Issakidis E, Jacquot J-P, Rouhier N (2006) Plant glutathione peroxidases are functional peroxiredoxins distributed in several subcellular compartments and regulated during biotic and abiotic stresses. Plant Physiol 142(4):1364–1379. https://doi.org/10.1104/pp.106.089458

Park AK, Kim I-S, Do H, Jeon BW, Lee CW, Roh SJ, Shin SC, Park H, Kim Y-S, Kim Y-H, Yoon H-S, Lee JH, Kim H-W (2016) Structure and catalytic mechanism of monodehydroascorbate reductase, MDHAR, from Oryza sativa L japonica. Sci Rep 6(1):33903. https://doi.org/10.1038/srep33903

Pessarakli M (2019) In: Pessarakli M (ed) Handbook of plant and crop stress, fourth edition. CRC Press. https://doi.org/10.1201/9781351104609

Praduman Y, Taruna Y, Sunil K, Babita R, Sandeep K, Veena J, Sarla PM (2014) Partial purification and characterization of ascorbate peroxidase from ripening ber (Ziziphus mauritiana L) fruits. Afr J Biotechnol 13(32):3323–3331. https://doi.org/10.5897/AJB2013.12193

Przybyło M, Langner M (2020) On the physiological and cellular homeostasis of ascorbate. Cell Mol Biol Lett 25(1):32. https://doi.org/10.1186/s11658-020-00223-y

Rahantaniaina M-S, Li S, Chatel-Innocenti G, Tuzet A, Mhamdi A, Vanacker H, Noctor G (2017) Glutathione oxidation in response to intracellular H2O2: key but overlapping roles for dehydroascorbate reductases. Plant Signal Behav 12(8):e1356531. https://doi.org/10.1080/15592324.2017.1356531

Rai GK, Kumar P, Choudhary SM, Singh H, Adab K, Kosser R, Magotra I, Kumar RR, Singh M, Sharma R, Corrado G, Rouphael Y (2023) Antioxidant potential of glutathione and crosstalk with phytohormones in enhancing abiotic stress tolerance in crop plants. Plants 12(5):1133. https://doi.org/10.3390/plants12051133

Raihan MT, Tanaka Y, Ishikawa T (2024) Characterization of chloroplastic thioredoxin dependent glutathione peroxidase like protein in *Euglena gracilis*: biochemical and functional perspectives. Biosci Biotechnol Biochem 88(9):1034–1046. https://doi.org/10.1093/bbb/zbae087

Rinne PLH, Kaikuranta PM, Van Der Schoot C (2001) The shoot apical meristem restores its symplasmic organization during chilling-induced release from dormancy. Plant J 26(3):249–264. https://doi.org/10.1046/j.1365-313X.2001.01022.x

Roy PC, Ojha DK, Sahu W, Al-Sadoon MK, Reddy KS, Chowdhary G (2024) Overexpression, purification and characterization of a heat-stable peroxisomal glutathione reductase from Oryza sativa involved in multifactorial abiotic stress tolerance. J King Saud Univ - Sci 36(11):103575. https://doi.org/10.1016/j.jksus.2024.103575

Sachdev S, Ansari SA, Ansari MI (2023) Antioxidant defensive mechanisms to regulate cellular redox homeostatic balance. In: Reactive oxygen species in plants. Springer Nature Singapore, pp 143–172. https://doi.org/10.1007/978-981-19-9884-3_9

Seregin IV, Kozhevnikova AD (2023) Phytochelatins: sulfur-containing metal(loid)-chelating ligands in plants. Int J Mol Sci 24(3):2430. https://doi.org/10.3390/ijms24032430

Sharma S, Singh V, Tanwar H, Mor VS, Kumar M, Punia RC, Dalal MS, Khan M, Sangwan S, Bhuker A, Dagar CS, Yashveer S, Singh J (2022) Impact of high temperature on germination, seedling growth and enzymatic activity of wheat. Agriculture 12(9):1500. https://doi.org/10.3390/agriculture12091500

Siddiqui MN, Mostofa MG, Rahman MM, Tahjib-Ul-Arif M, Das AK, Mohi-Ud-Din M, Rohman MM, Hafiz HR, Ansary MMU, Tran L-SP (2021) Glutathione improves rice tolerance to submergence: insights into its physiological and biochemical mechanisms. J Biotechnol 325:109–118. https://doi.org/10.1016/j.jbiotec.2020.11.011

Sidhu GK, Tuan PA, Renault S, Daayf F, Ayele BT (2020) Polyamine-mediated transcriptional regulation of enzymatic antioxidative response to excess soil moisture during early seedling growth in soybean. Biology 9(8):185. https://doi.org/10.3390/biology9080185

Sigel H (2024) Metal ions in biological systems. CRC Press. https://doi.org/10.1201/9781003573586

Singh AP, Dixit G, Mishra S, Dwivedi S, Tiwari M, Mallick S, Pandey V, Trivedi PK, Chakrabarty D, Tripathi RD (2015) Salicylic acid modulates arsenic toxicity by reducing its root to shoot translocation in rice (Oryza sativa L.). Front Plant Sci 6. https://doi.org/10.3389/fpls.2015.00340

Smirnoff N (2018) Ascorbic acid metabolism and functions: a comparison of plants and mammals. Free Radic Biol Med 122:116–129. https://doi.org/10.1016/j.freeradbiomed.2018.03.033

Smirnoff N, Wheeler GL (2024) The ascorbate biosynthesis pathway in plants is known, but there is a way to go with understanding control and functions. J Exp Bot 75(9):2604–2630. https://doi.org/10.1093/jxb/erad505

Song XS, Tiao CL, Shi K, Mao WH, Ogweno JO, Zhou YH, Yu JQ (2006) The response of antioxidant enzymes in cellular organelles in cucumber (Cucumis sativus L.) leaves to methyl viologen-induced photo-oxidative stress. Plant Growth Regul 49(1):85–93. https://doi.org/10.1007/s10725-006-0023-5

Sonkar KS, Achary VMM, Sahoo S, Reddy MK, Arockiasamy A (2023) Biochemical and structural characterization of a robust and thermostable ascorbate recycling monodehydroascorbate reductase (MDHAR) from stress adapted pearl millet. Biochem Biophys Res Commun 662:135–141. https://doi.org/10.1016/j.bbrc.2023.04.034

Sougrakpam Y, Babuta P, Deswal R (2023) Nitric oxide (NO) modulates low temperature-stress signalling via S-nitrosation, a NO PTM, inducing ethylene biosynthesis inhibition leading to enhanced post-harvest shelf-life of agricultural produce. Physiol Mol Biol Plants 29(12):2051–2065. https://doi.org/10.1007/s12298-023-01371-z

Strand Å, Hurry V, Henkes S, Huner N, Gustafsson P, Gardeström P, Stitt M (1999) Acclimation of Arabidopsis leaves developing at low temperatures. Increasing cytoplasmic volume accompanies increased activities of enzymes in the Calvin cycle and in the sucrose-biosynthesis pathway1. Plant Physiol 119(4):1387–1398. https://doi.org/10.1104/pp.119.4.1387

Tanaka M, Takahashi R, Hamada A, Terai Y, Ogawa T, Sawa Y, Ishikawa T, Maruta T (2021) Distribution and functions of monodehydroascorbate reductases in plants: comprehensive reverse genetic analysis of Arabidopsis thaliana enzymes. Antioxidants 10(11):1726. https://doi.org/10.3390/antiox10111726

Terai Y, Ueno H, Ogawa T, Sawa Y, Miyagi A, Kawai-Yamada M, Ishikawa T, Maruta T (2020) Dehydroascorbate reductases and glutathione set a threshold for high-light–induced ascorbate accumulation. Plant Physiol 183(1):112–122. https://doi.org/10.1104/pp.19.01556

Tripathi R, Singh B, Bisht S, Pandey J (2009) L-ascorbic acid in organic synthesis: an overview. Curr Org Chem 13(1):99–122. https://doi.org/10.2174/138527209787193792

Ullah S, Kolo Z, Egbichi I, Keyster M, Ludidi N (2016) Nitric oxide influences glycine betaine content and ascorbate peroxidase activity in maize. S Afr J Bot 105:218–225. https://doi.org/10.1016/j.sajb.2016.04.003

Veljović-Jovanović S, Vidović M, Morina F (2017) Ascorbate as a key player in plant abiotic stress response and tolerance. In: Ascorbic acid in plant growth, development and stress tolerance. Springer International Publishing, pp 47–109. https://doi.org/10.1007/978-3-319-74057-7_3

Wang HB, Liu YQ, Chen LL, Li XQ, Ha NH, Hoang TX, Li XH, Chen X (2023) The impact of cadmium stress on the ascorbate-glutathione pathway and ascorbate regeneration in tea plants. Biol Plant 67:45–53. https://doi.org/10.32615/bp.2023.002

Wei L, Zhang J, Wei S, Wang C, Deng Y, Hu D, Liu H, Gong W, Pan Y, Liao W (2022) Nitric oxide alleviates salt stress through protein S-nitrosylation and transcriptional regulation in tomato seedlings. Planta 256(6):101. https://doi.org/10.1007/s00425-022-04015-w

Wu S, Hu C, Tan Q, Nie Z, Sun X (2014) Effects of molybdenum on water utilization, antioxidative defense system and osmotic-adjustment ability in winter wheat (Triticum aestivum) under drought stress. Plant Physiol Biochem 83:365–374. https://doi.org/10.1016/j.plaphy.2014.08.022

Yamada A, Saitoh T, Mimura T, Ozeki Y (2002) Expression of mangrove allene oxide cyclase enhances salt tolerance in Escherichia coli, yeast, and tobacco cells. Plant Cell Physiol 43(8):903–910. https://doi.org/10.1093/pcp/pcf108

Yoshimura K, Ishikawa T (2017) Chemistry and metabolism of ascorbic acid in plants. In: Ascorbic acid in plant growth, development and stress tolerance. Springer International Publishing, pp 1–23. https://doi.org/10.1007/978-3-319-74057-7_1

Zhou F, Zheng B, Wang F, Cao A, Xie S, Chen X, Schick JA, Jin X, Li H (2021) Genome-wide analysis of MDHAR gene family in four cotton species provides insights into fiber development via regulating AsA redox homeostasis. Plants 10(2):227. https://doi.org/10.3390/plants10020227

MIX
Papier aus verantwortungsvollen Quellen
Paper from responsible sources
FSC® C105338

If you have any concerns about our products,
you can contact us on
**ProductSafety@springernature.com**

In case Publisher is established outside the EU,
the EU authorized representative is:
**Springer Nature Customer Service Center GmbH**
**Europaplatz 3, 69115 Heidelberg, Germany**

Printed by Libri Plureos GmbH
in Hamburg, Germany